PE Chemical Practice

Michael R. Lindeburg, PE

PPI®
PPI2PASS.COM
A KAPLAN COMPANY

Report Errors for This Book

PPI is grateful to every reader who notifies us of a possible error. Your feedback allows us to improve the quality and accuracy of our products. Report errata at **ppi2pass.com**.

PE CHEMICAL PRACTICE

Current release of this edition: 6

Release History

date	edition number	revision number	update
May 2020	1	4	Minor corrections.
Jan 2021	1	5	Minor corrections.
Jan 2023	1	6	Minor corrections.

Printed in the United States of America.

PPI
ppi2pass.com

ISBN: 978-1-59126-538-2

Topics

Table of Contents

Preface

The purpose of *PE Chemical Practice* is, along with the *PE Chemical Review*, to prepare you for the National Council of Examiners for Engineering and Surveying (NCEES) PE Chemical exam.

In 2017, the NCEES adopted revised specifications for the exam. The council also transitioned from a paper-based version of the exam to a computer-based testing (CBT) version. The PE Chemical exam now requires you to sit in front of a monitor, solve problems served up by the CBT system, access an electronic reference document called the *NCEES PE Chemical Reference Handbook* (*NCEES Handbook*), access an electronic reference document, and perform your scratch calculations on a reusable notepad. You may also use an on-screen calculator with which you will likely be unfamiliar.

Depending on the *NCEES Handbook* as your sole reference is not practical. It is my intention to lessen the challenge by providing ample guidance on how and when to employ the *NCEES Handbook* equations, table, and figures, to show them used in practice problems, and to refer to the more complete explanations in the *PE Chemical Review*.

In order to prepare you for the CBT PE Chemical exam, this book was created with the goal to familiarize you with the *NCEES Handbook*. Thus, for better or worse, this book duplicates the terms, variables, and formatting of the *NCEES Handbook* equations.

NCEES has selected what it believes to be all of the chemical engineering topics important to an early-career, minimally qualified chemical engineer, and has distilled them into its single reference, the *NCEES Handbook*. Personally, I cannot accept the premise that engineers learn and use so little engineering during their first few career years. However, regardless of whether you accept the NCEES subset of chemical engineering coverage, one thing is certain: In serving as your sole source of formulas, theory, methods, and data during the exam, the *NCEES Handbook* severely limits the types of problems that can be included in the PE Chemical exam. Based on the current PE Chemical exam specifications and the *NCEES Handbook*, the CBT PE Chemical exam is even more limited than the old paper-based exam. The number (breadth) of knowledge areas, the coverage (depth) of knowledge areas, the number of problems, and the duration of the exam are all significantly reduced. If you are only concerned about passing and/or "getting it over with" before graduation, these reductions are all in your favor. Your only deterrents will be the cost of the exam and the inconvenience of finding a time and place to take it.

Accepting that "it is what it is," I designed this book to guide you through the exam's body of knowledge. I have several admissions to make: (1) This book contains nothing magical or illicit. (2) This book, by itself, is only one part of a complete preparation. (3) This book stops well short of being perfect. What do I mean by those admissions?

First, this book does not contain anything magical. It's called a "practice problems" book, and though it will save you time in assembling hundreds of practice problems for your review, it will not learn the material for you. Merely owning it is not enough. You will have to put in the "practice" time to use it.

Similarly, there is nothing clandestine or unethical about this book. It does not contain any actual exam problems. It was written in a vacuum, based entirely on the NCEES PE Chemical exam specifications. This book is not based on feedback from actual examinees.

Truthfully, I expect that many exam problems will be similar to the problems I have used, because NCEES and I developed content with the same set of constraints. (If anything, NCEES is even more constrained when it comes to fringe, outlier, eccentric, or original topics.)

Any similarity between problems in this book and problems in the exam is easily attributed to the limited number of engineering formulas and concepts, the shallowness of coverage, and the need to keep the entire solution process (reading, researching, calculating, and responding) to less than three minutes for each problem.

To make it possible to be solved in the time allowed, any exam problem you see is likely to be more limited. In fact, most or all of the information you need to answer a problem will be given to you in its problem statement. If only the real world were so kind!

Second, by itself, this book is inadequate. It was never intended to define the entirety of your preparation activity. While it introduces essentially all of the exam knowledge areas and content in the *NCEES Handbook*, an introduction is only an introduction. To be a thorough review, this book needs augmentation.

By design, this book has four significant inadequacies.

1. This book has a limited number of pages, so it cannot contain enough of everything for everyone. The number of practice problems that can fit in it is also limited. The number of problems needed by you, personally, to come up to speed in a particular subject may be inadequate. So, additional exposure is inevitable if you want to be adequately prepared in every subject.
2. This book does not contain the *NCEES Handbook*. This book is limited in helping you become familiar with the idiosyncratic sequencing, formatting, variables, omissions, and presentation of topics in the *NCEES Handbook*. The only way to remedy this is to download your own PDF copy of the *NCEES Handbook* and use it in conjunction with your review.
3. This book does not contain a practice examination (mock exam, sample exam, etc.). With the advent of the CBT format, any sample exam in printed format is little more than another collection of practice problems. The actual PE Chemical exam is taken sitting in front of a computer using an online reference book, so the only way to practice is to sit in front of a computer while you answer problems. On the PPI Learning Hub (learn.ppi2pass.com), you'll find an online practice exam and other resources. Using an online reference is very different from the work environment experienced by most engineers, and it will take some getting used to.
4. This book does not contain explanatory background information, including figures and tables of data. Though all problems have associated step-by-step solutions, these solutions will not teach you the underlying engineering principles you need to solve the problems. Trying to extrapolate engineering principles from the solutions is like reading the ending of a book and then trying to guess at the "whos, whats, wheres, whens, and hows." In other words, reviewing solutions is only going to get you so far if you don't understand a topic. To truly understand how to solve practice problems in topics you're unfamiliar with, you'll need an actual review manual like the one PPI publishes, the *PE Chemical Review*. In it, you'll find all the "whos and whats" you were previously missing and these problems' "endings" will make much more sense.

Third, and finally, I reluctantly admit that I have never figured out how to write or publish a completely flawless first (or even subsequent) edition. The PPI staff comes pretty close to perfection in the areas of design, editing, typography, and illustrating. Subject matter experts help immensely with calculation checking, and beta testing before you see a book helps smooth out wrinkles. However, I still manage to muck up the content. So, I hope you will "let me have it" when you find my mistakes. PPI has established an easy way for you to report an error, as well as to review changes that resulted from errors that others have submitted. Just go to **ppi2pass.com/errata**. When you submit something, I'll receive it via email. When I answer it, you'll receive a response. We'll both benefit. Best wishes in your examination experience. Stay in touch!

Michael R. Lindeburg, PE

Acknowledgments

Developing a book specific to the computerized PE Chemical exam has been a monumental project. It involved the usual (from an author's and publisher's standpoint) activities of updating and repurposing existing content and writing new content. Many people helped accomplish these tasks.

Special thanks go to Corey Leclerc, PE, for developing new content related to the *NCEES Handbook*; and to calculation checker Anil Acharya.

Thanks to editorial project manager Rebecca Morgan managed the work with support from director of publishing services Grace Wong. Production services manager Cathy Schrott kept the process moving smoothly and swiftly, despite technical difficulties that seemed determined to stall the process at every opportunity. Steve Buehler, director of acquisitions, and Nicole Evans, acquisitions editor, arranged for the outside subject matter expert who was involved with this book. All the content was reviewed for consistency, PPI style, and accuracy by Jennifer Lindeburg King, editor-in-chief.

I would also like to thank Tom Bergstrom, technical illustrator; Robert Genevro, copy editor; Gabby Howitz and Scott Rutherford, typesetters; and Jenny Lindeburg King, editor-in-chief, for their support in producing this book. My thanks also to the information technology team who developed the PPI Learning Hub: Nik Heath, Kevin Milhoan, and Ahmed Osman, web developers; Alan Tu, associate product manager; Megan Synnestvedt, product manager; and Bryan Spaulding, chief technology officer. Thank you for your hard work and support in producing this new initiative for PPI.

Consistent with the past 40 years, I continue to thank my wife, Elizabeth, for accepting and participating in a writer's life that is full to overflowing. Even though our children have been out on their own for a long time, we seem to have even less time than we had before. As a corollary to Aristotle's "Nature abhors a vacuum," I propose: "Work expands to fill the void." Still, it's a great life with Elizabeth by my side and I'm thankful every day for her support and love. I couldn't do this without her.

To my daughters, Jenny and Katherine, I am so very proud of the strong, smart women you have become. You two are the greatest books I've ever written, and I'm so very proud to be your dad.

To my granddaughter Sydney, you bring joy and light to my life. Playing with you and your "brothers" Elmo and Grover is now, perhaps, my most important job. Thank you for showing me that life without play is no life at all. I look forward to your visits more than you will ever know.

I also appreciate the grant of permission to reproduce materials from several other publishers. In each case, attribution is provided where the material has been included. Neither PPI nor the publishers of the reproduced material make any representations or warranties as to the accuracy of the material, nor are they liable for any damages resulting from its use.

Thank you, everyone! I'm really proud of what you've accomplished. Your efforts will be pleasing to examinees and effective in preparing them for the PE Chemical exam.

Michael R. Lindeburg, PE

Codes and References Used to Prepare This Book

This book is based on the *NCEES PE Chemical Reference Handbook* (*NCEES Handbook*), v. 1.1 (July 2017 revision). The other documents, codes, and standards that were used to prepare this book were the most current available at the time.

NCEES does not specifically tie the PE Chemical exam to any edition (version) of any code or standard. Rather than make the PE Chemical exam subject to the vagaries of such codes and standards as are published by the American Chemical Society (ACS), the American Concrete Institute (ACI), the American Institute of Chemical Engineers (AIChE), the American Institute of Steel Construction (AISC), the American National Standards Institute (ANSI), the American Society of Civil Engineers (ASCE), the American Society of Heating, Refrigerating and Air-Conditioning Engineers (ASHRAE), the American Society of Mechanical Engineers (ASME), ASTM International (ASTM), the International Code Council (ICC), the Institute of Electrical and Electronic Engineers (IEEE), the National Fire Protection Association (NFPA), and so on, NCEES effectively writes its own "code," the *NCEES Handbook*.

Most surely, every standard- or code-dependent concept (e.g., flammability) in the *NCEES Handbook* can be traced back to some section of some edition of a standard or code (e.g., 29CFR). So, it would be logical to conclude that you need to be familiar with everything (the limitations, surrounding sections, and commentary) in the code related to that concept. However, that does not seem to be the case. The *NCEES Handbook* is a code unto itself, and you won't need to study the parent documents. Nor will you need to know anything pertaining to related, adjacent, similar, or parallel code concepts. Therefore, although methods and content in the *NCEES Handbook* can be ultimately traced back to some edition (version) of a relevant code, you don't need to know which. You don't need to know whether that content is current, limited in intended application, or relevant. You only need to use the content.

How to Use This Book

This book is written for one purpose, and one purpose only: to get you ready for the NCEES PE Chemical exam. Because it is a practice problems book, there are a few, but not many, ways to use it. Here's how this book was designed to be used.

Get the *NCEES PE Chemical Reference Handbook*

Get a copy of the *NCEES PE Chemical Reference Handbook* (*NCEES Handbook*). Use it as you solve the problems in this book. The *NCEES Handbook* is the only reference you can use during the exam, so you will want to know the sequence of its sections, what data are included, and the approximate locations of important figures and tables in the *NCEES Handbook*. You should also know the terminology (words and phrases) used in the *NCEES Handbook* to describe equations or subjects, because those are the terms you will have to look up during the exam.

Work Through Every Problem

NCEES has greatly reduced the number of subjects about which you are expected to be knowledgeable and has made nothing optional. Skipping your weakest subjects is no longer a viable preparation strategy. You should study all examination knowledge areas, not just your specialty areas. That means you solve every problem in this book and skip nothing. Do not limit the number of problems you solve in hopes of finding enough problems in your areas of expertise to pass the exam. To pinpoint subjects that will require additional study, use the diagnostic exams on the PPI Learning Hub (**learn.ppi2pass.com**). Your performance on these diagnostic exams will tell you which topics you need to devote more time to learning and which you can review more lightly.

Be Thorough

Being thorough means really doing the work. Some people think they can read a problem statement, think about it for 10 seconds, read the solution, and then say, "Yes, that's what I was thinking of, and that's what I would have done." Sadly, these people find out too late that the human brain doesn't learn very efficiently that way. Under pressure, they find they know and remember very little. For real learning, you'll have to spend some time with a stubby pencil.

There are so many places where you can get messed up solving a problem. Maybe it is in the use of your calculator, like pushing log instead of ln, or forgetting to set the angle to radians instead of degrees, and so on. Maybe it is rusty math. What is $\ln(ex)$ anyway? How do you factor a polynomial? Maybe it's in finding the data needed or the proper unit conversion. Maybe you're not familiar with the SI system of units. These things take time. And, you have to make the mistakes once so that you do not make them again.

If you do decide to get your hands dirty and actually work these problems, you will have to decide how much reliance you place on this book. It is tempting to turn to a solution when you get slowed down by details or stumped by the subject material. It is tempting to want to maximize the number of problems you solve by spending as little time as possible solving them. However, you need to struggle a little bit more than that to really learn the material.

Studying a new subject is analogous to using a machete to cut a path through a dense jungle. By doing the work, you develop pathways that weren't there before. It is a lot different than just looking at the route on a map. You actually get nowhere by looking at a map. But cut the path once, and you are in business until the jungle overgrowth closes in again. So do the problems—all of them. Do not look at the solutions until you have sweated a little.

1 Fluid Properties

PRACTICE PROBLEMS

(Use $g = 32.2$ ft/sec^2 (9.81 m/s^2) unless told otherwise.)

1. Atmospheric pressure is 14.7 lbf/in^2 (101.3 kPa). What is most nearly the absolute pressure in a tank if a gauge on the tank reads 8.7 lbf/in^2 (60 kPa) vacuum?

(A) 4 psia (27 kPa)

(B) 6 psia (41 kPa)

(C) 8 psia (55 kPa)

(D) 10 psia (68 kPa)

2. Air is considered to be an ideal gas with a specific gas constant of 53.3 ft-lbf/lbm-°R (287 J/kg·K). What is most nearly the kinematic viscosity of air at 80°F (27°C) and 70 psia (480 kPa)?

(A) 3.5×10^{-5} ft^2/sec (3.0×10^{-6} m^2/s)

(B) 4.0×10^{-5} ft^2/sec (4.0×10^{-6} m^2/s)

(C) 5.0×10^{-5} ft^2/sec (5.0×10^{-6} m^2/s)

(D) 6.0×10^{-5} ft^2/sec (6.0×10^{-6} m^2/s)

3. Three solutions of nitric acid are combined: one with 8% nitric acid by volume, one with 10% nitric acid by volume, and one with 20% nitric acid by volume. The combined solutions produce 100 mL of a solution that is 12% nitric acid by volume. The 8% solution contributes half of the total volume of nitric acid contributed by the 10% and 20% solutions. The volume of 10% solution in the 12% solution is most nearly

(A) 20 mL

(B) 30 mL

(C) 50 mL

(D) 80 mL

4. A 25% (by volume) mixture of ethylene glycol and water is used in a solar heating application. The components are nonreacting. The mixture is intended to operate at standard atmospheric pressure and an average temperature of 140°F. If the specific gravity of ethylene glycol at 140°F is 1.107 relative to water at 60°F, what is most nearly the specific gravity of the mixture referred to water at 60°F?

(A) 1.005

(B) 1.015

(C) 1.021

(D) 1.043

SOLUTIONS

1. *Customary U.S. Solution*

$$p_{\text{gage}} = -8.7\ \text{lbf/in}^2$$
$$p_{\text{atmospheric}} = 14.7\ \text{lbf/in}^2$$

The relationship between absolute, gage, and atmospheric pressure is in *NCEES Handbook* table "Conversion Table for the Most Commonly Used Units of Pressure."

$$\begin{aligned} p_{\text{absolute}} &= p_{\text{gage}} + p_{\text{atmospheric}} \\ &= -8.7\ \frac{\text{lbf}}{\text{in}^2} + 14.7\ \frac{\text{lbf}}{\text{in}^2} \\ &= 6\ \text{lbf/in}^2 \quad (6\ \text{psia}) \end{aligned}$$

SI Solution

$$p_{\text{gage}} = -60\ \text{kPa}$$
$$p_{\text{atmospheric}} = 101.3\ \text{kPa}$$

The relationship between absolute, gage, and atmospheric pressure is in *NCEES Handbook* table "Conversion Table for the Most Commonly Used Units of Pressure."

$$\begin{aligned} p_{\text{absolute}} &= p_{\text{gage}} + p_{\text{atmospheric}} = -60\ \text{kPa} + 101.3\ \text{kPa} \\ &= 41.3\ \text{kPa} \quad (41\ \text{kPa}) \end{aligned}$$

The answer is (B).

2. *Customary U.S. Solution*

From the *PE Chemical Handbook: Temperature-Dependent Properties of Air* table, the absolute viscosity of air at 14.7 psia can be found using linear interpolation between the data points of 32°F and 100°F. The viscosity of air at 80°F is

$$\mu(32°\text{F}) = 1.15 \times 10^{-5}\ \text{lbm/ft-sec}$$
$$\mu(100°\text{F}) = 1.28 \times 10^{-5}\ \text{lbm/ft-sec}$$

$$\begin{aligned} \mu(80°\text{F}) &= \mu(32°\text{F}) + (80°\text{F} - 32°\text{F})\frac{(\mu(100°\text{F}) - \mu(32°\text{F}))}{(100°\text{F} - 32°\text{F})} \\ &= \left(1.15 \times 10^{-5}\ \frac{\text{lbm}}{\text{ft-sec}}\right) + (48°\text{F})\left(\frac{(1.28 \times 10^{-5} - 1.15 \times 10^{-5})\frac{\text{lbm}}{\text{ft-sec}}}{68°\text{F}}\right) \\ &= 1.242 \times 10^{-5}\ \text{lbm/ft-sec} \end{aligned}$$

The density of the air at 70 psia and 80°F is

$$\begin{aligned} \rho &= \frac{p}{RT} = \frac{\left(70\ \frac{\text{lbf}}{\text{in}^2}\right)\left(12\ \frac{\text{in}}{\text{ft}}\right)^2}{\left(53.3\ \frac{\text{ft-lbf}}{\text{lbm-°R}}\right)(80°\text{F} + 460°)} \\ &= 0.350\ \text{lbm/ft}^3 \end{aligned}$$

The kinematic viscosity, ν, is in *NCEES Handbook: Viscosity and Fluid Properties*

$$\begin{aligned} \nu &= \frac{\mu}{\rho} = \frac{1.242 \times 10^{-5}\ \text{lbm/ft-sec}}{0.350\ \text{lbm/ft}^3} \\ &= 3.549 \times 10^{-5}\ \text{ft}^2/\text{sec} \quad (3.5 \times 10^{-5}\ \text{ft}^2/\text{sec}) \end{aligned}$$

SI Solution

From the *PE Chemical Handbook: Temperature-Dependent Properties of Air* table, the absolute viscosity of air at 0.1 MPa can be found using linear interpolation between the data points of 20°C and 40°C. The viscosity of air at 27°C is

$$\mu(20°\text{C}) = 18.2\mu\text{Pa·s}$$
$$\mu(40°\text{C}) = 19.1\mu\text{Pa·s}$$

$$\begin{aligned} \mu(27°\text{C}) &= \mu(20°\text{C}) + (27°\text{C} - 20°\text{C})\frac{(\mu(40°\text{C}) - \mu(20°\text{C}))}{(40°\text{C} - 20°\text{C})} \\ &= (1.82 \times 10^{-5}\ \text{Pa·s}) + (7°\text{C})\left(\frac{(1.91 \times 10^{-5} - 1.82 \times 10^{-5})\text{Pa·s}}{20°\text{C}}\right) \\ &= 1.852 \times 10^{-5}\ \text{Pa·s} \end{aligned}$$

The density of the air at 480 kPa and 27°C is

$$\begin{aligned} \rho &= \frac{p}{RT} = \frac{(480\ \text{kPa})\left(1000\ \frac{\text{Pa}}{\text{kPa}}\right)}{\left(287\ \frac{\text{J}}{\text{kg·K}}\right)(27°\text{C} + 273°)} \\ &= 5.575\ \text{kg/m}^3 \end{aligned}$$

The kinematic viscosity, ν, is in *NCEES Handbook: Viscosity and Fluid Properties*

$$\begin{aligned} \nu &= \frac{\mu}{\rho} = \frac{1.852 \times 10^{-5}\ \text{Pa·s}}{5.575\ \text{kg/m}^3} \\ &= 3.32 \times 10^{-6}\ \text{m}^2/\text{s} \quad (3.0 \times 10^{-6}\ \text{m}^2/\text{s}) \end{aligned}$$

The answer is (A).

3. Let

$$\begin{aligned} x &= \text{volume of 8\% solution} \\ 0.08x &= \text{volume of nitric acid contributed by 8\% solution} \\ y &= \text{volume of 10\% solution} \\ 0.10y &= \text{volume of nitric acid contributed by 10\% solution} \\ z &= \text{volume of 20\% solution} \\ 0.20z &= \text{volume of nitric acid contributed by 20\% solution} \end{aligned}$$

The three conditions that must be satisfied are

$$\begin{aligned} x + y + z &= 100 \text{ mL} \\ 0.08x + 0.10y + 0.20z &= (0.12)(100 \text{ mL}) = 12 \text{ mL} \\ 0.08x &= \left(\frac{1}{2}\right)(0.10y + 0.20z) \end{aligned}$$

Simplifying these equations,

$$\begin{aligned} x + y + z &= 100 \\ 4x + 5y + 10z &= 600 \\ 8x - 5y - 10z &= 0 \end{aligned}$$

Adding the second and third equations,

$$\begin{aligned} 12x &= 600 \\ x &= 50 \text{ mL} \end{aligned}$$

Working with the first two equations,

$$\begin{aligned} y + z &= 100 - 50 = 50 \\ 5y + 10z &= 600 - (4)(50) = 400 \end{aligned}$$

Multiplying the top equation by −5 and adding to the bottom equation,

$$\begin{aligned} 5z &= 150 \\ z &= 30 \text{ mL} \end{aligned}$$

Using the first equation, the volume of 10% solution in the 12% solution is

$$y = 20 \text{ mL}$$

The answer is (A).

4. From App. 1A (also *NCEES Handbook* table "Physical Properties of Liquid Water"), the density of 140°F water is 61.38 lbm/ft^3. Consider 100 ft^3 of mixture. Since the mixture percentages are volumetric, there is 25 ft^3 of ethylene glycol and 75 ft^3 of water. The weight of 100 ft^3 is

$$\begin{aligned} m &= \rho_{\text{water}} V_{\text{water}} + \rho_{\text{glycol}} V_{\text{glycol}} \\ &= \left(61.38 \ \frac{\text{lbm}}{\text{ft}^3}\right)(75 \text{ ft}^3) \\ &\quad + (1.107)\left(62.37 \ \frac{\text{lbm}}{\text{ft}^3}\right)(25 \text{ ft}^3) \\ &= 6329.59 \text{ lbm} \end{aligned}$$

The specific gravity of the mixture is

$$\begin{aligned} \text{SG} &= \frac{\rho}{\rho_{\text{ref}}} = \frac{m}{V\rho_{\text{ref}}} = \frac{6329.59 \text{ lbm}}{(100 \text{ ft}^3)\left(62.37 \ \frac{\text{lbm}}{\text{ft}^3}\right)} \\ &= 1.0148 \quad (1.015) \end{aligned}$$

The answer is (B).

2 Fluid Statics

PRACTICE PROBLEMS

(Use $g = 32.2\ \text{ft/sec}^2$ or $9.81\ \text{m/s}^2$ unless told otherwise.)

1. A 4000 lbm (1800 kg) blimp contains 10,000 lbm (4500 kg) of hydrogen (specific gas constant = 766.5 ft-lbf/lbm-°R (4124 J/kg·K)) at 56°F (13°C) and 30.2 in Hg (770 mm Hg). If the hydrogen and air are in thermal and pressure equilibrium, what is most nearly the blimp's lift (lifting force)?

(A) 7.6×10^3 lbf (3.4×10^4 N)

(B) 1.2×10^4 lbf (5.3×10^4 N)

(C) 1.3×10^5 lbf (5.7×10^5 N)

(D) 1.7×10^5 lbf (7.7×10^5 N)

2. A hollow 6 ft (1.8 m) diameter sphere floats half-submerged in seawater. The mass of concrete (150 lbm/ft³ or 2400 kg/m³) that is required as an external anchor to just submerge the sphere completely is most nearly

(A) 2700 lbm (1200 kg)

(B) 4200 lbm (1900 kg)

(C) 5500 lbm (2500 kg)

(D) 6300 lbm (2700 kg)

3. Water removed from Lake Superior (elevation, 601 ft above mean sea level; water density, 62.4 lbm/ft³) is transported by tanker ship to the Atlantic Ocean (elevation, 0 ft) through 16 Seaway locks. A tanker's displacement is 32,000 tonnes when loaded, and 5100 tonnes when empty. Each lock is 766 ft long and 80 ft wide. Water pumped from each lock flows to the Atlantic Ocean. Compared to a passage from Lake Superior to the Atlantic Ocean when empty, what is most nearly the change in water loss from Lake Superior when a ship passes through the locks fully loaded?

(A) 27,000 tonnes less loss

(B) 27,000 tonnes additional loss

(C) 54,000 tonnes additional loss

(D) no change in loss

SOLUTIONS

1. *Customary U.S. Solution*

The lift (lifting force) of the hydrogen-filled blimp, F_{lift}, is equal to the difference between the buoyant force, F_b, and the weight of the hydrogen contained in the blimp, W_{H}.

$$F_{\text{lift}} = F_b - W_{\text{H}} - W_{\text{blimp}}$$

The weight of the hydrogen is calculated from the mass of hydrogen.

$$W_{\text{H}} = \frac{m_{\text{H}}g}{g_c} = \frac{(10{,}000\ \text{lbm})\left(32.2\ \frac{\text{ft}}{\text{sec}^2}\right)}{32.2\ \frac{\text{lbm-ft}}{\text{lbf-sec}^2}} = 10{,}000\ \text{lbf}$$

The buoyant force is equal to the weight of the displaced air. The volume of the air displaced is equal to the volume of hydrogen enclosed in the blimp.

The absolute temperature of the hydrogen is

$$T = 56°\text{F} + 460° = 516°\text{R}$$

The pressure of the hydrogen is

$$p = \frac{(30.2\ \text{in Hg})\left(12\ \frac{\text{in}}{\text{ft}}\right)^2}{2.036\ \frac{\text{in Hg}}{\frac{\text{lbf}}{\text{in}^2}}} = 2136\ \text{lbf/ft}^2$$

Compute the volume of hydrogen from the ideal gas law.

$$V_{\text{H}} = \frac{m_{\text{H}}RT}{p} = \frac{(10{,}000\ \text{lbm})\left(766.5\ \frac{\text{ft-lbf}}{\text{lbm-°R}}\right)(516°\text{R})}{2136\ \frac{\text{lbf}}{\text{ft}^2}} = 1.85 \times 10^6\ \text{ft}^3$$

Since the volume of the hydrogen contained in the blimp is equal to the air displaced, the air displaced can be computed from the ideal gas equation. Since the air and hydrogen are in thermal and pressure equilibrium, the temperature and pressure are equal to the values given for the hydrogen.

For air, $R = 53.35$ ft-lbf/lbm-°R.

$$m_{air} = \frac{pV_H}{RT} = \frac{\left(2136\ \frac{\text{lbf}}{\text{ft}^2}\right)(1.85\times 10^6\ \text{ft}^3)}{\left(53.35\ \frac{\text{ft-lbf}}{\text{lbm-°R}}\right)(516\text{°R})}$$
$$= 1.435\times 10^5\ \text{lbm}$$

The buoyant force is equal to the weight of the air.

$$F_b = W_{air} = \frac{m_{air}g}{g_c} = \frac{(1.435\times 10^5\ \text{lbm})\left(32.2\ \frac{\text{ft}}{\text{sec}^2}\right)}{32.2\ \frac{\text{lbm-ft}}{\text{lbf-sec}^2}}$$
$$= 1.435\times 10^5\ \text{lbf}$$

The lift (lifting force) is

$$F_{lift} = F_b - W_H - W_{blimp}$$
$$= 1.435\times 10^5\ \text{lbf} - 10{,}000\ \text{lbf} - 4000\ \text{lbf}$$
$$= 1.295\times 10^5\ \text{lbf}\quad (1.3\times 10^5\ \text{lbf})$$

SI Solution

The lift (lifting force) of the hydrogen-filled blimp, F_{lift}, is equal to the difference between the buoyant force, F_b, and the weight of the hydrogen contained in the blimp, W_H.

$$F_{lift} = F_b - W_H - W_{blimp}$$

The weight of the hydrogen is calculated from the mass of hydrogen.

$$W_H = m_H g$$
$$= (4500\ \text{kg})\left(9.81\ \frac{\text{m}}{\text{s}^2}\right)$$
$$= 44\,145\ \text{N}$$

The buoyant force is equal to the weight of the displaced air. The volume of the air displaced is equal to the volume of hydrogen enclosed in the blimp.

The absolute temperature of the hydrogen is

$$T = 13\text{°C} + 273° = 286\text{K}$$

The absolute pressure of the hydrogen is

$$p = \frac{(770\ \text{mm Hg})\left(133.4\ \frac{\text{kPa}}{\text{m}}\right)}{1000\ \frac{\text{mm}}{\text{m}}}$$
$$= 102.7\ \text{kPa}$$

Compute the volume of hydrogen from the ideal gas law.

$$V_H = \frac{m_H RT}{p}$$
$$= \frac{(4500\ \text{kg})\left(4124\ \frac{\text{J}}{\text{kg·K}}\right)(286\text{K})}{(102.7\ \text{kPa})\left(1000\ \frac{\text{Pa}}{\text{kPa}}\right)}$$
$$= 5.168\times 10^4\ \text{m}^3$$

Since the volume of the hydrogen contained in the blimp is equal to the air displaced, the air displaced can be computed from the ideal gas equation. Since the air and hydrogen are assumed to be in thermal and pressure equilibrium, the temperature and pressure are equal to the values given for the hydrogen.

For air, $R = 287.03$ J/kg·K.

$$m_{air} = \frac{pV_H}{RT}$$
$$= \frac{(102.7\ \text{kPa})\left(1000\ \frac{\text{Pa}}{\text{kPa}}\right)(5.168\times 10^4\ \text{m}^3)}{\left(287.03\ \frac{\text{J}}{\text{kg·K}}\right)(286\text{K})}$$
$$= 6.465\times 10^4\ \text{kg}$$

The buoyant force is equal to the weight of the air.

$$F_b = W_{air} = m_{air}g$$
$$= (6.465\times 10^4\ \text{kg})\left(9.81\ \frac{\text{m}}{\text{s}^2}\right)$$
$$= 6.34\times 10^5\ \text{N}$$

The lift (lifting force) is

$$F_{lift} = F_b - W_H - W_{blimp}$$
$$= 6.34\times 10^5\ \text{N} - 44\,145\ \text{N} - (1800\ \text{kg})\left(9.81\ \frac{\text{m}}{\text{s}^2}\right)$$
$$= 5.7\times 10^5\ \text{N}$$

The answer is (C).

2. *Customary U.S. Solution*

The weight of the sphere is equal to the weight of the displaced volume of water when floating.

The buoyant force is given by

$$F_b = \frac{\rho g V_{\text{displaced}}}{g_c}$$

Since the sphere is half submerged,

$$W_{\text{sphere}} = \frac{1}{2}\left(\frac{\rho g V_{\text{sphere}}}{g_c}\right)$$

For seawater, $\rho = 64.0\ \text{lbm/ft}^3$.

The volume of the sphere is

$$\begin{aligned} V_{\text{sphere}} &= \frac{\pi}{6}d^3 = \left(\frac{\pi}{6}\right)(6\ \text{ft})^3 \\ &= 113.1\ \text{ft}^3 \end{aligned}$$

The weight of the sphere is

$$\begin{aligned} W_{\text{sphere}} &= \frac{1}{2}\left(\frac{\rho g V_{\text{sphere}}}{g_c}\right) \\ &= \left(\frac{1}{2}\right)\left(\frac{\left(64.0\ \frac{\text{lbm}}{\text{ft}^3}\right)\left(32.2\ \frac{\text{ft}}{\text{sec}^2}\right)(113.1\ \text{ft}^3)}{32.2\ \frac{\text{lbm-ft}}{\text{lbf-sec}^2}}\right) \\ &= 3619\ \text{lbf} \end{aligned}$$

The equilibrium equation for a fully submerged sphere and anchor can be solved for the concrete volume.

$$W_{\text{sphere}} + W_{\text{concrete}} = (V_{\text{sphere}} + V_{\text{concrete}})\rho_{\text{water}}$$

$$W_{\text{sphere}} + \rho_{\text{concrete}} V_{\text{concrete}}\left(\frac{g}{g_c}\right) = (V_{\text{sphere}} + V_{\text{concrete}})\rho_{\text{water}}\left(\frac{g}{g_c}\right)$$

$$\begin{aligned} &3619\ \text{lbf} + \left(150\ \frac{\text{lbm}}{\text{ft}^3}\right) \\ &\quad \times V_{\text{concrete}}\left(\frac{32.2\ \frac{\text{ft}}{\text{sec}^2}}{32.2\ \frac{\text{lbm-ft}}{\text{lbf-sec}^2}}\right) = (113.1\ \text{ft}^3 + V_{\text{concrete}}) \\ &\quad \times\left(64.0\ \frac{\text{lbm}}{\text{ft}^3}\right) \\ &\quad \times\left(\frac{32.2\ \frac{\text{ft}}{\text{sec}^2}}{32.2\ \frac{\text{lbm-ft}}{\text{lbf-sec}^2}}\right) \end{aligned}$$

$$\begin{aligned} V_{\text{concrete}} &= 42.09\ \text{ft}^3 \\ m_{\text{concrete}} &= \rho_{\text{concrete}} V_{\text{concrete}} \\ &= \left(150\ \frac{\text{lbm}}{\text{ft}^3}\right)(42.09\ \text{ft}^3) \\ &= 6314\ \text{lbm} \quad (6300\ \text{lbm}) \end{aligned}$$

SI Solution

The weight of the sphere is equal to the weight of the displaced volume of water when floating.

The buoyant force is given by

$$F_b = \rho g V_{\text{displaced}}$$

Since the sphere is half submerged,

$$W_{\text{sphere}} = \frac{1}{2}\rho g V_{\text{sphere}}$$

For seawater, $\rho = 1025\ \text{kg/m}^3$.

The volume of the sphere is

$$V_{\text{sphere}} = \frac{\pi}{6}d^3 = \left(\frac{\pi}{6}\right)(1.8\ \text{m})^3 = 3.054\ \text{m}^3$$

The weight of the sphere required is

$$\begin{aligned} W_{\text{sphere}} &= \frac{1}{2}\rho g V_{\text{sphere}} \\ &= \left(\frac{1}{2}\right)\left(1025\ \frac{\text{kg}}{\text{m}^3}\right)\left(9.81\ \frac{\text{m}}{\text{s}^2}\right)(3.054\ \text{m}^3) \\ &= 15\,354\ \text{N} \end{aligned}$$

The equilibrium equation for a fully submerged sphere and anchor can be solved for the concrete volume.

$$W_{\text{sphere}} + W_{\text{concrete}} = g(V_{\text{sphere}} + V_{\text{concrete}})\rho_{\text{water}}$$

$$W_{\text{sphere}} + \rho_{\text{concrete}} g V_{\text{concrete}} = g(V_{\text{sphere}} + V_{\text{concrete}})\rho_{\text{water}}$$

$$\begin{aligned}15\,354\ \text{N} + \left(2400\ \frac{\text{kg}}{\text{m}^3}\right) & \\ \times\left(9.81\ \frac{\text{m}}{\text{s}^2}\right)V_{\text{concrete}} &= (3.054\ \text{m}^3 + V_{\text{concrete}}) \\ &\quad \times\left(1025\ \frac{\text{kg}}{\text{m}^3}\right)\left(9.81\ \frac{\text{m}}{\text{s}^2}\right)\end{aligned}$$

$$\begin{aligned}V_{\text{concrete}} &= 1.138\ \text{m}^3 \\ m_{\text{concrete}} &= \rho_{\text{concrete}} V_{\text{concrete}} \\ &= \left(2400\ \frac{\text{kg}}{\text{m}^3}\right)(1.138\ \text{m}^3) \\ &= 2731\ \text{kg} \quad (2700\ \text{kg})\end{aligned}$$

The answer is (D).

3. From Archimedes' principle, each tonne of water carried in the tanker displaces a tonne of water in the lock. So, each tonne of water transported out of the lake results in a tonne less of lock loss. Compared to an empty tanker, the net result is zero.

The answer is (D).

3 Fluid Flow Parameters

PRACTICE PROBLEMS

Use the following values unless told otherwise.

$$g = 32.2 \text{ ft/sec}^2 \ (9.81 \text{ m/s}^2)$$
$$\rho_{\text{water}} = 62.4 \text{ lbm/ft}^3 \ (1000 \text{ kg/m}^3)$$
$$p_{\text{atmospheric}} = 14.7 \text{ psia} \ (101.3 \text{ kPa})$$

1. A 10 in (25 cm) composition pipe is compressed by a tree root into an elliptical cross section until its inside height is only 7.2 in (18 cm). What is most nearly the hydraulic radius when flowing half full?

(A) 2.2 in (5.5 cm)

(B) 2.7 in (6.9 cm)

(C) 3.2 in (8.1 cm)

(D) 4.5 in (11.4 cm)

2. A pipe with an inside diameter of 18.812 in contains water to a depth of 15.7 in. What is most nearly the hydraulic radius? (Work in customary U.S. units only.)

(A) 4.4 in

(B) 5.1 in

(C) 5.7 in

(D) 6.5 in

SOLUTIONS

1. *Customary U.S. Solution*

The perimeter of the pipe, as in *NCEES Handbook:* Planar Geometry—Area and Perimeter, is

$$p = \pi d = \pi(10 \text{ in}) = 31.42 \text{ in}$$

If the pipe is flowing half full, the wetted perimeter is

$$\text{wetted perimeter} = \tfrac{1}{2}p = \left(\frac{1}{2}\right)(31.42 \text{ in}) = 15.71 \text{ in}$$

The ellipse will have a minor axis, b, equal to one-half the height of the compressed pipe, or

$$b = \frac{7.2 \text{ in}}{2} = 3.6 \text{ in}$$

When the pipe is compressed, the perimeter of the pipe remains constant. The perimeter of an ellipse, as in *NCEES Handbook:* Planar Geometry—Area and Perimeter, is

$$p \approx 2\pi\sqrt{\tfrac{1}{2}(a^2 + b^2)}$$

Solve for the major axis.

$$\begin{aligned} a &= \sqrt{2\left(\frac{p}{2\pi}\right)^2 - b^2} = \sqrt{(2)\left(\frac{31.42 \text{ in}}{2\pi}\right)^2 - (3.6 \text{ in})^2} \\ &= 6.09 \text{ in} \end{aligned}$$

The flow area or area of the ellipse, as in *NCEES Handbook:* Planar Geometry—Area and Perimeter, is

$$\begin{aligned} \text{flow area} &= \tfrac{1}{2}\pi ab \\ &= \left(\frac{1}{2}\pi\right)(6.09 \text{ in})(3.6 \text{ in}) \\ &= 34.4 \text{ in}^2 \end{aligned}$$

The hydraulic radius is

$$r_h = \frac{\text{area in flow}}{\text{wetted perimeter}} = \frac{34.4 \text{ in}^2}{15.71 \text{ in}} = 2.19 \text{ in} \quad (2.2 \text{ in})$$

SI Solution

The perimeter of the pipe, as in *NCEES Handbook:* Planar Geometry—Area and Perimeter, is

$$p = \pi d = \pi(25\text{ cm}) = 78.54\text{ cm}$$

If the pipe is flowing half full, the wetted perimeter is

$$\text{wetted perimeter} = \tfrac{1}{2}p = \left(\frac{1}{2}\right)(78.54\text{ cm}) = 39.27\text{ cm}$$

The ellipse will have a minor axis, b, equal to one-half the height of the compressed pipe, or

$$b = \frac{18\text{ cm}}{2} = 9\text{ cm}$$

When the pipe is compressed, the perimeter of the pipe remains constant. The perimeter of an ellipse, as in *NCEES Handbook:* Planar Geometry—Area and Perimeter, is

$$p \approx 2\pi\sqrt{\tfrac{1}{2}(a^2 + b^2)}$$

Solve for the major axis.

$$\begin{aligned} a &= \sqrt{2\left(\frac{p}{2\pi}\right)^2 - b^2} = \sqrt{(2)\left(\frac{78.54\text{ cm}}{2\pi}\right)^2 - (9\text{ cm})^2} \\ &= 15.2\text{ cm} \end{aligned}$$

The flow area or area of the ellipse, as in *NCEES Handbook:* Planar Geometry—Area and Perimeter, is

$$\begin{aligned} \text{flow area} &= \tfrac{1}{2}\pi ab = \left(\frac{1}{2}\pi\right)(15.2\text{ cm})(9\text{ cm}) \\ &= 214.9\text{ cm}^2 \end{aligned}$$

The hydraulic radius is

$$\begin{aligned} r_h &= \frac{\text{area in flow}}{\text{wetted perimeter}} = \frac{214.9\text{ cm}^2}{39.27\text{ cm}} \\ &= 5.47\text{ cm} \quad (5.5\text{ cm}) \end{aligned}$$

The answer is (A).

2. As in *NCEES Handbook:* Planar Geometry—Area and Perimeter, for a circular segment,

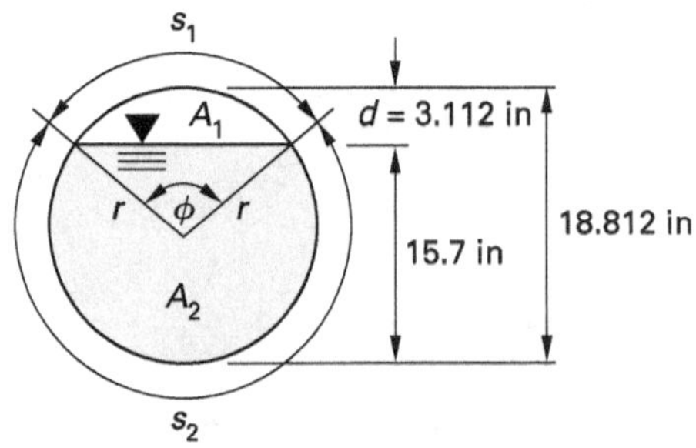

$$\begin{aligned} r &= \frac{d}{2} = \frac{18.812\text{ in}}{2} = 9.406\text{ in} \\ \phi &= 2\arccos\frac{r - d}{r} \\ &= 2\arccos\frac{9.406\text{ in} - 3.112\text{ in}}{9.406\text{ in}} \\ &= 1.675\text{ rad} \end{aligned}$$

$$\begin{aligned} \sin\phi &= 0.9946 \\ A_1 &= \tfrac{1}{2}r^2(\phi - \sin\phi) \\ &= \left(\frac{1}{2}\right)(9.406\text{ in})^2(1.675 - 0.9946) \\ &= 30.1\text{ in}^2 \\ A_{\text{total}} &= A_1 + A_2 = \frac{\pi}{4}d^2 = \left(\frac{\pi}{4}\right)(18.812\text{ in})^2 \\ &= 277.95\text{ in}^2 \\ A_2 &= A_{\text{total}} - A_1 = 277.95\text{ in}^2 - 30.1\text{ in}^2 \\ &= 247.85\text{ in}^2 \\ s_1 &= r\phi = (9.406\text{ in})(1.675) = 15.76\text{ in} \\ s_{\text{total}} &= s_1 + s_2 = \pi d = \pi(18.812\text{ in}) \\ &= 59.1\text{ in} \\ s_2 &= s_{\text{total}} - s_1 = 59.1\text{ in} - 15.76\text{ in} = 43.34\text{ in} \\ r_h &= \frac{A_2}{s_2} = \frac{247.85\text{ in}^2}{43.34\text{ in}} = 5.719\text{ in} \quad (5.7\text{ in}) \end{aligned}$$

The answer is (C).

4 Fluid Dynamics

PRACTICE PROBLEMS

(Use $g = 32.2\ \text{ft/sec}^2$ ($9.81\ \text{m/s}^2$) unless told otherwise.)

1. The capacity of an old municipal pressurized water supply pipe system must be doubled without increasing the average velocity. A new replacement pipe will have 40% less friction than the existing pipe. What is the approximate required increase in pipe diameter?

(A) 20%

(B) 30%

(C) 40%

(D) 100%

2. A pressurized water pipe system will be modified such that the pipe diameter will be doubled and the average flow velocity halved. Fluid properties are unchanged. When these changes are made, the Reynolds number will

(A) halve

(B) double

(C) quadruple

(D) remain the same

3. 5 ft^3/sec (130 L/s) of water flows through a schedule-40 steel pipe that changes gradually in diameter from 6 in at point A to 18 in at point B. Point B is 15 ft (4.6 m) higher than point A. The respective pressures at points A and B are 10 psia (70 kPa) and 7 psia (48.3 kPa). All minor losses are insignificant. The velocity and direction of flow at point A are most nearly

(A) 3.2 ft/sec (1 m/s); from A to B

(B) 25 ft/sec (7 m/s); from A to B

(C) 3.2 ft/sec (1 m/s); from B to A

(D) 25 ft/sec (7 m/s); from B to A

4. Points A and B are separated by 3000 ft of new 6 in schedule-40 steel pipe. 750 gal/min of 60°F water flows from point A to point B. Point B is 60 ft above point A. Approximately what must be the pressure at point A if the pressure at B must be 50 psig?

(A) 90 psig

(B) 100 psig

(C) 120 psig

(D) 170 psig

5. A pipe network connects junctions A, B, C, and D as shown. All pipe sections have a Hazen-Williams C-value of 150. Water can be added and removed at any of the junctions to achieve the flows listed. Water flows from point A to point D. No flows are backward. All minor losses are insignificant. For simplicity, use the nominal piple diameter listed.

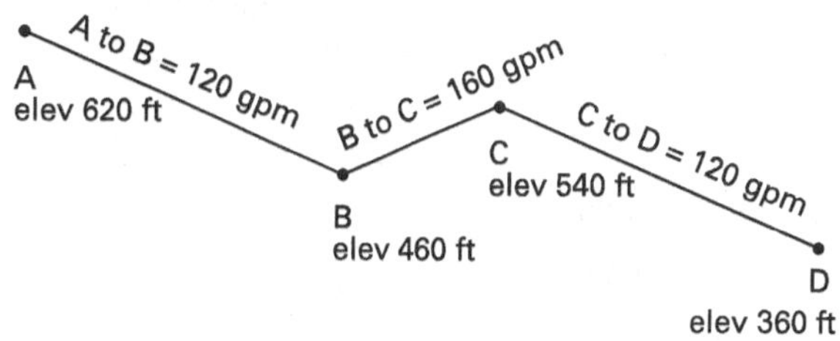

The Hazen-Williams frictional head loss is

$$h_f = \frac{10.44 L_{\text{ft}} Q_{\text{gpm}}^{1.85}}{C^{1.85} d_{\text{inch}}^{4.87}}$$

The lengths and the nominal pipe diameters of the pipe network are as shown.

path	length	nominal diameter
A to B	20,000 ft	6 in
B to C	10,000 ft	6 in
C to D	30,000 ft	4 in

If the minimum static pressure anywhere in the system is 20 psig, the pressure at point A is most nearly

(A) 14 psig

(B) 17 psig

(C) 24 psig

(D) 30 psig

6. Based on the formulas commonly used by engineers, is friction head loss ever proportional to velocity, instead of velocity squared?

(A) yes, in laminar flow

(B) yes, for non-Newtonian fluids

(C) yes, in smooth pipes

(D) no, never

7. Water flows at 10 ft/s through a pipe network consisting of 50 ft of new 3 in diameter, schedule-40 steel pipe, four 45° standard elbows, and a fully open gate valve. All fittings are flanged. The elevation of the network discharge is 20 ft higher than the elevation of the inlet. The element that contributes the most to specific energy loss is the

(A) four elbows

(B) gate valve

(C) pipe friction (excluding the fittings)

(D) elevation change

8. 1.5 ft^3/sec (40 L/s) of 70°F (20°C) water flows through 1200 ft (355 m) of 6 in (nominal) diameter new schedule-40 steel pipe. The friction loss is most nearly

(A) 4 ft (1.2 m)

(B) 18 ft (5.2 m)

(C) 36 ft (9.5 m)

(D) 70 ft (21 m)

9. 500 gal/min (30 L/s) of 100°F (40°C) water flows through 300 ft (90 m) of 6 in schedule-40 pipe. The pipe contains two 6 in flanged steel elbows, two gate valves, a 90° angle valve, and a swing check valve. The discharge is located 20 ft (6 m) higher than the entrance. The pressure difference between the two ends of the pipe is most nearly

(A) 12 psi (78 kPa)

(B) 21 psi (140 kPa)

(C) 45 psi (310 kPa)

(D) 87 psi (600 kPa)

10. 70°F (20°C) air is flowing at 60 ft/sec (18 m/s) through 300 ft (90 m) of 6 in schedule-40 pipe. The pipe contains two 6 in (0.15 m) flanged steel elbows, two full-open gate valves, a 90° angle valve, and a swing check valve. The discharge is located 20 ft (6 m) higher than the entrance, and the average air density is 0.075 lbm/ft^3. The pressure difference between the two ends of the pipe is most nearly

(A) 0.26 psi (1.8 kPa)

(B) 0.52 psi (3.4 kPa)

(C) 1.5 psi (10 kPa)

(D) 13 psi (90 kPa)

11. Three reservoirs (A, B, and C) are interconnected with a common junction (point D) at elevation 25 ft above an arbitrary reference point. The water levels for reservoirs A, B, and C are at elevations of 50 ft, 40 ft, and 22 ft, respectively. The pipe from reservoir A to the junction is 800 ft of 3 in (nominal) steel pipe. The pipe from reservoir B to the junction is 500 ft of 10 in (nominal) steel pipe. The pipe from reservoir C to the junction is 1000 ft of 4 in (nominal) steel pipe. All pipes are schedule-40 with a friction factor of 0.02. All minor losses and velocity heads can be neglected. The direction of flow and the pressure at point D are most nearly

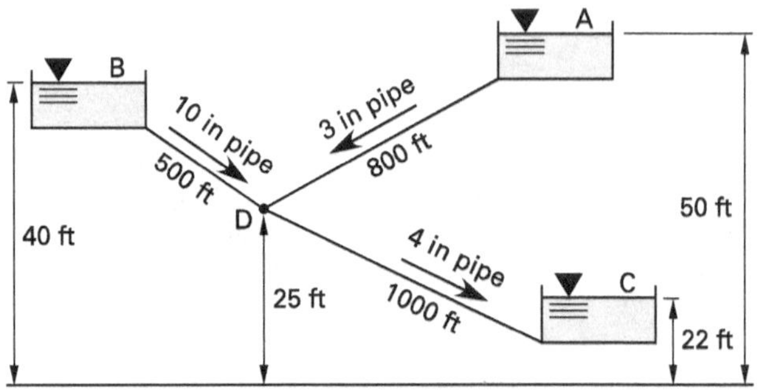

(A) out of reservoir B; 500 psf

(B) out of reservoir B; 930 psf

(C) into reservoir B; 1100 psf

(D) into reservoir B; 1260 psf

12. 8 MGD (millions of gallons per day) (350 L/s) of 70°F (20°C) water flows into the new schedule-40 steel pipe network shown. Minor losses are insignificant.

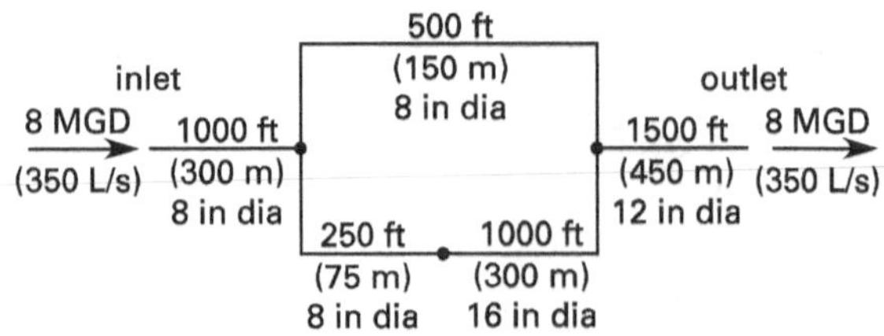

The quantity of water flowing in the upper branch is most nearly

(A) 1.2 ft^3/sec (0.034 m^3/s)

(B) 2.9 ft^3/sec (0.081 m^3/s)

(C) 4.1 ft^3/sec (0.11 m^3/s)

(D) 5.4 ft^3/sec (0.15 m^3/s)

13. Water flows through the horizontal nozzle shown so that the incoming velocity, v_1, is 4 m/s. Neglect any losses through the nozzle and use specific weight of 9800 N/m^3 and kinematic viscosity equal to 10^{-6} m^2/s for water. The horizontal section of the pipe is cast iron with a specific roughness of 0.00024 m.

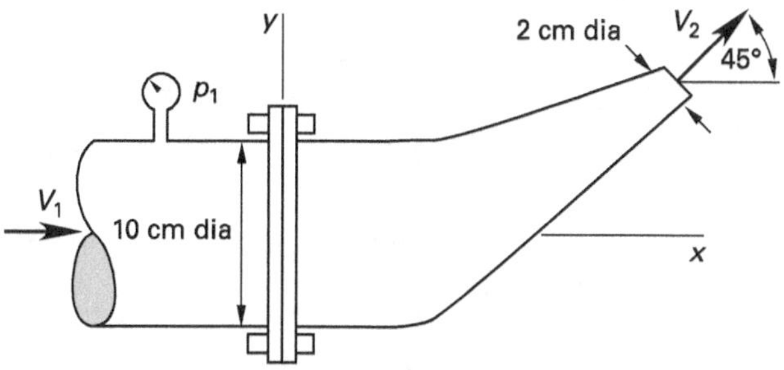

The difference between the pressure at a section 100 meters upstream of the bolted joint and the inlet pressure, p_1, is most nearly

(A) 190 kPa

(B) 200 kPa

(C) 290 kPa

(D) 420 kPa

14. A full cylindrical tank that is 40 ft (12 m) high has a constant diameter of 20 ft. The tank has a 4 in (100 mm) diameter hole in its bottom. The coefficient of discharge for the hole is 0.98. Approximately how long will it take for the water level to drop from 40 ft to 20 ft (12 m to 6 m)?

(A) 950 sec

(B) 1200 sec

(C) 1450 sec

(D) 1700 sec

15. A 24 in diameter siphon is used to transfer irrigation water from a water distribution canal to an irrigation ditch for a field of row crops below. The elevation of the water in the canal is 320 ft MSL (relative to mean sea level), and the elevation of the stilling basin for the row crops is 305 ft MSL. Counting fittings and minor losses, the siphon has a total equivalent length of 42 ft. The siphon is constructed of corrugated metal pipe with a standard galvanized surface. Neglecting entrance and exit losses, what is most nearly the total head loss experienced by the water?

(A) 7.0 ft

(B) 11 ft

(C) 15 ft

(D) 23 ft

16. A venturi meter with an 8 in diameter throat is installed in a 12 in diameter water line. The venturi is perfectly smooth, so that the discharge coefficient is 1.00. An attached mercury manometer registers a 4 in differential. The volumetric flow rate is most nearly

(A) 1.7 ft^3/sec

(B) 5.2 ft^3/sec

(C) 6.4 ft^3/sec

(D) 18 ft^3/sec

17. 60°F (15°C) benzene (specific gravity at 60°F (15°C) of 0.885) flows through an 8 in/3.5 in (200 mm/90 mm) venturi meter whose coefficient of discharge is 0.99. A mercury manometer indicates a 4 in difference in the heights of the mercury columns. The volumetric flow rate of the benzene is most nearly

(A) 1.2 ft^3/sec (34 L/s)

(B) 9.1 ft^3/sec (250 L/s)

(C) 13 ft^3/sec (360 L/s)

(D) 27 ft^3/sec (760 L/s)

18. A sharp-edged orifice meter with a 0.2 ft diameter opening is installed in a 1 ft diameter pipe. 70°F water approaches the orifice at 2 ft/sec. The indicated pressure drop across the orifice meter is most nearly

(A) 5.9 psi

(B) 13 psi

(C) 22 psi

(D) 47 psi

19. A mercury manometer is used to measure a pressure difference across an orifice meter in a water line. The difference in mercury levels is 7 in (17.8 cm). The pressure differential is most nearly

(A) 1.7 psi (12 kPa)

(B) 3.2 psi (22 kPa)

(C) 7.9 psi (55 kPa)

(D) 23 psi (160 kPa)

20. A sharp-edged ISA orifice is used in a schedule-40 steel 12 in (300 mm inside diameter) water line. The water temperature is 70°F (20°C), and the flow rate is 10 ft^3/sec (250 L/s). The differential pressure change across the orifice (to the vena contracta) should be approximately 25 ft (7.5 m). The smallest orifice that can be used is most nearly

(A) 5.5 in (14 cm)

(B) 7.3 in (19 cm)

(C) 8.1 in (20 cm)

(D) 8.9 in (23 cm)

21. A pipe necks down from 24 in at point A to 12 in at point B. 8 ft^3/sec of 60°F water flows from point A to point B. The pressure head at point A is 20 ft. Friction is insignificant over the distance between points A and B. The magnitude and direction of the resultant force on the water are most nearly

(A) 2900 lbf, toward A

(B) 3500 lbf, toward A

(C) 2900 lbf, toward B

(D) 3500 lbf, toward B

22. 2000 gal/min (125 L/s) of brine with a specific gravity of 1.2 passes through an 85% efficient pump. The centerlines of the pump's 12 in inlet and 8 in outlet are at the same elevation. The inlet suction gauge indicates 6 in (150 mm) of mercury below atmospheric. The discharge pressure gauge is located 4 ft (1.2 m) above the centerline of the pump's outlet and indicates 20 psig (138 kPa). All pipes are schedule-40. The input power to the pump is most nearly

(A) 12 hp (8.9 kW)

(B) 36 hp (26 kW)

(C) 52 hp (39 kW)

(D) 87 hp (65 kW)

23. 100 ft^3/sec (2.6 m^3/s) of water passes through a horizontal turbine. The water's pressure is reduced from 30 psig (210 kPa) to 5 psig (35 kPa) vacuum. Disregarding friction, velocity, and other factors, the power generated is most nearly

(A) 110 hp (82 kW)

(B) 380 hp (280 kW)

(C) 730 hp (540 kW)

(D) 920 hp (640 kW)

24. A refrigeration truck is driven at 65 mph into a 15 mph headwind. The frontal area of the truck is 100 ft^2. The coefficient of drag is 0.5. When calculating the drag force on the truck, the velocity that should be used is most nearly

(A) 50 mph

(B) 65 mph

(C) 73 mph

(D) 80 mph

25. A dish-shaped antenna faces directly into a 60 mph wind. The projected area of the antenna is 0.8 ft^2; the coefficient of drag, C_D, is 1.2; and the density of air is 0.076 lbm/ft^3. The total amount of drag force experienced by the antenna is most nearly

(A) 9.0 lbf

(B) 10 lbf

(C) 14 lbf

(D) 16 lbf

26. A car traveling through 70°F (20°C) air has the following characteristics.

frontal area	28 ft^2 (2.6 m^2)
mass	3300 lbm (1500 kg)
drag coefficient	0.42
rolling resistance	1% of weight
engine thermal efficiency	28%
fuel heating value	115,000 Btu/gal (32 MJ/L)

Assume the car is traveling at 55 mi/hr (90 km/h). Considering only the drag and rolling resistance, the fuel consumption of the car is most nearly

(A) 0.026 gal/mi (0.062 L/km)

(B) 0.038 gal/mi (0.087 L/km)

(C) 0.051 gal/mi (0.12 L/km)

(D) 0.13 gal/mi (0.30 L/km)

27. A 1/20th airplane model is tested in a wind tunnel at full velocity and temperature. What is the approximate ratio of the wind tunnel pressure to normal ambient pressure?

(A) 5

(B) 10

(C) 20

(D) 40

28. 68°F (20°C) castor oil (kinematic viscosity at 68°F (20°C) of 1110×10^{-5} ft²/sec (103×10^{-5} m²/s)) flows through a pump whose impeller turns at 1000 rpm. A similar pump twice the first pump's size is tested with 68°F (20°C) air. Theoretically, what should be the approximate speed of the second pump's impeller to ensure similarity?

(A) 3.6 rpm

(B) 88 rpm

(C) 250 rpm

(D) 1600 rpm

SOLUTIONS

1. The decrease in friction will affect pumping power required, and increase the volumetric flow, which will increase the velocity. Since there are no details provided about a specific model of pump and system curve, capacity can be estimated as a function of velocity and area. The velocity is unchanged. The volumetric flow rate can be found in *NCEES Handbook:* Conservation of Mass.

$$\frac{Q_2}{Q_1} = \frac{\mathrm{v}A_2}{\mathrm{v}A_1} = \frac{\frac{\pi \mathrm{v} d_2^2}{4}}{\frac{\pi \mathrm{v} d_1^2}{4}} = \frac{d_2^2}{d_1^2}$$
$$= 2$$
$$d_2 = \sqrt{2}\, d_1 \approx 1.41 d_1 \quad \text{(40\% increase)}$$

It is worth noting that the pump should be modified by either changing the impeller or reducing the speed, if possible for this type of pump, to prevent the average velocity from exceeding the desired value.

The answer is (C).

2. As in *NCEES Handbook:* Similitude, the initial Reynolds number is

$$\mathrm{Re}_1 = \frac{D\mathrm{v}}{\nu}$$

After modifications, the Reynolds number is

$$\mathrm{Re}_2 = \frac{(2D)\left(\frac{\mathrm{v}}{2}\right)}{\nu} = \frac{D\mathrm{v}}{\nu} = \mathrm{Re}_1$$

The Reynolds number will remain the same.

The answer is (D).

3. *Customary U.S. Solution*

As in *NCEES Handbook* table "Pipe Dimensions and Weights," for schedule-40 pipe,

$$D_\mathrm{A} = 0.5054 \text{ ft}$$
$$D_\mathrm{B} = 1.4063 \text{ ft}$$

Let point A be at zero elevation.

As in *NCEES Handbook:* The Bernoulli Equation, the total energy at point A is

$$E_{t,\mathrm{A}} = E_p + E_\mathrm{v} + E_z = \frac{p_\mathrm{A}}{\rho} + \frac{\mathrm{v}_\mathrm{A}^2}{2g_c} + \frac{z_\mathrm{A} g}{g_c}$$

At point A, from App. 3C, the diameter is 6 in. As in *NCEES Handbook:* Conservation of Mass, the velocity at point A is

$$\dot{V} = \mathrm{v_A} A_\mathrm{A} = \mathrm{v_A}\left(\frac{\pi}{4}\right)D_\mathrm{A}^2$$

$$\mathrm{v_A} = \left(\frac{4}{\pi}\right)\left(\frac{\dot{V}}{D_\mathrm{A}^2}\right) = \left(\frac{4}{\pi}\right)\left(\frac{5\ \frac{\mathrm{ft}^3}{\mathrm{sec}}}{(0.5054\ \mathrm{ft})^2}\right)$$

$$= 24.9\ \mathrm{ft/sec}$$

$$p_\mathrm{A} = \left(10\ \frac{\mathrm{lbf}}{\mathrm{in}^2}\right)\left(12\ \frac{\mathrm{in}}{\mathrm{ft}}\right)^2 = 1440\ \mathrm{lbf/ft^2}$$

$$z_\mathrm{A} = 0$$

As in *NCEES Handbook* table "Physical Properties of Liquid Water (U.S. Units)," for water, $\rho \approx 62.4\ \mathrm{lbm/ft^3}$.

$$E_{t,\mathrm{A}} = \frac{p_\mathrm{A}}{\rho} + \frac{\mathrm{v_A^2}}{2g_c} + \frac{z_\mathrm{A} g}{g_c}$$

$$= \frac{1440\ \frac{\mathrm{lbf}}{\mathrm{ft}^2}}{62.4\ \frac{\mathrm{lbm}}{\mathrm{ft}^3}} + \frac{\left(24.9\ \frac{\mathrm{ft}}{\mathrm{sec}}\right)^2}{(2)\left(32.2\ \frac{\mathrm{lbm\text{-}ft}}{\mathrm{lbf\text{-}sec^2}}\right)} + 0$$

$$= 32.7\ \mathrm{ft\text{-}lbf/lbm}$$

Similarly, the total energy at point B is

$$\mathrm{v_B} = \left(\frac{4}{\pi}\right)\left(\frac{\dot{V}}{D_\mathrm{B}^2}\right) = \left(\frac{4}{\pi}\right)\left(\frac{5\ \frac{\mathrm{ft}^3}{\mathrm{sec}}}{(1.4063\ \mathrm{ft})^2}\right)$$

$$= 3.22\ \mathrm{ft/sec}$$

$$p_\mathrm{B} = \left(7\ \frac{\mathrm{lbf}}{\mathrm{in}^2}\right)\left(12\ \frac{\mathrm{in}}{\mathrm{ft}}\right)^2 = 1008\ \mathrm{lbf/ft^2}$$

$$z_\mathrm{B} = 15\ \mathrm{ft}$$

$$E_{t,\mathrm{B}} = \frac{p_\mathrm{B}}{\rho} + \frac{\mathrm{v_B^2}}{2g_c} + \frac{z_\mathrm{B} g}{g_c}$$

$$= \frac{1008\ \frac{\mathrm{lbf}}{\mathrm{ft}^2}}{62.4\ \frac{\mathrm{lbm}}{\mathrm{ft}^3}} + \frac{\left(3.22\ \frac{\mathrm{ft}}{\mathrm{sec}}\right)^2}{(2)\left(32.2\ \frac{\mathrm{lbm\text{-}ft}}{\mathrm{lbf\text{-}sec^2}}\right)} + \frac{(15\ \mathrm{ft})\left(32.2\ \frac{\mathrm{ft}}{\mathrm{sec}^2}\right)}{32.2\ \frac{\mathrm{lbm\text{-}ft}}{\mathrm{lbf\text{-}sec^2}}}$$

$$= 31.3\ \mathrm{ft\text{-}lbf/lbm}$$

Since $E_{t,\mathrm{A}} > E_{t,\mathrm{B}}$, the flow is from point A to point B.

SI Solution

Let point A be at zero elevation.

As in *NCEES Handbook:* The Bernoulli Equation, the total energy at point A is

$$E_{t,\mathrm{A}} = E_p + E_\mathrm{v} + E_z = \frac{p_\mathrm{A}}{\rho} + \frac{\mathrm{v_A^2}}{2} + z_\mathrm{A} g$$

At point A, the diameter is 154 mm (0.154 m). The velocity at point A is

$$\dot{V} = \mathrm{v_A} A_\mathrm{A} = \mathrm{v_A}\left(\frac{\pi}{4}\right)D_\mathrm{A}^2$$

$$\mathrm{v_A} = \left(\frac{4}{\pi}\right)\left(\frac{\dot{V}}{D_\mathrm{A}^2}\right) = \left(\frac{4}{\pi}\right)\left(\frac{130\ \frac{\mathrm{L}}{\mathrm{s}}}{(0.154\ \mathrm{m})^2\left(1000\ \frac{\mathrm{L}}{\mathrm{m}^3}\right)}\right)$$

$$= 6.98\ \mathrm{m/s}\quad (7\ \mathrm{m/s})$$

$$p_\mathrm{A} = 70\ \mathrm{kPa}\quad (70\,000\ \mathrm{Pa})$$

$$z_\mathrm{A} = 0$$

As in *NCEES Handbook* table "Physical Properties of Liquid Water (SI Units)," for water, $\rho \approx 1000\ \mathrm{kg/m^3}$.

$$E_{t,\mathrm{A}} = \frac{p_\mathrm{A}}{\rho} + \frac{\mathrm{v_A^2}}{2} + z_\mathrm{A} g$$

$$= \frac{70\,000\ \mathrm{Pa}}{1000\ \frac{\mathrm{kg}}{\mathrm{m}^3}} + \frac{\left(6.98\ \frac{\mathrm{m}}{\mathrm{s}}\right)^2}{2} + 0$$

$$= 94.36\ \mathrm{J/kg}$$

Similarly, at B, the diameter is 429 mm. The total energy at point B is

$$\begin{aligned} \mathrm{v}_B &= \left(\frac{4}{\pi}\right)\left(\frac{\dot{V}}{D_B^2}\right) \\ &= \left(\frac{4}{\pi}\right)\left(\frac{130\ \frac{\text{L}}{\text{s}}}{(0.429\ \text{m})^2\left(1000\ \frac{\text{L}}{\text{m}^3}\right)}\right) \\ &= 0.90\ \text{m/s} \\ p_B &= 48.3\ \text{kPa} \quad (48\,300\ \text{Pa}) \\ z_B &= 4.6\ \text{m} \end{aligned}$$

$$\begin{aligned} E_{t,B} &= \frac{p_B}{\rho} + \frac{\mathrm{v}_B^2}{2} + z_B g \\ &= \frac{48\,300\ \text{Pa}}{1000\ \frac{\text{kg}}{\text{m}^3}} + \frac{\left(0.90\ \frac{\text{m}}{\text{s}}\right)^2}{2} \\ &\quad + (4.6\ \text{m})\left(9.81\ \frac{\text{m}}{\text{s}^2}\right) \\ &= 93.8\ \text{J/kg} \end{aligned}$$

Since $E_{t,A} > E_{t,B}$, the flow is from point A to point B.

The answer is (B).

4.

$$\begin{aligned} \dot{V} &= \frac{750\ \frac{\text{gal}}{\text{min}}}{\left(7.4805\ \frac{\text{gal}}{\text{ft}^3}\right)\left(60\ \frac{\text{sec}}{\text{min}}\right)} \\ &= 1.671\ \text{ft}^3/\text{sec} \end{aligned}$$

From App. 3B (also *NCEES Handbook* table "Pipe Dimensions and Weights"), $D = 0.5054$ ft, and $A = 0.2006\ \text{ft}^2$.

$$\mathrm{v} = \frac{\dot{V}}{A} = \frac{1.671\ \frac{\text{ft}^3}{\text{sec}}}{0.2006\ \text{ft}^2} = 8.33\ \text{ft/sec}$$

For 60°F water, from App. 1A (also *NCEES Handbook* table "Physical Properties of Liquid Water (U.S. Units)"),

$$\begin{aligned} \rho &= 62.37\ \text{lbm/ft}^3 \\ \nu &= 1.217 \times 10^{-5}\ \text{ft}^2/\text{sec} \end{aligned}$$

As in *NCEES Handbook:* Similitude, the Reynolds number is

$$\text{Re} = \frac{\mathrm{v}D}{\nu} = \frac{\left(8.33\ \frac{\text{ft}}{\text{sec}}\right)(0.5054\ \text{ft})}{1.217 \times 10^{-5}\ \frac{\text{ft}^2}{\text{sec}}} = 3.46 \times 10^5$$

The specific weight is

$$\gamma = \frac{\rho g}{g_c} = \frac{\left(62.37\ \frac{\text{lbm}}{\text{ft}^3}\right)\left(32.2\ \frac{\text{ft}}{\text{sec}^2}\right)}{32.2\ \frac{\text{lbm-ft}}{\text{lbf-sec}^2}} = 62.37\ \text{lbf/ft}^3$$

As in *NCEES Handbook:* Absolute Roughness and Relative Roughness, for steel,

$$\begin{aligned} \epsilon &= 0.0002 \\ \frac{\epsilon}{D} &= \frac{0.0002\ \text{ft}}{0.5054\ \text{ft}} \approx 0.0004 \\ f &= 0.0175 \end{aligned}$$

The friction factor is found by following the curved line for relative roughness 0.0004 back to the intersection of the vertical line for Reynolds number 346,000. From this intersection point, travel straight left to the primary y-axis to determine the friction factor.

From Eq. 4.22 (also *NCEES Handbook:* Head Loss in Pipe or Conduit),

$$\begin{aligned} h_f &= \frac{fL\mathrm{v}^2}{2Dg} = \frac{(0.0175)(3000\ \text{ft})\left(8.33\ \frac{\text{ft}}{\text{sec}}\right)^2}{(2)(0.5054\ \text{ft})\left(32.2\ \frac{\text{ft}}{\text{sec}^2}\right)} \\ &= 111.9\ \text{ft} \end{aligned}$$

Use the Bernoulli equation (see *NCEES Handbook:* The Bernoulli Equation). Since velocity is the same at points A and B, it may be omitted.

$$\frac{p_1}{\gamma_1} = \frac{p_2}{\gamma_2} + (z_2 - z_1) + h_f$$

$$\frac{\left(12\ \frac{\text{in}}{\text{ft}}\right)^2 p_1}{62.37\ \frac{\text{lbf}}{\text{ft}^3}} = \frac{\left(50\ \frac{\text{lbf}}{\text{in}^2}\right)\left(12\ \frac{\text{in}}{\text{ft}}\right)^2}{62.37\ \frac{\text{lbf}}{\text{ft}^3}} + 60\ \text{ft} + 111.9\ \text{ft}$$

$$p_1 = 124.5\ \text{lbf/in}^2 \quad (120\ \text{psig})$$

The answer is (C).

5. From Eq. 4.29, the friction loss from A to B is

$$\begin{aligned} h_{f,\text{ft,A-B}} &= \frac{10.44 L_{\text{ft}} Q_{\text{gpm}}^{1.85}}{C^{1.85} d_{\text{in}}^{4.87}} \\ &= \frac{(10.44)(20{,}000\ \text{ft})\left(120\ \frac{\text{gal}}{\text{min}}\right)^{1.85}}{(150)^{1.85}(6\ \text{in})^{4.87}} \\ &= 22.4\ \text{ft} \end{aligned}$$

Calculate the velocity head.

$$\begin{aligned} \text{v} = \frac{\dot{V}}{A} &= \frac{120\ \frac{\text{gal}}{\text{min}}}{\left(\frac{\pi}{4}\right)\left(\frac{6\ \text{in}}{12\ \frac{\text{in}}{\text{ft}}}\right)^2 \left(7.4805\ \frac{\text{gal}}{\text{ft}^3}\right)\left(60\ \frac{\text{sec}}{\text{min}}\right)} \\ &= 1.36\ \text{ft/sec} \end{aligned}$$

$$\begin{aligned} h_{\text{v}} = \frac{\text{v}^2}{2g} &= \frac{\left(1.36\ \frac{\text{ft}}{\text{sec}}\right)^2}{(2)\left(32.2\ \frac{\text{ft}}{\text{sec}^2}\right)} \\ &= 0.029\ \text{ft} \end{aligned}$$

Velocity heads are low and can be disregarded. The friction loss from B to C is

$$\begin{aligned} h_{f,\text{B-C}} &= \frac{(10.44)(10{,}000\ \text{ft})\left(160\ \frac{\text{gal}}{\text{min}}\right)^{1.85}}{(150)^{1.85}(6\ \text{in})^{4.87}} \\ &= 19.10\ \text{ft} \end{aligned}$$

For C to D,

$$\begin{aligned} h_{f,\text{C-D}} &= \frac{(10.44)(30{,}000\ \text{ft})\left(120\ \frac{\text{gal}}{\text{min}}\right)^{1.85}}{(150)^{1.85}(4\ \text{in})^{4.87}} \\ &= 242.4\ \text{ft} \end{aligned}$$

Assume a pressure of 20 psig at point A.

$$\begin{aligned} h_{p,\text{A}} &= \frac{\left(20\ \frac{\text{lbf}}{\text{in}^2}\right)\left(12\ \frac{\text{in}}{\text{ft}}\right)^2}{62.4\ \frac{\text{lbf}}{\text{ft}^3}} \\ &= 46.2\ \text{ft} \end{aligned}$$

From the Bernoulli equation, ignoring velocity head,

$$\begin{aligned} h_{p,\text{A}} + z_{\text{A}} &= h_{p,\text{B}} + z_{\text{B}} + h_{f,\text{A-B}} \\ 46.2\ \text{ft} + 620\ \text{ft} &= h_{p,\text{B}} + 460\ \text{ft} + 22.4\ \text{ft} \\ h_{p,\text{B}} &= 183.8\ \text{ft} \end{aligned}$$

$$\begin{aligned} p_{\text{B}} = \gamma h_{p,\text{B}} &= \frac{\left(62.4\ \frac{\text{lbf}}{\text{ft}^3}\right)(183.8\ \text{ft})}{\left(12\ \frac{\text{in}}{\text{ft}}\right)^2} \\ &= 79.6\ \text{lbf/in}^2 \end{aligned}$$

For B to C,

$$\begin{aligned} h_{p,\text{B}} + z_{\text{B}} &= h_{p,\text{C}} + z_{\text{C}} + h_{f,\text{B-C}} \\ 183.8\ \text{ft} + 460\ \text{ft} &= h_{p,\text{C}} + 540\ \text{ft} + 19.10\ \text{ft} \\ h_{p,\text{C}} &= 84.7\ \text{ft} \end{aligned}$$

$$\begin{aligned} p_{\text{C}} = \gamma h_{p,\text{C}} &= \frac{\left(62.4\ \frac{\text{lbf}}{\text{ft}^3}\right)(84.7\ \text{ft})}{\left(12\ \frac{\text{in}}{\text{ft}}\right)^2} \\ &= 36.7\ \text{lbf/in}^2 \end{aligned}$$

For C to D,

$$h_{p,\mathrm{C}} + z_\mathrm{C} = h_{p,\mathrm{D}} + z_\mathrm{D} + h_{f,\mathrm{C\text{-}D}}$$
$$84.7 \text{ ft} + 540 \text{ ft} = h_{p,\mathrm{D}} + 360 \text{ ft} + 242.4 \text{ ft}$$
$$h_{p,\mathrm{D}} = 22.3 \text{ ft}$$
$$p_\mathrm{D} = \gamma h_{p,\mathrm{D}} = \frac{\left(62.4 \ \frac{\text{lbf}}{\text{ft}^3}\right)(22.3 \text{ ft})}{\left(12 \ \frac{\text{in}}{\text{ft}}\right)^2} = 9.7 \text{ lbf/in}^2$$

$p_\mathrm{D} = 9.7$ lbf/in^2 is too low; therefore, add 20 lbf/in^2 − 9.7 lbf/in^2 = 10.3 lbf/in^2 (psig) to each point.

$$p_\mathrm{A} = 20.0 \ \frac{\text{lbf}}{\text{in}^2} + 10.3 \ \frac{\text{lbf}}{\text{in}^2} = 30.3 \text{ lbf/in}^2 \quad (30 \text{ psig})$$

The answer is (D).

6. As in *NCEES Handbook:* Head Loss in Pipe or Conduit, the Darcy equation is applicable to fluids in the laminar and turbulent regions.

$$h_f = \frac{fL\mathrm{v}^2}{2Dg}$$

From *NCEES Handbook:* Friction Factors for Laminar Flow, for laminar flow in circular pipes, the friction factor is

$$f = \frac{64}{\mathrm{Re}} = \frac{64\nu}{D\mathrm{v}}$$

Combining these two equations,

$$h_f = \frac{fL\mathrm{v}^2}{2Dg} = \frac{\left(\frac{64\nu}{D\mathrm{v}}\right)L\mathrm{v}^2}{2Dg} = \frac{32\nu L\mathrm{v}}{D^2 g}$$

Friction head loss in laminar flow is proportional to velocity.

The answer is (A).

7. As in *NCEES Handbook:* Turbulent Flow, the equivalent length of four 45° standard elbows with 3 in diameter is

$$\frac{L}{D} = 16$$
$$L = 16D = \frac{(16)\left(3 \ \frac{\text{in}}{\text{elbow}}\right)(4 \text{ elbows})}{12 \ \frac{\text{in}}{\text{ft}}} = 16 \text{ ft}$$

The equivalent length of the fully open gate valve is

$$\frac{L}{D} = 10$$
$$L = 10D = \frac{(10)(3 \text{ in})}{12 \ \frac{\text{in}}{\text{ft}}} = 2.5 \text{ ft}$$

The elevation change is 20 ft, but this distance is not an equivalent length of pipe. Taking terms from the Bernoulli equation,

$$\Delta z = \frac{fL_e\mathrm{v}^2}{2Dg}$$
$$L_e = \frac{2Dg\Delta z}{f\mathrm{v}^2}$$

Use a value of 0.02, which is appropriate for steel pipe and turbulent flow. Using nominal values for a quick estimate, the equivalent length of pipe equal to the elevation increase is

$$L_e = \frac{2Dg\Delta z}{f\mathrm{v}^2} = \frac{(2)(3 \text{ in})\left(32.2 \ \frac{\text{ft}}{\text{sec}^2}\right)(20 \text{ ft})}{(0.02)\left(10 \ \frac{\text{ft}}{\text{sec}}\right)^2\left(12 \ \frac{\text{in}}{\text{ft}}\right)} = 161 \text{ ft}$$

The answer is (D).

8. *Customary U.S. Solution*

As in *NCEES Handbook* table "Pipe Dimensions and Weights," for 6 in schedule-40 pipe, the internal diameter, D, is 0.5054 ft. The internal area is 0.2006 ft^2.

As in *NCEES Handbook:* Conservation of Mass, the velocity, v, is calculated from the volumetric flow, $\dot{V}$, and the flow area, A, by

$$\begin{aligned} v &= \frac{\dot{V}}{A} = \frac{1.5\ \frac{\text{ft}^3}{\text{sec}}}{0.2006\ \text{ft}^2} \\ &= 7.48\ \text{ft/sec} \end{aligned}$$

Use App. 1A (also *NCEES Handbook* table "Physical Properties of Liquid Water (U.S. Units)"). For water at 70°F, the density is $\rho = 62.299$ lbm/ft^3, and the viscosity is $\mu = 6.552 \times 10^{-4}$ lbm/ft-sec.

As in *NCEES Handbook* table "Dimensionless Numbers," the Reynolds number is

$$\begin{aligned} \text{Re} &= \frac{Dv\rho}{\mu} \\ &= \frac{(0.5054\ \text{ft})\left(7.48\ \frac{\text{ft}}{\text{sec}}\right)\left(62.299\ \frac{\text{lbm}}{\text{ft}^3}\right)}{6.552 \times 10^{-4}\ \frac{\text{lbm}}{\text{ft-sec}}} \\ &= 3.59 \times 10^5 \end{aligned}$$

Since Re > 2100, the flow is turbulent. As in *NCEES Handbook:* Friction Factors for Turbulent Flow, the friction loss coefficient can be determined from the Moody diagram.

As in *NCEES Handbook:* Absolute Roughness and Relative Roughness, for new steel pipe, the specific roughness, ϵ, is 0.0002 ft.

The relative roughness is

$$\frac{\epsilon}{D} = \frac{0.0002\ \text{ft}}{0.5054\ \text{ft}} = 0.0004$$

From the Moody diagram with Re = 3.57×10^5 and $\epsilon/D = 0.0004$, the friction factor, f, is 0.0174.

As in *NCEES Handbook:* Head Loss in Pipe or Conduit, use Darcy's equation to compute the frictional loss.

$$\begin{aligned} h_f &= \frac{fLv^2}{2Dg} = \frac{(0.0174)(1200\ \text{ft})\left(7.48\ \frac{\text{ft}}{\text{sec}}\right)^2}{(2)(0.5054\ \text{ft})\left(32.2\ \frac{\text{ft}}{\text{sec}^2}\right)} \\ &= 35.9\ \text{ft} \quad (36\ \text{ft}) \end{aligned}$$

SI Solution

As in *NCEES Handbook* table "Pipe Dimensions and Weights," for 6 in pipe, the internal diameter is 154.1 mm, and the internal area is 186.5×10^{-4} m^2.

As in *NCEES Handbook:* Conservation of Mass, the velocity, v, is calculated from the volumetric flow, $\dot{V}$, and the flow area, A, by

$$\begin{aligned} v &= \frac{\dot{V}}{A} = \frac{40\ \frac{\text{L}}{\text{s}}}{(186.5 \times 10^{-4}\ \text{m}^2)\left(1000\ \frac{\text{L}}{\text{m}^3}\right)} \\ &= 2.145\ \text{m/s} \end{aligned}$$

From App. 1B (also *NCEES Handbook* table "Physical Properties of Liquid Water (SI Units)"), for water at 20°C, the kinematic viscosity is

$$\begin{aligned} \nu &= \frac{\mu}{\rho} = \frac{1.0050 \times 10^{-3}\ \text{Pa·s}}{998.23\ \frac{\text{kg}}{\text{m}^3}} \\ &= 1.007 \times 10^{-6}\ \text{m}^2/\text{s} \end{aligned}$$

Calculate the Reynolds number (see *NCEES Handbook* table "Dimensionless Numbers"),

$$\begin{aligned} \text{Re} &= \frac{Dv}{\nu} = \frac{(154.1\ \text{mm})\left(2.145\ \frac{\text{m}}{\text{s}}\right)}{\left(1.007 \times 10^{-6}\ \frac{\text{m}^2}{\text{s}}\right)\left(1000\ \frac{\text{mm}}{\text{m}}\right)} \\ &= 3.282 \times 10^5 \end{aligned}$$

Since Re > 2100, the flow is turbulent. The friction loss coefficient can be determined from the Moody diagram.

From *NCEES Handbook:* Absolute Roughness and Relative Roughness, for new steel pipe, the specific roughness, ϵ, is 6.0×10^{-5} m.

The relative roughness is

$$\frac{\epsilon}{D} = \frac{6.0 \times 10^{-5}\ \text{m}}{0.1541\ \text{m}} = 0.0004$$

From the Moody diagram with Re = 3.28×10^5 and $\epsilon/D = 0.0004$, the friction factor, f, is 0.0175.

As in *NCEES Handbook:* Head Loss in Pipe or Conduit, use Darcy's equation to compute the frictional loss.

$$h_f = \frac{fL\text{v}^2}{2Dg} = \frac{(0.0175)(355 \text{ m})\left(2.145 \ \frac{\text{m}}{\text{s}}\right)^2}{(2)(0.1541 \text{ m})\left(9.81 \ \frac{\text{m}}{\text{s}^2}\right)}$$
$$= 9.45 \text{ m} \quad (9.5 \text{ m})$$

The answer is (C).

9. *Customary U.S. Solution*

As in *NCEES Handbook* table "Pipe Dimensions and Weights," for 6 in schedule-40 pipe, the internal diameter, D, is 0.5054 ft. The internal area is 0.2006 ft^2.

Convert the volumetric flow rate from gal/min to ft^3/sec.

$$\dot{V} = \frac{500 \ \frac{\text{gal}}{\text{min}}}{\left(60 \ \frac{\text{sec}}{\text{min}}\right)\left(7.4805 \ \frac{\text{gal}}{\text{ft}^3}\right)} = 1.114 \text{ ft}^3/\text{sec}$$

As in *NCEES Handbook:* Conservation of Mass, the velocity is

$$\text{v} = \frac{\dot{V}}{A} = \frac{1.114 \ \frac{\text{ft}^3}{\text{sec}}}{0.2006 \text{ ft}^2} = 5.55 \text{ ft/sec}$$

Use App. 1A (also *NCEES Handbook* table "Physical Properties of Liquid Water (U.S. Units)"). For water at 100°F, the kinematic viscosity, ν, is 0.739×10^{-5} ft^2/sec, and the density is 62.00 lbm/ft^2.

As in *NCEES Handbook:* Similitude, the Reynolds number is

$$\text{Re} = \frac{D\text{v}}{\nu} = \frac{(0.5054 \text{ ft})\left(5.55 \ \frac{\text{ft}}{\text{sec}}\right)}{0.739 \times 10^{-5} \ \frac{\text{ft}^2}{\text{sec}}}$$
$$= 3.80 \times 10^5$$

Since Re > 2100, the flow is turbulent. The friction loss coefficient can be determined from the Moody diagram.

For new steel pipe, the specific roughness, ϵ, is 0.0002 ft.

From *NCEES Handbook:* Absolute Roughness and Relative Roughness, the relative roughness is

$$\frac{\epsilon}{D} = \frac{0.0002 \text{ ft}}{0.5054 \text{ ft}} = 0.0004$$

From the Moody diagram in *NCEES Handbook:* Friction Factors for Turbulent Flow, with Re $= 3.80 \times 10^5$ and $\epsilon/D = 0.0004$, the friction factor, f, is 0.0173.

Use *NCEES Handbook:* Head Loss in Pipe or Conduit. The equivalent length of two 6 in steel flanged elbows is

$$\frac{L}{D} = 20$$
$$L = 20D$$
$$= \frac{(20)\left(6 \ \frac{\text{in}}{\text{elbow}}\right)(2 \text{ elbows})}{12 \ \frac{\text{in}}{\text{ft}}}$$
$$= 20 \text{ ft}$$

The equivalent length of two fully open gate valves is

$$\frac{L}{D} = 10$$
$$L = 10D$$
$$= \frac{(10)\left(6 \ \frac{\text{in}}{\text{valve}}\right)(2 \text{ valves})}{12 \ \frac{\text{in}}{\text{ft}}}$$
$$= 10 \text{ ft}$$

The equivalent length of a fully open 90° angle valve is

$$\frac{L}{D} = 165$$
$$L = 165D$$
$$= \frac{(165)(6 \text{ in})}{12 \ \frac{\text{in}}{\text{ft}}}$$
$$= 82.5 \text{ ft}$$

The equivalent length of a swing check valve is

$$\frac{L}{D} = 125$$
$$L = 125D$$
$$= \frac{(125)(6 \text{ in})}{12 \ \frac{\text{in}}{\text{ft}}}$$
$$= 62.5 \text{ ft}$$

The total is

$$20 \text{ ft} + 10 \text{ ft} + 82.5 \text{ ft} + 62.5 \text{ ft} = 175 \text{ ft}$$

The equivalent pipe length is the sum of the straight run of pipe and the equivalent length of pipe for the valves and fittings.

$$\begin{aligned} L_e &= L + L_{\text{fittings}} = 300\ \text{ft} + 175\ \text{ft} \\ &= 475\ \text{ft} \end{aligned}$$

Using Darcy's equation (see *NCEES Handbook:* Head Loss in Pipe or Conduit), the frictional loss is

$$\begin{aligned} h_f &= \frac{fL_e \text{v}^2}{2Dg} \\ &= \frac{(0.0173)(475\ \text{ft})\left(5.55\ \frac{\text{ft}}{\text{sec}}\right)^2}{(2)(0.5054\ \text{ft})\left(32.2\ \frac{\text{ft}}{\text{sec}^2}\right)} \\ &= 7.78\ \text{ft} \end{aligned}$$

The head loss is the sum of the head losses through the pipe, valves, and fittings and the change in elevation.

$$\begin{aligned} \Delta h &= h_f + \Delta z = 7.78\ \text{ft} + 20\ \text{ft} \\ &= 27.78\ \text{ft} \end{aligned}$$

The pressure difference between the entrance and discharge is

$$\begin{aligned} \Delta p &= \gamma \Delta h = \rho \Delta h \times \frac{g}{g_c} \\ &= \left(\frac{\left(62.0\ \frac{\text{lbm}}{\text{ft}^3}\right)(27.78\ \text{ft})}{\left(12\ \frac{\text{in}}{\text{ft}}\right)^2}\right)\left(\frac{32.2\ \frac{\text{ft}}{\text{sec}^2}}{32.2\ \frac{\text{lbm-ft}}{\text{lbf-sec}^2}}\right) \\ &= 12\ \text{lbf/in}^2\quad (12\ \text{psi}) \end{aligned}$$

SI Solution

As in *NCEES Handbook* table "Pipe Dimensions and Weights," for 6 in pipe, the internal diameter is 154.1 mm (0.1541 m). The internal area is $186.5 \times 10^{-4}\ \text{m}^2$.

As in *NCEES Handbook:* Conservation of Mass, the velocity, v, is

$$\begin{aligned} \text{v} &= \frac{\dot{V}}{A} = \frac{30\ \frac{\text{L}}{\text{s}}}{(186.5 \times 10^{-4}\ \text{m}^2)\left(1000\ \frac{\text{L}}{\text{m}^3}\right)} \\ &= 1.61\ \text{m/s} \end{aligned}$$

Use App. 1B (also *NCEES Handbook* table "Physical Properties of Liquid Water (SI Units)"). For water at 40°C, the kinematic viscosity, ν, is $6.611 \times 10^{-7}\ \text{m}^2/\text{s}$, and the density is $992.25\ \text{kg/m}^3$.

As in *NCEES Handbook:* Similitude, the Reynolds number is

$$\begin{aligned} \text{Re} &= \frac{D\text{v}}{\nu} = \frac{(0.1541\ \text{m})\left(1.61\ \frac{\text{m}}{\text{s}}\right)}{6.611 \times 10^{-7}\ \frac{\text{m}^2}{\text{s}}} \\ &= 3.75 \times 10^5 \end{aligned}$$

Since Re > 2100, the flow is turbulent. The friction loss coefficient can be determined from the Moody diagram.

From *NCEES Handbook:* Absolute Roughness and Relative Roughness, for new steel pipe, the specific roughness, ϵ, is 6.0×10^{-5} m.

The relative roughness is

$$\frac{\epsilon}{D} = \frac{6.0 \times 10^{-5}\ \text{m}}{0.1541\ \text{m}} = 0.0004$$

From the Moody diagram in *NCEES Handbook:* Friction Factors for Turbulent Flow, with Re = 3.75×10^5 and $\epsilon/D = 0.0004$, the friction factor, *f*, is 0.0173.

Use App. 4D (also *NCEES Handbook:* Head Loss in Pipe or Conduit). Since 6 in is 0.1524 m, the equivalent length of two 6 in steel flanged elbows is

$$\begin{aligned} \frac{L}{D} &= 20 \\ &= 20D \\ &= (20)\left(0.1524\ \frac{\text{m}}{\text{valve}}\right)(2\ \text{elbows}) \\ &= 6.1\ \text{m} \end{aligned}$$

The equivalent length of two fully open gate valves is

$$\begin{aligned} \frac{L}{D} &= 10 \\ L &= 10D \\ &= (10)\left(0.1524\ \frac{\text{m}}{\text{valve}}\right)(2\ \text{valves}) \\ &= 3.0\ \text{m} \end{aligned}$$

The equivalent length of a fully open 90° angle valve is

$$\begin{aligned} \frac{L}{D} &= 165 \\ L &= 165D \\ &= (165)(0.1524\ \text{m}) \\ &= 25.1\ \text{m} \end{aligned}$$

The equivalent length of a swing check valve is

$$\begin{aligned}\frac{L}{D} &= 125\\ L &= 125D\\ &= (125)(0.1524\ \text{m})\\ &= 19.1\ \text{m}\end{aligned}$$

The total is

$$6.1\ \text{m} + 3.0\ \text{m} + 25.1\ \text{m} + 19.1\ \text{m} = 53.3\ \text{m}$$

The equivalent pipe length is the sum of the straight run of pipe and the equivalent length of pipe for the valves and fittings.

$$\begin{aligned}L_e &= L + L_{\text{fittings}} = 90\ \text{m} + 53.3\ \text{m}\\ &= 143.3\ \text{m}\end{aligned}$$

Using Darcy's equation (see *NCEES Handbook:* Head Loss in Pipe or Conduit), the frictional loss is

$$\begin{aligned}h_f &= \frac{fL_e\text{v}^2}{2Dg} = \frac{(0.0173)(143.3\ \text{m})\left(1.61\ \frac{\text{m}}{\text{s}}\right)^2}{(2)(0.1541\ \text{m})\left(9.81\ \frac{\text{m}}{\text{s}^2}\right)}\\ &= 2.12\ \text{m}\end{aligned}$$

The total head loss is the sum of the head losses through the pipe, valves, and fittings and the change in elevation.

$$\begin{aligned}\Delta h &= h_f + \Delta z = 2.12\ \text{m} + 6\ \text{m}\\ &= 8.12\ \text{m}\end{aligned}$$

The pressure difference between the entrance and discharge is

$$\begin{aligned}\Delta p &= \rho\Delta hg = \left(992.25\ \frac{\text{kg}}{\text{m}^3}\right)(8.12\ \text{m})\left(9.81\ \frac{\text{m}}{\text{s}^2}\right)\\ &= 79\,040\ \text{Pa}\quad (78\ \text{kPa})\end{aligned}$$

The answer is (A).

10. *Customary U.S. Solution*

As in *NCEES Handbook* table "Pipe Dimensions and Weights," for 6 in schedule-40 pipe, the internal diameter, D, is 0.5054 ft. The internal area is 0.2006 ft^2.

Use App. 1C (also *NCEES Handbook:* Temperature-Dependent Properties of Air (U.S. Customary Units)). For air at 70°F and atmospheric pressure, the kinematic viscosity is $16.39 \times 10^{-5}\ \text{ft}^2/\text{sec}$.

As in *NCEES Handbook:* Similitude, the Reynolds number is

$$\begin{aligned}\text{Re} &= \frac{D\text{v}}{\nu} = \frac{(0.5054\ \text{ft})\left(60\ \frac{\text{ft}}{\text{sec}}\right)}{16.39 \times 10^{-5}\ \frac{\text{ft}^2}{\text{sec}}}\\ &= 1.85 \times 10^5\end{aligned}$$

Since Re > 2100, the flow is turbulent. The friction loss coefficient can be determined from the Moody diagram.

For new steel pipe, the specific roughness, ϵ, is 0.0002 ft.

As in *NCEES Handbook:* Absolute Roughness and Relative Roughness, the relative roughness is

$$\frac{\epsilon}{D} = \frac{0.0002\ \text{ft}}{0.5054\ \text{ft}} = 0.0004$$

From the Moody diagram in *NCEES Handbook:* Friction Factors for Turbulent Flow, the friction factor, f, is 0.0184 when $\text{Re} = 1.85 \times 10^5$ and $\epsilon/D = 0.0004$.

Use *NCEES Handbook:* Head Loss in Pipe or Conduit. The equivalent length of two 6 in steel flanged elbows is

$$\begin{aligned}\frac{L}{D} &= 20\\ L &= 20D\\ &= \frac{(20)\left(6\ \frac{\text{in}}{\text{elbow}}\right)(2\ \text{elbows})}{12\ \frac{\text{in}}{\text{ft}}}\\ &= 20\ \text{ft}\end{aligned}$$

Use App. 4D. The equivalent length of two fully open gate valves is

$$\begin{aligned}\frac{L}{D} &= 10\\ L &= 10D\\ &= \frac{(10)\left(6\ \frac{\text{in}}{\text{valve}}\right)(2\ \text{valves})}{12\ \frac{\text{in}}{\text{ft}}}\\ &= 10\ \text{ft}\end{aligned}$$

The equivalent length of a fully open 90° angle valve is

$$\begin{aligned} \frac{L}{D} &= 165 \\ L &= 165D \\ &= \frac{(165)(6\ \text{in})}{12\ \dfrac{\text{in}}{\text{ft}}} \\ &= 82.5\ \text{ft} \end{aligned}$$

The equivalent length of a swing check valve is

$$\begin{aligned} \frac{L}{D} &= 125 \\ L &= 125D \\ &= \frac{(125)(6\ \text{in})}{12\ \dfrac{\text{in}}{\text{ft}}} \\ &= 62.5\ \text{ft} \end{aligned}$$

The total is

$$20\ \text{ft} + 10\ \text{ft} + 82.5\ \text{ft} + 62.5\ \text{ft} = 175\ \text{ft}$$

The equivalent pipe length is the sum of the straight run of pipe and the equivalent lengths of pipe for the valves and fittings.

$$\begin{aligned} L_e &= L + L_{\text{fittings}} = 300\ \text{ft} + 175\ \text{ft} \\ &= 475\ \text{ft} \end{aligned}$$

Using Darcy's equation (see *NCEES Handbook:* Head Loss in Pipe or Conduit), the frictional loss is

$$\begin{aligned} h_f &= \frac{fL_e\text{v}^2}{2Dg} = \frac{(0.0184)(450.2\ \text{ft})\left(60\ \dfrac{\text{ft}}{\text{sec}}\right)^2}{(2)(0.5054\ \text{ft})\left(32.2\ \dfrac{\text{ft}}{\text{sec}^2}\right)} \\ &= 966.7\ \text{ft} \end{aligned}$$

The head loss is the sum of the head losses through the pipe, valves, and fittings and the change in elevation.

$$\begin{aligned} \Delta h &= h_f + \Delta z = 966.7\ \text{ft} + 20\ \text{ft} \\ &= 986.7\ \text{ft} \end{aligned}$$

The pressure difference between the entrance and discharge is

$$\begin{aligned} \Delta p &= \gamma \Delta h = \rho \Delta h \times \frac{g}{g_c} \\ &= \left(\frac{\left(0.075\ \dfrac{\text{lbm}}{\text{ft}^3}\right)(986.7\ \text{ft})}{\left(12\ \dfrac{\text{in}}{\text{ft}}\right)^2}\right)\left(\frac{32.2\ \dfrac{\text{ft}}{\text{sec}^2}}{32.2\ \dfrac{\text{lbm-ft}}{\text{lbf-sec}^2}}\right) \\ &= 0.52\ \text{lbf/in}^2 \quad (0.52\ \text{psi}) \end{aligned}$$

SI Solution

Use App. 3C (also *NCEES Handbook* table "Pipe Dimensions and Weights"). For 6 in pipe, the internal diameter, D, is 154.1 mm (0.1541 m), and the internal area is $186.5 \times 10^{-4}\ \text{m}^2$.

Use App. 1D (also *NCEES Handbook:* Temperature-Dependent Properties of Air (SI Units)). For air at 20°C, the kinematic viscosity, ν, is $1.512 \times 10^{-5}\ \text{m}^2/\text{s}$.

As in *NCEES Handbook:* Similitude, the Reynolds number is

$$\begin{aligned} \text{Re} &= \frac{D\text{v}}{\nu} = \frac{(0.1541\ \text{m})\left(18\ \dfrac{\text{m}}{\text{s}}\right)}{1.512 \times 10^{-5}\ \dfrac{\text{m}^2}{\text{s}}} \\ &= 1.83 \times 10^5 \end{aligned}$$

Since Re > 2100, the flow is turbulent. The friction loss coefficient can be determined from the Moody diagram.

From *NCEES Handbook:* Absolute Roughness and Relative Roughness, for new steel pipe, the specific roughness, ϵ, is 6.0×10^{-5} m.

The relative roughness is

$$\frac{\epsilon}{D} = \frac{6.0 \times 10^{-5}\ \text{m}}{0.1541\ \text{m}} = 0.0004$$

From the Moody diagram in *NCEES Handbook:* Friction Factors for Turbulent Flow, with Re = 1.83×10^5 and $\epsilon/D = 0.0004$, the friction factor, f, is 0.0185.

Compute the equivalent lengths of the valves and fittings. (Convert from App. 4D (also *NCEES Handbook:* Head Loss in Pipe or Conduit).)

Since 6 in is 0.1524 m, the equivalent length of two 6 in steel flanged elbows is

$$\begin{aligned}\frac{L}{D} &= 20\\ L &= 20D\\ &= (20)\left(0.1524\ \frac{\text{m}}{\text{elbow}}\right)(2\text{ elbows})\\ &= 6.1\text{ m}\end{aligned}$$

The equivalent length of two fully open gate valves is

$$\begin{aligned}\frac{L}{D} &= 10\\ L &= 10D\\ &= (10)\left(0.1524\ \frac{\text{m}}{\text{valve}}\right)(2\text{ valves})\\ &= 3.0\text{ m}\end{aligned}$$

The equivalent length of a fully open 90° angle valve is

$$\begin{aligned}\frac{L}{D} &= 165\\ L &= 165D\\ &= (165)(0.1524\text{ m})\\ &= 25.1\text{ m}\end{aligned}$$

The equivalent length of a swing check valve is

$$\begin{aligned}\frac{L}{D} &= 125\\ L &= 125D\\ &= (125)(0.1524\text{ m})\\ &= 19.1\text{ m}\end{aligned}$$

The total is

$$6.1\text{ m} + 3.0\text{ m} + 25.1\text{ m} + 19.1\text{ m} = 53.3\text{ m}$$

The equivalent pipe length is the sum of the straight run of pipe and the equivalent lengths of pipe for the valves and fittings.

$$\begin{aligned}L_e &= L + L_{\text{fittings}} = 90\text{ m} + 53.3\text{ m}\\ &= 143.3\text{ m}\end{aligned}$$

As in *NCEES Handbook:* Head Loss in Pipe or Conduit, use Darcy's equation to compute the frictional loss.

$$h_f = \frac{fL_e\text{v}^2}{2Dg} = \frac{(0.0185)(143.3\text{ m})\left(18\ \frac{\text{m}}{\text{s}}\right)^2}{(2)(0.1541\text{ m})\left(9.81\ \frac{\text{m}}{\text{s}^2}\right)} = 284.1\text{ m}$$

The head loss is the sum of the head losses through the pipe, valves, and fittings and the change in elevation.

$$\begin{aligned}\Delta h &= h_f + \Delta z = 284.1\text{ m} + 6\text{ m}\\ &= 290.1\text{ m}\end{aligned}$$

The density of the air, ρ, is approximately 1.20 kg/m^3.

The pressure difference between the entrance and discharge is

$$\begin{aligned}\Delta p = \rho\Delta hg &= \left(1.20\ \frac{\text{kg}}{\text{m}^3}\right)(290.1\text{ m})\left(9.81\ \frac{\text{m}}{\text{s}^2}\right)\\ &= 3415\text{ Pa}\quad(3.4\text{ kPa})\end{aligned}$$

The answer is (B).

11. Assume that flows from reservoirs A and B are toward D and then toward C. From continuity,

$$\begin{aligned}\dot{V}_{\text{A-D}} + \dot{V}_{\text{B-D}} &= \dot{V}_{\text{D-C}}\\ A_\text{A}\text{v}_{\text{A-D}} + A_\text{B}\text{v}_{\text{B-D}} - A_\text{C}\text{v}_{\text{D-C}} &= 0\end{aligned}$$

From App. 3B (also *NCEES Handbook* table "Pipe Dimensions and Weights"), for schedule-40 pipe,

$$\begin{aligned}A_\text{A} &= 0.05134\text{ ft}^2 & D_\text{A} &= 0.2557\text{ ft}\\ A_\text{B} &= 0.5476\text{ ft}^2 & D_\text{B} &= 0.8350\text{ ft}\\ A_\text{C} &= 0.08841\text{ ft}^2 & D_\text{C} &= 0.3355\text{ ft}\end{aligned}$$

$$0.05134\text{v}_{\text{A-D}} + 0.5476\text{v}_{\text{B-D}} - 0.08841\text{v}_{\text{D-C}} = 0\quad[\text{Eq. I}]$$

Ignoring the velocity heads, the conservation of energy equation between A and D is

$$z_\text{A} = \frac{p_\text{D}}{\gamma} + z_\text{D} + h_{f,\text{A-D}}$$

$$50\text{ ft} = \frac{p_\text{D}}{62.4\ \frac{\text{lbf}}{\text{ft}^3}} + 25\text{ ft} + \frac{(0.02)(800\text{ ft})\text{v}_{\text{A-D}}^2}{(2)(0.2557\text{ ft})\left(32.2\ \frac{\text{ft}}{\text{sec}^2}\right)}$$

$$\text{v}_{\text{A-D}} = \sqrt{25.73 - 0.0165p_\text{D}}\quad[\text{Eq. II}]$$

Similarly, for B–D,

$$40\text{ ft} = \frac{p_\text{D}}{62.4\ \frac{\text{lbf}}{\text{ft}^3}} + 25\text{ ft} + \frac{(0.02)(500\text{ ft})\text{v}_{\text{B-D}}^2}{(2)(0.8350\text{ ft})\left(32.2\ \frac{\text{ft}}{\text{sec}^2}\right)}$$

$$\text{v}_{\text{B-D}} = \sqrt{80.66 - 0.0862p_\text{D}}\quad[\text{Eq. III}]$$

For D–C,

$$22\ \text{ft} = \frac{p_D}{62.4\ \frac{\text{lbf}}{\text{ft}^3}} + 25\ \text{ft} - \frac{(0.02)(1000\ \text{ft})v_{\text{D-C}}^2}{(2)(0.3355\ \text{ft})\left(32.2\ \frac{\text{ft}}{\text{sec}^2}\right)}$$

$$v_{\text{D-C}} = \sqrt{3.24 + 0.0173p_D} \quad \text{[Eq. IV]}$$

Equations I, II, III, and IV must be solved simultaneously. To do this, assume a value for p_D. This value then determines all three velocities in Eqs. II, III, and IV. These velocities are substituted into Eq. I. A trial-and-error solution yields

$$\begin{aligned} v_{\text{A-D}} &= 3.21\ \text{ft/sec} \\ v_{\text{B-D}} &= 0.408\ \text{ft/sec} \\ v_{\text{D-C}} &= 4.40\ \text{ft/sec} \\ p_D &= 933.8\ \text{lbf/ft}^2 \end{aligned}$$

Flow is from B to D.

The answer is (B).

12. *Customary U.S. Solution*

First, it is necessary to collect data on schedule-40 pipe and water. The fluid viscosity, pipe dimensions, and other parameters can be found in Chap. 1 and Chap. 3 (also *NCEES Handbook* tables "Pipe Dimensions and Weights" and "Physical Properties of Liquid Water (U.S. Units)"). At 70°F water, $\nu = 1.059 \times 10^{-5}\ \text{ft}^2/\text{sec}$.

From Table 4.2 (also *NCEES Handbook:* Absolute Roughness and Relative Roughness), $\epsilon = 0.0002$ ft. From App. 3B (also *NCEES Handbook* table "Pipe Dimensions and Weights"),

8 in pipe	$D = 0.6651$ ft
	$A = 0.3474\ \text{ft}^2$
12 in pipe	$D = 0.9948$ ft
	$A = 0.7773\ \text{ft}^2$
16 in pipe	$D = 1.25$ ft
	$A = 1.2272\ \text{ft}^2$

The flow quantity is converted from gallons per minute to cubic feet per second.

$$\begin{aligned} \dot{V} &= \frac{(8\ \text{MGD})\left(10^6\ \frac{\frac{\text{gal}}{\text{day}}}{\text{MGD}}\right)}{\left(24\ \frac{\text{hr}}{\text{day}}\right)\left(60\ \frac{\text{min}}{\text{hr}}\right)\left(7.4805\ \frac{\text{gal}}{\text{ft}^3}\right)\left(60\ \frac{\text{sec}}{\text{min}}\right)} \\ &= 12.378\ \text{ft}^3/\text{sec} \end{aligned}$$

As in *NCEES Handbook:* Conservation of Mass, the velocity for the inlet pipe is

$$v = \frac{\dot{V}}{A} = \frac{12.378\ \frac{\text{ft}^3}{\text{sec}}}{0.3474\ \text{ft}^2} = 35.63\ \text{ft/sec}$$

As in *NCEES Handbook:* Similitude, the Reynolds number is

$$\begin{aligned} \text{Re} &= \frac{Dv}{\nu} = \frac{(0.6651\ \text{ft})\left(35.63\ \frac{\text{ft}}{\text{sec}}\right)}{1.059 \times 10^{-5}\ \frac{\text{ft}^2}{\text{sec}}} \\ &= 2.24 \times 10^6 \end{aligned}$$

The relative roughness is

$$\frac{\epsilon}{D} = \frac{0.0002\ \text{ft}}{0.6651\ \text{ft}} = 0.0003$$

From the Moody diagram in *NCEES Handbook:* Friction Factors for Turbulent Flow, $f = 0.015$.

Using Eq. 4.23(b) (also *NCEES Handbook:* Head Loss in Pipe or Conduit), the frictional energy loss is

$$\begin{aligned} E_{f,1} &= h_f \times \frac{g}{g_c} = \frac{fLv^2}{2Dg_c} \\ &= \frac{(0.015)(1000\ \text{ft})\left(35.63\ \frac{\text{ft}}{\text{sec}}\right)^2}{(2)(0.6651\ \text{ft})\left(32.2\ \frac{\text{lbm-ft}}{\text{lbf-sec}^2}\right)} \\ &= 444.6\ \text{ft-lbf/lbm} \end{aligned}$$

As in *NCEES Handbook:* Conservation of Mass, the velocity for the outlet pipe is

$$v = \frac{\dot{V}}{A} = \frac{12.378\ \frac{\text{ft}^3}{\text{sec}}}{0.7773\ \text{ft}^2} = 15.92\ \text{ft/sec}$$

As in *NCEES Handbook:* Similitude, the Reynolds number is

$$\mathrm{Re} = \frac{D\mathrm{v}}{\nu} = \frac{(0.9948\ \mathrm{ft})\left(15.92\ \frac{\mathrm{ft}}{\mathrm{sec}}\right)}{1.059\times 10^{-5}\ \frac{\mathrm{ft}^2}{\mathrm{sec}}} = 1.5\times 10^6$$

The relative roughness is

$$\frac{\epsilon}{D} = \frac{0.0002\ \mathrm{ft}}{0.9948\ \mathrm{ft}} = 0.0002$$

From the Moody diagram in *NCEES Handbook:* Friction Factors for Turbulent Flow, $f = 0.014$.

Using Eq. 4.23(b) (also *NCEES Handbook:* Head Loss in Pipe or Conduit), the frictional energy loss is

$$\begin{aligned} E_{f,2} &= h_f \times \frac{g}{g_c} = \frac{fL\mathrm{v}^2}{2Dg_c} \\ &= \frac{(0.014)(1500\ \mathrm{ft})\left(15.92\ \frac{\mathrm{ft}}{\mathrm{sec}}\right)^2}{(2)(0.9948\ \mathrm{ft})\left(32.2\ \frac{\mathrm{lbm\text{-}ft}}{\mathrm{lbf\text{-}sec}^2}\right)} \\ &= 83.1\ \mathrm{ft\text{-}lbf/lbm} \end{aligned}$$

Assume a 50% split through the two branches. As in *NCEES Handbook:* Conservation of Mass, for the upper branch, the velocity is

$$\mathrm{v} = \frac{\dot{V}}{A} = \frac{\left(\frac{1}{2}\right)\left(12.378\ \frac{\mathrm{ft}^3}{\mathrm{sec}}\right)}{0.3474\ \mathrm{ft}^2} = 17.82\ \mathrm{ft/sec}$$

As in *NCEES Handbook:* Similitude, the Reynolds number is

$$\mathrm{Re} = \frac{D\mathrm{v}}{\nu} = \frac{(0.6651\ \mathrm{ft})\left(17.82\ \frac{\mathrm{ft}}{\mathrm{sec}}\right)}{1.059\times 10^{-5}\ \frac{\mathrm{ft}^2}{\mathrm{sec}}} = 1.1\times 10^6$$

The relative roughness is

$$\frac{\epsilon}{D} = \frac{0.0002\ \mathrm{ft}}{0.6651\ \mathrm{ft}} = 0.0003$$

From the Moody diagram in *NCEES Handbook:* Friction Factors for Turbulent Flow, $f = 0.015$.

As in *NCEES Handbook:* Conservation of Mass, for the 16 in pipe in the lower branch, the velocity is

$$\mathrm{v} = \frac{\dot{V}}{A} = \frac{\left(\frac{1}{2}\right)\left(12.378\ \frac{\mathrm{ft}^3}{\mathrm{sec}}\right)}{1.2272\ \mathrm{ft}^2} = 5.04\ \mathrm{ft/sec}$$

As in *NCEES Handbook:* Similitude, the Reynolds number is

$$\mathrm{Re} = \frac{D\mathrm{v}}{\nu} = \frac{(1.25\ \mathrm{ft})\left(5.04\ \frac{\mathrm{ft}}{\mathrm{sec}}\right)}{1.059\times 10^{-5}\ \frac{\mathrm{ft}^2}{\mathrm{sec}}} = 5.95\times 10^5$$

The relative roughness is

$$\frac{\epsilon}{D} = \frac{0.0002\ \mathrm{ft}}{1.25\ \mathrm{ft}} = 0.00016$$

From the Moody diagram in *NCEES Handbook:* Friction Factors for Turbulent Flow, $f = 0.015$.

These values of f for the two branches are fairly insensitive to changes in $\dot{V}$, so they will be used for the rest of the problem in both branches.

Use Eq. 4.23(b) (also *NCEES Handbook:* Head Loss in Pipe or Conduit) to find the frictional energy loss in the upper branch.

$$\begin{aligned} E_{f,\mathrm{upper}} &= h_f \times \frac{g}{g_c} = \frac{fL\mathrm{v}^2}{2Dg_c} \\ &= \frac{(0.015)(500\ \mathrm{ft})\left(17.81\ \frac{\mathrm{ft}}{\mathrm{sec}}\right)^2}{(2)(0.6651\ \mathrm{ft})\left(32.2\ \frac{\mathrm{lbm\text{-}ft}}{\mathrm{lbf\text{-}sec}^2}\right)} \\ &= 55.5\ \mathrm{ft\text{-}lbf/lbm} \end{aligned}$$

To calculate a loss for any other flow in the upper branch,

$$E_{f,\text{upper}2} = E_{f,\text{upper}}\left(\frac{\dot{V}}{\left(\frac{1}{2}\right)\left(12.378\ \frac{\text{ft}^3}{\text{sec}}\right)}\right)^2$$

$$= \left(55.5\ \frac{\text{ft-lbf}}{\text{lbm}}\right)\left(\frac{\dot{V}}{6.189\ \frac{\text{ft}^3}{\text{sec}}}\right)^2$$

$$= 1.45\dot{V}^2$$

Similarly, for the lower branch, in the 8 in section,

$$E_{f,\text{lower,8 in}} = h_f \times \frac{g}{g_c} = \frac{fL\text{v}^2}{2Dg_c}$$

$$= \frac{(0.015)(250\ \text{ft})\left(17.81\ \frac{\text{ft}}{\text{sec}}\right)^2}{(2)(0.6651\ \text{ft})\left(32.2\ \frac{\text{lbm-ft}}{\text{lbf-sec}^2}\right)}$$

$$= 27.8\ \text{ft-lbf/lbm}$$

For the lower branch, in the 16 in section,

$$E_{f,\text{lower,16 in}} = h_f \times \frac{g}{g_c} = \frac{fL\text{v}^2}{2Dg_c}$$

$$= \frac{(0.015)(1000\ \text{ft})\left(5.04\ \frac{\text{ft}}{\text{sec}}\right)^2}{(2)(1.25\ \text{ft})\left(32.2\ \frac{\text{lbm-ft}}{\text{lbf-sec}^2}\right)}$$

$$= 4.7\ \text{ft-lbf/lbm}$$

The total loss in the lower branch is

$$E_{f,\text{lower}} = E_{f,\text{lower,8 in}} + E_{f,\text{lower,16 in}}$$

$$= 27.8\ \frac{\text{ft-lbf}}{\text{lbm}} + 4.7\ \frac{\text{ft-lbf}}{\text{lbm}}$$

$$= 32.5\ \text{ft-lbf/lbm}$$

To calculate a loss for any other flow in the lower branch,

$$E_{f,\text{lower}2} = E_{f,\text{lower}}\left(\frac{\dot{V}}{\left(\frac{1}{2}\right)\left(12.378\ \frac{\text{ft}^3}{\text{sec}}\right)}\right)^2$$

$$= \left(32.5\ \frac{\text{ft-lbf}}{\text{lbm}}\right)\left(\frac{\dot{V}}{6.189\ \frac{\text{ft}^3}{\text{sec}}}\right)^2$$

$$= 0.85\dot{V}^2$$

Let x be the fraction flowing in the upper branch. Since the friction losses are equal,

$$E_{f,\text{upper}2} = E_{f,\text{lower}2}$$

$$1.45x^2 = (0.85)(1-x)^2$$

$$x = 0.434$$

$$\dot{V}_{\text{upper}} = (0.434)\left(12.378\ \frac{\text{ft}^3}{\text{sec}}\right)$$

$$= 5.372\ \text{ft}^3/\text{sec} \quad \left(5.4\ \text{ft}^3/\text{sec}\right)$$

SI Solution

First, it is necessary to collect data on schedule-40 pipe and water. The fluid viscosity, pipe dimensions, and other parameters can be found in various appendices in Chap. 1 and Chap. 3 (also *NCEES Handbook* tables "Pipe Dimensions and Weights" and "Physical Properties of Liquid Water (SI Units)"). At 20°C water, $\nu = 1.007 \times 10^{-6}\ \text{m}^2/\text{s}$.

From *NCEES Handbook:* Absolute Roughness and Relative Roughness, $\epsilon = 6 \times 10^{-5}$ m. As in *NCEES Handbook* table "Pipe Dimensions and Weights,"

8 in pipe	$D = 202.7$ mm
	$A = 322.75\ \text{cm}^2$
12 in pipe	$D = 303.2$ mm
	$A = 721.9 \times 10^{-4}\ \text{m}^2$
16 in pipe	$D = 381$ mm
	$A = 1140 \times 10^{-4}\ \text{m}^2$

As in *NCEES Handbook:* Conservation of Mass, the velocity for the inlet pipe is

$$\mathrm{v} = \frac{\dot{V}}{A} = \frac{\left(350\ \frac{\mathrm{L}}{\mathrm{s}}\right)\left(100\ \frac{\mathrm{cm}}{\mathrm{m}}\right)^2}{(322.75\ \mathrm{cm}^2)\left(1000\ \frac{\mathrm{L}}{\mathrm{m}^3}\right)} = 10.85\ \mathrm{m/s}$$

As in *NCEES Handbook:* Similitude, the Reynolds number is

$$\mathrm{Re} = \frac{D\mathrm{v}}{\nu} = \frac{(0.2027\ \mathrm{m})\left(10.85\ \frac{\mathrm{m}}{\mathrm{s}}\right)}{1.007 \times 10^{-6}\ \frac{\mathrm{m}^2}{\mathrm{s}}}$$
$$= 2.18 \times 10^6$$

The relative roughness is

$$\frac{\epsilon}{D} = \frac{6 \times 10^{-5}\ \mathrm{m}}{0.2027\ \mathrm{m}} = 0.0003$$

From the Moody diagram in *NCEES Handbook:* Friction Factors for Turbulent Flow, $f = 0.015$.

Using Eq. 4.23(a) (also *NCEES Handbook:* Head Loss in Pipe or Conduit), the frictional energy loss is

$$E_{f,1} = h_f g = \frac{fL\mathrm{v}^2}{2D}$$
$$= \frac{(0.015)(300\ \mathrm{m})\left(10.85\ \frac{\mathrm{m}}{\mathrm{s}}\right)^2}{(2)(0.2027\ \mathrm{m})}$$
$$= 1307\ \mathrm{J/kg}$$

As in *NCEES Handbook:* Conservation of Mass, the velocity for the outlet pipe is

$$\mathrm{v} = \frac{\dot{V}}{A} = \frac{350\ \frac{\mathrm{L}}{\mathrm{s}}}{(721.9 \times 10^{-4}\ \mathrm{m}^2)\left(1000\ \frac{\mathrm{L}}{\mathrm{m}^3}\right)} = 4.848\ \mathrm{m/s}$$

As in *NCEES Handbook:* Similitude, the Reynolds number is

$$\mathrm{Re} = \frac{D\mathrm{v}}{\nu} = \frac{(0.3032\ \mathrm{m})\left(4.848\ \frac{\mathrm{m}}{\mathrm{s}}\right)}{1.007 \times 10^{-6}\ \frac{\mathrm{m}^2}{\mathrm{s}}}$$
$$= 1.46 \times 10^6$$

The relative roughness is

$$\frac{\epsilon}{D} = \frac{6 \times 10^{-5}\ \mathrm{m}}{0.3032\ \mathrm{m}} = 0.0002$$

From the Moody diagram in *NCEES Handbook:* Friction Factors for Turbulent Flow, $f = 0.014$.

Using Eq. 4.23(a) (also *NCEES Handbook:* Head Loss in Pipe or Conduit), the frictional energy loss is

$$E_{f,2} = h_f g = \frac{fL\mathrm{v}^2}{2D}$$
$$= \frac{(0.014)(450\ \mathrm{m})\left(4.848\ \frac{\mathrm{m}}{\mathrm{s}}\right)^2}{(2)(0.3032\ \mathrm{m})}$$
$$= 244.2\ \mathrm{J/kg}$$

Assume a 50% split through the two branches. As in *NCEES Handbook:* Conservation of Mass, in the upper branch, the velocity is

$$\mathrm{v} = \frac{\dot{V}}{A} = \frac{\left(\frac{1}{2}\right)\left(350\ \frac{\mathrm{L}}{\mathrm{s}}\right)}{(322.7 \times 10^{-4}\ \mathrm{m}^2)\left(1000\ \frac{\mathrm{L}}{\mathrm{m}^3}\right)} = 5.423\ \mathrm{m/s}$$

As in *NCEES Handbook:* Similitude, the Reynolds number is

$$\mathrm{Re} = \frac{D\mathrm{v}}{\nu} = \frac{(0.2027\ \mathrm{m})\left(5.423\ \frac{\mathrm{m}}{\mathrm{s}}\right)}{1.007 \times 10^{-6}\ \frac{\mathrm{m}^2}{\mathrm{s}}}$$
$$= 1.1 \times 10^6$$

The relative roughness is

$$\frac{\epsilon}{D} = \frac{6 \times 10^{-5}\ \mathrm{m}}{0.2027\ \mathrm{m}} = 0.0003$$

From the Moody diagram in *NCEES Handbook:* Friction Factors for Turbulent Flow, $f = 0.015$.

As in *NCEES Handbook:* Conservation of Mass, for the 16 in pipe in the lower branch, the velocity is

$$\mathrm{v} = \frac{\dot{V}}{A} = \frac{\left(\frac{1}{2}\right)\left(350\ \frac{\mathrm{L}}{\mathrm{s}}\right)}{(1140 \times 10^{-4}\ \mathrm{m}^2)\left(1000\ \frac{\mathrm{L}}{\mathrm{m}^3}\right)} = 1.535\ \mathrm{m/s}$$

As in *NCEES Handbook:* Similitude, the Reynolds number is

$$\text{Re} = \frac{D\text{v}}{\nu} = \frac{(0.381\ \text{m})\left(1.535\ \frac{\text{m}}{\text{s}}\right)}{1.007\times 10^{-6}\ \frac{\text{m}^2}{\text{s}}}$$
$$= 5.81\times 10^5$$

The relative roughness is

$$\frac{\epsilon}{D} = \frac{6\times 10^{-5}\ \text{m}}{0.381\ \text{m}} = 0.00016$$

From the Moody diagram in *NCEES Handbook:* Friction Factors for Turbulent Flow, $f = 0.015$.

These values of f for the two branches are fairly insensitive to changes in $\dot{V}$, so they will be used for the rest of the problem in both branches.

Using Eq. 4.23(a) (also *NCEES Handbook:* Head Loss in Pipe or Conduit), the frictional energy loss in the upper branch is

$$E_{f,\text{upper}} = h_f g = \frac{fL\text{v}^2}{2D}$$
$$= \frac{(0.015)(150\ \text{m})\left(5.423\ \frac{\text{m}}{\text{s}}\right)^2}{(2)(0.2027\ \text{m})}$$
$$= 163.2\ \text{J/kg}$$

To calculate a loss for any other flow in the upper branch,

$$E_{f,\text{upper2}} = E_{f,\text{upper}}\left(\frac{\dot{V}}{\left(\frac{1}{2}\right)\left(0.350\ \frac{\text{m}^3}{\text{s}}\right)}\right)^2$$
$$= \left(163.2\ \frac{\text{J}}{\text{kg}}\right)\left(\frac{\dot{V}}{0.175\ \frac{\text{m}^3}{\text{s}}}\right)^2$$
$$= 5329\dot{V}^2$$

Similarly, for the lower branch, in the 8 in section,

$$E_{f,\text{lower},8\,\text{in}} = h_f g = \frac{fL\text{v}^2}{2D} = \frac{(0.015)(75\ \text{m})\left(5.423\ \frac{\text{m}}{\text{s}}\right)^2}{(2)(0.2027\ \text{m})}$$
$$= 81.61\ \text{J/kg}$$

For the lower branch, in the 16 in section,

$$E_{f,\text{lower},16\,\text{in}} = h_f g = \frac{fL\text{v}^2}{2D} = \frac{(0.015)(300\ \text{m})\left(1.585\ \frac{\text{m}}{\text{s}}\right)^2}{(2)(0.381\ \text{m})}$$
$$= 14.84\ \text{J/kg}$$

The total loss in the lower branch is

$$E_{f,\text{lower}} = E_{f,\text{lower},8\,\text{in}} + E_{f,\text{lower},16\,\text{in}}$$
$$= 81.61\ \frac{\text{J}}{\text{kg}} + 14.84\ \frac{\text{J}}{\text{kg}}$$
$$= 96.45\ \text{J/kg}$$

To calculate a loss for any other flow in the lower branch,

$$E_{f,\text{lower2}} = E_{f,\text{lower}}\left(\frac{\dot{V}}{\left(\frac{1}{2}\right)\left(0.350\ \frac{\text{m}^3}{\text{s}}\right)}\right)^2$$
$$= \left(96.45\ \frac{\text{J}}{\text{kg}}\right)\left(\frac{\dot{V}}{0.175\ \frac{\text{m}^3}{\text{s}}}\right)^2$$
$$= 3149\dot{V}^2$$

Let x be the fraction flowing in the upper branch. Since the friction losses are equal,

$$E_{f,\text{upper}2} = E_{f,\text{lower}2}$$
$$5329x^2 = (3149)(1-x)^2$$
$$x = 0.435$$

$$\dot{V}_{\text{upper}} = (0.435)\left(0.350\ \frac{\text{m}^3}{\text{s}}\right)$$
$$= 0.15\ \text{m}^3/\text{s}$$

The answer is (D).

13. First, find the Reynolds number from the velocity (v), flow diameter (D), and kinematic viscosity (ν).

$$\begin{aligned} \text{Re} &= \frac{\text{v}D}{\nu} \\ &= \frac{\left(4\ \frac{\text{m}}{\text{s}}\right)(0.1\ \text{m})}{10^{-6}\ \frac{\text{m}^2}{\text{s}}} \\ &= 4 \times 10^5 \end{aligned}$$

Next, find the relative roughness. For plain cast iron, the specific roughness is usually taken as 0.00024 m.

$$\frac{\epsilon}{D} = \frac{0.00024\ \text{m}}{0.1\ \text{m}} = 0.0024$$

With the value of the Reynolds number and the relative roughness, use a Moody diagram to find the friction factor, f.

$$f = 0.025$$

Multiply Darcy's equation by the specific weight of water to find the pressure difference, Δp. L is length and g is gravitational acceleration.

$$\begin{aligned} h_f &= \frac{fL\text{v}^2}{2Dg} \\ \Delta p &= \gamma h_f \\ &= \gamma \frac{fL\text{v}^2}{2Dg} \end{aligned}$$

$$\begin{aligned} \Delta p &= \left(9800\ \frac{\text{N}}{\text{m}^3}\right) \frac{(0.025)(100\ \text{m})\left(4\ \frac{\text{m}}{\text{s}}\right)^2}{(2)(0.1\ \text{m})\left(9.8\ \frac{\text{m}}{\text{s}^2}\right)} \\ &= 2.0 \times 10^5\ \text{Pa} \quad (200\ \text{kPa}) \end{aligned}$$

The answer is (B).

14.

$$A_o = \left(\frac{\pi}{4}\right)\left(\frac{4\ \text{in}}{12\ \frac{\text{in}}{\text{ft}}}\right)^2 = 0.08727\ \text{ft}^2$$

$$A_t = \left(\frac{\pi}{4}\right)(20\ \text{ft})^2 = 314.16\ \text{ft}^2$$

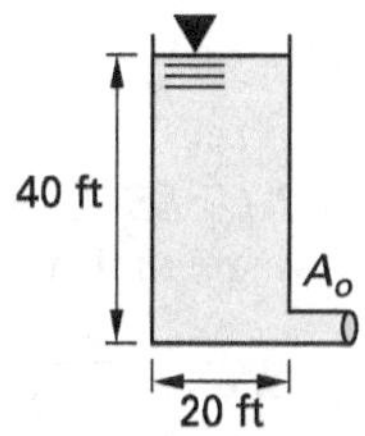

Using Torricelli's equation in Eq. 4.83 (also *NCEES Handbook*: Orifice Discharging Freely into Atmosphere), the time it takes to drop from 40 ft to 20 ft is

$$\begin{aligned} t &= \frac{2A_t(\sqrt{z_1} - \sqrt{z_2})}{C_d A_o \sqrt{2g}} \\ &= \frac{(2)(314.16\ \text{ft}^2)(\sqrt{40\ \text{ft}} - \sqrt{20\ \text{ft}})}{(0.98)(0.08727\ \text{ft}^2)\sqrt{(2)\left(32.2\ \frac{\text{ft}}{\text{sec}^2}\right)}} \\ &= 1696\ \text{sec} \quad (1700\ \text{sec}) \end{aligned}$$

The answer is (D).

15. For low velocities, the total energy head of the water in the canal is 320 ft. Similarly, the total energy head of the water in the irrigation ditch is 305 ft. The head loss is 320 ft − 305 ft = 15 ft.

The answer is (C).

16. From *NCEES Handbook:* Orifice, Nozzle, and Venturi Meters, the orifice coefficient is $C_{\text{orifice}} = 1.00$.

$$C = \frac{C_{\text{orifice}}}{\sqrt{1-\beta^4}} = \frac{C_{\text{orifice}}}{\sqrt{1 - \left(\frac{D_2}{D_1}\right)^4}}$$

$$A_2 = \left(\frac{\pi}{4}\right)\left(\frac{8\ \text{in}}{12\ \frac{\text{in}}{\text{ft}}}\right)^2 = 0.3491\ \text{ft}^2$$

The specific weight of mercury is 0.491 lbf/in^3; the specific weight of water is 0.0361 lbf/in^3.

$$\begin{aligned} p_1 - p_2 &= \Delta(\gamma h) \\ &= \left(\begin{array}{c} \left(0.491\ \frac{\text{lbf}}{\text{in}^3}\right)(4\ \text{in}) \\ -\left(0.0361\ \frac{\text{lbf}}{\text{in}^3}\right)(4\ \text{in}) \end{array} \right) \left(12\ \frac{\text{in}}{\text{ft}}\right)^2 \\ &= 262.0\ \text{lbf/ft}^2 \end{aligned}$$

From Eq. 4.156 (also *NCEES Handbook:* Square-Edge Orifice Meter (Vena Contracta Taps)),

$$\begin{aligned} \dot{V} &= CA_{\text{venturi}}\sqrt{\frac{2g_c(p_1 - p_2)}{p}} \\ &= (1.116)(1)(0.3491\ \text{ft}^2) \\ &\quad \times \sqrt{\frac{(2)\left(32.2\ \frac{\text{ft}}{\text{sec}^2}\right)\left(262\ \frac{\text{lbf}}{\text{ft}^2}\right)}{62.4\ \frac{\text{lbf}}{\text{ft}^3}}} \\ &= 6.406\ \text{ft}^3/\text{sec} \quad (6.4\ \text{ft}^3/\text{sec}) \end{aligned}$$

The answer is (C).

17. *Customary U.S. Solution*

As in *NCEES Handbook:* Square-Edge Orifice Meter (Vena Contracta Taps), the volumetric flow rate of benzene through the venturi meter is given by

$$\dot{V} = C_f A_2 \sqrt{\frac{2g(\rho_m - \rho)h}{\rho}}$$

As in *NCEES Handbook* table "Standard Values," the density of mercury, ρ_m, at 60°F is approximately 848 lbm/ft^3.

As in *NCEES Handbook:* Density Definitions, the density of the benzene at 60°F is

$$\rho = (\text{SG})\rho_{\text{water}} = (0.885)\left(62.4\ \frac{\text{lbm}}{\text{ft}^3}\right) = 55.22\ \text{lbm/ft}^3$$

The throat area is

$$A_2 = \frac{\pi D_2^2}{4} = \frac{\pi\left(\frac{3.5\ \text{in}}{12\ \frac{\text{in}}{\text{ft}}}\right)^2}{4} = 0.0668\ \text{ft}^2$$

$$\beta = \frac{3.5\ \text{in}}{8\ \text{in}} = 0.4375$$

As in *NCEES Handbook:* Venturi Flow Nozzle Meter,

$$C_f = \frac{C_d}{\sqrt{1 - \beta^4}} = \frac{0.99}{\sqrt{1 - (0.4375)^4}} = 1.00865$$

Find the volumetric flow of benzene.

$$\begin{aligned} \dot{V} &= C_f A_2 \sqrt{\frac{2g(\rho_m - \rho)h}{\rho}} \\ &= (1.00865)(0.0668\ \text{ft}^2) \\ &\quad \times \sqrt{\frac{(2)\left(32.2\ \frac{\text{ft}}{\text{sec}^2}\right)\left(848\ \frac{\text{lbm}}{\text{ft}^3} - 55.22\ \frac{\text{lbm}}{\text{ft}^3}\right) \times (4\ \text{in})}{\left(55.22\ \frac{\text{lbm}}{\text{ft}^3}\right)\left(12\ \frac{\text{in}}{\text{ft}}\right)}} \\ &= 1.181\ \text{ft}^3/\text{sec} \quad (1.2\ \text{ft}^3/\text{sec}) \end{aligned}$$

SI Solution

As in *NCEES Handbook:* Square-Edge Orifice Meter (Vena Contracta Taps), the volumetric flow rate of benzene through the venturi meter is

$$\dot{V} = C_f A_2 \sqrt{\frac{2g(\rho_m - \rho)h}{\rho}}$$

As in *NCEES Handbook* table "Standard Values," ρ_m is the density of mercury at 15°C; ρ_m is approximately 13 600 kg/m^3.

As in *NCEES Handbook:* Density Definitions, the density of the benzene at 15°C is

$$\begin{aligned} \rho &= (\text{SG})\rho_{\text{water}} = (0.885)\left(1000\ \frac{\text{kg}}{\text{m}^3}\right) \\ &= 885\ \text{kg/m}^3 \end{aligned}$$

The throat area is

$$A_2 = \frac{\pi D_2^2}{4} = \frac{\pi(0.09\ \text{m})^2}{4} = 0.0064\ \text{m}^2$$

$$\beta = \frac{9\ \text{cm}}{20\ \text{cm}} = 0.45$$

As in *NCEES Handbook:* Venturi Flow Nozzle Meter,

$$C_f = \frac{C_d}{\sqrt{1-\beta^4}} = \frac{0.99}{\sqrt{1-(0.45)^4}} = 1.01094$$

Find the volumetric flow of benzene.

$$\begin{aligned}\dot{V} &= C_f A_2 \sqrt{\frac{2g(\rho_m-\rho)h}{\rho}} \\ &= (1.01094)(0.0064\ \text{m}^2) \\ &\quad \times \sqrt{\frac{(2)\left(9.81\ \frac{\text{m}}{\text{s}^2}\right) \times \left(13\,600\ \frac{\text{kg}}{\text{m}^3} - 885\ \frac{\text{kg}}{\text{m}^3}\right)(0.1\ \text{m})}{885\ \frac{\text{kg}}{\text{m}^3}}} \\ &= 0.0344\ \text{m}^3/\text{s} \quad (34\ \text{L/s})\end{aligned}$$

The answer is (A).

18. From App. 1A (also *NCEES Handbook* table "Physical Properties of Liquid Water (U.S. Units)"), for 70°F water,

$$\nu = 1.059 \times 10^{-5}\ \text{ft}^2/\text{sec}$$

$$\gamma = 62.3\ \text{lbf/ft}^3$$

$$D_o = 0.2\ \text{ft}$$

$$\begin{aligned}\text{v}_o &= \text{v}\left(\frac{D}{D_o}\right)^2 = \left(2\ \frac{\text{ft}}{\text{sec}}\right)\left(\frac{1\ \text{ft}}{0.2\ \text{ft}}\right)^2 \\ &= 50\ \text{ft/sec}\end{aligned}$$

As in *NCEES Handbook:* Similitude,

$$\text{Re} = \frac{D_o \text{v}_o}{\nu} = \frac{(0.2\ \text{ft})\left(50\ \frac{\text{ft}}{\text{sec}}\right)}{1.059 \times 10^{-5}\ \frac{\text{ft}^2}{\text{sec}}} = 9.44 \times 10^5$$

$$A_o = \left(\frac{\pi}{4}\right)(0.2\ \text{ft})^2 = 0.0314\ \text{ft}^2$$

$$A = \left(\frac{\pi}{4}\right)(1\ \text{ft})^2 = 0.7854\ \text{ft}^2$$

$$\frac{A_o}{A} = \frac{0.0314\ \text{ft}^2}{0.7854\ \text{ft}^2} = 0.040$$

From Fig. 4.28 (also *NCEES Handbook:* Square-Edge Orifice Meter (Vena Contracta Taps)),

$$C_f \approx 0.60$$

$$\dot{V} = A\text{v} = (0.7854\ \text{ft}^2)\left(2\ \frac{\text{ft}}{\text{sec}}\right) = 1.571\ \text{ft}^3/\text{sec}$$

From Eq. 4.162(b) (also *NCEES Handbook:* Square-Edge Orifice Meter (Vena Contracta Taps)), substituting $\gamma = \rho g/g_c$,

$$\begin{aligned}p_p - p_o &= \left(\frac{\gamma}{2g}\right)\left(\frac{\dot{V}}{C_f A_o}\right)^2 \\ &= \frac{\left(\frac{62.3\ \frac{\text{lbf}}{\text{ft}^3}}{(2)\left(32.2\ \frac{\text{ft}}{\text{sec}^2}\right)}\right)\left(\frac{1.571\ \frac{\text{ft}^3}{\text{sec}}}{(0.60)(0.0314\ \text{ft}^2)}\right)^2}{\left(12\ \frac{\text{in}}{\text{ft}}\right)^2} \\ &= 46.7\ \text{lbf/in}^2 \quad (47\ \text{psi})\end{aligned}$$

The answer is (D).

19. *Customary U.S. Solution*

As in *NCEES Handbook* table "Standard Values," the densities of mercury and water are

$$\rho_{\text{mercury}} = 848\ \text{lbm/ft}^3$$

$$\rho_{\text{water}} = 62.4\ \text{lbm/ft}^3$$

The manometer tube is filled with water above the mercury column. The pressure differential across the orifice meter is

$$\Delta p = p_1 - p_2 = (\rho_{\text{mercury}} - \rho_{\text{water}})h \times \frac{g}{g_c}$$

$$= \left(\frac{\left(848\ \frac{\text{lbm}}{\text{ft}^3} - 62.4\ \frac{\text{lbm}}{\text{ft}^3}\right)(7\ \text{in})}{12\ \frac{\text{in}}{\text{ft}}}\right)\left(\frac{32.2\ \frac{\text{ft}}{\text{sec}^2}}{32.2\ \frac{\text{lbm-ft}}{\text{lbf-sec}^2}}\right)$$

$$= 458.3\ \text{lbf/ft}^2 \quad (3.2\ \text{psi})$$

SI Solution

As in *NCEES Handbook* table "Standard Values," the densities of mercury and water are

$$\rho_{\text{mercury}} = 13\,600\ \text{kg/m}^3$$
$$\rho_{\text{water}} = 1000\ \text{kg/m}^3$$

The manometer tube is filled with water above the mercury column. The pressure differential across the orifice meter is given by

$$\Delta p = p_1 - p_2 = (\rho_{\text{mercury}} - \rho_{\text{water}})hg$$
$$= \left(13\,600\ \frac{\text{kg}}{\text{m}^3} - 1000\ \frac{\text{kg}}{\text{m}^3}\right)(0.178\ \text{m})\left(9.81\ \frac{\text{m}}{\text{s}^2}\right)$$
$$= 22\,002\ \text{Pa} \quad (22\ \text{kPa})$$

The answer is (B).

20. *Customary U.S. Solution*

Use App. 3B (also *NCEES Handbook* table "Pipe Dimensions and Weights"). For 12 in pipe,

$$D = 0.99483\ \text{ft}$$
$$A = 0.7773\ \text{ft}^2$$

As in *NCEES Handbook:* Conservation of Mass, the velocity is

$$\text{v} = \frac{\dot{V}}{A} = \frac{10\ \frac{\text{ft}^3}{\text{sec}}}{0.7773\ \text{ft}^2} = 12.87\ \text{ft/sec}$$

As in *NCEES Handbook* table "Physical Properties of Liquid Water (U.S. Units)," for water at 70°F, $\nu = 1.059 \times 10^{-5}\ \text{ft}^2\text{/sec}$.

As in *NCEES Handbook:* Similitude, the Reynolds number in the pipe is

$$\text{Re} = \frac{\text{v}D}{\nu} = \frac{\left(12.87\ \frac{\text{ft}}{\text{sec}}\right)(0.99483\ \text{ft})}{1.059 \times 10^{-5}\ \frac{\text{ft}^2}{\text{sec}}}$$
$$= 1.21 \times 10^6 \quad \text{[fully turbulent]}$$

Flow through the orifice will have a higher Reynolds number and will also be turbulent.

The volumetric flow rate through a sharp-edged orifice can be found in *NCEES Handbook:* Square-Edge Orifice Meter (Vena Contracta Taps).

$$\dot{V} = CA_o\sqrt{\frac{2g(\rho_m - \rho)h}{\rho}} = CA_o\sqrt{\frac{2g_c(p_1 - p_2)}{\rho}}$$

Rearranging,

$$CA_o = \frac{\dot{V}}{\sqrt{\frac{2g_c(p_1 - p_2)}{\rho}}}$$

The maximum head loss must not exceed 25 ft.

$$\frac{\frac{g_c}{g} \times (p_1 - p_2)}{\rho} = 25\ \text{ft}$$
$$\frac{g_c(p_1 - p_2)}{\rho} = (25\ \text{ft})g$$

Substituting,

$$CA_o = \frac{10\ \frac{\text{ft}^3}{\text{sec}}}{\sqrt{(2)\left(32.2\ \frac{\text{ft}}{\text{sec}^2}\right)(25\ \text{ft})}} = 0.249\ \text{ft}^2$$

Both C and A_o depend on the orifice diameter. For a 7 in diameter orifice,

$$A_o = \frac{\pi D_o^2}{4} = \frac{\pi\left(\frac{7\ \text{in}}{12\ \frac{\text{in}}{\text{ft}}}\right)^2}{4} = 0.267\ \text{ft}^2$$
$$\frac{A_o}{A_1} = \frac{0.267\ \text{ft}^2}{0.7773\ \text{ft}^2} = 0.343$$

From Fig. 4.28 (also *NCEES Handbook* figure "Discharge Coefficient C_{orifice} for Square-Edge Orifice Meters") for $d_2/d_1 = 7/12$ and fully turbulent flow,

$$\begin{aligned} C &= \frac{C_{\text{orifice}}}{\sqrt{1-\beta^4}} = \frac{C_{\text{orifice}}}{\sqrt{1-\left(\frac{D_2}{D_1}\right)^4}} \\ &= \frac{0.605}{\sqrt{1-\left(\frac{7\text{ in}}{12\text{ in}}\right)^4}} \\ &= 0.645 \\ CA_o &= (0.645)(0.267\text{ ft}^2) = 0.172\text{ ft}^2 < 0.249\text{ ft}^2 \end{aligned}$$

Therefore, a 7 in diameter orifice is too small.

Try a 9 in diameter orifice.

$$A_o = \frac{\pi D_o^2}{4} = \frac{\pi\left(\frac{9\text{ in}}{12\ \frac{\text{in}}{\text{ft}}}\right)^2}{4} = 0.442\text{ ft}^2$$

$$\frac{A_o}{A_1} = \frac{0.442\text{ ft}^2}{0.7773\text{ ft}^2} = 0.569$$

From Fig. 4.28 (also *NCEES Handbook* figure "Discharge Coefficient C_{orifice} for Square-Edge Orifice Meters"), for $d_2/d_1 = 0.75$ and fully turbulent flow,

$$\begin{aligned} C &= \frac{C_{\text{orifice}}}{\sqrt{1-\beta^4}} = \frac{C_{\text{orifice}}}{\sqrt{1-\left(\frac{D_2}{D_1}\right)^4}} \\ &= \frac{0.615}{\sqrt{1-\left(\frac{9\text{ in}}{12\text{ in}}\right)^4}} \\ &= 0.73 \\ CA_o &= (0.73)(0.442\text{ ft}^2) = 0.323\text{ ft}^2 > 0.249\text{ ft}^2 \end{aligned}$$

Therefore, a 9 in orifice is too large.

Interpolating gives

$$\begin{aligned} D_o &= 7\text{ in} + \frac{(9\text{ in} - 7\text{ in})(0.249\text{ ft}^2 - 0.172\text{ ft}^2)}{0.323\text{ ft}^2 - 0.172\text{ ft}^2} \\ &= 8.0\text{ in} \end{aligned}$$

Further iterations yield

$$\begin{aligned} D_o &\approx 8.1\text{ in} \\ CA_o &= 0.243\text{ ft}^2 \end{aligned}$$

SI Solution

For 300 mm inside diameter pipe, $D = 0.3$ m.

As in *NCEES Handbook:* Conservation of Mass, the velocity is

$$\begin{aligned} \text{v} &= \frac{\dot{V}}{A} = \frac{\dot{V}}{\frac{\pi D^2}{4}} = \frac{250\ \frac{\text{L}}{\text{s}}}{\left(\frac{\pi(0.3\text{ m})^2}{4}\right)\left(1000\ \frac{\text{L}}{\text{m}^3}\right)} \\ &= 3.54\text{ m/s} \end{aligned}$$

From App. 1B (also *NCEES Handbook* table "Physical Properties of Liquid Water (SI Units)"), for water at 20°C, $\nu = 1.007 \times 10^{-6}\ \text{m}^2/\text{s}$.

As in *NCEES Handbook:* Similitude, the Reynolds number in the pipe is

$$\begin{aligned} \text{Re} &= \frac{\text{v}D}{\nu} = \frac{\left(3.54\ \frac{\text{m}}{\text{s}}\right)(0.3\text{ m})}{1.007 \times 10^{-6}\ \frac{\text{m}^2}{\text{s}}} \\ &= 1.05 \times 10^6 \quad \text{[fully turbulent]} \end{aligned}$$

Flow through the orifice will have a higher Reynolds number and also be turbulent.

As in *NCEES Handbook:* Square-Edge Orifice Meter (Vena Contracta Taps), the volumetric flow rate through a sharp-edged orifice is

$$\dot{V} = CA_o\sqrt{\frac{2g(\rho_m - \rho)h}{\rho}} = CA_o\sqrt{\frac{2(p_1 - p_2)}{\rho}}$$

Rearranging,

$$CA_o = \frac{\dot{V}}{\sqrt{\frac{2(p_1 - p_2)}{\rho}}}$$

The maximum head loss must not exceed 7.5 m.

$$\begin{aligned} \frac{p_1 - p_2}{g\rho} &= 7.5\text{ m} \\ \frac{p_1 - p_2}{\rho} &= (7.5\text{ m})g \end{aligned}$$

Substituting,

$$CA_o = \frac{0.25 \ \frac{\text{m}^3}{\text{s}}}{\sqrt{(2)\left(9.81 \ \frac{\text{m}}{\text{s}^2}\right)(7.5 \text{ m})}} = 0.021 \text{ m}^2$$

Both C and A_o depend on the orifice diameter. For an 18 cm diameter orifice,

$$A_o = \frac{\pi D_o^2}{4} = \frac{\pi(0.18 \text{ m})^2}{4} = 0.0254 \text{ m}^2$$

$$\frac{A_o}{A_1} = \frac{0.0254 \text{ m}^2}{0.0707 \text{ m}^2} = 0.359$$

From Fig. 4.28 (also *NCEES Handbook* figure "Discharge Coefficient C_{orifice} for Square-Edge Orifice Meters"), for $d_2/d_1 =$ 18 cm/30 cm and fully turbulent flow,

$$\begin{aligned} C &= \frac{C_{\text{orifice}}}{\sqrt{1-\beta^4}} = \frac{C_{\text{orifice}}}{\sqrt{1-\left(\frac{D_2}{D_1}\right)^4}} \\ &= \frac{0.605}{\sqrt{1-\left(\frac{18 \text{ cm}}{30 \text{ cm}}\right)^4}} \\ &= 0.65 \\ CA_o &= (0.65)(0.0254 \text{ m}^2) = 0.0165 \text{ m}^2 < 0.021 \text{ m}^2 \end{aligned}$$

Therefore, an 18 cm diameter orifice is too small.

Try a 23 cm diameter orifice.

$$A_o = \frac{\pi D_o^2}{4} = \frac{\pi(0.23 \text{ m})^2}{4} = 0.0415 \text{ m}^2$$

$$\frac{A_o}{A_1} = \frac{0.0415 \text{ m}^2}{0.0707 \text{ m}^2} = 0.587$$

From Fig. 4.28 (also *NCEES Handbook* figure "Discharge Coefficient C_{orifice} for Square-Edge Orifice Meters"), for $d_2/d_1 =$ 23 cm/30 cm and fully turbulent flow,

$$\begin{aligned} C &= \frac{C_{\text{orifice}}}{\sqrt{1-\beta^4}} = \frac{C_{\text{orifice}}}{\sqrt{1-\left(\frac{D_2}{D_1}\right)^4}} \\ &= \frac{0.615}{\sqrt{1-\left(\frac{23 \text{ cm}}{30 \text{ cm}}\right)^4}} \\ &= 0.73 \\ CA_o &= (0.73)(0.0415 \text{ m}^2) = 0.0303 \text{ m}^2 > 0.021 \text{ m}^2 \end{aligned}$$

Therefore, a 23 cm orifice is too large.

Interpolating gives

$$\begin{aligned} D_o &= 18 \text{ cm} \\ &\quad +(23 \text{ cm} - 18 \text{ cm})\left(\frac{0.021 \text{ m}^2 - 0.0165 \text{ m}^2}{0.0303 \text{ m}^2 - 0.0165 \text{ m}^2}\right) \\ &= 19.6 \text{ cm} \quad (19 \text{ cm}) \end{aligned}$$

Further iteration yields

$$\begin{aligned} D_o &= 20 \text{ cm} \\ C &= 0.675 \\ CA_o &= 0.021 \text{ m}^2 \end{aligned}$$

The answer is (C).

21.

$$A_A = \left(\frac{\pi}{4}\right)\left(\frac{24 \text{ in}}{12 \ \frac{\text{in}}{\text{ft}}}\right)^2 = 3.142 \text{ ft}^2$$

$$A_B = \left(\frac{\pi}{4}\right)\left(\frac{12 \text{ in}}{12 \ \frac{\text{in}}{\text{ft}}}\right)^2 = 0.7854 \text{ ft}^2$$

As in *NCEES Handbook:* Conservation of Mass,

$$\text{v}_A = \frac{\dot{V}}{A} = \frac{8 \ \frac{\text{ft}^3}{\text{sec}}}{3.142 \text{ ft}^2} = 2.546 \text{ ft/sec}$$

$$p_A = \gamma h = \left(62.4 \ \frac{\text{lbf}}{\text{ft}^3}\right)(20 \text{ ft}) = 1248 \text{ lbf/ft}^2$$

$$\text{v}_B = \frac{\dot{V}}{A} = \frac{8 \ \frac{\text{ft}^3}{\text{sec}}}{0.7854 \text{ ft}^2} = 10.19 \text{ ft/sec}$$

Use the Bernoulli equation (see *NCEES Handbook:* The Bernoulli Equation) to solve for p_B.

$$\begin{aligned}
p_B &= p_A + \left(\frac{v_A^2}{2g} - \frac{v_B^2}{2g}\right)\gamma \\
&= 1248\ \frac{\text{lbf}}{\text{ft}^2} + \left(\frac{\left(2.546\ \frac{\text{ft}}{\text{sec}}\right)^2 - \left(10.19\ \frac{\text{ft}}{\text{sec}}\right)^2}{(2)\left(32.2\ \frac{\text{ft}}{\text{sec}^2}\right)}\right) \\
&\quad \times \left(62.4\ \frac{\text{lbf}}{\text{ft}^3}\right) \\
&= 1153.7\ \text{lbf/ft}^2
\end{aligned}$$

With $\theta = 0°$, from Eq. 4.202(b),

$$\begin{aligned}
F_x &= p_B A_B - p_A A_A + \frac{\dot{m}(v_B - v_A)}{g_c} \\
&= \left(1153.7\ \frac{\text{lbf}}{\text{ft}^2}\right)(0.7854\ \text{ft}^2) - \left(1248\ \frac{\text{lbf}}{\text{ft}^2}\right)(3.142\ \text{ft}^2) \\
&\quad + \frac{\left(\left(8\ \frac{\text{ft}^3}{\text{sec}}\right)\left(62.4\ \frac{\text{lbm}}{\text{ft}^3}\right)\right)\left(10.19\ \frac{\text{ft}}{\text{sec}} - 2.546\ \frac{\text{ft}}{\text{sec}}\right)}{32.2\ \frac{\text{lbm-ft}}{\text{lbf-sec}^2}} \\
&= -2897\ \text{lbf}\ (2900\ \text{lbf})\ \text{on the fluid (toward A)}
\end{aligned}$$

$$F_y = 0$$

The answer is (A).

22. *Customary U.S. Solution*

As in *NCEES Handbook* table "Pipe Dimensions and Weights," for schedule-40 pipe,

$$D_i = 0.9948\ \text{ft}$$
$$A_i = 0.7773\ \text{ft}^2$$

As in *NCEES Handbook:* Conservation of Mass,

$$\begin{aligned}
v &= \frac{\dot{V}}{A_i} = \frac{2000\ \frac{\text{gal}}{\text{min}}}{(0.7773\ \text{ft}^2)\left(7.4805\ \frac{\text{gal}}{\text{ft}^3}\right)\left(60\ \frac{\text{sec}}{\text{min}}\right)} \\
&= 5.73\ \text{ft/sec}
\end{aligned}$$

The pressures are in terms of gage pressure. As in *NCEES Handbook* table "Standard Values," the density of mercury is 0.491 lbm/in.

$$\begin{aligned}
p_i &= \left(14.7\ \frac{\text{lbf}}{\text{in}^2} - \frac{(6\ \text{in})\left(0.491\ \frac{\text{lbm}}{\text{in}^3}\right)\left(32.2\ \frac{\text{ft}}{\text{sec}^2}\right)}{32.2\ \frac{\text{lbm-ft}}{\text{lbf-sec}^2}}\right) \\
&\quad \times \left(12\ \frac{\text{in}}{\text{ft}}\right)^2 \\
&= 1692.6\ \text{lbf/ft}^2
\end{aligned}$$

$$E_{ti} = \frac{p_i}{\rho} + \frac{v_i^2}{2g_c} + \frac{z_i g}{g_c}$$

The inlet pump head can be calculated from a variation of the Bernoulli equation (also *NCEES Handbook:* The Bernoulli Equation).

$$E_{ti} = \frac{p_i}{\rho} + \frac{v_i^2}{2g_c} + \frac{z_i g}{g_c}$$

Since the pump inlet and outlet are at the same elevation, use $\Delta z = 0$. $\rho = (\text{SG})\rho_{\text{water}}$.

$$\begin{aligned}
E_{ti} &= \frac{p_i}{(\text{SG})\rho_{\text{water}}} + \frac{v_i^2}{2g_c} + 0 \\
&= \frac{1692.6\ \frac{\text{lbf}}{\text{ft}^2}}{(1.2)\left(62.4\ \frac{\text{lbm}}{\text{ft}^3}\right)} + \frac{\left(5.73\ \frac{\text{ft}}{\text{sec}}\right)^2}{(2)\left(32.2\ \frac{\text{lbm-ft}}{\text{lbf-sec}^2}\right)} \\
&= 23.11\ \text{ft-lbf/lbm}
\end{aligned}$$

Calculate the total head at the inlet.

$$\begin{aligned}
h_{ti} &= E_{ti} \times \frac{g_c}{g} = \left(23.11\ \frac{\text{ft-lbf}}{\text{lbm}}\right)\left(\frac{32.2\ \frac{\text{lbm-ft}}{\text{lbf-sec}^2}}{32.2\ \frac{\text{ft}}{\text{sec}^2}}\right) \\
&= 23.11\ \text{ft}
\end{aligned}$$

Refer to *NCEES Handbook* table "Pipe Dimensions and Weights." At the outlet side of the pump,

$$D_o = 0.6651 \text{ ft}$$
$$A_o = 0.3474 \text{ ft}^2$$

$$\text{v}_o = \frac{Q}{A_o} = \frac{2000\ \frac{\text{gal}}{\text{min}}}{(0.3474 \text{ ft}^2)\left(7.4805\ \frac{\text{gal}}{\text{ft}^3}\right)\left(60\ \frac{\text{sec}}{\text{min}}\right)}$$
$$= 12.83 \text{ ft/sec}$$

The pressures are in terms of gage pressure. The gauge is located 4 ft above the pump outlet, which adds 4 ft of pressure head at the pump outlet.

$$p_o = \left(14.7\ \frac{\text{lbf}}{\text{in}^2} + 20\ \frac{\text{lbf}}{\text{in}^2}\right)\left(12\ \frac{\text{in}}{\text{ft}}\right)^2 + (4 \text{ ft})\left(\frac{(1.2)\left(62.4\ \frac{\text{lbm}}{\text{ft}^3}\right)\left(32.2\ \frac{\text{ft}}{\text{sec}^2}\right)}{32.2\ \frac{\text{lbm-ft}}{\text{lbf-sec}^2}}\right)$$
$$= 5296 \text{ lbf/ft}^2$$

Calculate the outlet pressure head using the Bernoulli equation.

$$E_{to} = \frac{p_o}{\rho} + \frac{\text{v}_o^2}{2g_c} + \frac{z_o g}{g_c}$$

Since the pump inlet and outlet are at the same elevation, $\Delta z = 0$. $\rho = (\text{SG})\rho_{\text{water}}$.

$$E_{to} = \frac{p_o}{(\text{SG})\rho_{\text{water}}} + \frac{\text{v}_o^2}{2g_c} + 0$$
$$= \frac{5296\ \frac{\text{lbf}}{\text{ft}^2}}{(1.2)\left(62.4\ \frac{\text{lbm}}{\text{ft}^3}\right)} + \frac{\left(12.83\ \frac{\text{ft}}{\text{sec}}\right)^2}{(2)\left(32.2\ \frac{\text{lbm-ft}}{\text{lbf-sec}^2}\right)}$$
$$= 73.28 \text{ ft-lbf/lbm}$$

Calculate the total head at the outlet.

$$h_{to} = E_{to} \times \frac{g_c}{g} = \left(73.28\ \frac{\text{ft-lbf}}{\text{lbm}}\right)\left(\frac{32.2\ \frac{\text{lbm-ft}}{\text{lbf-sec}^2}}{32.2\ \frac{\text{ft}}{\text{sec}^2}}\right)$$
$$= 73.28 \text{ ft}$$

Compute the total head across the pump.

$$\Delta h = h_{to} - h_{ti} = 73.28 \text{ ft} - 23.11 \text{ ft} = 50.17 \text{ ft}$$

The mass flow rate is

$$\dot{m} = \rho\dot{V} = (\text{SG})\rho_{\text{water}}\dot{V}$$
$$= \frac{(1.2)\left(62.4\ \frac{\text{lbm}}{\text{ft}^3}\right)\left(2000\ \frac{\text{gal}}{\text{min}}\right)}{\left(7.4805\ \frac{\text{gal}}{\text{ft}^3}\right)\left(60\ \frac{\text{sec}}{\text{min}}\right)}$$
$$= 333.7 \text{ lbm/sec}$$

As in *NCEES Handbook:* Pump Power, the power input to the pump is

$$P = \frac{\Delta h\dot{m} \times \frac{g}{g_c}}{\eta}$$
$$= \frac{(50.17 \text{ ft})\left(333.7\ \frac{\text{lbm}}{\text{sec}}\right)\left(\frac{32.2\ \frac{\text{ft}}{\text{sec}^2}}{32.2\ \frac{\text{lbm-ft}}{\text{lbf-sec}^2}}\right)}{(0.85)\left(550\ \frac{\text{ft-lbf}}{\text{hp-sec}}\right)}$$
$$= 35.8 \text{ hp} \quad (36 \text{ hp})$$

(It is not necessary to use absolute pressures as has been done in this solution.)

SI Solution

As in *NCEES Handbook* table "Pipe Dimensions and Weights," for schedule-40 pipe,

$$D_i = 303.2 \text{ mm}$$
$$A_i = 0.0722 \text{ m}^2$$

As in *NCEES Handbook:* Conservation of Mass,

$$\text{v} = \frac{\dot{V}}{A_i} = \frac{0.125\ \frac{\text{m}^3}{\text{s}}}{0.0722 \text{ m}^2} = 1.73 \text{ m/s}$$

The pressures are in terms of gage pressure, and the density of mercury is 13 600 kg/m^3.

$$p_i = 1.013 \times 10^5 \text{ Pa} - (0.15 \text{ m})\left(13\,600\ \frac{\text{kg}}{\text{m}^3}\right)\left(9.81\ \frac{\text{m}}{\text{s}^2}\right)$$
$$= 8.13 \times 10^4 \text{ Pa}$$

The inlet pump head can be calculated from a variation of the Bernoulli equation (also *NCEES Handbook:* The Bernoulli Equation).

$$E_{ti} = \frac{p}{\rho} + \frac{v_i^2}{2} + z_i g$$

Since the pump inlet and outlet are at the same elevation, $\Delta z = 0$. $\rho = (\text{SG})\rho_{\text{water}}$.

$$\begin{aligned} E_{ti} &= \frac{p}{(\text{SG})\rho_{\text{water}}} + \frac{v_i^2}{2} + 0 \\ &= \frac{8.13 \times 10^4 \text{ Pa}}{(1.2)\left(1000 \frac{\text{kg}}{\text{m}^3}\right)} + \frac{\left(1.73 \frac{\text{m}}{\text{s}}\right)^2}{2} \\ &= 69.2 \text{ J/kg} \end{aligned}$$

The total head at the inlet is

$$h_{ti} = \frac{E_{ti}}{g} = \frac{69.2 \frac{\text{J}}{\text{kg}}}{9.81 \frac{\text{m}}{\text{s}^2}} = 7.05 \text{ m}$$

Assume the pipe nominal diameter is equal to the internal diameter. As in *NCEES Handbook* table "Pipe Dimensions and Weights," for the outlet side of the pump,

$$\begin{aligned} D_i &= 202.7 \text{ mm} \\ A_o &= 0.0323 \text{ m}^2 \end{aligned}$$

As in *NCEES Handbook:* Conservation of Mass,

$$v_o = \frac{\dot{V}}{A_o} = \frac{0.125 \frac{\text{m}^3}{\text{s}}}{0.0323 \text{ m}^2} = 3.87 \text{ m/s}$$

The pressures are in terms of gage pressure. The gauge is located 1.2 m above the pump outlet, which adds 1.2 m of pressure head at the pump outlet.

$$\begin{aligned} p_o &= 1.013 \times 10^5 \text{ Pa} + 138 \times 10^3 \text{ Pa} \\ &\quad + (1.2 \text{ m})\left((1.2)\left(1000 \frac{\text{kg}}{\text{m}^3}\right)\left(9.81 \frac{\text{m}}{\text{s}^2}\right)\right) \\ &= 2.53 \times 10^5 \text{ Pa} \end{aligned}$$

Calculate the outlet pressure head using the Bernoulli equation.

$$E_{to} = \frac{p_o}{\rho} + \frac{v_o^2}{2} + z_o g$$

Since the pump inlet and outlet are at the same elevation, $\Delta z = 0$. $\rho = (\text{SG})\rho_{\text{water}}$.

$$\begin{aligned} E_{to} &= \frac{p_o}{(\text{SG})\rho_{\text{water}}} + \frac{v_o^2}{2} + 0 \\ &= \frac{2.53 \times 10^5 \text{ Pa}}{(1.2)\left(1000 \frac{\text{kg}}{\text{m}^3}\right)} + \frac{\left(3.87 \frac{\text{m}}{\text{s}}\right)^2}{2} \\ &= 218.3 \text{ J/kg} \end{aligned}$$

The total head at the outlet is

$$h_{to} = \frac{E_{to}}{g} = \frac{218.3 \frac{\text{J}}{\text{kg}}}{9.81 \frac{\text{m}}{\text{s}^2}} = 22.25 \text{ m}$$

The total head across the pump is

$$\Delta h = h_{to} - h_{ti} = 22.25 \text{ m} - 7.05 \text{ m} = 15.2 \text{ m}$$

The mass flow rate is

$$\begin{aligned} \dot{m} &= \rho \dot{V} = (\text{SG})\rho_{\text{water}} Q \\ &= (1.2)\left(1000 \frac{\text{kg}}{\text{m}^3}\right)\left(0.125 \frac{\text{m}^3}{\text{s}}\right) \\ &= 150 \text{ kg/s} \end{aligned}$$

As in *NCEES Handbook:* Pump Power, the power input to the pump is

$$\begin{aligned} P &= \frac{\Delta h \dot{m} g}{\eta} = \frac{(15.2 \text{ m})\left(150 \frac{\text{kg}}{\text{s}}\right)\left(9.81 \frac{\text{m}}{\text{s}^2}\right)}{0.85} \\ &= 26\,314 \text{ W} \quad (26 \text{ kW}) \end{aligned}$$

(It is not necessary to use absolute pressures as has been done in this solution.)

The answer is (B).

23. *Customary U.S. Solution*

The mass flow rate is

$$\dot{m} = \dot{V}\rho = \left(100\ \frac{\text{ft}^3}{\text{sec}}\right)\left(62.4\ \frac{\text{lbm}}{\text{ft}^3}\right) = 6240\ \text{lbm/sec}$$

Refer to *NCEES Handbook:* Similitude. The head loss across the horizontal turbine can be found using a modified version of the Bernoulli equation.

$$h_{\text{loss}} = \frac{\Delta p}{\rho} \times \frac{g_c}{g} = \left(\frac{\left(30\ \frac{\text{lbf}}{\text{in}^2} - \left(-5\ \frac{\text{lbf}}{\text{in}^2}\right)\right)\left(12\ \frac{\text{in}}{\text{ft}}\right)^2}{62.4\ \frac{\text{lbm}}{\text{ft}^3}}\right)\left(\frac{32.2\ \frac{\text{lbm-ft}}{\text{lbf-sec}^2}}{32.2\ \frac{\text{ft}}{\text{sec}^2}}\right) = 80.77\ \text{ft}$$

From Table 5.5 (also *NCEES Handbook:* Pump Power), the power developed by the turbine is

$$P = \frac{h_{\text{loss}}\dot{m}}{550} \times \frac{g}{g_c} = \left(\frac{\left(6240\ \frac{\text{lbm}}{\text{sec}}\right)(80.77\ \text{ft})}{550\ \frac{\text{ft-lbf}}{\text{hp-sec}}}\right)\left(\frac{32.2\ \frac{\text{ft}}{\text{sec}^2}}{32.2\ \frac{\text{lbm-ft}}{\text{lbf-sec}^2}}\right) = 916\ \text{hp} \quad (920\ \text{hp})$$

SI Solution

The mass flow rate is

$$\dot{m} = \dot{V}\rho = \left(2.6\ \frac{\text{m}^3}{\text{s}}\right)\left(1000\ \frac{\text{kg}}{\text{m}^3}\right) = 2600\ \text{kg/s}$$

The head loss across the horizontal turbine is

$$h_{\text{loss}} = \frac{\Delta p}{\rho g} = \frac{(210\ \text{kPa} - (-35\ \text{kPa}))\left(1000\ \frac{\text{Pa}}{\text{kPa}}\right)}{\left(1000\ \frac{\text{kg}}{\text{m}^3}\right)\left(9.81\ \frac{\text{m}}{\text{s}^2}\right)} = 25.0\ \text{m}$$

From Table 5.6, the power developed by the turbine is

$$P = 9.81 h_{\text{loss}}\dot{m} = \frac{\left(9.81\ \frac{\text{m}}{\text{s}^2}\right)(25.0\ \text{m})\left(2600\ \frac{\text{kg}}{\text{s}}\right)}{1000\ \frac{\text{W}}{\text{kW}}} = 638\ \text{W} \quad (640\ \text{kW})$$

The answer is (D).

24. The speed of the truck relative to the ground is 65 mph. However, power is also required to overcome oncoming wind. The speed of the truck relative to the wind is 65 mph + 15 mph = 80 mph. If the truck were stationary, the wind would exert a force on the truck that would push the truck backward, relative to the ground. The frictional forces between the tires and ground, as well as within the parking brake systems, perform work while preventing this motion.

The answer is (D).

25. From Eq. 4.218(b) (also *NCEES Handbook:* Particle Flow), the drag force on the antenna is

$$F_D = \frac{C_D A \rho \text{v}^2}{2 g_c} = \frac{(1.2)(0.8\ \text{ft}^2)\left(0.076\ \frac{\text{lbm}}{\text{ft}^3}\right) \times \left(\frac{\left(60\ \frac{\text{mi}}{\text{hr}}\right)\left(5280\ \frac{\text{ft}}{\text{mi}}\right)}{3600\ \frac{\text{sec}}{\text{hr}}}\right)^2}{(2)\left(32.2\ \frac{\text{lbm-ft}}{\text{lbf-sec}^2}\right)} = 8.77\ \text{lbf} \quad (9.0\ \text{lbf})$$

The answer is (A).

26. *Customary U.S. Solution*

For air at 70°F,

$$\rho = \frac{p}{RT} = \frac{\left(14.7\ \frac{\text{lbf}}{\text{in}^2}\right)\left(12\ \frac{\text{in}}{\text{ft}}\right)^2}{\left(53.35\ \frac{\text{ft-lbf}}{\text{lbm-°R}}\right)(70\text{°F} + 460\text{°})}$$
$$= 0.0749\ \text{lbm/ft}^3$$

$$\text{v} = \frac{\left(55\ \frac{\text{mi}}{\text{hr}}\right)\left(5280\ \frac{\text{ft}}{\text{mi}}\right)}{3600\ \frac{\text{sec}}{\text{hr}}}$$
$$= 80.67\ \text{ft/sec}$$

The drag on the car is

$$F_D = \frac{C_D A \rho \text{v}^2}{2g_c}$$
$$= \frac{(0.42)(28\ \text{ft}^2)\left(0.0749\ \frac{\text{lbm}}{\text{ft}^3}\right)\left(80.67\ \frac{\text{ft}}{\text{sec}}\right)^2}{(2)\left(32.2\ \frac{\text{lbm-ft}}{\text{lbf-sec}^2}\right)}$$
$$= 89.0\ \text{lbf}$$

The total resisting force is

$$F = F_D + \text{rolling resistance}$$
$$= 89.0\ \text{lbf} + (0.01)(3300\ \text{lbm}) \times \frac{g}{g_c}$$
$$= 89.0\ \text{lbf} + (0.01)(3300\ \text{lbm})\left(\frac{32.2\ \frac{\text{ft}}{\text{sec}^2}}{32.2\ \frac{\text{lbm-ft}}{\text{lbf-sec}^2}}\right)$$
$$= 122.0\ \text{lbf}$$

The power manifested by virtue of the car's velocity is

$$P = F\text{v} = \frac{(122.0\ \text{lbf})\left(80.67\ \frac{\text{ft}}{\text{sec}}\right)}{778\ \frac{\text{ft-lbf}}{\text{Btu}}}$$
$$= 12.65\ \text{Btu/sec}$$

The energy available from the fuel is

$$E_A = (\text{engine thermal efficiency})(\text{fuel heating value})$$
$$= (0.28)\left(115{,}000\ \frac{\text{Btu}}{\text{gal}}\right)$$
$$= 32{,}2000\ \text{Btu/gal}$$

The fuel consumption at 55 mi/hr is

$$\frac{P}{E_A\text{v}} = \frac{\left(12.65\ \frac{\text{Btu}}{\text{sec}}\right)\left(3600\ \frac{\text{sec}}{\text{hr}}\right)}{\left(32{,}200\ \frac{\text{Btu}}{\text{gal}}\right)\left(55\ \frac{\text{mi}}{\text{hr}}\right)}$$
$$= 0.0257\ \text{gal/mi} \quad (0.026\ \text{gal/mi})$$

SI Solution

For air at 20°C,

$$\rho = \frac{p}{RT} = \frac{1.013 \times 10^5\ \text{Pa}}{\left(287.03\ \frac{\text{J}}{\text{kg·K}}\right)(20\text{°C} + 273\text{°})}$$
$$= 1.205\ \text{kg/m}^3$$

$$\text{v} = \frac{\left(90\ \frac{\text{km}}{\text{h}}\right)\left(1000\ \frac{\text{m}}{\text{km}}\right)}{3600\ \frac{\text{s}}{\text{h}}}$$
$$= 25.0\ \text{m/s}$$

The drag on the car is

$$F_D = \frac{C_D A \rho \text{v}^2}{2} = \frac{(0.42)(2.6\ \text{m}^2)\left(1.205\ \frac{\text{kg}}{\text{m}^3}\right)\left(25.0\ \frac{\text{m}}{\text{s}}\right)^2}{2}$$
$$= 411.2\ \text{N}$$

The total resisting force is

$$F = F_D + \text{rolling resistance} \times g$$
$$= 411.2\ \text{N} + (0.01)(1500\ \text{kg})g$$
$$= 411.2\ \text{N} + (0.01)(1500\ \text{kg})\left(9.81\ \frac{\text{m}}{\text{s}^2}\right)$$
$$= 558.4\ \text{N}$$

The power required is

$$P = F\text{v} = (558.4\ \text{N})\left(25\ \frac{\text{m}}{\text{s}}\right)$$
$$= 13\,960\ \text{W}$$

The energy available from the fuel is

$$\begin{aligned} E_A &= (\text{engine thermal efficiency})(\text{fuel heating value}) \\ &= (0.28)\left(32 \times 10^6 \ \frac{\text{J}}{\text{L}}\right) \\ &= 8.96 \times 10^6 \ \text{J/L} \end{aligned}$$

The fuel consumption at 90 km/h is

$$\begin{aligned} \frac{P}{E_A \text{v}} &= \frac{(13\,960 \ \text{W})\left(3600 \ \frac{\text{s}}{\text{h}}\right)}{\left(8.96 \times 10^6 \ \frac{\text{J}}{\text{L}}\right)\left(90 \ \frac{\text{km}}{\text{h}}\right)} \\ &= 0.0623 \ \text{L/km} \quad (0.062 \ \text{L/km}) \end{aligned}$$

The answer is (A).

27. To ensure similarity between the model and the true conditions of the full-scale airplane, the Reynolds numbers must be equal.

$$\left(\frac{\text{v}L}{\nu}\right)_{\text{model}} = \left(\frac{\text{v}L}{\nu}\right)_{\text{true}}$$

Use the absolute viscosity.

$$\mu = \frac{\rho\nu}{g_c}$$

$$\nu = \frac{\mu g_c}{\rho}$$

$$\left(\frac{\text{v}L\rho}{\mu g_c}\right)_{\text{model}} = \left(\frac{\text{v}L\rho}{\mu g_c}\right)_{\text{true}}$$

Recall that the absolute viscosity is independent of pressure, so $\mu_{\text{model}} = \mu_{\text{true}}$.

Since g_c is a constant,

$$\left(\frac{\text{v}L\rho}{\mu}\right)_{\text{model}} = \left(\frac{\text{v}L\rho}{\mu}\right)_{\text{true}}$$

Assume the air behaves as an ideal gas.

$$\rho = \frac{p}{RT}$$

$$\left(\frac{\text{v}Lp}{\mu RT}\right)_{\text{model}} = \left(\frac{\text{v}Lp}{\mu RT}\right)_{\text{true}}$$

Since the tunnel operates with air at true velocity and temperature,

$$R_{\text{model}} = R_{\text{true}}$$

$$\text{v}_{\text{model}} = \text{v}_{\text{true}}$$

$$T_{\text{model}} = T_{\text{true}}$$

$$\left(\frac{Lp}{\mu}\right)_{\text{model}} = \left(\frac{Lp}{\mu}\right)_{\text{true}}$$

Therefore,

$$(Lp)_{\text{model}} = (Lp)_{\text{true}}$$

Since the scale of the model is 1/20,

$$L_{\text{model}} = \frac{L_{\text{true}}}{20}$$

Substituting gives

$$\begin{aligned} (Lp)_{\text{model}} &= (Lp)_{\text{true}} \\ \left(\frac{L_{\text{true}}}{20}\right)p_{\text{model}} &= L_{\text{true}}p_{\text{true}} \\ p_{\text{model}} &= 20p_{\text{true}} \end{aligned}$$

The answer is (C).

28. To ensure similarity between the two impellers, the Reynolds numbers, Re, must be equal.

$$\text{Re}_{\text{oil}} = \text{Re}_{\text{air}}$$

$$\frac{\text{v}_{\text{oil}}D_{\text{oil}}}{\nu_{\text{oil}}} = \frac{\text{v}_{\text{air}}D_{\text{air}}}{\nu_{\text{air}}}$$

$$\frac{\text{v}_{\text{oil}}}{\text{v}_{\text{air}}} = \left(\frac{\nu_{\text{oil}}}{\nu_{\text{air}}}\right)\left(\frac{D_{\text{air}}}{D_{\text{oil}}}\right)$$

v is the tangential velocity, and D is the impeller diameter.

$$\text{v} \propto nD$$

$$\frac{\text{v}_{\text{oil}}}{\text{v}_{\text{air}}} = \left(\frac{n_{\text{oil}}}{n_{\text{air}}}\right)\left(\frac{D_{\text{oil}}}{D_{\text{air}}}\right)$$

Therefore,

$$\left(\frac{\nu_{\text{oil}}}{\nu_{\text{air}}}\right)\left(\frac{D_{\text{air}}}{D_{\text{oil}}}\right) = \left(\frac{n_{\text{oil}}}{n_{\text{air}}}\right)\left(\frac{D_{\text{oil}}}{D_{\text{air}}}\right)$$

$$n_{\text{air}} = n_{\text{oil}}\left(\frac{\nu_{\text{air}}}{\nu_{\text{oil}}}\right)\left(\frac{D_{\text{oil}}}{D_{\text{air}}}\right)^2$$

Since the air impeller is twice the size of the oil impeller, $D_{\text{air}} = 2D_{\text{oil}}$.

$$\begin{aligned} n_{\text{air}} &= n_{\text{oil}}\left(\frac{\nu_{\text{air}}}{\nu_{\text{oil}}}\right)\left(\frac{D_{\text{oil}}}{D_{\text{air}}}\right)^2 \\ &= n_{\text{oil}}\left(\frac{\nu_{\text{air}}}{\nu_{\text{oil}}}\right)\left(\frac{D_{\text{oil}}}{2D_{\text{oil}}}\right)^2 \\ &= \tfrac{1}{4}n_{\text{oil}}\left(\frac{\nu_{\text{air}}}{\nu_{\text{oil}}}\right) \end{aligned}$$

Customary U.S. Solution

From App. 1C (also *NCEES Handbook:* Temperature-Dependent Properties of Air (U.S. Customary Units)), for air at 68°F, $\nu = 15.72 \times 10^{-5}$ ft²/sec.

For castor oil at 68°F, $\nu = 1110 \times 10^{-5}$ ft²/sec (given).

$$\begin{aligned} n_{\text{air}} &= \tfrac{1}{4}n_{\text{oil}}\left(\frac{\nu_{\text{air}}}{\nu_{\text{oil}}}\right) \\ &= \left(\frac{1}{4}\right)\left(1000\ \frac{\text{rev}}{\text{min}}\right)\left(\frac{15.72 \times 10^{-5}\ \frac{\text{ft}^2}{\text{sec}}}{1110 \times 10^{-5}\ \frac{\text{ft}^2}{\text{sec}}}\right) \\ &= 3.54\ \text{rpm} \quad (3.6\ \text{rpm}) \end{aligned}$$

SI Solution

From App. 1D (also *NCEES Handbook:* Temperature-Dependent Properties of Air (SI Units)), for air at 20°C, $\nu = 1.512 \times 10^{-5}$ m²/s.

For castor oil at 20°C, $\nu = 103 \times 10^{-5}$ m²/s (given),

$$\begin{aligned} n_{\text{air}} &= \tfrac{1}{4}n_{\text{oil}}\left(\frac{\nu_{\text{air}}}{\nu_{\text{oil}}}\right) \\ &= \left(\frac{1}{4}\right)\left(1000\ \frac{\text{rev}}{\text{min}}\right)\left(\frac{1.512 \times 10^{-5}\ \frac{\text{m}^2}{\text{s}}}{103 \times 10^{-5}\ \frac{\text{m}^2}{\text{s}}}\right) \\ &= 3.67\ \text{rpm} \quad (3.6\ \text{rpm}) \end{aligned}$$

The answer is (A).

5 Hydraulic Machines

PRACTICE PROBLEMS

1. Two centrifugal pumps used in a water pumping application have the characteristic curves shown. The pumps operate in parallel and discharge into a common header against a head of 40 ft. What is most nearly the discharge rate of the pumps operating in parallel?

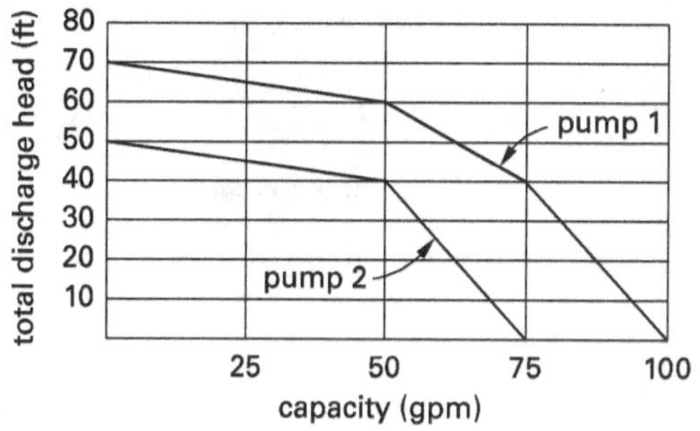

(A) 30 gpm

(B) 50 gpm

(C) 75 gpm

(D) 130 gpm

2. An electric motor drives a pump in a gasoline transfer network. The system and pump curves are defined by the points in the given table. The gasoline has a specific gravity of 0.7. What is most nearly the horsepower for the motor?

	volume	
head(ft)	system curve (gpm)	pump curve (gpm)
10	0	1500
20	500	1200
30	1000	1000
40	1200	500
50	1500	0

(A) 4.0 hp

(B) 5.3 hp

(C) 5.5 hp

(D) 7.6 hp

3. Two centrifugal pumps used in a water pumping application have the characteristic curves shown. The pumps operate in series and have a combined discharge of 50 gpm. What is most nearly the total discharge head?

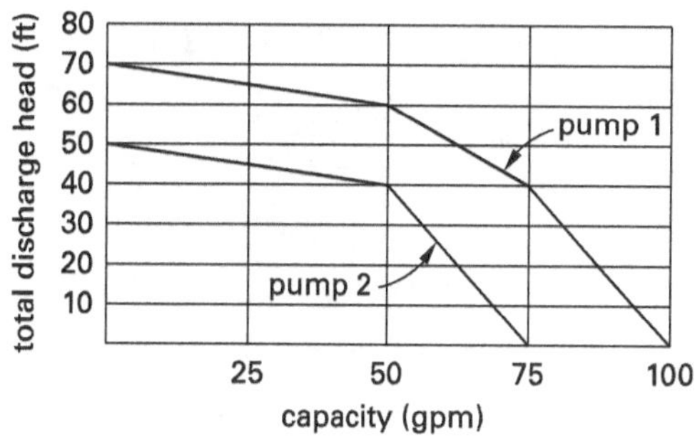

(A) 24 ft

(B) 50 ft

(C) 80 ft

(D) 100 ft

4. 2000 gal/min of 60°F thickened sludge with a specific gravity of 1.2 flows through a pump with an inlet diameter of 12 in and an outlet diameter of 8 in. The centerlines of the inlet and outlet are at the same elevation. The inlet pressure is 8 in of mercury (vacuum). A discharge pressure gauge located 4 ft above the pump discharge centerline reads 20 psig. The pump efficiency is 85%. All pipes are schedule-40. The input power of the pump is most nearly

(A) 26 hp

(B) 31 hp

(C) 37 hp

(D) 53 hp

5. 1.25 ft^3/sec (35 L/s) of 70°F (21°C) water is pumped from the bottom of a tank through 700 ft (230 m) of 4 in (102.3 mm) schedule-40 steel pipe. The line includes a 50 ft (15 m) rise in elevation, two right-angle elbows, a wide-open gate valve, and a swing check valve. All fittings and valves are regular screwed. The inlet pressure is 50 psig (345 kPa), and a working pressure of 20 psig (140 kPa) is needed at the end of the pipe. The hydraulic power for this pumping application is most nearly

(A) 16 hp (12 kW)

(B) 23 hp (17 kW)

(C) 49 hp (37 kW)

(D) 66 hp (50 kW)

6. 80 gal/min (5 L/s) of 80°F (27°C) water is lifted 12 ft (4 m) vertically by a pump through a total length of 50 ft (15 m) of a 2 in (5.1 cm) diameter smooth rubber hose. The discharge end of the hose is submerged in 8 ft (2.5 m) of water as shown.

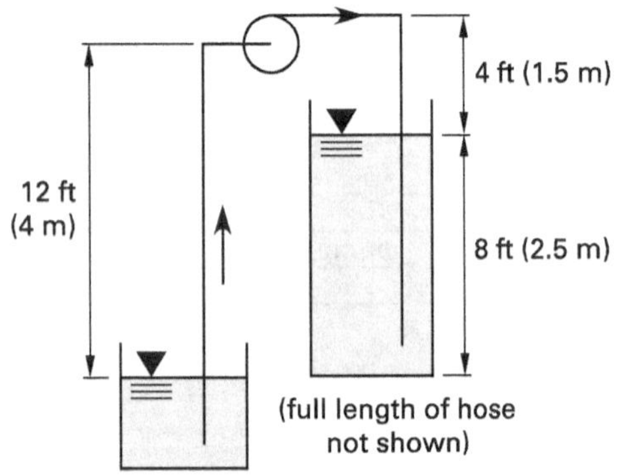

The head added by the pump is most nearly

(A) 10 ft (3.0 m)

(B) 13 ft (4.0 m)

(C) 22 ft (6.6 m)

(D) 31 ft (9.3 m)

7. A 20 hp motor drives a centrifugal pump. The pump discharges 60°F (16°C) water at a velocity of 12 ft/sec (4 m/s) into a 6 in (15.2 cm) steel schedule-40 line. The inlet is 8 in (20.3 cm) schedule-40 steel pipe. The pump suction is 5 psig (35 kPa) below standard atmospheric pressure. The friction and fitting head loss in the system is 10 ft (3.3 m). The pump efficiency is 70%. The suction and discharge lines are at the same elevation. The maximum height above the pump inlet that water is available with that velocity at standard atmospheric pressure is most nearly

(A) 28 ft (6.9 m)

(B) 37 ft (11 m)

(C) 49 ft (15 m)

(D) 81 ft (25 m)

8. An electrically driven pump is used to fill a tank on a hill from a lake below. The flow rate is 10,000 gal/hr (10.5 L/s) of 60°F (16°C) water. The atmospheric pressure is 14.7 psia (101 kPa). The pump is 12 ft (4 m) above the lake, and the tank surface level is 350 ft (115 m) above the pump. The suction and discharge lines are 4 in (10.2 cm) diameter schedule-40 steel pipe. The equivalent length of the inlet line between the lake and the pump is 300 ft (100 m). The total equivalent length between the lake and the tank is 7000 ft (2300 m), including all fittings, bends, screens, and valves. The cost of electricity is \$0.04 per kW·h. The overall efficiency of the pump and motor set is 70%.

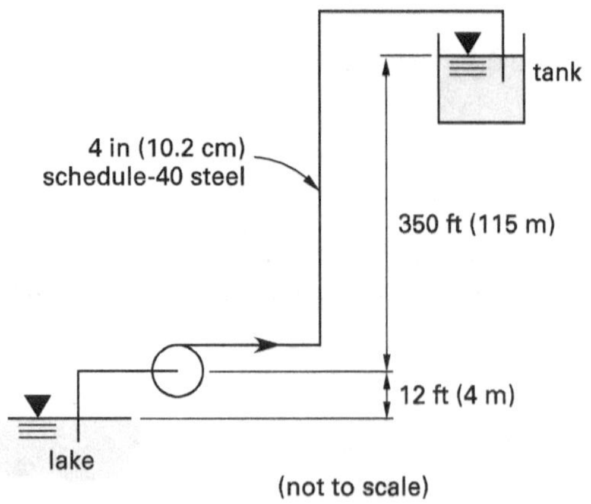

The cost to operate the pump for one hour is most nearly

(A) \$0.1

(B) \$1

(C) \$3

(D) \$6

9. A pump transfers 3.5 MGD of filtered water from the clear well of a 10 ft wide by 20 ft long rapid sand filter to a higher elevation. The pump efficiency is 85%, and the motor driving the pump has an efficiency of 90%. Minor losses are insignificant. Refer to the illustration shown for additional information.

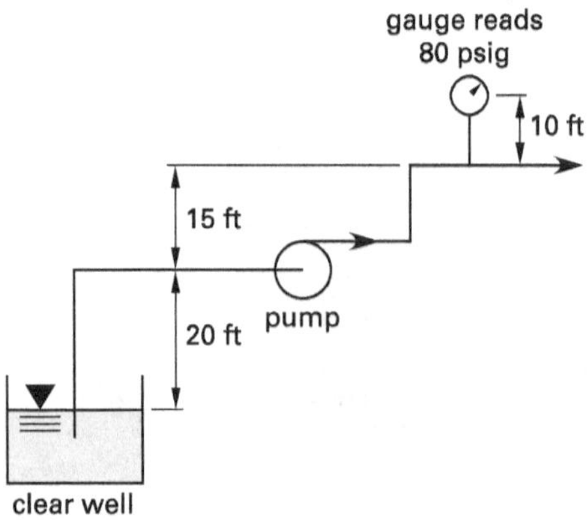

The required motor power is most nearly

(A) 50 hp

(B) 100 hp

(C) 150 hp

(D) 200 hp

10. Gasoline with a specific gravity of 0.7 and kinematic viscosity of 6×10^{-6} ft^2/sec (5.6×10^{-10} m^2/s) is transferred from a tanker to a storage tank. The interior of the storage tank is maintained at atmospheric pressure by a vapor-recovery system. The free surface in the storage tank is 60 ft (20 m) above the tanker's free surface. The pipe consists of 500 ft (170 m) of 3 in (7.62 cm) schedule-40 steel pipe with six flanged elbows and two wide-open gate valves. The pump and motor both have individual efficiencies of 88%. Electricity costs \$0.045 per kW·h. The pump's performance data (based on cold, clear water) are known.

flow rate (gpm (L/s))	head (ft (m))
0 (0)	127 (42)
100 (6.3)	124 (41)
200 (12)	117 (39)
300 (18)	108 (36)
400 (24)	96 (32)
500 (30)	80 (27)
600 (36)	55 (18)

The total cost of operating the pump for one hour is most nearly

(A) \$0.20

(B) \$0.80

(C) \$1.30

(D) \$2.70

11. The pressure of 37 gal/min (65 L/s) of 80°F (27°C) SAE 40 oil is increased from 1 atm to 40 psig (275 kPa). The hydraulic power required is most nearly

(A) 0.45 hp (9 kW)

(B) 0.9 hp (18 kW)

(C) 1.8 hp (36 kW)

(D) 3.6 hp (72 kW)

12. A double-suction water pump moving 300 gal/sec (1.1 kL/s) turns at 900 rpm. The pump adds 20 ft (7 m) of head to the water. The specific speed is most nearly

(A) 3000 rpm (52 rpm)

(B) 6000 rpm (100 rpm)

(C) 9000 rpm (160 rpm)

(D) 12,000 rpm (210 rpm)

13. 100 gal/min (6.3 L/s) of pressurized hot water at 281°F and 80 psia (138°C and 550 kPa) is drawn through 30 ft (10 m) of 1.5 in (3.81 cm) schedule-40 steel pipe into a tank pressurized to a constant 2 psig (14 kPa). The inlet and outlet are both 20 ft (6 m) below the surface of the water when the tank is full. The inlet line contains a square mouth inlet, two wide-open gate valves, and two long-radius elbows. All components are regular screwed. The pump's NPSHR is 10 ft (3 m) for this application. The kinematic viscosity of 281°F (138°C) water is 0.239×10^{-5} ft^2/sec (0.222×10^{-6} m^2/s), and the vapor pressure is 50.02 psia (3.431 bar). Will the pump cavitate?

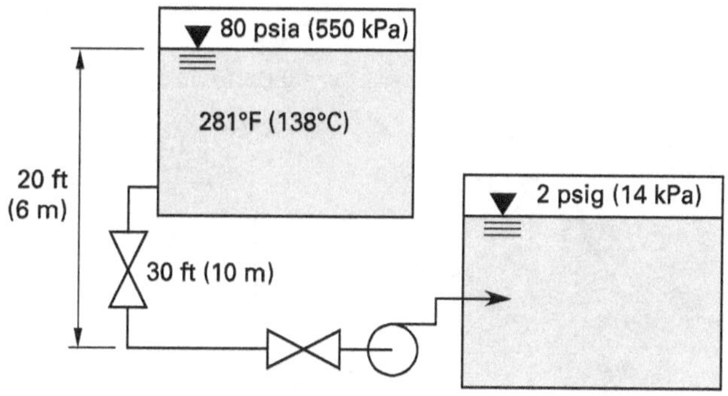

(A) yes; NPSHA = 4 ft (1.2 m)

(B) yes; NPSHA = 9 ft (2.7 m)

(C) no; NPSHA = 24 ft (7.2 m)

(D) no; NPSHA = 68 ft (20.6 m)

14. The velocity of the tip of a marine propeller is 4.2 times the velocity of the boat. The propeller is located 8 ft (3 m) below the surface. The temperature of the seawater is 68°F (20°C). The density of seawater is approximately 64.0 lbm/ft^3 (1024 kg/m^3), and the salt content is 2.5% by weight. The practical maximum boat velocity, as limited strictly by cavitation, is most nearly

(A) 9.1 ft/sec (2.7 m/s)

(B) 12 ft/sec (3.8 m/s)

(C) 15 ft/sec (4.5 m/s)

(D) 22 ft/sec (6.6 m/s)

15. The inlet of a centrifugal water pump is 7 ft (2.3 m) above the free surface from which it draws. The suction point is a submerged pipe. The suction line consists of 12 ft (4 m) of 2 in (5.08 cm) schedule-40 steel pipe and contains one long-radius 90° elbow and one swing check valve. The discharge line is 2 in (5.08 cm) schedule-40 steel pipe and includes two long radius elbows and an 80 ft (27 m) run. The discharge is 20 ft (6.3 m) above the free surface and is a jet to the open atmosphere. All components are regular screwed. The water temperature is 70°F (21°C). Use the following pump curve data.

flow rate (gpm (L/s))	head (ft (m))
0 (0)	110 (37)
10 (0.6)	108 (36)
20 (1.2)	105 (35)
30 (1.8)	102 (34)
40 (2.4)	98 (33)
50 (3.2)	93 (31)
60 (3.6)	87 (29)
70 (4.4)	79 (26)
80 (4.8)	66 (22)
90 (5.7)	50 (17)

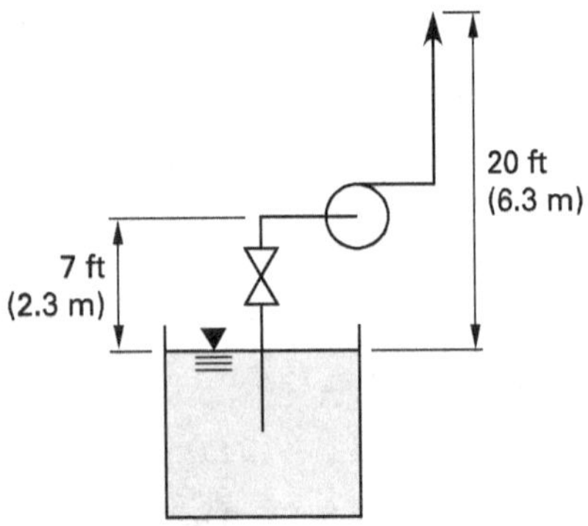

The flow rate is most nearly

(A) 44 gal/min (2.9 L/s)

(B) 69 gal/min (4.5 L/s)

(C) 82 gal/min (5.5 L/s)

(D) 95 gal/min (6.2 L/s)

16. A pump was intended to run at 1750 rpm when driven by a 0.5 hp (0.37 kW) motor. The required power rating of a motor that will turn the pump at 2000 rpm is most nearly

(A) 0.25 hp (0.19 kW)

(B) 0.45 hp (0.34 kW)

(C) 0.65 hp (0.49 kW)

(D) 0.75 hp (0.55 kW)

17. A centrifugal pump running at 1400 rpm has the curve shown and an efficiency of 86%. The pump will convey water in an existing pipeline with known head requirements given by the formula $H = 30 + 2Q^2$ where H is the system head in feet of water, and Q is the flow rate in cubic feet per second (cfs).

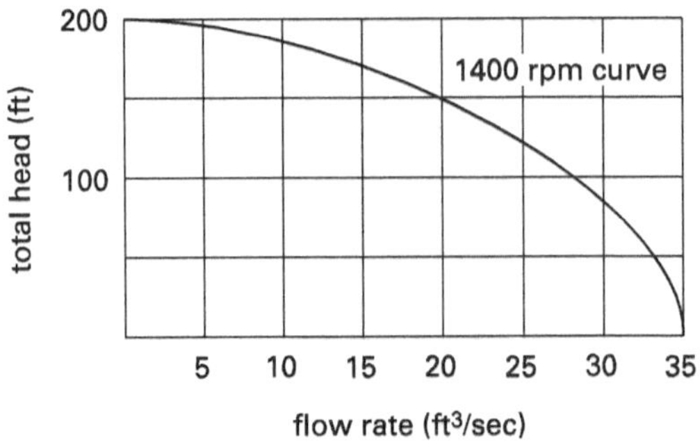

The power required at the pump shaft is most nearly

(A) 190 hp

(B) 210 hp

(C) 230 hp

(D) 260 hp

18. A horizontal turbine reduces 100 ft^3/sec of water from 30 psia to 5 psia. Friction is negligible. The power developed is most nearly

(A) 350 hp

(B) 500 hp

(C) 650 hp

(D) 800 hp

19. 1000 ft^3/sec of 60°F water flows from a high reservoir through a hydroelectric turbine installation, exiting 625 ft lower. The head loss due to friction is 58 ft. The turbine efficiency is 89%. The power developed in the turbines is most nearly

(A) 40 kW

(B) 18 MW

(C) 43 MW

(D) 71 MW

20. A Francis-design hydraulic reaction turbine with 22 in (560 mm) diameter blades runs at 610 rpm. The turbine develops 250 hp (185 kW) when 25 ft^3/sec (700 L/s) of water flow through it. The pressure head at the turbine entrance is 92.5 ft (28.2 m). The elevation of the turbine above the tailwater level is 5.26 ft (1.75 m). The inlet velocity is 12 ft/s (3.6 m/s) greater than the outlet velocity. The overall turbine efficiency is most nearly

(A) 81%

(B) 88%

(C) 93%

(D) 96%

SOLUTIONS

1. Theoretically, when operating in parallel, each pump performs as if the other pump is not present. The capacities of each pump at a 40 ft discharge head are cumulative: 50 gpm for pump 2 and 75 gpm for pump 1.

$$\begin{aligned} Q_{\text{parallel}} &= Q_2 + Q_1 = 50\ \frac{\text{gal}}{\text{min}} + 75\ \frac{\text{gal}}{\text{min}} \\ &= 125\ \text{gpm} \quad (130\ \text{gpm}) \end{aligned}$$

The answer is (D).

2. The system and pump curves intersect at 1000 gpm and 30 ft. From Table 5.5 (also *NCEES Handbook:* Pump Power), the hydraulic horsepower is

$$\begin{aligned} \text{WHP} &= \frac{h_A Q(\text{SG})}{3956} = \frac{(30\ \text{ft})\left(1000\ \frac{\text{gal}}{\text{min}}\right)(0.7)}{3956\ \frac{\text{ft-gal}}{\text{hp-min}}} \\ &= 5.31\ \text{hp} \quad (5.3\ \text{hp}) \end{aligned}$$

This is the minimum power that the electric motor can produce.

The answer is (B).

3. When operated in series, the second pump receives water at the rate of the first pump's discharge, so both pumps experience the same flow rate. The second pump adds pressure head to the first pump's pressurization, so the discharge heads are cumulative. At 50 gpm, the discharge heads are 60 ft for pump 1 and 40 ft for pump 2, respectively.

$$h_{A,\text{series}} = h_{A,1} + h_{A,2} = 60\ \text{ft} + 40\ \text{ft} = 100\ \text{ft}$$

The answer is (D).

4. The flow rate is

$$\dot{V} = \frac{2000\ \frac{\text{gal}}{\text{min}}}{\left(7.4805\ \frac{\text{gal}}{\text{ft}^3}\right)\left(60\ \frac{\text{sec}}{\text{min}}\right)} = 4.456\ \text{ft}^3/\text{sec}$$

From App. 3B (also *NCEES Handbook* table "Pipe Dimensions and Weights"),

$$\begin{aligned} &12\ \text{in:}\ D_1 = 0.99483\ \text{ft} \qquad A_1 = 0.7773\ \text{ft}^2 \\ &8\ \text{in:}\ D_2 = 0.6651\ \text{ft} \qquad A_2 = 0.3474\ \text{ft}^2 \end{aligned}$$

$$p_1 = \left(14.7\ \frac{\text{lbf}}{\text{in}^2} - (8\ \text{in})\left(0.491\ \frac{\text{lbf}}{\text{in}^3}\right)\right)\left(12\ \frac{\text{in}}{\text{ft}}\right)^2$$
$$= 1551.2\ \text{lbf/ft}^2$$
$$p_2 = \left(14.7\ \frac{\text{lbf}}{\text{in}^2} + 20\ \frac{\text{lbf}}{\text{in}^2}\right)\left(12\ \frac{\text{in}}{\text{ft}}\right)^2 + (4\ \text{ft})(1.2)\left(62.4\ \frac{\text{lbf}}{\text{ft}^3}\right)$$
$$= 5296.3\ \text{lbf/ft}^2$$

As in *NCEES Handbook:* Conservation of Mass,

$$\text{v}_1 = \frac{\dot{V}}{A_1} = \frac{4.456\ \frac{\text{ft}^3}{\text{sec}}}{0.7773\ \text{ft}^2} = 5.73\ \text{ft/sec}$$

$$\text{v}_2 = \frac{\dot{V}}{A_2} = \frac{4.456\ \frac{\text{ft}^3}{\text{sec}}}{0.3474\ \text{ft}^2} = 12.83\ \text{ft/sec}$$

Using a modified Bernoulli equation from Eq. 5.8 (also *NCEES Handbook:* The Bernoulli Equation), the total heads (in feet of sludge) at points 1 and 2 are

$$h_{t,1} = h_{t,s} = \frac{p_1}{\gamma} + \frac{\text{v}_1^2}{2g}$$
$$= \frac{1551.2\ \frac{\text{lbf}}{\text{ft}^2}}{\left(62.4\ \frac{\text{lbf}}{\text{ft}^3}\right)(1.2)} + \frac{\left(5.73\ \frac{\text{ft}}{\text{sec}}\right)^2}{(2)\left(32.2\ \frac{\text{ft}}{\text{sec}^2}\right)}$$
$$= 21.23\ \text{ft}$$

$$h_{t,2} = h_{t,d} = \frac{p_2}{\gamma} + \frac{\text{v}_2^2}{2g}$$
$$= \frac{5296.3\ \frac{\text{lbf}}{\text{ft}^2}}{\left(62.4\ \frac{\text{lbf}}{\text{ft}^3}\right)(1.2)} + \frac{\left(12.83\ \frac{\text{ft}}{\text{sec}}\right)^2}{(2)\left(32.2\ \frac{\text{ft}}{\text{sec}^2}\right)}$$
$$= 73.29\ \text{ft}$$

The pump must add 73.29 ft − 21.23 ft = 52.06 ft of head (sludge head).

The power required is given in Table 5.5 (also *NCEES Handbook:* Pump Power).

$$P_{\text{ideal}} = \frac{\Delta p\dot{V}}{550} = \frac{\Delta h\gamma\dot{V}}{550} = \frac{\Delta h(\text{SG})\gamma_w\dot{V}}{550}$$
$$= \frac{(52.06\ \text{ft})(1.2)\left(62.4\ \frac{\text{lbf}}{\text{ft}^3}\right)\left(4.456\ \frac{\text{ft}^3}{\text{sec}}\right)}{550\ \frac{\text{ft-lbf}}{\text{hp-sec}}}$$
$$= 31.58\ \text{hp}$$

The input horsepower is

$$P_{\text{in}} = \frac{P_{\text{ideal}}}{\eta} = \frac{31.58\ \text{hp}}{0.85} = 37.15\ \text{hp}\quad (37\ \text{hp})$$

The answer is (C).

5. *Customary U.S. Solution*

From App. 3B (also *NCEES Handbook* table "Pipe Dimensions and Weights"), data for 4 in schedule-40 steel pipe are

$$D = 0.3355\ \text{ft}$$
$$A = 0.08841\ \text{ft}^2$$

The velocity in the pipe is in *NCEES Handbook:* Conservation of Mass.

$$\text{v} = \frac{\dot{V}}{A} = \frac{1.25\ \frac{\text{ft}^3}{\text{sec}}}{0.08841\ \text{ft}^2} = 14.139\ \text{ft/sec}$$

From App. 4D (also *NCEES Handbook:* Head Loss in Pipe or Conduit), typical equivalent lengths for schedule-40, screwed steel fittings for 4 in pipes are

90° elbow: 13 ft
gate valve: 2.5 ft
check valve: 38 ft

The total equivalent length is

$$(2)(13\ \text{ft}) + (1)(2.5\ \text{ft}) + (1)(38\ \text{ft}) = 66.5\ \text{ft}$$

From App. 1A (also *NCEES Handbook* table "Physical Properties of Liquid Water (U.S. Units)"), the density of water is 62.3 lbm/ft^3, and the kinematic viscosity of water, ν, at 70°F is $1.059 \times 10^{-5}\ \text{ft}^2/\text{sec}$. The Reynolds number is from *NCEES Handbook:* Similitude.

$$\text{Re} = \frac{D\text{v}}{\nu} = \frac{(0.3355\ \text{ft})\left(14.139\ \frac{\text{ft}}{\text{sec}}\right)}{1.059 \times 10^{-5}\ \frac{\text{ft}^2}{\text{sec}}}$$
$$= 4.479 \times 10^5$$

From App. 4A (also *NCEES Handbook:* Absolute Roughness and Relative Roughness), for steel, $\epsilon = 0.0002$ ft.

So,

$$\frac{\epsilon}{D} = \frac{0.0002\ \text{ft}}{0.3355\ \text{ft}} \approx 0.0006$$

Interpolating from App. 4B (also *NCEES Handbook:* Friction Factors for Turbulent Flow), the friction factor is $f = 0.01835$.

The friction head is given by Eq. 5.5 (also *NCEES Handbook:* Head Loss in Pipe or Conduit).

$$h_f = \frac{fL\text{v}^2}{2Dg}$$
$$= \frac{(0.01835)(700\ \text{ft} + 66.5\ \text{ft})\left(14.139\ \frac{\text{ft}}{\text{sec}}\right)^2}{(2)(0.3355\ \text{ft})\left(32.2\ \frac{\text{ft}}{\text{sec}^2}\right)}$$
$$= 130.1\ \text{ft}$$

The total dynamic head is given by Eq. 5.8 (also *NCEES Handbook:* Pump Similitude). Point 1 is taken as the bottom of the supply tank. Point 2 is taken as the end of the discharge pipe.

$$h = \frac{(p_2 - p_1)g_c}{\rho g} + \frac{\text{v}_2^2 - \text{v}_1^2}{2g} + z_2 - z_1$$
$$\text{v}_1 \approx 0$$
$$z_2 - z_1 = 50\ \text{ft} \quad [\text{given as rise in elevation}]$$

The outlet and inlet pressures are

$$p_2 = 20\ \text{psig}$$
$$p_1 = 50\ \text{psig}$$

The pressure head added by the pump is

$$h = \frac{(p_2 - p_1)g_c}{\rho g} + \frac{\text{v}_2^2 - \text{v}_1^2}{2g} + z_2 - z_1$$
$$= \frac{\left(20\ \frac{\text{lbf}}{\text{in}^2} - 50\ \frac{\text{lbf}}{\text{in}^2}\right)\left(12\ \frac{\text{in}}{\text{ft}}\right)^2\left(32.2\ \frac{\text{lbm-ft}}{\text{lbf-sec}^2}\right)}{\left(62.3\ \frac{\text{lbm}}{\text{ft}^3}\right)\left(32.2\ \frac{\text{ft}}{\text{sec}^2}\right)}$$
$$+ \frac{\left(14.139\ \frac{\text{ft}}{\text{sec}}\right)^2}{(2)\left(32.2\ \frac{\text{ft}}{\text{sec}^2}\right)} + 50\ \text{ft}$$
$$= -16.2\ \text{ft}$$

The head added is

$$h_A = h + h_f$$
$$= -16.2\ \text{ft} + 130.1\ \text{ft}$$
$$= 113.9\ \text{ft}$$

The mass flow rate is

$$\dot{m} = \rho\dot{V}$$
$$= \left(62.3\ \frac{\text{lbm}}{\text{ft}^3}\right)\left(1.25\ \frac{\text{ft}^3}{\text{sec}}\right)$$
$$= 77.875\ \text{lbm/sec}$$

From Table 5.5 (also *NCEES Handbook:* Pump Power), the hydraulic horsepower is

$$\text{WHP} = \frac{h_A\dot{m}}{550} \times \frac{g}{g_c}$$
$$= \left(\frac{(113.9\ \text{ft})\left(77.875\ \frac{\text{lbm}}{\text{sec}}\right)}{550\ \frac{\text{ft-lbf}}{\text{hp-sec}}}\right)\left(\frac{32.2\ \frac{\text{ft}}{\text{sec}^2}}{32.2\ \frac{\text{lbm-ft}}{\text{lbf-sec}^2}}\right)$$
$$= 16.13\ \text{hp} \quad (16\ \text{hp})$$

SI Solution

From App. 3C (also *NCEES Handbook* table "Pipe Dimensions and Weights"), data for 4 in schedule-40 steel pipe are

$$D = 102.26\ \text{mm}$$
$$A = 82.30\ \text{cm}^2$$

As in *NCEES Handbook:* Conservation of Mass, the velocity in the pipe is

$$\text{v} = \frac{\dot{V}}{A} = \frac{\left(35\ \frac{\text{L}}{\text{s}}\right)\left(100\ \frac{\text{cm}}{\text{m}}\right)^2}{(82.30\ \text{cm}^2)\left(1000\ \frac{\text{L}}{\text{m}^3}\right)} = 4.25\ \text{m/s}$$

From App. 4D (also *NCEES Handbook:* Head Loss in Pipe or Conduit), typical equivalent lengths for schedule-40, screwed steel fittings for 4 in pipes are

90° elbow: 13.0 ft
gate valve: 2.5 ft
check valve: 38.0 ft

The total equivalent length is

$$(2)(13.0\ \text{ft}) + (1)(2.5\ \text{ft}) + (1)(38.0\ \text{ft}) = 66.5\ \text{ft}$$

$$(66.5\ \text{ft})\left(0.3048\ \frac{\text{m}}{\text{ft}}\right) = 20.27\ \text{m}$$

At 21°C, from App. 1B (also *NCEES Handbook* table "Physical Properties of Liquid Water (SI Units)"), the water properties are

$$\rho = 998\ \text{kg/m}^3$$

$$\mu = 0.9827 \times 10^{-3}\ \text{Pa·s}$$

$$\nu = \frac{\mu}{\rho} = \frac{0.9827 \times 10^{-3}\ \text{Pa·s}}{998\ \frac{\text{kg}}{\text{m}^3}} = 9.85 \times 10^{-7}\ \text{m}^2/\text{s}$$

As in *NCEES Handbook:* Similitude, the Reynolds number is

$$\text{Re} = \frac{D\text{v}}{\nu} = \frac{(102.26\ \text{mm})\left(4.25\ \frac{\text{m}}{\text{s}}\right)}{\left(9.85 \times 10^{-7}\ \frac{\text{m}^2}{\text{s}}\right)\left(1000\ \frac{\text{mm}}{\text{m}}\right)} = 4.417 \times 10^5$$

From Table 4.2 (also *NCEES Handbook:* Absolute Roughness and Relative Roughness), for steel, $\epsilon = 6.0 \times 10^{-5}$ m.

$$\frac{\epsilon}{D} = \frac{(6.0 \times 10^{-5}\ \text{m})\left(1000\ \frac{\text{mm}}{\text{m}}\right)}{102.26\ \text{mm}} = 0.0006$$

Interpolating from App. 4B (also *NCEES Handbook:* Friction Factors for Turbulent Flow), the friction factor is $f = 0.01836$.

From Eq. 5.5 (also *NCEES Handbook:* Head Loss in Pipe or Conduit), the friction head is

$$h_f = \frac{fL\text{v}^2}{2Dg} = \frac{(0.01836)(230\ \text{m} + 20.27\ \text{m}) \times \left(4.25\ \frac{\text{m}}{\text{s}}\right)^2\left(1000\ \frac{\text{mm}}{\text{m}}\right)}{(2)(102.26\ \text{mm})\left(9.81\ \frac{\text{m}}{\text{s}^2}\right)} = 41.4\ \text{m}$$

The total dynamic head is given by Eq. 5.8 (also *NCEES Handbook:* Pump Similitude). Point 1 is taken as the bottom of the supply tank. Point 2 is taken as the end of the discharge pipe.

$$h = \frac{p_2 - p_1}{\rho g} + \frac{\text{v}_2^2 - \text{v}_1^2}{2g} + z_2 - z_1$$

$$\text{v}_1 \approx 0$$

$$z_2 - z_1 = 15\ \text{m} \quad \text{[given as rise in elevation]}$$

The difference between outlet and inlet pressure is

$$p_2 - p_1 = 140\ \text{kPa} - 345\ \text{kPa} = -205\ \text{kPa}$$

$$h = \frac{(-205\ \text{kPa})\left(1000\ \frac{\text{Pa}}{\text{kPa}}\right)}{\left(998\ \frac{\text{kg}}{\text{m}^3}\right)\left(9.81\ \frac{\text{m}}{\text{s}^2}\right)} + \frac{\left(4.25\ \frac{\text{m}}{\text{s}}\right)^2}{(2)\left(9.81\ \frac{\text{m}}{\text{s}^2}\right)} + 15\ \text{m} = -5.0\ \text{m}$$

The head added by the pump is

$$h_A = h + h_f = -5.0\ \text{m} + 41.4\ \text{m} = 36.4\ \text{m}$$

The mass flow rate is

$$\dot{m} = \rho\dot{V} = \frac{\left(998\ \frac{\text{kg}}{\text{m}^3}\right)\left(35\ \frac{\text{L}}{\text{s}}\right)}{1000\ \frac{\text{L}}{\text{m}^3}} = 34.93\ \text{kg/s}$$

From Table 5.6 (also *NCEES Handbook:* Pump Power), the hydraulic power is

$$\text{WkW} = \frac{9.81 h_A \dot{m}}{1000} = \frac{\left(9.81\ \frac{\text{m}}{\text{s}^2}\right)(36.4\ \text{m})\left(34.93\ \frac{\text{kg}}{\text{s}}\right)}{1000\ \frac{\text{W}}{\text{kW}}}$$
$$= 12.46\ \text{kW} \quad (12\ \text{kW})$$

The answer is (A).

6. *Customary U.S. Solution*

The area of the rubber hose is

$$A = \frac{\pi D^2}{4} = \frac{\pi\left(\frac{2\ \text{in}}{12\ \frac{\text{in}}{\text{ft}}}\right)^2}{4} = 0.0218\ \text{ft}^2$$

As in *NCEES Handbook:* Conservation of Mass, the velocity of water in the hose is

$$\text{v} = \frac{\dot{V}}{A} = \frac{80\ \frac{\text{gal}}{\text{min}}}{(0.0218\ \text{ft}^2)\left(7.4805\ \frac{\text{gal}}{\text{ft}^3}\right)\left(60\ \frac{\text{sec}}{\text{min}}\right)}$$
$$= 8.176\ \text{ft/sec}$$

At 80°F, as in *NCEES Handbook* table "Physical Properties of Liquid Water (U.S. Units)," the kinematic viscosity of water is $\nu = 0.930 \times 10^{-5}\ \text{ft}^2/\text{sec}$.

As in *NCEES Handbook:* Similitude, the Reynolds number is

$$\text{Re} = \frac{\text{v}D}{\nu} = \frac{\left(8.176\ \frac{\text{ft}}{\text{sec}}\right)(2\ \text{in})}{\left(0.930 \times 10^{-5}\ \frac{\text{ft}^2}{\text{sec}}\right)\left(12\ \frac{\text{in}}{\text{ft}}\right)}$$
$$= 1.47 \times 10^5$$

Since the rubber hose is smooth, from *NCEES Handbook:* Friction Factors for Turbulent Flow, the friction factor is $f = 0.0166$.

From Eq. 5.5 (also *NCEES Handbook:* Head Loss in Pipe or Conduit), the friction head is

$$h_f = \frac{fL\text{v}^2}{2Dg} = \frac{(0.0166)(50\ \text{ft})\left(8.176\ \frac{\text{ft}}{\text{sec}}\right)^2\left(12\ \frac{\text{in}}{\text{ft}}\right)}{(2)(2\ \text{in})\left(32.2\ \frac{\text{ft}}{\text{sec}^2}\right)}$$
$$= 5.17\ \text{ft}$$

Neglecting entrance and exit losses, the head added by the pump is

$$h_A = h_f + h_z = 5.17\ \text{ft} + (12\ \text{ft} - 4\ \text{ft})$$
$$= 13.17\ \text{ft} \quad (13\ \text{ft})$$

Note that the change in elevation only considers the surface of the two reservoirs, therefore the exact elevation of the inlet and discharge pipes are not needed.

SI Solution

The area of the rubber hose is

$$A = \frac{\pi D^2}{4} = \frac{\pi(5.1\ \text{cm})^2}{(4)\left(100\ \frac{\text{cm}}{\text{m}}\right)^2} = 0.00204\ \text{m}^2$$

As in *NCEES Handbook:* Conservation of Mass, the velocity of water in the hose is

$$\text{v} = \frac{\dot{V}}{A} = \frac{5\ \frac{\text{L}}{\text{s}}}{(0.00204\ \text{m}^2)\left(1000\ \frac{\text{L}}{\text{m}^3}\right)} = 2.45\ \text{m/s}$$

At 27°C, from App. 1B (also *NCEES Handbook* table "Physical Properties of Liquid Water (SI Units)"), the kinematic viscosity of water is $\nu = 0.854 \times 10^{-6}\ \text{m}^2/\text{s}$.

As in *NCEES Handbook:* Similitude, the Reynolds number is

$$\text{Re} = \frac{\text{v}D}{\nu} = \frac{\left(2.45\ \frac{\text{m}}{\text{s}}\right)(5.1\ \text{cm})}{\left(0.854 \times 10^{-6}\ \frac{\text{m}^2}{\text{s}}\right)\left(100\ \frac{\text{cm}}{\text{m}}\right)} = 1.46 \times 10^5$$

Since the rubber hose is smooth, from App. 4B (also *NCEES Handbook:* Friction Factors for Turbulent Flow), the friction factor is $f = 0.0166$.

From Eq. 5.5 (also *NCEES Handbook:* Head Loss in Pipe or Conduit), the friction head is

$$h_f = \frac{fL\text{v}^2}{2Dg} = \frac{(0.0166)(15\text{ m})\left(2.45\ \frac{\text{m}}{\text{s}}\right)^2\left(100\ \frac{\text{cm}}{\text{m}}\right)}{(2)(5.1\text{ cm})\left(9.81\ \frac{\text{m}}{\text{s}^2}\right)}$$
$$= 1.49\text{ m}$$

Neglecting entrance and exit losses, the head added by the pump is

$$\begin{aligned} h_A &= h_f + h_z \\ &= 1.49\text{ m} + 4\text{ m} - 1.5\text{ m} \\ &= 3.99\text{ m} \quad (4.0\text{ m}) \end{aligned}$$

Note that the change in elevation only considers the surface of the two reservoirs, therefore the exact elevation of the inlet and discharge pipes are not needed.

The answer is (B).

7. *Customary U.S. Solution*

From App. 3B (also *NCEES Handbook* table "Pipe Dimensions and Weights"), the diameters (inside) for 8 in and 6 in schedule-40 steel pipe are

$$\begin{aligned} D_1 &= 7.981\text{ in} \\ D_2 &= 6.065\text{ in} \end{aligned}$$

At 60°F, from App. 1A, the density of water is 62.37 lbm/ft^3.

The mass flow rate through 6 in pipe is from *NCEES Handbook:* Conservation of Mass.

$$\dot{m} = A_2\text{v}_2\rho = \frac{\pi\left(\frac{(6.065\text{ in})^2}{4}\right)\left(12\ \frac{\text{ft}}{\text{sec}}\right)\left(62.37\ \frac{\text{lbm}}{\text{ft}^3}\right)}{\left(12\ \frac{\text{in}}{\text{ft}}\right)^2}$$
$$= 150.2\text{ lbm/sec}$$

The inlet (suction) pressure is

$$(14.7\text{ psia} - 5\text{ psig})\left(12\ \frac{\text{in}}{\text{ft}}\right)^2 = 1397\text{ lbf/ft}^2$$

From Table 5.5 (also *NCEES Handbook:* Pump Power), the head added by the pump is

$$\begin{aligned} h_A &= \frac{550(\text{BHP})\eta}{\dot{m}} \times \frac{g_c}{g} \\ &= \left(\frac{\left(550\ \frac{\text{ft-lbf}}{\text{hp-sec}}\right)(20\text{ hp})(0.70)}{150.2\ \frac{\text{lbm}}{\text{sec}}}\right)\left(\frac{32.2\ \frac{\text{lbm-ft}}{\text{lbf-sec}^2}}{32.2\ \frac{\text{ft}}{\text{sec}^2}}\right) \\ &= 51.26\text{ ft} \end{aligned}$$

At 1 (pump inlet),

$$\begin{aligned} p_1 &= 1397\text{ lbf/ft}^2 \quad [\text{absolute}] \\ z_1 &= 0 \\ \text{v}_1 &= \frac{\text{v}_2A_2}{A_1} = \text{v}_2\left(\frac{D_2}{D_1}\right)^2 = \left(12\ \frac{\text{ft}}{\text{sec}}\right)\left(\frac{6.065\text{ in}}{7.981\text{ in}}\right)^2 \\ &= 6.93\text{ ft/sec} \end{aligned}$$

At 2 (pump outlet),

$$\begin{aligned} p_2 &\quad [\text{unknown}] \\ \text{v}_2 &= 12\text{ ft/sec} \quad [\text{given}] \\ z_2 &= z_1 = 0 \\ h_{f,1-2} &= 0 \end{aligned}$$

Let z_3 be the additional head above atmospheric. As in *NCEES Handbook:* Pump Head, the head added by the pump is

$$h_A = \frac{(p_2 - p_1)g_c}{\rho g} + \frac{\text{v}_2^2 - \text{v}_1^2}{2g} + z_2 - z_1 + h_{f,1-2}$$

$$\begin{aligned} 51.26\text{ ft} &= \frac{\left(p_2 - 1397\ \frac{\text{lbf}}{\text{ft}^2}\right)\left(32.2\ \frac{\text{lbm-ft}}{\text{lbf-sec}^2}\right)}{\left(62.37\ \frac{\text{lbm}}{\text{ft}^3}\right)\left(32.2\ \frac{\text{ft}}{\text{sec}^2}\right)} \\ &\quad + \frac{\left(12\ \frac{\text{ft}}{\text{sec}}\right)^2 - \left(6.93\ \frac{\text{ft}}{\text{sec}}\right)^2}{(2)\left(32.2\ \frac{\text{ft}}{\text{sec}^2}\right)} \\ &\quad + 0\text{ ft} - 0\text{ ft} + 0\text{ ft} \\ p_2 &= 4501.2\text{ lbf/ft}^2 \end{aligned}$$

At 3 (discharge),

$$p_3 = \left(14.7\ \frac{\text{lbf}}{\text{in}^2}\right)\left(12\ \frac{\text{in}}{\text{ft}}\right)^2 = 2117\ \text{lbf/ft}^2$$
$$v_3 = 12\ \text{ft/sec}\quad [\text{given}]$$
$$z_3\quad [\text{unknown}]$$
$$h_{f,2-3} = 10\ \text{ft}$$
$$h_A = 0\quad [\text{no pump between points 2 and 3}]$$
$$h_A = \frac{(p_3 - p_2)g_c}{\rho g} + \frac{v_3^2 - v_2^2}{2g} + z_3 - z_2 + h_{f,2-3}$$
$$0 = \frac{\left(2117\ \frac{\text{lbf}}{\text{ft}^2} - 4501.2\ \frac{\text{lbf}}{\text{ft}^2}\right)\left(32.2\ \frac{\text{lbm-ft}}{\text{lbf-sec}^2}\right)}{\left(62.37\ \frac{\text{lbm}}{\text{ft}^3}\right)\left(32.2\ \frac{\text{ft}}{\text{sec}^2}\right)} + \frac{\left(12\ \frac{\text{ft}}{\text{sec}}\right)^2 - \left(12\ \frac{\text{ft}}{\text{sec}}\right)^2}{(2)\left(32.2\ \frac{\text{ft}}{\text{sec}^2}\right)} + z_3 - 0\ \text{ft} + 10\ \text{ft}$$
$$z_3 = 28.2\ \text{ft}\quad (28\ \text{ft})$$

z_3 could have been found directly without determining the intermediate pressure, p_2. This method is illustrated in the SI solution.

SI Solution

From App. 3C (also *NCEES Handbook* table "Pipe Dimensions and Weights"), the inside diameters for 8 in and 6 in steel schedule-40 pipe are

$$D_1 = 202.717\ \text{mm}$$
$$D_2 = 154.051\ \text{mm}$$

At 16°C, from App. 1B (also *NCEES Handbook* table "Physical Properties of Liquid Water (SI Units)"), the density of water is 998.95 kg/m^3.

As in *NCEES Handbook:* Conservation of Mass, the mass flow rate through the 6 in pipe is

$$\dot{m} = A_2 v_2 \rho = \frac{\pi\left(\frac{(154.051\ \text{mm})^2}{4}\right)\left(4\ \frac{\text{m}}{\text{s}}\right)\left(998.95\ \frac{\text{kg}}{\text{m}^3}\right)}{\left(1000\ \frac{\text{mm}}{\text{m}}\right)^2}$$
$$= 74.5\ \text{kg/s}$$

The inlet (suction) pressure is

$$101.3\ \text{kPa} - 35\ \text{kPa} = 66.3\ \text{kPa}$$

From Table 5.6 (also *NCEES Handbook:* Pump Power), the head added by the pump is

$$h_A = \frac{1000(\text{BkW})\eta}{9.81\dot{m}} = \frac{\left(1000\ \frac{\text{W}}{\text{kW}}\right)(20\ \text{hp})\left(0.7457\ \frac{\text{kW}}{\text{hp}}\right)(0.70)}{\left(9.81\ \frac{\text{m}}{\text{s}^2}\right)\left(74.5\ \frac{\text{kg}}{\text{s}}\right)}$$
$$= 14.29\ \text{m}$$

At 1 (pump inlet),

$$p_1 = 66.3\ \text{kPa}$$
$$z_1 = 0$$
$$v_1 = v_2\left(\frac{A_2}{A_1}\right) = v_2\left(\frac{D_2}{D_1}\right)^2 = \left(4\ \frac{\text{m}}{\text{s}}\right)\left(\frac{154.051\ \text{mm}}{202.717\ \text{mm}}\right)^2$$
$$= 2.31\ \text{m/s}$$

At 2 (pump outlet),

$$p_2 = 101.3\ \text{kPa}$$
$$v_2 = 4\ \text{m/s}\quad [\text{given}]$$

As in *NCEES Handbook:* Pump Head, the head added by the pump is

$$h_A = \frac{p_2 - p_1}{\rho g} + \frac{v_2^2 - v_1^2}{2g} + z_2 - z_1 + h_f + z_3$$
$$14.29\ \text{m} = \frac{(101.3\ \text{kPa} - 66.3\ \text{kPa})\left(1000\ \frac{\text{Pa}}{\text{kPa}}\right)}{\left(998.95\ \frac{\text{kg}}{\text{m}^3}\right)\left(9.81\ \frac{\text{m}}{\text{s}^2}\right)} + \frac{\left(4\ \frac{\text{m}}{\text{s}}\right)^2 - \left(2.31\ \frac{\text{m}}{\text{s}}\right)^2}{(2)\left(9.81\ \frac{\text{m}}{\text{s}^2}\right)} + 0\ \text{m} - 0\ \text{m} + 3.3\ \text{m} + z_3$$
$$z_3 = 6.87\ \text{m}\quad (6.9\ \text{m})$$

The answer is (A).

8. *Customary U.S. Solution*

The flow rate is

$$\dot{V} = \left(10{,}000\ \frac{\text{gal}}{\text{hr}}\right)\left(0.1337\ \frac{\text{ft}^3}{\text{gal}}\right) = 1337\ \text{ft}^3/\text{hr}$$

From App. 3B (also *NCEES Handbook* table "Pipe Dimensions and Weights"), for 4 in schedule-40 steel pipe,

$$D = 0.3355 \text{ ft}$$
$$A = 0.08841 \text{ ft}^2$$

As in *NCEES Handbook:* Conservation of Mass, the velocity in the pipe is

$$\text{v} = \frac{\dot{V}}{A} = \frac{1337 \ \frac{\text{ft}^3}{\text{hr}}}{(0.08841 \text{ ft}^2)\left(3600 \ \frac{\text{sec}}{\text{hr}}\right)} = 4.20 \text{ ft/sec}$$

From App. 1A (also *NCEES Handbook* table "Physical Properties of Liquid Water (U.S. Units)"), the kinematic viscosity of water at 60°F is

$$\nu = 1.217 \times 10^{-5} \text{ ft}^2\text{/sec}$$
$$\rho = 62.37 \text{ lbm/ft}^3$$

As in *NCEES Handbook:* Similitude, the Reynolds number is

$$\text{Re} = \frac{D\text{v}}{\nu} = \frac{(0.3355 \text{ ft})\left(4.20 \ \frac{\text{ft}}{\text{sec}}\right)}{1.217 \times 10^{-5} \ \frac{\text{ft}^2}{\text{sec}}} = 1.16 \times 10^5$$

From App. 4A (also *NCEES Handbook:* Absolute Roughness and Relative Roughness), for welded and seamless steel, $\epsilon = 0.0002$ ft.

$$\frac{\epsilon}{D} = \frac{0.0002 \text{ ft}}{0.3355 \text{ ft}} \approx 0.0006$$

From App. 4B (also *NCEES Handbook:* Friction Factors for Turbulent Flow), the friction factor, f, is 0.0205. The 7000 ft of equivalent length includes the pipe between the lake and the pump. Using *NCEES Handbook:* Head Loss in Pipe or Conduit, the friction head is

$$h_f = \frac{fL\text{v}^2}{2Dg} = \frac{(0.0205)(7000 \text{ ft})\left(4.20 \ \frac{\text{ft}}{\text{sec}}\right)^2}{(2)(0.3355 \text{ ft})\left(32.2 \ \frac{\text{ft}}{\text{sec}^2}\right)} = 117.2 \text{ ft}$$

The head added by the pump is

$$h_A = h_f + h_z = 117.2 \text{ ft} + (12 \text{ ft} + 350 \text{ ft}) = 479.2 \text{ ft}$$

From Table 5.5 (also *NCEES Handbook:* Pump Power), the hydraulic horsepower is

$$\text{WP} = \frac{h_A Q(\text{SG})}{3956} = \frac{(479.2 \text{ ft})\left(10{,}000 \ \frac{\text{gal}}{\text{hr}}\right)(1)}{\left(3956 \ \frac{\text{ft-gal}}{\text{hp-min}}\right)\left(60 \ \frac{\text{min}}{\text{hr}}\right)} = 20.2 \text{ hp}$$

From Eq. 5.15 (also *NCEES Handbook:* Pump Power), the electrical horsepower is

$$\text{EHP} = \frac{\text{WP}}{\eta} = \frac{20.2 \text{ hp}}{0.7} = 28.9 \text{ hp}$$

At \$0.04/kW-hr, power costs for 1 hr are

$$(28.9 \text{ hp})\left(0.7457 \ \frac{\text{kW}}{\text{hp}}\right)(1 \text{ hr})\left(0.04 \ \frac{\$}{\text{kW-hr}}\right) = \$0.86 \text{ per hour} \quad (\$1 \text{ per hour})$$

SI Solution

From App. 3C (also *NCEES Handbook* table "Pipe Dimensions and Weights"), for 4 in schedule-40 steel pipe,

$$D = 102.26 \text{ mm}$$
$$A = 82.30 \times 10^{-4} \text{ m}^2$$

As in *NCEES Handbook:* Conservation of Mass, the velocity in the pipe is

$$\text{v} = \frac{\dot{V}}{A} = \frac{10.5 \ \frac{\text{L}}{\text{s}}}{(82.30 \times 10^{-4} \text{ m}^2)\left(1000 \ \frac{\text{L}}{\text{m}^3}\right)} = 1.28 \text{ m/s}$$

From App. 1B (also *NCEES Handbook* table "Physical Properties of Liquid Water (SI Units)"), at 16°C the water data are

$$\rho = 998.95 \text{ kg/m}^3$$
$$\mu = 1.1081 \times 10^{-3} \text{ Pa·s}$$

As in *NCEES Handbook:* Similitude, the Reynolds number is

$$\text{Re} = \frac{\rho v D}{\mu} = \frac{\left(998.95\ \frac{\text{kg}}{\text{m}^3}\right)\left(1.28\ \frac{\text{m}}{\text{s}}\right)(102.26\ \text{mm})}{(1.1081 \times 10^{-3}\ \text{Pa·s})\left(1000\ \frac{\text{mm}}{\text{m}}\right)}$$

$$= 1.18 \times 10^5$$

From Table 4.2 (also *NCEES Handbook:* Absolute Roughness and Relative Roughness), for welded and seamless steel, $\epsilon = 6.0 \times 10^{-5}$ m.

$$\frac{\epsilon}{D} = \frac{(6.0 \times 10^{-5}\ \text{m})\left(1000\ \frac{\text{mm}}{\text{m}}\right)}{102.26\ \text{mm}} \approx 0.0006$$

From App. 4B (also *NCEES Handbook:* Friction Factors for Turbulent Flow), the friction factor is $f = 0.0205$.

From Eq. 5.5 (also *NCEES Handbook:* Head Loss in Pipe or Conduit), the friction head is

$$h_f = \frac{fL\text{v}^2}{2Dg} = \frac{(0.0205)(2300\ \text{m})\left(1.28\ \frac{\text{m}}{\text{s}}\right)^2\left(1000\ \frac{\text{mm}}{\text{m}}\right)}{(2)(102.26\ \text{mm})\left(9.81\ \frac{\text{m}}{\text{s}^2}\right)}$$

$$= 38.5\ \text{m}$$

The head added by the pump is

$$h_A = h_f + h_z = 38.5\ \text{m} + 4\ \text{m} + 115\ \text{m} = 157.5\ \text{m}$$

From Table 5.6 (also *NCEES Handbook:* Pump Power), the hydraulic power is

$$\text{WkW} = \frac{9.81 h_A Q(\text{SG})}{1000} = \frac{\left(9.81\ \frac{\text{m}}{\text{s}^2}\right)(157.5\ \text{m})\left(10.5\ \frac{\text{L}}{\text{s}}\right)(1)}{1000\ \frac{\text{W·L}}{\text{kW·kg}}}$$

$$= 16.22\ \text{kW}$$

From Eq. 5.15 (also *NCEES Handbook:* Pump Power), the electrical power is

$$\text{EHP} = \frac{\text{WkW}}{\eta} = \frac{16.22\ \text{kW}}{0.7} = 23.2\ \text{kW}$$

At \$0.04/kW·h, power costs for 1 h are

$$(23.2\ \text{kW})(1\ \text{h})\left(0.04\ \frac{\$}{\text{kW·h}}\right) = \$0.93\quad(\$1)$$

The answer is (B).

9. The static suction lift, $h_{p(s)}$, is 20 ft.

The static discharge head, $h_{p(d)}$, is 15 ft.

There is no pipe size specified, so h_v cannot be calculated. Even so, v is typically in the 5–10 ft/sec range, and $h_v \approx 0$. Since pipe lengths are not given, assume $h_f \approx 0$.

$$20\ \text{ft} + 15\ \text{ft} + \frac{\left(80\ \frac{\text{lbf}}{\text{in}^2}\right)\left(12\ \frac{\text{in}}{\text{ft}}\right)^2}{62.4\ \frac{\text{lbf}}{\text{ft}^3}} + 10\ \text{ft}$$

$$= 229.6\ \text{ft of water}$$

The mass flow rate is

$$\frac{(3.5\ \text{MGD})\left(62.4\ \frac{\text{lbm}}{\text{ft}^3}\right)\left(10^6\ \frac{\text{gal}}{\text{MG}}\right)}{\left(7.4805\ \frac{\text{gal}}{\text{ft}^3}\right)\left(24\ \frac{\text{hr}}{\text{day}}\right) \times \left(60\ \frac{\text{min}}{\text{hr}}\right)\left(60\ \frac{\text{sec}}{\text{min}}\right)} = 337.9\ \text{lbm/sec}$$

The rated motor output power does not depend on the motor efficiency. The motor produces what it is rated to produce. From Table 5.5 (also *NCEES Handbook:* Pump Power),

$$P = \frac{h_A \dot{m}}{550\eta_{\text{pump}}} \times \frac{g}{g_c}$$

$$= \left(\frac{(229.6\ \text{ft})\left(337.9\ \frac{\text{lbm}}{\text{sec}}\right)}{\left(550\ \frac{\text{ft-lbf}}{\text{hp-sec}}\right)(0.85)}\right)\left(\frac{32.2\ \frac{\text{ft}}{\text{sec}^2}}{32.2\ \frac{\text{lbm-ft}}{\text{lbf-sec}^2}}\right)$$

$$= 166.0\ \text{hp}$$

Use a 200 hp motor.

The answer is (D).

10. *Customary U.S. Solution*

From the App. 3B (also *NCEES Handbook* table "Pipe Dimensions and Weights"), for 3 in schedule-40 steel pipe,

$$D = 0.2557 \text{ ft}$$
$$A = 0.05134 \text{ ft}^2$$

As in *NCEES Handbook:* Head Loss in Pipe or Conduit, the equivalent lengths for various fittings are

flanged elbow, $L_e = (20)(3 \text{ in}) = 60 \text{ in} = 5 \text{ ft}$

wide-open gate valve, $L_e = (10)(3 \text{ in}) = 30 \text{ in} = 2.5 \text{ ft}$

The total equivalent length of pipe and fittings is

$$\begin{aligned} L_e &= 500 \text{ ft} + (6)(5 \text{ ft}) + (2)(2.5 \text{ ft}) \\ &= 535 \text{ ft} \end{aligned}$$

The flow rate is 100 gal/min.

As in *NCEES Handbook:* Conservation of Mass, the velocity in the pipe is

$$\begin{aligned} \text{v} = \frac{\dot{V}}{A} &= \frac{100 \ \frac{\text{gal}}{\text{min}}}{(0.05134 \text{ ft}^2)\left(7.4805 \ \frac{\text{gal}}{\text{ft}^3}\right)\left(60 \ \frac{\text{sec}}{\text{min}}\right)} \\ &= 4.34 \text{ ft/sec} \end{aligned}$$

As in *NCEES Handbook:* Similitude, the Reynolds number is

$$\text{Re} = \frac{\text{v}D}{\nu} = \frac{\left(4.34 \ \frac{\text{ft}}{\text{sec}}\right)(0.2557 \text{ ft})}{6 \times 10^{-6} \ \frac{\text{ft}^2}{\text{sec}}} = 1.85 \times 10^5$$

From App. 4A (also *NCEES Handbook:* Absolute Roughness and Relative Roughness), $\epsilon = 0.0002$ ft.

$$\frac{\epsilon}{D} = \frac{0.0002 \text{ ft}}{0.2557 \text{ ft}} \approx 0.0008$$

From the friction factor table, $f \approx 0.0204$.

For higher flow rates, f approaches 0.0186. Since the chosen flow rate was almost the lowest, $f = 0.0186$ should be used.

From Eq. 5.5 (also *NCEES Handbook:* Friction Factors for Turbulent Flow), the friction head loss is

$$\begin{aligned} h_f = \frac{fL\text{v}^2}{2Dg} &= \frac{(0.0186)(535 \text{ ft})\left(4.34 \ \frac{\text{ft}}{\text{sec}}\right)^2}{(2)(0.2557 \text{ ft})\left(32.2 \ \frac{\text{ft}}{\text{sec}^2}\right)} \\ &= 11.4 \text{ ft of gasoline} \end{aligned}$$

The friction head loss neglects the small velocity head. The other system points can be found using Eq. 5.44 (also *NCEES Handbook:* Pump Similitude).

$$\frac{h_{f_1}}{h_{f_2}} = \left(\frac{Q_1}{Q_2}\right)^2$$

$$\begin{aligned} h_{f_2} = h_{f_1}\left(\frac{Q_2}{100 \ \frac{\text{gal}}{\text{min}}}\right)^2 &= (11.4 \text{ ft})\left(\frac{Q_2}{100 \ \frac{\text{gal}}{\text{min}}}\right)^2 \\ &= 0.00114 Q_2^2 \end{aligned}$$

Q (gal/min)	h_f (ft)	$h_f + 60$ (ft)
100	11.4	71.3
200	45.2	105.2
300	101.7	161.7
400	180.8	240.8
500	282.5	342.5
600	406.8	466.8

Plot the system and pump curves. The pump's characteristic curve is independent of the liquid's specific gravity.

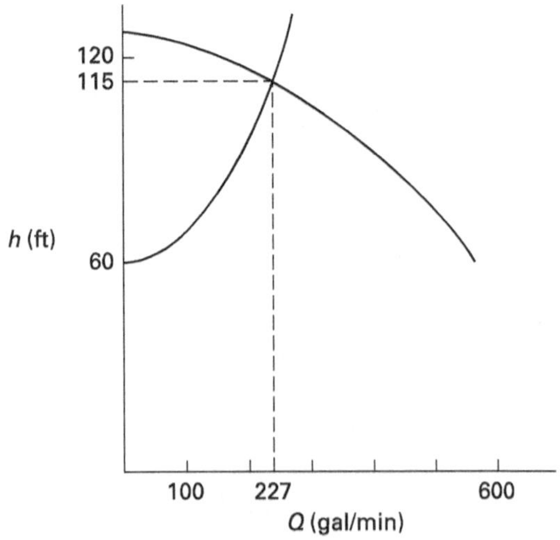

$$h = 115 \text{ ft}$$

The transfer rate is

$$Q = 227 \text{ gal/min}$$

(This value could be used to determine a new friction factor.)

From Table 5.5 (also *NCEES Handbook:* Pump Power), the electrical power supplied to the motor is

$$\text{EHP} = \frac{h_A Q(\text{SG})}{3956\eta_{\text{pump}}\eta_{\text{motor}}} = \frac{(115 \text{ ft})\left(227 \ \frac{\text{gal}}{\text{min}}\right)(0.7)}{\left(3956 \ \frac{\text{ft-gal}}{\text{hp-min}}\right)(0.88)(0.88)}$$
$$= 5.96 \text{ hp}$$

The cost per hour is

$$(5.96 \text{ hp})\left(0.7457 \ \frac{\text{kW}}{\text{hp}}\right)(1 \text{ hr})\left(0.045 \ \frac{\$}{\text{kW·h}}\right) = \$0.20$$

SI Solution

From App. 3C (also *NCEES Handbook* table "Pipe Dimensions and Weights"), for 3 in schedule-40 pipe,

$$D = 77.92 \text{ mm}$$
$$A = 47.69 \times 10^{-4} \text{ m}^2$$

From App. 4D (also *NCEES Handbook:* Head Loss in Pipe or Conduit), the equivalent lengths for various fittings are

flanged elbow: $L_e = (20)(77.92 \text{ mm}) = 1558.4 \text{ mm} = 1.6 \text{ m}$

wide-open gate valve: $L_e = (10)(77.92 \text{ mm}) = 779.2 \text{ mm} = 0.8 \text{ m}$

The total equivalent length of pipe and fittings is

$$L_e = 170 \text{ m} + \big((6)(1.6\text{m}) + (2)(0.8 \text{ m})\big)$$
$$= 181.2 \text{ m}$$

The flow rate is 6.3 L/s. As in *NCEES Handbook:* Conservation of Mass, the velocity in the pipe is

$$\text{v} = \frac{\dot{V}}{A} = \frac{6.3 \ \frac{\text{L}}{\text{s}}}{(47.69 \times 10^{-4} \text{ m}^2)\left(1000 \ \frac{\text{L}}{\text{m}^3}\right)}$$
$$= 1.32 \text{ m/s}$$

As in *NCEES Handbook:* Similitude, the Reynolds number is

$$\text{Re} = \frac{\text{v}D}{\nu} = \frac{\left(1.32 \ \frac{\text{m}}{\text{s}}\right)(77.92 \text{ mm})}{\left(5.6 \times 10^{-7} \ \frac{\text{m}^2}{\text{s}}\right)\left(1000 \ \frac{\text{mm}}{\text{m}}\right)} = 1.84 \times 10^5$$

From Table 4.2 (also *NCEES Handbook:* Absolute Roughness and Relative Roughness), $\epsilon = 6.0 \times 10^{-5}$ m.

$$\frac{\epsilon}{D} = \frac{(6.0 \times 10^{-5} \text{ m})\left(1000 \ \frac{\text{mm}}{\text{m}}\right)}{77.92 \text{ mm}} \approx 0.0008$$

From App. 4B (also *NCEES Handbook:* Friction Factors for Turbulent Flow), $f = 0.0204$.

For higher flow rates, f approaches 0.0186. Since the chosen flow rate was almost the lowest, $f = 0.0186$ should be used.

From Eq. 5.5 (also *NCEES Handbook:* Head Loss in Pipe or Conduit), the friction head loss is

$$h_f = \frac{fL\text{v}^2}{2Dg} = \frac{(0.0186)(181.2 \text{ m})\left(1.32 \ \frac{\text{m}}{\text{s}}\right)^2\left(1000 \ \frac{\text{mm}}{\text{m}}\right)}{(2)(77.92 \text{ mm})\left(9.81 \ \frac{\text{m}}{\text{s}^2}\right)}$$
$$= 3.80 \text{ m of gasoline}$$

The friction head loss neglects the small velocity head. The other system points can be found using Eq. 5.44 (also *NCEES Handbook:* Pump Similitude).

$$\frac{h_{f1}}{h_{f2}} = \left(\frac{Q_1}{Q_2}\right)^2$$

$$h_{f2} = h_{f1}\left(\frac{Q_2}{Q_1}\right)^2 = (3.80 \text{ m})\left(\frac{Q_2}{6.3 \ \frac{\text{L}}{\text{s}}}\right)^2$$
$$= 0.0957Q_2^2$$

Q (L/s)	h_f (m)	$h_f + 20$(m)
6.3	3.80	23.80
12	13.78	33.78
18	31.0	51.0
24	55.1	75.1
30	86.1	106.1
36	124.0	144.0

Plot the system and pump curves. The pump's characteristic curve is independent of the liquid's specific gravity.

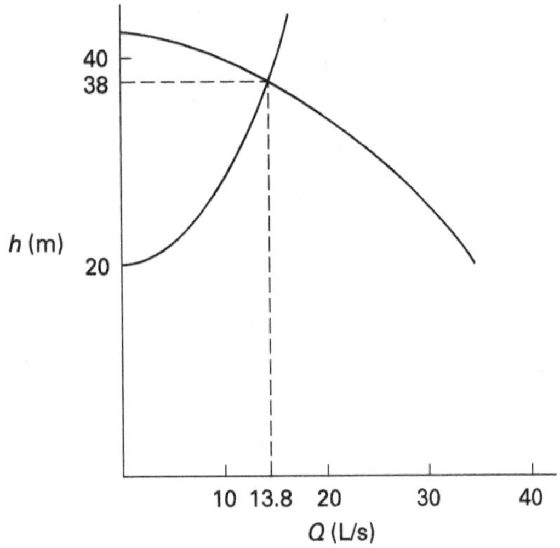

$$h = 38.0 \text{ m}$$

The transfer rate is

$$Q = 13.8 \text{ L/s}$$

(This value could be used to determine a new friction factor.)

From Table 5.6 (also *NCEES Handbook:* Pump Power), the electric power delivered to the motor is

$$\begin{aligned}\text{EkW} &= \frac{9.81 h_A Q(\text{SG})}{1000\eta_{\text{pump}}\eta_{\text{motor}}} \\ &= \frac{\left(9.81 \ \frac{\text{m}}{\text{s}^2}\right)(38.0 \text{ m})\left(13.8 \ \frac{\text{L}}{\text{s}}\right)(0.7)}{\left(1000 \ \frac{\text{W}}{\text{kW}}\right)(0.88)(0.88)} \\ &= 4.65 \text{ kW}\end{aligned}$$

The cost per hour is

$$(4.65 \text{ kW})(1 \text{ h})\left(0.045 \ \frac{\$}{\text{kW·h}}\right) = \$0.21 \quad (\$0.20)$$

The answer is (A).

11. *Customary U.S. Solution*

At base form,

$$Power = \Delta p Q$$

However, unit conversion is necessary to take a pressure and flow rate in the units of psi and gpm and achieve horsepower.

$$WHP = k\left(\frac{lb_f}{in^2}\right)\left(\frac{gal}{min}\right)$$

$$1 \ hp = 746.04 \ W = 746.04 \ \frac{kg \ m^2}{s^3}$$

$$1 \ \frac{lb_f}{in^2} = 6{,}894.64 \ \frac{N}{m^2} = 6{,}894.64 \ \frac{kg}{m \ s^2}$$

$$1 \ \frac{gal}{min} = \frac{3.785 \ m^3}{60 \ s} = 6.31 \ x \ 10^{-5} \ \frac{m^3}{s}$$

$$k = \frac{\left(6{,}894.64 \ \frac{kg}{m \ s^2}\right)\left(6.31 \ x \ 10^{-5} \ \frac{m^3}{s}\right)}{746.04 \ \frac{kg \ m^2}{s^3}} = 5.834 \ x \ 10^{-4} \ \frac{psi * gpm}{hp} = \frac{1}{1714} \ \frac{psi * gpm}{hp}$$

$$\text{WHP} = \frac{\Delta p Q}{1714}$$
$$\Delta p = p_d - p_s$$

The absolute pressures are

$$\begin{aligned}p_d &= 40 \text{ psig} + 14.7 \text{ psia} = 54.7 \text{ psia} \\ p_s &= 1 \text{ atm} = 14.7 \text{ psia} \\ \Delta p &= p_d - p_s = 54.7 \text{ psia} - 14.7 \text{ psia} = 40 \text{ psia}\end{aligned}$$

$$\text{WHP} = \frac{\left(40 \ \frac{\text{lbf}}{\text{in}^2}\right)\left(37 \ \frac{\text{gal}}{\text{min}}\right)}{1714 \ \frac{\text{lbf-gal}}{\text{in}^2\text{-min-hp}}} = 0.863 \text{ hp} \quad (0.9 \text{ hp})$$

SI Solution

From Table 5.6 (also *NCEES Handbook:* Pump Power), the hydraulic kilowatts are

$$\text{WkW} = \frac{\Delta p Q}{1000}$$
$$\Delta p = p_d - p_s$$

The absolute pressures are

$$p_d = 275 \text{ kPa} + 101.3 \text{ kPa} = 376.3 \text{ kPa}$$
$$p_s = 1 \text{ atm} = 101.3 \text{ kPa}$$
$$\Delta p = p_d - p_s = 376.3 \text{ kPa} - 101.3 \text{ kPa} = 275 \text{ kPa}$$
$$\text{WkW} = \frac{(275 \text{ kPa})\left(65 \ \frac{\text{L}}{\text{s}}\right)}{1000 \ \frac{\text{W}}{\text{kW}}} = 17.88 \text{ kW} \quad (18 \text{ kW})$$

The answer is (B).

12. *Customary U.S. Solution*

From Eq. 5.28(b) (also *NCEES Handbook:* Specific Speed (N_s) at the BEP), the specific speed is

$$N_s = \frac{n\sqrt{Q}}{h_A^{0.75}}$$

For a double-suction pump, Q in the preceding equation is half of the full flow rate.

$$N_s = \frac{\left(900 \ \frac{\text{rev}}{\text{min}}\right)\sqrt{\left(\frac{1}{2}\right)\left(300 \ \frac{\text{gal}}{\text{sec}}\right)\left(60 \ \frac{\text{sec}}{\text{min}}\right)}}{(20 \text{ ft})^{0.75}}$$
$$= 9028 \text{ rpm} \quad (9000 \text{ rpm})$$

SI Solution

From Eq. 5.28(a) (also *NCEES Handbook:* Specific Speed (N_s) at the BEP), the specific speed is

$$N_s = \frac{n\sqrt{\dot{V}}}{h_A^{0.75}}$$

For a double-suction pump, $\dot{V}$ in the preceding equation is half of the full flow rate.

$$N_s = \frac{\left(900 \ \frac{\text{rev}}{\text{min}}\right)\sqrt{\left(\frac{1}{2}\right)\left(1.1 \ \frac{\text{kL}}{\text{s}}\right)\left(1 \ \frac{\text{m}^3}{\text{kL}}\right)}}{(7 \text{ m})^{0.75}}$$
$$= 155.1 \text{ rpm} \quad (160 \text{ rpm})$$

The answer is (C).

13. *Customary U.S. Solution*

From App. 3B (also *NCEES Handbook* table "Pipe Dimensions and Weights"), for 1.5 in schedule-40 steel pipe,

$$D = 0.1342 \text{ ft}$$
$$A = 0.01414 \text{ ft}^2$$

As in *NCEES Handbook:* Conservation of Mass, The velocity in the pipe is

$$\text{v} = \frac{\dot{V}}{A} = \frac{100 \ \frac{\text{gal}}{\text{min}}}{(0.01414 \text{ ft}^2)\left(7.4805 \ \frac{\text{gal}}{\text{ft}^3}\right)\left(60 \ \frac{\text{sec}}{\text{min}}\right)}$$
$$= 15.76 \text{ ft/sec}$$

From App. 4A (also *NCEES Handbook:* Absolute Roughness and Relative Roughness), for steel, ϵ = 0.0002 ft.

$$\frac{\epsilon}{D} = \frac{0.0002 \text{ ft}}{0.1342 \text{ ft}} = 0.0015$$

At 281°F, $\nu = 0.239 \times 10^{-5}$ ft²/sec. As in *NCEES Handbook:* Similitude, the Reynolds number is

$$\text{Re} = \frac{D\text{v}}{\nu} = \frac{(0.1342 \text{ ft})\left(15.76 \ \frac{\text{ft}}{\text{sec}}\right)}{0.239 \times 10^{-5} \ \frac{\text{ft}^2}{\text{sec}}}$$
$$= 8.85 \times 10^5$$

From App. 4B (also *NCEES Handbook:* Friction Factors for Turbulent Flow), the friction factor is $f = 0.022$.

From App. 4D (also *NCEES Handbook:* Head Loss in Pipe or Conduit and *NCEES Handbook:* Loss Coefficients for Pipe Entrance and Exit), the equivalent lengths for screwed steel fittings are

inlet (square mouth):

$$L_e = \left(\frac{D}{f}\right)\left(\frac{160}{\text{Re}} + (0.5)\left(1 + \frac{1}{D_{\text{inches}}}\right)\right)$$
$$= \left(\frac{1.5 \text{ in}}{0.022}\right) \times \left(\frac{160}{8.85 \times 10^5} + (0.5)\left(1 + \frac{1}{1.5}\right)\right)$$
$$= 56.8 \text{ in}$$
$$= 4.7 \text{ ft}$$

long radius 90° elbow: $L_e = (16)(1.5 \text{ in}) = 24 \text{ in} = 2 \text{ ft}$

wide-open gate valves: $L_e = (10)(1.5 \text{ in}) = 15 \text{ in} = 1.3 \text{ ft}$

The total equivalent length is

$$30 \text{ ft} + 4.7 \text{ ft} + (2)(2 \text{ ft}) + (2)(1.3 \text{ ft}) = 41.3 \text{ ft}$$

From Eq. 5.5 (also *NCEES Handbook:* Head Loss in Pipe or Conduit), the friction head is

$$\begin{aligned} h_f &= \frac{fL\text{v}^2}{2Dg} = \frac{(0.022)(41.3 \text{ ft})\left(15.76 \ \frac{\text{ft}}{\text{sec}}\right)^2}{(2)(0.1342 \text{ ft})\left(32.2 \ \frac{\text{ft}}{\text{sec}^2}\right)} \\ &= 26.11 \text{ ft} \end{aligned}$$

The density of the liquid is the reciprocal of the specific volume, taken from App. 10B (also *NCEES Handbook* table "Physical Properties of Liquid Water (U.S. Units)"), at 281°F.

$$\rho = \frac{1}{\upsilon_f} = \frac{1}{0.01727 \ \frac{\text{ft}^3}{\text{lbm}}} = 57.9 \text{ lbm/ft}^3$$

Extracted from Eq. 5.4(b) (also *NCEES Handbook:* The Bernoulli Equation), the vapor pressure head is

$$\begin{aligned} h_{\text{vp}} &= \frac{p_{\text{vapor}}}{\rho} \times \frac{g_c}{g} \\ &= \left(\frac{\left(50.06 \ \frac{\text{lbf}}{\text{in}^2}\right)\left(12 \ \frac{\text{in}}{\text{ft}}\right)^2}{57.9 \ \frac{\text{lbm}}{\text{ft}^3}}\right)\left(\frac{32.2 \ \frac{\text{lbm-ft}}{\text{lbf-sec}^2}}{32.2 \ \frac{\text{ft}}{\text{sec}^2}}\right) \\ &= 124.5 \text{ ft} \end{aligned}$$

Extracted from Eq. 5.4(b) (also *NCEES Handbook:* The Bernoulli Equation), the pressure head is

$$\begin{aligned} h_p &= \frac{p}{\rho} \times \frac{g_c}{g} = \left(\frac{\left(80 \ \frac{\text{lbf}}{\text{in}^2}\right)\left(12 \ \frac{\text{in}}{\text{ft}}\right)^2}{57.9 \ \frac{\text{lbm}}{\text{ft}^3}}\right)\left(\frac{32.2 \ \frac{\text{lbm-ft}}{\text{lbf-sec}^2}}{32.2 \ \frac{\text{ft}}{\text{sec}^2}}\right) \\ &= 199.0 \text{ ft} \end{aligned}$$

From Eq. 5.30 (also *NCEES Handbook:* Net-Positive Suction Head (NPSH)), the NPSHA is

$$\begin{aligned} \text{NPSHA} &= h_p + h_{z(s)} - h_{f(s)} - h_{\text{vp}} \\ &= 199.0 \text{ ft} + 20 \text{ ft} - 26.11 \text{ ft} - 124.5 \text{ ft} \\ &= 68.5 \text{ ft} \quad (68 \text{ ft}) \end{aligned}$$

Since NPSHR = 10 ft, the pump will not cavitate.

(A pump may not be needed in this configuration.)

SI Solution

From App. 3C (also *NCEES Handbook* table "Pipe Dimensions and Weights"), for 1.5 in schedule-40 steel pipe,

$$\begin{aligned} D &= 40.89 \text{ mm} \\ A &= 13.13 \times 10^{-4} \text{ m}^2 \end{aligned}$$

As in *NCEES Handbook:* Conservation of Mass, the velocity in the pipe is

$$\begin{aligned} \text{v} &= \frac{\dot{V}}{A} = \frac{6.3 \ \frac{\text{L}}{\text{s}}}{(13.13 \times 10^{-4} \text{ m}^2)\left(1000 \ \frac{\text{L}}{\text{m}^3}\right)} \\ &= 4.80 \text{ m/s} \end{aligned}$$

From Table 4.2 (also *NCEES Handbook:* Absolute Roughness and Relative Roughness), for steel, $\epsilon = 6.0 \times 10^{-5}$ m.

$$\begin{aligned} \frac{\epsilon}{D} &= \frac{(6.0 \times 10^{-5} \text{ m})\left(1000 \ \frac{\text{mm}}{\text{m}}\right)}{40.89 \text{ mm}} \\ &\approx 0.0015 \end{aligned}$$

Using *NCEES Handbook* table "Physical Properties of Liquid Water (SI Units)," at 138°C, $\nu = 0.222 \times 10^{-6}$ m²/s.

As in *NCEES Handbook:* Similitude, the Reynolds number is

$$\begin{aligned} \text{Re} &= \frac{D\text{v}}{\nu} = \frac{(40.89 \text{ mm})\left(4.80 \ \frac{\text{m}}{\text{s}}\right)}{\left(0.222 \times 10^{-6} \ \frac{\text{m}^2}{\text{s}}\right)\left(1000 \ \frac{\text{mm}}{\text{m}}\right)} \\ &= 8.84 \times 10^5 \end{aligned}$$

From App. 4B (also *NCEES Handbook:* Friction Factors for Turbulent Flow), the friction factor is $f = 0.022$.

From App. 4D (also *NCEES Handbook:* Head Loss in Pipe or Conduit and *NCEES Handbook:* Loss Coefficients for Pipe Entrance and Exit), the equivalent lengths for screwed steel fittings are

inlet (square mouth):

$$\begin{aligned} L_e &= \left(\frac{D}{f}\right)\left(\frac{160}{\text{Re}} + (0.5)\left(1+\frac{1}{D_{\text{inches}}}\right)\right) \\ &= \left(\frac{40.89 \text{ mm}}{0.022}\right) \\ &\quad \times\left(\frac{160}{8.85\times 10^5} + (0.5)\left(1+\frac{1}{1.5}\right)\right) \\ &= 1548 \text{ mm} \\ &= 1.5 \text{ m} \end{aligned}$$

long radius 90° elbow: $L_e = (16)(40.89 \text{ mm}) = 654 \text{ mm} = 0.7 \text{ m}$

wide-open gate valves: $L_e = (10)(40.89 \text{ mm}) = 409 \text{ mm} = 0.4 \text{ m}$

The total equivalent length is

$$10 \text{ m}+1.5 \text{ m}+(2)(0.7 \text{ m})+(2)(0.4 \text{ m}) = 13.7 \text{ m}$$

As in *NCEES Handbook:* Head Loss in Pipe or Conduit, the friction head is

$$\begin{aligned} h_f &= \frac{fL\text{v}^2}{2Dg} \\ &= \frac{(0.022)(13.7 \text{ m})\left(4.80 \frac{\text{m}}{\text{s}}\right)^2\left(1000 \frac{\text{mm}}{\text{m}}\right)}{(2)(40.89 \text{ mm})\left(9.81 \frac{\text{m}}{\text{s}^2}\right)} \\ &= 8.66 \text{ m} \end{aligned}$$

The density of the liquid is the reciprocal of the specific volume. From App. 10B (also *NCEES Handbook* table "Physical Properties of Liquid Water (SI Units)"), at 138°C,

$$\rho = \frac{1}{\upsilon_f} = \frac{(1)\left(3.281 \frac{\text{ft}}{\text{m}}\right)^3}{\left(0.01727 \frac{\text{ft}^3}{\text{lbm}}\right)\left(2.205 \frac{\text{lbm}}{\text{kg}}\right)} = 927.5 \text{ kg/m}^3$$

Extracted from Eq. 5.4(a) (also *NCEES Handbook:* The Bernoulli Equation), the vapor pressure head is

$$\begin{aligned} h_{\text{vp}} &= \frac{p_{\text{vapor}}}{\rho g} = \frac{(3.431 \text{ bar})\left(1\times 10^5 \frac{\text{Pa}}{\text{bar}}\right)}{\left(927.5 \frac{\text{kg}}{\text{m}^3}\right)\left(9.81 \frac{\text{m}}{\text{s}^2}\right)} \\ &= 37.71 \text{ m} \end{aligned}$$

Extracted from Eq. 5.4(a) (also *NCEES Handbook:* The Bernoulli Equation), the pressure head is

$$\begin{aligned} h_p &= \frac{p}{\rho g} = \frac{(550 \text{ kPa})\left(1000 \frac{\text{Pa}}{\text{kPa}}\right)}{\left(927.5 \frac{\text{kg}}{\text{m}^3}\right)\left(9.81 \frac{\text{m}}{\text{s}^2}\right)} \\ &= 60.45 \text{ m} \end{aligned}$$

From Eq. 5.30 (also *NCEES Handbook:* Net-Positive Suction Head (NPSH)), the NPSHA is

$$\begin{aligned} \text{NPSHA} &= h_p + h_{z(s)} - h_{f(s)} - h_{\text{vp}} \\ &= 60.45 \text{ m} + 6 \text{ m} - 8.66 \text{ m} - 37.71 \text{ m} \\ &= 20.08 \text{ m} \quad (20.6 \text{ m}) \end{aligned}$$

Since NPSHR is 3 m, the pump will not cavitate.

(A pump may not be needed in this configuration.)

The answer is (D).

14. The solvent is the freshwater, and the solution is the seawater. Since seawater contains approximately 2½% salt (NaCl) by weight, 100 lbm of seawater will yield 2.5 lbm salt and 97.5 lbm water. The molecular weight of salt is 23.0 + 35.5 = 58.5 lbm/lbmol. The number of moles of salt in 100 lbm of seawater is

$$n_{\text{salt}} = \frac{m}{\text{MW}} = \frac{2.5 \text{ lbm}}{58.5 \frac{\text{lbm}}{\text{lbmol}}} = 0.043 \text{ lbmol}$$

Similarly, water's molecular weight is 18.016 lbm/lbmol. The number of moles of water is

$$n_{\text{water}} = \frac{97.5 \text{ lbm}}{18.016 \frac{\text{lbm}}{\text{lbmol}}} = 5.412 \text{ lbmol}$$

The mole fraction of water is

$$\frac{5.412 \text{ lbmol}}{5.412 \text{ lbmol} + 0.043 \text{ lbmol}} = 0.992$$

Customary U.S. Solution

Cavitation will occur when

$$h_{atm} - h_v < h_{vp}$$

The density of seawater is 64.0 lbm/ft³.

Extracted from Eq. 5.4(b) (also *NCEES Handbook:* The Bernoulli Equation), the atmospheric head is

$$h_{atm} = \frac{p}{\rho} \times \frac{g_c}{g} = \left(\frac{\left(14.7\ \frac{\text{lbf}}{\text{in}^2}\right)\left(12\ \frac{\text{in}}{\text{ft}}\right)^2}{64.0\ \frac{\text{lbm}}{\text{ft}^3}} \right) \left(\frac{32.2\ \frac{\text{lbm-ft}}{\text{lbf-sec}^2}}{32.2\ \frac{\text{ft}}{\text{sec}^2}} \right)$$

$$= 33.075\ \text{ft} \quad [\text{ft of seawater}]$$

$$h_{depth} = 8\ \text{ft} \quad [\text{given}]$$

Extracted from Eq. 5.6 (also *NCEES Handbook:* The Bernoulli Equation), the velocity head is

$$h_v = \frac{v^2_{propeller}}{2g} = \frac{(4.2v_{boat})^2}{(2)\left(32.2\ \frac{\text{ft}}{\text{sec}^2}\right)} = 0.2739v^2_{boat}\ \text{sec}^2/\text{ft}$$

From App. 10A (also *NCEES Handbook* table "Physical Properties of Liquid Water (U.S. Units)"), the vapor pressure of freshwater at 68°F is $p_{vp} = 0.3393$ psia.

From App. 1A (also *NCEES Handbook* table "Physical Properties of Liquid Water (U.S. Units)"), the density of water at 68°F is 62.32 lbm/ft³. Raoult's law predicts the actual vapor pressure of the solution.

$$p_{vapor,solution} = \begin{pmatrix} \text{mole fraction} \\ \text{of the solvent} \end{pmatrix} p_{vapor,solvent}$$

$$p_{vapor,seawater} = (0.992)\left(0.3393\ \frac{\text{lbf}}{\text{in}^2}\right) = 0.3366\ \text{lbf/in}^2$$

Extracted from Eq. 5.4(b) (also *NCEES Handbook:* The Bernoulli Equation), the vapor pressure head is

$$h_{vapor,seawater} = \frac{p}{\rho} \times \frac{g_c}{g} = \frac{\left(0.3366\ \frac{\text{lbf}}{\text{in}^2}\right)\left(12\ \frac{\text{in}}{\text{ft}}\right)^2}{\left(64.0\ \frac{\text{lbm}}{\text{ft}^3}\right)} \left(\frac{32.2\ \frac{\text{lbm-ft}}{\text{lbf-sec}^2}}{32.2\ \frac{\text{ft}}{\text{sec}^2}} \right)$$

$$= 0.7574\ \text{ft}$$

Solve for the boat velocity.

SI Solution

Cavitation will occur when

$$h_{atm} - h_v < h_{vp}$$

The density of seawater is 1024 kg/m³.

Extracted from Eq. 5.4(a) (also *NCEES Handbook:* The Bernoulli Equation), the atmospheric head is

$$h_{atm} = \frac{p}{\rho g} = \frac{(101.3\ \text{kPa})\left(1000\ \frac{\text{Pa}}{\text{kPa}}\right)}{\left(1024\ \frac{\text{kg}}{\text{m}^3}\right)\left(9.81\ \frac{\text{m}}{\text{s}^2}\right)} = 10.08\ \text{m}$$

$$h_{depth} = 3\ \text{m} \quad [\text{given}]$$

Extracted from Eq. 5.6 (also *NCEES Handbook:* The Bernoulli Equation), the velocity head is

$$h_v = \frac{v^2_{propeller}}{2g} = \frac{(4.2v_{boat})^2}{(2)\left(9.81\ \frac{\text{m}}{\text{s}^2}\right)} = 0.899v^2_{boat}\ \text{sec}^2/\text{m}$$

As in *NCEES Handbook* table "Physical Properties of Liquid Water (U.S. Units)," the vapor pressure of 20°C freshwater is

$$p_{vp} = (0.02339\ \text{bar})\left(100\ \frac{\text{kPa}}{\text{bar}}\right) = 2.339\ \text{kPa}$$

From App. 1B (also *NCEES Handbook* table "Physical Properties of Liquid Water (U.S. Units)"), the density of water at 20°C is 998.23 kg/m³. Raoult's law predicts the actual vapor pressure of the solution.

$$p_{vapor,solution} = p_{vapor,solvent} \begin{pmatrix} \text{mole fraction} \\ \text{of the solvent} \end{pmatrix}$$

The solvent is the freshwater and the solution is the seawater.

The mole fraction of water is 0.992.

$$p_{vapor,seawater} = (2.339\ \text{kPa})(0.992) = 2.320\ \text{kPa}$$

Extracted from Eq. 5.4(a) (also *NCEES Handbook:* The Bernoulli Equation), the vapor pressure head is

$$h_{vapor,seawater} = \frac{p}{\rho g} = \frac{(2.320\ \text{kPa})\left(1000\ \frac{\text{Pa}}{\text{kPa}}\right)}{\left(1024\ \frac{\text{kg}}{\text{m}^3}\right)\left(9.81\ \frac{\text{m}}{\text{s}^2}\right)} = 0.231\ \text{m}$$

Solve for the boat velocity.

$$3 \text{ m} + 10.08 \text{ m} - 0.899\text{v}_{\text{boat}}^2 \text{ sec}^2/\text{m} = 0.231 \text{ m}$$

$$\text{v}_{\text{boat}} = 3.78 \text{ m/s} \quad (3.8 \text{ m/s})$$

The answer is (B).

15. *Customary U.S. Solution*

From App. 3C (also *NCEES Handbook* table "Pipe Dimensions and Weights"), for 2 in schedule-40 pipe,

$$D = 0.1723 \text{ ft}$$
$$A = 0.02330 \text{ ft}^2$$

Since the flow rate is unknown, it must be assumed in order to find velocity. Estimate 90 gal/min.

$$\dot{V} = \frac{90 \ \frac{\text{gal}}{\text{min}}}{\left(7.4805 \ \frac{\text{gal}}{\text{ft}^3}\right)\left(60 \ \frac{\text{sec}}{\text{min}}\right)} = 0.2005 \text{ ft}^3/\text{sec}$$

As in *NCEES Handbook:* Conservation of Mass, the velocity is

$$\text{v} = \frac{\dot{V}}{A} = \frac{0.2005 \ \frac{\text{ft}^3}{\text{sec}}}{0.02330 \text{ ft}^2} = 8.605 \text{ ft/sec}$$

From App. 1B (also *NCEES Handbook* table "Physical Properties of Liquid Water (U.S. Units)"), the kinematic viscosity of water at 70°F is $\nu = 1.059 \times 10^{-5} \text{ ft}^2/\text{sec}$.

As in *NCEES Handbook:* Similitude, the Reynolds number is

$$\text{Re} = \frac{D\text{v}}{\nu} = \frac{(0.1723 \text{ ft})\left(8.605 \ \frac{\text{ft}}{\text{sec}}\right)}{1.059 \times 10^{-5} \ \frac{\text{ft}^2}{\text{sec}}} = 1.4 \times 10^5$$

From Table 4.2 (also *NCEES Handbook:* Absolute Roughness and Relative Roughness), the specific roughness of steel pipe is $\epsilon = 0.0002$ ft.

$$\frac{\epsilon}{D} = \frac{0.0002 \text{ ft}}{0.1723 \text{ ft}} \approx 0.0012$$

From App. 4B (also *NCEES Handbook:* Friction Factors for Turbulent Flow), $f = 0.022$.

From App. 4D (also *NCEES Handbook:* Head Loss in Pipe or Conduit and *NCEES Handbook:* Loss Coefficients for Pipe Entrance and Exit), the equivalent lengths of various screwed steel fittings are

$$\text{inlet (reentrant): } L_e = \left(\frac{D}{f}\right)\left(\frac{160}{\text{Re}} + (1.0)\left(1 + \frac{1}{D_{\text{inches}}}\right)\right)$$
$$= \left(\frac{2 \text{ in}}{0.022}\right) \times \left(\frac{160}{1.4 \times 10^5} + (1.0)\left(1 + \frac{1}{2 \text{ in}}\right)\right)$$
$$= 136.5 \text{ in}$$
$$= 11.4 \text{ ft}$$

swing check valve: $L_e = (125)(2 \text{ in}) = 250 \text{ in} = 20.8 \text{ ft}$

long radius elbows: $L_e = (16)(2 \text{ in}) = 32 \text{ in} = 2.7 \text{ ft}$

The total equivalent length of the 2 in line is

$$L_e = 12 \text{ ft} + 11.4 \text{ ft} + 20.8 \text{ ft} + (3)(2.7 \text{ ft}) + 80 \text{ ft}$$
$$= 132.3 \text{ ft}$$

At 90 gal/min, the friction loss in the line from *NCEES Handbook:* Head Loss in Pipe or Conduit is

$$h_f = \frac{fL\text{v}^2}{2Dg} = \frac{(0.022)(132.3 \text{ ft})\left(8.605 \ \frac{\text{ft}}{\text{sec}}\right)^2}{(2)(0.1723 \text{ ft})\left(32.2 \ \frac{\text{ft}}{\text{sec}^2}\right)} = 19.4 \text{ ft}$$

Extracted from Eq. 5.6 (also *NCEES Handbook:* The Bernoulli Equation), the velocity head at 90 gal/min is

$$h_\text{v} = \frac{\text{v}^2}{2g} = \frac{\left(8.605 \ \frac{\text{ft}}{\text{sec}}\right)^2}{(2)\left(32.2 \ \frac{\text{ft}}{\text{sec}^2}\right)} = 1.1 \text{ ft}$$

In general, the friction head and velocity head are proportional to v^2 and Q^2 (see*NCEES Handbook:* Pump Similitude).

$$h_f = (19.4 \text{ ft})\left(\frac{Q_2}{90 \ \frac{\text{gal}}{\text{min}}}\right)^2$$

$$h_\text{v} = (1.1 \text{ ft})\left(\frac{Q_2}{90 \ \frac{\text{gal}}{\text{min}}}\right)^2$$

(The 7 ft suction lift is included in the 20 ft static discharge head.)

The equation for the total system head is

$$h = h_z + h_v + h_f = 20\ \text{ft} + (1.1\ \text{ft} + 19.4\ \text{ft})\left(\frac{Q_2}{90\ \frac{\text{gal}}{\text{min}}}\right)^2$$

Q_2 (gal/min)	system head, h (ft)
0	20.0
10	20.2
20	21.0
30	22.2
40	24.0
50	26.2
60	29.0
70	32.2
80	36.0
90	40.2
100	44.9
110	50.2

The intersection point of the system curve and the pump curve defines the operating flow rate.

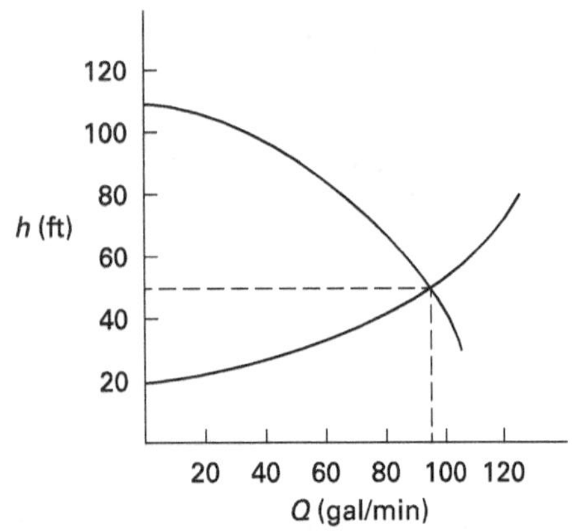

The flow rate is 95 gal/min.

SI Solution

As in *NCEES Handbook* table "Pipe Dimensions and Weights," for 2 in schedule-40 pipe,

$$D = 52.501\ \text{mm}$$
$$A = 21.648 \times 10^{-4}\ \text{m}^2$$

Since the flow rate is unknown, it must be assumed in order to find velocity. Estimate 6 L/s.

$$\dot{V} = \frac{6\ \frac{\text{L}}{\text{s}}}{1000\ \frac{\text{L}}{\text{m}^3}} = 6 \times 10^{-3}\ \text{m}^3/\text{s}$$

As in *NCEES Handbook:* Conservation of Mass, the velocity is

$$\text{v} = \frac{\dot{V}}{A} = \frac{6 \times 10^{-3}\ \frac{\text{m}^3}{\text{s}}}{21.648 \times 10^{-4}\ \text{m}^2} = 2.77\ \text{m/s}$$

As in *NCEES Handbook* table "Physical Properties of Liquid Water (SI Units)," the kinematic viscosity of water at 21°C is approximately $\nu = 9.849 \times 10^{-7}\ \text{m}^2/\text{s}$.

As in *NCEES Handbook:* Similitude, the Reynolds number is

$$\text{Re} = \frac{\text{v}D}{\nu} = \frac{\left(2.77\ \frac{\text{m}}{\text{s}}\right)(52.501\ \text{mm})}{\left(9.849 \times 10^{-7}\ \frac{\text{m}^2}{\text{s}}\right)\left(1000\ \frac{\text{mm}}{\text{m}}\right)}$$
$$= 1.48 \times 10^5$$

As in *NCEES Handbook:* Absolute Roughness and Relative Roughness, the specific roughness of steel pipe is $\epsilon = 6.0 \times 10^{-5}$ m.

$$\frac{\epsilon}{D} = \frac{(6.0 \times 10^{-5}\ \text{m})\left(1000\ \frac{\text{mm}}{\text{m}}\right)}{52.501\ \text{mm}} \approx 0.0011$$

As in *NCEES Handbook:* Friction Factors for Turbulent Flow, $f = 0.022$. Use the equivalent lengths of various screwed steel fittings from *NCEES Handbook:* Head Loss in Pipe or Conduit. The total equivalent length of 5.08 cm schedule-40 pipe is

$$\text{inlet (reentrant): } L_e = \left(\frac{D}{f}\right)\left(\frac{160}{\text{Re}} + (1.0)\left(1 + \frac{1}{D_{\text{inches}}}\right)\right)$$
$$= \left(\frac{5.08\ \text{cm}}{0.022}\right) \times \left(\frac{160}{1.4 \times 10^5} + (1.0)\left(1 + \frac{1}{2\ \text{in}}\right)\right)$$
$$= 346.7\ \text{cm}$$
$$= 3.5\ \text{m}$$

$$\text{swing check valve: } L_e = (125)(5.08\ \text{cm}) = 635\ \text{cm} = 6.4\ \text{m}$$

long radius elbows: $L_e = (16)(5.08 \text{ cm}) = 81.3 \text{ cm} = 0.8 \text{ m}$

$$\begin{aligned} L_e &= 4 \text{ m} + 3.5 \text{ m} + 6.4 \text{ m} + (3)(0.8 \text{ m}) + 27 \text{ m} \\ &= 43.3 \text{ m} \end{aligned}$$

At 6 L/s, the friction loss in the line from *NCEES Handbook:* Head Loss in Pipe or Conduit is

$$\begin{aligned} h_f &= \frac{fL\text{v}^2}{2Dg} \\ &= \frac{(0.022)(43.3 \text{ m})\left(2.77 \ \frac{\text{m}}{\text{s}}\right)^2\left(1000 \ \frac{\text{mm}}{\text{m}}\right)}{(2)(52.501 \text{ mm})\left(9.81 \ \frac{\text{m}}{\text{s}^2}\right)} \\ &= 7.1 \text{ m} \end{aligned}$$

At 6 L/s, the velocity head is (see also *NCEES Handbook:* The Bernoulli Equation),

$$h_\text{v} = \frac{\text{v}^2}{2g} = \frac{\left(2.77 \ \frac{\text{m}}{\text{s}}\right)^2}{(2)\left(9.81 \ \frac{\text{m}}{\text{s}^2}\right)} = 0.39 \text{ m}$$

In general, the friction head and velocity head are proportional to v^2 and Q^2.

$$h_f = (7.1 \text{ m})\left(\frac{Q_2}{6 \ \frac{\text{L}}{\text{s}}}\right)^2$$

$$h_\text{v} = (0.39 \text{ m})\left(\frac{Q_2}{6 \ \frac{\text{L}}{\text{s}}}\right)^2$$

(The 2.3 m suction lift is included in the 6.3 m static discharge head.)

The equation for the total system head is

$$\begin{aligned} h &= h_z + h_\text{v} + h_f \\ &= 6.3 \text{ m} + (0.39 \text{ m} + 7.1 \text{ m})\left(\frac{Q_2}{6 \ \frac{\text{L}}{\text{s}}}\right)^2 \end{aligned}$$

Q_2 (L/s)	h (m)
0	6.3
0.6	6.37
1.2	6.60
1.8	6.96
2.4	7.48
3.2	8.40
3.6	8.96
4.4	10.27
4.8	11.03
5.7	12.97
6.0	13.69
6.5	14.98
7.0	16.36
7.5	17.85

The intersection point of the system curve and the pump curve defines the operating flow rate.

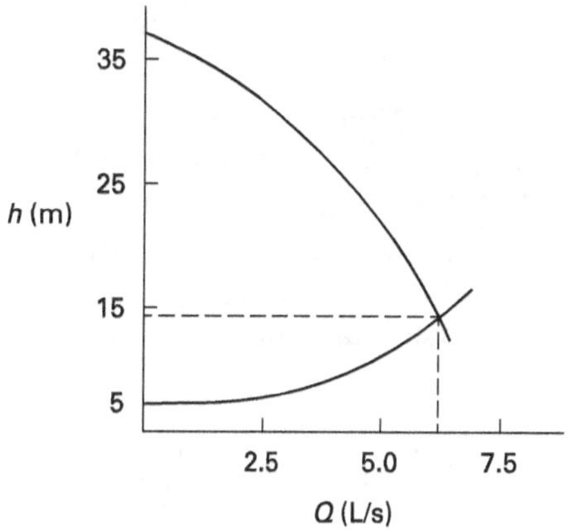

The flow rate is 6.2 L/s.

The answer is (D).

16. As in *NCEES Handbook:* Pump Similitude,

$$P_2 = P_1\left(\frac{\rho_2 n_2^3 D_2^5}{\rho_1 n_1^3 D_1^5}\right) = P_1\left(\frac{n_2}{n_1}\right)^3 \quad [\rho_2 = \rho_1 \text{ and } D_2 = D_1]$$

Customary U.S. Solution

$$P_2 = (0.5 \text{ hp})\left(\frac{2000 \ \frac{\text{rev}}{\text{min}}}{1750 \ \frac{\text{rev}}{\text{min}}}\right)^3 = 0.75 \text{ hp}$$

SI Solution

$$P_2 = (0.37\ \text{kW})\left(\frac{2000\ \frac{\text{rev}}{\text{min}}}{1750\ \frac{\text{rev}}{\text{min}}}\right)^3 = 0.55\ \text{kW}$$

The answer is (D).

17. The operating point for this pump occurs at the intersection of the 1400 rpm pump curve provided and the system (head) curve defined the equation $H = 30 + 2\ Q^2$. Using the guess and check method

1. Assume $Q = 1$ cfs. $H = 30 + (2)(1^2) = 32$ ft, but the pump curve is at 200 ft. Need to increase flow.
2. Assume $Q = 5$ cfs. $H = 30 + (2)(5^2) = 80$ ft, but the pump curve is at approximately 195 ft. Need to increase flow.
3. Assume $Q = 10$ cfs. $H = 30 + (2)(10^2) = 230$ ft, but the pump curve is at approximately 200 ft. Need to decrease flow.
4. Assume $Q = 9$ cfs. $H = 30 + (2)(9^2) = 192$ ft, and the pump curve is at approximately 192 ft. Therefore, the operating point occurs at $Q = 9$ cfs and $H = 192$ ft.

$$\begin{aligned} Q &= \left(9\ \frac{\text{ft}^3}{\text{sec}}\right)\left(7.4805\ \frac{\text{gal}}{\text{ft}^3}\right)\left(60\ \frac{\text{sec}}{\text{min}}\right) \\ &= 4039\ \text{gal/min} \end{aligned}$$

From the *NCEES Handbook*: Pump Power, the water power (WP) is

$$\begin{aligned} \text{WP [hp]} &= \frac{Q\ [\text{gpm}] \times H\ [\text{ft}] \times \rho\ [\text{lbm/ft}^3]}{246{,}780} \\ &= \frac{(4039)(192)(62.4)}{246{,}780} \\ &= 196\ \text{hp} \end{aligned}$$

Therefore, the power required at the pump shaft, also known as brake power (BP), is

$$\text{BP} = \frac{\text{WP}}{\eta_{\text{pump}}} = \frac{196\ \text{hp}}{0.86} = 228\ \text{hp} \quad (230\ \text{hp})$$

The answer is (C).

18. Since turbines are essentially pumps running backward, use *NCEES Handbook:* Pump Power.

$$\begin{aligned} \Delta p &= \left(30\ \frac{\text{lbf}}{\text{in}^2} - 5\ \frac{\text{lbf}}{\text{in}^2}\right)\left(12\ \frac{\text{in}}{\text{ft}}\right)^2 \\ &= 3600\ \text{lbf/ft}^2 \end{aligned}$$

$$\begin{aligned} P &= \frac{\Delta p \dot{V}}{550} = \frac{\left(3600\ \frac{\text{lbf}}{\text{ft}^2}\right)\left(100\ \frac{\text{ft}^3}{\text{sec}}\right)}{550\ \frac{\text{ft-lbf}}{\text{hp-sec}}} \\ &= 654.5\ \text{hp} \quad (650\ \text{hp}) \end{aligned}$$

The answer is (C).

19. The flow rate is

$$\begin{aligned} \dot{m} = \rho\dot{V} &= \left(62.4\ \frac{\text{lbm}}{\text{ft}^3}\right)\left(1000\ \frac{\text{ft}^3}{\text{sec}}\right) \\ &= 6.24 \times 10^4\ \text{lbm/sec} \end{aligned}$$

The head available for work is

$$\Delta h = 625\ \text{ft} - 58\ \text{ft} = 567\ \text{ft}$$

Use Table 5.5 (also *NCEES Handbook:* Pump Power). The turbine efficiency is 89%. So, the power is

$$\begin{aligned} P &= \frac{h_A \dot{m}}{550} \times \frac{g}{g_c} \\ &= \frac{(0.89)\left(6.24 \times 10^4\ \frac{\text{lbm}}{\text{sec}}\right)(567\ \text{ft})\left(32.2\ \frac{\text{ft}}{\text{sec}^2}\right)}{\left(550\ \frac{\text{ft-lbf}}{\text{hp-sec}}\right)\left(32.2\ \frac{\text{lbm-ft}}{\text{lbf-sec}^2}\right)} \\ &= 57{,}253\ \text{hp} \end{aligned}$$

Convert from hp to kW.

$$\begin{aligned} P &= (57{,}253\ \text{hp})\left(0.7457\ \frac{\text{kW}}{\text{hp}}\right) \\ &= 4.27 \times 10^4\ \text{kW} \quad (43\ \text{MW}) \end{aligned}$$

The answer is (C).

20. *Customary U.S. Solution*

The total effective head is due to the pressure head, velocity head, and tailwater head.

$$h_{\text{eff}} = h_p + h_v - h_{z,\text{tailwater}} = h_p + \frac{v^2}{2g} - h_{z,\text{tailwater}}$$

$$= 92.5 \text{ ft} + \frac{\left(12 \ \frac{\text{ft}}{\text{sec}}\right)^2}{(2)\left(32.2 \ \frac{\text{ft}}{\text{sec}^2}\right)} - (-5.26 \text{ ft})$$

$$= 100 \text{ ft}$$

From Table 5.5 (also *NCEES Handbook:* Fans), the theoretical hydraulic horsepower is

$$P_{\text{th}} = \frac{h_A \dot{V}(\text{SG})}{8.814} = \frac{(100 \text{ ft})\left(25 \ \frac{\text{ft}^3}{\text{sec}}\right)(1)}{8.814 \ \frac{\text{ft}^4}{\text{hp-sec}}}$$

$$= 283.6 \text{ hp}$$

The overall turbine efficiency is

$$\eta = \frac{P_{\text{brake}}}{P_{\text{th}}} = \frac{250 \text{ hp}}{283.6 \text{ hp}} = 0.882 \quad (88\%)$$

SI Solution

The total effective head is due to the pressure head, velocity head, and tailwater head.

$$h_{\text{eff}} = h_p + h_v - h_{z,\text{tailwater}} = h_p + \frac{v^2}{2g} - h_{z,\text{tailwater}}$$

$$= 28.2 \text{ m} + \frac{\left(3.6 \ \frac{\text{m}}{\text{s}}\right)^2}{(2)\left(9.81 \ \frac{\text{m}}{\text{s}^2}\right)} - (-1.75 \text{ m})$$

$$= 30.61 \text{ m} \quad (31 \text{ m})$$

From Table 5.6 (also *NCEES Handbook:* Fans), the theoretical hydraulic kilowatts are

$$P_{\text{th}} = \frac{9.81 h_A Q(\text{SG})}{1000}$$

$$= \frac{\left(9.81 \ \frac{\text{m}}{\text{s}^2}\right)(30.61 \text{ m})\left(700 \ \frac{\text{L}}{\text{s}}\right)(1)}{1000 \ \frac{\text{W}}{\text{kW}}}$$

$$= 210.2 \text{ kW}$$

The overall turbine efficiency is

$$\eta = \frac{P_{\text{brake}}}{P_{\text{th}}} = \frac{185 \text{ kW}}{210.2 \text{ kW}} = 0.88 \quad (88\%)$$

The answer is (B).

6 Special Fluid Topics

PRACTICE PROBLEMS

1. A 3 ft diameter column is packed with 1 in Intalox® saddles. 5000 lbm/hr of air at an average pressure of 1 atm and an average temperature of 70°F flows through the tower. The viscosity and density of air are 0.044 lbm/ft-hr and 0.075 lbm/ft^3, respectively. Use the Ergun equation to estimate the pressure drop through the dry packing.

(A) 0.29 psf/ft

(B) 1.00 psf/ft

(C) 1.44 psf/ft

(D) 3.00 psf/ft

SOLUTIONS

1. As in *NCEES Handbook:* Liquid-Liquid Extraction Equipment: Packed Columns,

$$\epsilon = 0.73$$
$$a_p = 256\ \text{m}^2/\text{m}^3 = 78\ \text{ft}^2/\text{ft}^3$$
$$D_p = \frac{6(1-\epsilon)}{a_p} = \frac{(6)(1-0.73)}{78\ \frac{\text{ft}^2}{\text{ft}^3}} = 0.0207\ \text{ft}$$

The superficial gas velocity is

$$G_0 = \frac{\dot{m}_g}{A} = \frac{5000\ \frac{\text{lbm}}{\text{hr}}}{\left(\frac{\pi}{4}\right)(3\ \text{ft})^2} = 707\ \text{lbm/hr-ft}^2$$

Equation 6.1(b) is used to estimate the pressure drop in dry packing.

$$\frac{\Delta p}{L} = \left(\frac{1-\epsilon}{\epsilon^3}\right)\left(\frac{G_0^2}{D_p g_c \rho_g}\right)\left(\frac{150(1-\epsilon)\mu_g'}{D_p G_0} + 1.75\right)$$

$$= \left(\frac{1-0.73}{(0.73)^3}\right)\left(\frac{\left(707\ \frac{\text{lbm}}{\text{hr-ft}^2}\right)^2}{(0.0207\ \text{ft})\left(32.2\ \frac{\text{ft-lbm}}{\text{lbf-sec}^2}\right) \times\left(0.075\ \frac{\text{lbm}}{\text{ft}^3}\right)\left(3600\ \frac{\text{sec}}{\text{hr}}\right)^2}\right)$$
$$\times\left(\frac{(150)(1-0.73)\left(0.044\ \frac{\text{lbm}}{\text{ft-hr}}\right)}{(0.0207\ \text{ft})\left(707\ \frac{\text{lbm}}{\text{hr-ft}^2}\right)} + 1.75\right)$$
$$= 1.00\ \text{psf/ft}$$

The answer is (B).

7 Inorganic Chemistry

PRACTICE PROBLEMS

1. The gravimetric analysis of a compound is 40% carbon, 6.7% hydrogen, and 53.3% oxygen. The simplest formula for the compound is most nearly

(A) HCO

(B) HCO_2

(C) CH_2O

(D) CHO_2

2. When used together, which of the following metal pairs is LEAST likely to experience galvanic corrosion in sea water?

(A) zinc and platinum

(B) zinc and 316/317 stainless steel

(C) 316/317 stainless steel and lead

(D) brass and copper

3. When use of dissimilar metals is unavoidable, which of the following CANNOT be used to prevent or reduce the likelihood of galvanic corrosion?

(A) inert spacers between the dissimilar metals

(B) sacrificial anodes

(C) paints, coatings, and natural oxide buildups

(D) layers of cadmium, nickel, chromium, or zinc plating

4. A sacrificial galvanic protection system is proposed for a water storage tank made of low-alloy steel. Which of the following materials is most suitable as the sacrificial anode?

(A) brass

(B) nickel

(C) stainless steel

(D) zinc

5. A linear actuator mechanism incorporates a naval brass threaded rod that passes through a threaded metallic collar. The mechanism is exposed to both fresh and saltwater. To limit galvanic corrosion, which material would be best for the mechanism's collar?

(A) zinc-plated carbon steel

(B) nickel-plated carbon steel

(C) 316 stainless steel

(D) titanium

Thermodynamics

SOLUTIONS

1. Calculate the relative mole ratios of the atoms by assuming there are 100 g of sample.

For 100 g of sample,

substance	mass	$\frac{m}{AW}$ = no. moles	relative mole ratio
C	40 g	$\frac{40}{12} = 3.33$	1
H	6.7 g	$\frac{6.7}{1} = 6.7$	2
O	53.3 g	$\frac{53.3}{16} = 3.33$	1

The empirical formula is CH_2O.

The answer is (C).

2. Galvanic corrosion (galvanic action) results from a difference in oxidation potentials of metallic ions. The greater the difference in oxidation potentials, the greater the likelihood of galvanic action. To determine which of the options are least likely to experience galvanic action, use the galvanic series chart, in Table 7.6 (also the *NCEES Handbook* table titled "Galvanic Corrosion"). The closer the pair of metals are, the less likely they are to experience galvanic corrosion. Therefore, of the options, brass and copper are least likely to experience galvanic corrosion.

The answer is (D).

3. While cadmium, nickel, chromium, and zinc are often used as protective deposits on steel, porosities in the surfaces can act as small galvanic cells, resulting in invisible subsurface corrosion.

The answer is (D).

4. From Table 7.6 (also *NCEES Handbook* table "Galvanic Corrosion"), brass, nickel, and stainless steel have lower potentials than low-alloy steel; therefore, they will be cathodic, not anodic. Zinc has a higher potential than low-alloy steel, so it can be used as the sacrificial anode.

The answer is (D).

5. From Table 7.6 (also *NCEES Handbook* table "Galvanic Corrosion"), nickel is the closest to brass of the options provided, so nickel-plated carbon steel should be used to make the actuator's threaded collar.

The answer is (B).

8 Combustion and Incineration

PRACTICE PROBLEMS

1. Methane (MW = 16.043) with a heating value of 24,000 Btu/lbm (55.8 MJ/kg) is burned with a 50% efficiency. If the heat of vaporization of any water vapor formed is recovered, approximately how much water (specific heat of 1 Btu/lbm-°F (4.1868 kJ/kg·°C)) can be heated from 60°F to 200°F (15°C to 95°C) when 7 ft^3 (200 L) of methane at 60°F and 14.73 psia (288.9K and 101.51 kPa) are burned?

(A) 25 lbm (11 kg)

(B) 35 lbm (16 kg)

(C) 50 lbm (23 kg)

(D) 95 lbm (43 kg)

2. 15 lbm/hr (6.8 kg/h) of propane (C_3H_8, MW = 44.097) is burned stoichiometrically in air. Approximately what volume of dry carbon dioxide (CO_2, MW = 44.011) is formed after cooling to 70°F (21°C) and 14.7 psia (101 kPa)?

(A) 180 ft^3/hr (5.0 m^3/h)

(B) 270 ft^3/hr (7.6 m^3/h)

(C) 390 ft^3/hr (11 m^3/h)

(D) 450 ft^3/hr (13 m^3/h)

3. In a particular installation, 30% excess air at 15 psia (103 kPa) and 100°F (40°C) is needed for the combustion of methane. Approximately how much nitrogen (MW = 28.016) passes through the furnace if methane is burned at the rate of 4000 ft^3/hr (31 L/s)?

(A) 270 lbm/hr (0.033 kg/s)

(B) 930 lbm/hr (0.11 kg/s)

(C) 1800 lbm/hr (0.22 kg/s)

(D) 2700 lbm/hr (0.34 kg/s)

4. Approximately how much air is required to completely burn one unit mass of a fuel that is 84% carbon, 15.3% hydrogen, 0.3% sulfur, and 0.4% nitrogen by weight?

(A) 9 lbm air/lbm fuel (9 kg air/kg fuel)

(B) 12 lbm air/lbm fuel (12 kg air/kg fuel)

(C) 15 lbm air/lbm fuel (15 kg air/kg fuel)

(D) 18 lbm air/lbm fuel (18 kg air/kg fuel)

5. Propane (C_3H_8) is burned with 20% excess air. The gravimetric percentage of carbon dioxide in the flue gas is most nearly

(A) 8%

(B) 12%

(C) 15%

(D) 22%

6. The ultimate analysis of a coal is 80% carbon, 4% hydrogen, 2% oxygen, and the rest ash. The flue gases are 60°F and 14.7 psia (15.6°C and 101.3 kPa) when sampled, and are 80% nitrogen, 12% carbon dioxide, 7% oxygen, and 1% carbon monoxide by volume. The air required to burn 1 lbm (1 kg) of coal under these conditions is most nearly

(A) 11 lbm (11 kg)

(B) 15 lbm (15 kg)

(C) 19 lbm (19 kg)

(D) 23 lbm (23 kg)

7. The ultimate analysis of a coal is 75% carbon, 5% hydrogen, 3% oxygen, 2% nitrogen, and the rest ash. Atmospheric air is 60°F (16°C) and at standard pressure. Assuming the lower heating value of this fuel is 12,957 Btu/lbm (30.14 MJ/kg), the theoretical temperatures of the combustion product is most nearly

(A) 3500°F (1900°C)

(B) 4000°F (2200°C)

(C) 4500°F (2500°C)

(D) 5000°F (2800°C)

8. An anhydrous fuel oil has the following ultimate analysis: 85.43% carbon, 11.31% hydrogen, 2.7% oxygen, 0.34% sulfur, and 0.22% nitrogen. The oil is burned with 60% excess air. The mass of wet flue gas produced per kg of fuel burned is most nearly

(A) 18 kg

(B) 23 kg

(C) 30 kg

(D) 38 kg

9. The ultimate analysis of a coal is 51.45% carbon, 16.69% ash, 15.71% moisture, 7.28% oxygen, 4.02% hydrogen, 3.92% sulfur, and 0.93% nitrogen. 15,395 lbm (6923 kg) of the coal are burned, and 2816 lbm (1267 kg) of ash containing 20.9% carbon (by weight) are recovered. 13.3 lbm (kg) of dry gases are produced per pound (kilogram) of fuel burned. The air used per unit mass of fuel is most nearly

(A) 13 lbm air/lbm fuel (13 kg air/kg fuel)

(B) 14 lbm air/lbm fuel (14 kg air/kg fuel)

(C) 15 lbm air/lbm fuel (15 kg air/kg fuel)

(D) 17 lbm air/lbm fuel (17 kg air/kg fuel)

10. A coal is 65% carbon by weight. During combustion, 3% of the coal is lost in the ash pit. Combustion uses 9.87 lbm (kg) of air per pound (kg) of fuel. The flue gas analysis is 81.5% nitrogen, 9.5% carbon dioxide, and 9% oxygen. The percentage of excess air by mass is most nearly

(A) 10%

(B) 30%

(C) 70%

(D) 140%

11. A coal has an ultimate analysis of 67.34% carbon, 4.91% oxygen, 4.43% hydrogen, 4.28% sulfur, 1.08% nitrogen, and the rest ash. 3% of the carbon is lost during combustion. The flue gases are 81.9% nitrogen, 15.5% carbon dioxide, 1.6% carbon monoxide, and 1% oxygen by volume. The heat produced from the combustion of carbon monoxide is most nearly

(A) 600 Btu/lbm

(B) 800 Btu/lbm

(C) 1000 Btu/lbm

(D) 1200 Btu/lbm

12. A natural gas is 93% methane, 3.73% nitrogen, 1.82% hydrogen, 0.45% carbon monoxide, 0.35% oxygen, 0.25% ethylene, 0.22% carbon dioxide, and 0.18% hydrogen sulfide by volume. The gas is burned with 40% excess air. Atmospheric air is 60°F and at standard atmospheric pressure. The theoretical air requirements are most nearly

(A) 5 ft^3 air/ft^3 fuel

(B) 7 ft^3 air/ft^3 fuel

(C) 9 ft^3 air/ft^3 fuel

(D) 13 ft^3 air/ft^3 fuel

13. A coal has a gravimetric analysis of 83% carbon, 5% hydrogen, 5% oxygen, and 7% noncombustible matter. 10% of the fired coal mass, including all of the noncombustible matter, is recovered in the ash pit. 26 lbm (26 kg) of air at 70°F (21°C) and 1 atmosphere are used per lbm (kg) of coal burned. When loaded into the furnace, the coal temperature is 60°F (15.6°C). The stack gas from coal combustion has a temperature of 550°F (290°C). The percentage of the combustion heat carried away by the stack gases, based on the coal's as-delivered properties, is most nearly

(A) 18%

(B) 24%

(C) 37%

(D) 49%

14. A utility boiler burns coal with an ultimate analysis of 76.56% carbon, 7.7% oxygen, 6.1% silicon, 5.5% hydrogen, 2.44% sulfur, and 1.7% nitrogen. 410 lbm/hr of refuse are removed with a composition of 30% carbon and 0% sulfur. All the sulfur and the remaining carbon are burned. The power plant has the following characteristics.

- coal feed rate: 15,300 lbm/hr
- electric power rating: 17 MW
- generator efficiency: 95%
- steam generator efficiency: 86%
- cooling water rate: 225 ft^3/sec

The emission rate of solid particulates in lbm/hr is most nearly

(A) 23 lbm/hr

(B) 150 lbm/hr

(C) 810 lbm/hr

(D) 1700 lbm/hr

15. 250 SCFM (118 L/s) of propane are mixed with an oxidizer consisting of 60% oxygen and 40% nitrogen by volume in a proportion resulting in 40% excess oxygen by weight. The maximum velocity for the two reactants when combined is 400 ft/min (2 m/s) at 14.7 psia (101 kPa) and 80°F (27°C). Maximum velocity for the products is 800 ft/min (4 m/s) at 8 psia (55 kPa) and 460°F (240°C). The minimum size of the inlet pipe is most nearly

(A) 9 ft^2 (0.9 m^2)

(B) 11 ft^2 (1.1 m^2)

(C) 14 ft^2 (1.3 m^2)

(D) 17 ft^2 (1.6 m^2)

16. An industrial process uses hot gas at 3600°R (1980°C) and 14.7 psia (101 kPa). It is proposed that propane be burned stoichiometrically in a mixture of nitrogen and oxygen. After passing through the process, gas will be exhausted through a duct, being cooled slowly to 100°F (38°C) and 14.7 psia (101 kPa) before discharge. The following data are available.

- The enthalpies of formation (at the standard reference temperature) are

$C_3H_8(g)$:	$\Delta H_f = +28{,}800$ Btu/lbmol	(+67.0 GJ/kmol)
$CO_2(g)$:	$\Delta H_f = -169{,}300$ Btu/lbmol	(−393.8 GJ/kmol)
$H_2O(g)$:	$\Delta H_f = -104{,}040$ Btu/lbmol	(−242 GJ/kmol)

- The enthalpy increases from the standard reference temperature to 3600°R (1980°C) are

CO_2:	39,791 Btu/lbmol	(92.6 GJ/kmol)
H_2O:	31,658 Btu/lbmol	(73.6 GJ/kmol)
N_2:	24,471 Btu/lbmol	(56.9 GJ/kmol)

What are most nearly the combining weights of the nitrogen and oxygen, respectively, per pound-mole of propane?

(A) 760 lbm/lbmol, 128 lbm/lbmol (760 kg/kmol, 128 kg/kmol)

(B) 810 lbm/lbmol, 160 lbm/lbmol (810 kg/kmol, 160 kg/kmol)

(C) 845 lbm/lbmol, 160 lbm/lbmol (845 kg/kmol, 160 kg/kmol)

(D) 850 lbm/lbmol, 160 lbm/lbmol (850 kg/kmol, 160 kg/kmol)

17. An electrical power-generating plant burns refuse-derived fuel (RDF). After sorting, incoming refuse is shredded and compressed before being fed into the combustor. The raw refuse averages 7% by weight incombustible solids. 5000 lbm/hr of processed RDF produces 20,070 lbm/hr of saturated steam at 200 lbf/in^2. The combustion products are used to heat incoming feedwater to a saturated temperature of 160°F before entering the combustor. 2000 lbm/hr of water vapor condense in the feedwater heater at a partial pressure of 4 lbf/in^2 and are removed. All thermal losses are to be disregarded. What is most nearly the higher heating value of the RDF?

(A) 4300 Btu/lbm

(B) 4700 Btu/lbm

(C) 4900 Btu/lbm

(D) 5100 Btu/lbm

Thermodynamics

SOLUTIONS

1. *Customary U.S. Solution*

As in *NCEES Handbook:* Temperature,

$$T = 60°F + 460° = 520°R$$
$$p = 14.73 \text{ psia}$$

The molecular weight of methane can be found in *NCEES Handbook:* Physical Properties of Liquids and Gases—Temperature-Independent Properties (U.S. Units). The mass is

$$m = \frac{pVM}{RT} = \frac{\left(14.73 \frac{\text{lbf}}{\text{in}^2}\right)\left(12 \frac{\text{in}}{\text{ft}}\right)^2 (7 \text{ ft}^3)\left(16.043 \frac{\text{lbm}}{\text{lbmol}}\right)}{\left(1545.35 \frac{\text{ft-lbf}}{\text{lbm-°R}}\right)(520°R)}$$
$$= 0.296 \text{ lbm}$$

The combustion energy available from methane is

$$Q = \eta m(\text{HHV}) = (0.5)(0.296 \text{ lbm})\left(24{,}000 \frac{\text{Btu}}{\text{lbm}}\right)$$
$$= 3552 \text{ Btu}$$

As in *NCEES Handbook:* Heat Capacity/Specific Heat (c_p), this energy is used to heat water from 60°F to 200°F.

$$Q = m_{\text{water}} c_p (T_2 - T_1)$$
$$m_{\text{water}} = \frac{3552 \text{ Btu}}{\left(1 \frac{\text{Btu}}{\text{lbm-°F}}\right)(200°F - 60°F)}$$
$$= 25.37 \text{ lbm} \quad (25 \text{ lbm})$$

SI Solution

The molecular weight of methane can be found in *NCEES Handbook:* Physical Properties of Liquids and Gases—Temperature-Independent Properties (SI Units). The mass is

$$m = \frac{pVM}{RT}$$
$$= \frac{(101.51 \text{ kPa})\left(1000 \frac{\text{Pa}}{\text{kPa}}\right)(200 \text{ L})\left(16.043 \frac{\text{g}}{\text{mol}}\right)}{\left(8314.5 \frac{\text{J}}{\text{kmol·K}}\right)(288.9\text{K})\left(1000 \frac{\text{L}}{\text{m}^3}\right)}$$
$$= 0.136 \text{ kg}$$

The combustion energy available from methane is

$$Q = \eta m(\text{HHV})$$
$$= (0.5)(0.136 \text{ kg})\left(55.8 \frac{\text{MJ}}{\text{kg}}\right)\left(1000 \frac{\text{kJ}}{\text{MJ}}\right)$$
$$= 3794 \text{ kJ}$$

As in *NCEES Handbook:* Heat Capacity/Specific Heat (c_p), this energy is used to heat water from 15°C to 95°C.

$$Q = m_{\text{water}} c_p (T_2 - T_1)$$
$$m_{\text{water}} = \frac{3794 \text{ kJ}}{\left(4.1868 \frac{\text{kJ}}{\text{kg·°C}}\right)(95°C - 15°C)}$$
$$= 11.33 \text{ kg} \quad (11 \text{ kg})$$

The answer is (A).

2. *Customary U.S. Solution*

The molecular weights can be found in Table 10.7 (also *NCEES Handbook:* Physical Properties of Liquids and Gases—Temperature-Independent Properties). The combustion reaction should lead to CO_2 and H_2O as the products. Use the balanced chemical reaction equation from Table 8.7. When properly balanced,

$$\begin{array}{lccc} & C_3H_8 + & 5O_2 \rightarrow & 3CO_2 + 4H_2O \\ \text{MW} & 44.097 & (5)(32) & (3)(44.011) \\ & 44.097 & 160 & 132.033 \end{array}$$

The amount of carbon dioxide produced is 132.033 lbm/44.097 lbm propane. For 15 lbm/hr of propane, the amount of carbon dioxide produced is

$$\left(\frac{132.033 \text{ lbm}}{44.097 \text{ lbm}}\right)\left(15 \frac{\text{lbm}}{\text{hr}}\right) = 44.91 \text{ lbm/hr}$$

As in *NCEES Handbook:* Temperature,

$$T = 70°F + 460° = 530°R$$

$$\dot{V} = \frac{\dot{m}RT}{pM} = \frac{\left(44.91 \frac{\text{lbm}}{\text{h}}\right)\left(1545.35 \frac{\text{ft-lbf}}{\text{lbmol-°R}}\right)(530°R)}{\left(14.7 \frac{\text{lbf}}{\text{in}^2}\right)\left(12 \frac{\text{in}}{\text{ft}}\right)^2\left(44.011 \frac{\text{lbm}}{\text{lbmol}}\right)}$$
$$= 394.8 \text{ ft}^3/\text{hr} \quad (390 \text{ ft}^3/\text{hr})$$

SI Solution

The molecular weights can be found in Table 8.7 (also *NCEES Handbook* table "Temperature-Independent Properties of Liquids and Gases (SI Units)"). The combustion reaction should lead to CO_2 and H_2O as the products. Use the balanced chemical reaction equation from Table 8.7. When properly balanced,

$$\begin{array}{lcccc} & C_3H_8 & + \; 5O_2 & \rightarrow & 3CO_2 \; + 4H_2O \\ \text{MW} & 44.097 & (5)(32) & & (3)(44.011) \\ & 44.097 & 160 & & 132.033 \end{array}$$

The amount of carbon dioxide produced is 132.033 kg/44.097 kg propane. For 6.8 kg/h of propane, the amount of carbon dioxide produced is

$$\left(\frac{132.033 \text{ kg}}{44.097 \text{ kg}}\right)\left(6.8 \; \frac{\text{kg}}{\text{h}}\right) = 20.36 \text{ kg/h}$$

As in *NCEES Handbook:* Temperature,

$$T = 21°\text{C} + 273° = 294\text{K}$$

$$\dot{V} = \frac{\dot{m}RT}{pM} = \frac{\left(20.36 \; \frac{\text{kg}}{\text{h}}\right)\left(8314.5 \; \frac{\text{J}}{\text{kmol}\cdot\text{K}}\right)(294\text{K})}{(101\,000 \text{ Pa})\left(44.011 \; \frac{\text{g}}{\text{mol}}\right)}$$
$$= 11.20 \text{ m}^3\text{/h} \quad (11 \text{ m}^3\text{/h})$$

The answer is (C).

3. The molecular weights can be found in *NCEES Handbook* table "Temperature-Independent Properties of Liquids and Gases (U.S. Units)." The combustion reaction should lead to CO_2 and H_2O as the products where Ar and gases other than oxygen are treated as N_2, as is common for combustion analysis to simplify to a binary mixture. When properly balanced,

$$CH_4 + 2O_2 \rightarrow CO_2 + 2H_2O$$

Use Table 8.6 and Table 8.7 (also *NCEES Handbook:* Combustion Reactions and *NCEES Handbook:* Dry Atmospheric Air Composition). With 30% excess air and considering that there are 3.773 volumes of nitrogen for every volume of oxygen, the reaction equation is

$$\begin{aligned} &CH_4 + (1.3)(2)O_2 + (1.3)(2)(3.773)N_2 \\ &\quad \rightarrow CO_2 + 2H_2O + (1.3)(2)(3.773)N_2 + 0.6O_2 \\ &CH_4 + 2.6O_2 + 9.81N_2 \\ &\quad \rightarrow CO_2 + 2H_2O + 9.81N_2 + 0.6O_2 \end{aligned}$$

Customary U.S. Solution

The volume of nitrogen that accompanies 4000 ft^3/hr of entering methane is

$$\begin{aligned} V_{N_2} &= \left(9.81 \; \frac{\text{ft}^3 \; N_2}{\text{ft}^3 \; CH_4}\right)\left(4000 \; \frac{\text{ft}^3 \; CH_4}{\text{hr}}\right) \\ &= 39{,}240 \text{ ft}^3 \; N_2\text{/hr} \end{aligned}$$

This is the "partial volume" of nitrogen in the input stream. As in *NCEES Handbook:* Temperature, the absolute temperature is

$$T = 100°\text{F} + 460° = 560°\text{R}$$

$$\begin{aligned} \dot{m} &= \frac{p\dot{V}M}{RT} \\ &= \frac{\left(15 \; \frac{\text{lbf}}{\text{in}^2}\right)\left(12 \; \frac{\text{in}}{\text{ft}}\right)^2\left(39240 \; \frac{\text{ft}^3}{\text{hr}}\right)\left(28.016 \; \frac{\text{lbm}}{\text{lbmol}}\right)}{\left(1545.35 \; \frac{\text{ft-lbf}}{\text{lbmol-°R}}\right)(560°\text{R})} \\ &= 2744 \text{ lbm/hr} \quad (2700 \text{ lbm/hr}) \end{aligned}$$

SI Solution

The volume of nitrogen that accompanies 31 L/s of entering methane is

$$\frac{\left(9.81 \; \frac{\text{m}^3 \; N_2}{\text{m}^3 \; CH_4}\right)\left(31 \; \frac{\text{L} \; CH_4}{\text{s}}\right)}{1000 \; \frac{\text{L}}{\text{m}^3}} = 0.3041 \text{ m}^3\text{/s}$$

This is the "partial volume" of nitrogen in the input stream. As in *NCEES Handbook:* Temperature, the absolute temperature is

$$T = 40°\text{C} + 273° = 313\text{K}$$

$$\begin{aligned} \dot{m} &= \frac{p\dot{V}M}{RT} \\ &= \frac{(103 \text{ kPa})\left(1000 \; \frac{\text{Pa}}{\text{kPa}}\right)\left(0.3041 \; \frac{\text{m}^3}{\text{s}}\right)\left(28.016 \; \frac{\text{g}}{\text{mol}}\right)}{\left(8314.5 \; \frac{\text{J}}{\text{kmol}\cdot\text{K}}\right)(313\text{K})} \\ &= 0.337 \text{ kg/s} \quad (0.34 \text{ kg/s}) \end{aligned}$$

The answer is (D).

Thermodynamics

4. The molecular weights can be found in *NCEES Handbook* table "Temperature-Independent Properties of Liquids and Gases (U.S. Units)." The combustion reaction should lead to CO_2 and H_2O as the products. From Table 8.7, when properly balanced,

$$\begin{array}{lccccc} & C & + & O_2 & \rightarrow & CO_2 \\ MW & 12 & & 32 & & \end{array}$$

The mass of oxygen required per unit mass of carbon is

$$\frac{32}{12} = 2.67$$

$$\begin{array}{lccccc} & 2H_2 & + & O_2 & \rightarrow & 2H_2O \\ MW & (2)(2) & & 32 & & \end{array}$$

The mass of oxygen required per unit mass of hydrogen is

$$\frac{32}{(2)(2)} = 8.0$$

$$\begin{array}{lccccc} & S & + & O_2 & \rightarrow & SO_2 \\ MW & 32.1 & & 32 & & \end{array}$$

The mass of oxygen required per unit mass of sulfur is

$$\frac{32}{32.1} = 1.0$$

Nitrogen does not burn.

The mass of oxygen required per unit mass of fuel is

$$\begin{aligned}&(0.84)(2.67) + (0.153)(8) + (0.003)(1)\\ &\quad = 3.47 \text{ units of mass of } O_2/\text{unit mass fuel}\end{aligned}$$

From Table 8.6 (also *NCEES Handbook:* Dry Atmospheric Air Composition), air is 0.2315 O_2/unit mass.

Customary U.S. Solution

The air required is

$$\frac{3.47}{0.2315} = 14.99 \quad (15 \text{ lbm air/lbm fuel})$$

SI Solution

The air required is

$$\frac{3.47}{0.2315} = 14.99 \quad (15 \text{ kg air/kg fuel})$$

The answer is (C).

5. The molecular weights can be found in *NCEES Handbook* table "Temperature-Independent Properties of Liquids and Gases (U.S. Units)." The combustion reaction should lead to CO_2 and H_2O as the products. From Table 8.7, when properly balanced,

$$C_3H_8 + 5O_2 \rightarrow 3CO_2 + 4H_2O$$

With 20% excess air, the oxygen volume is $(1.2)(5) = 6$.

$$C_3H_8 + 6O_2 \rightarrow 3CO_2 + 4H_2O + O_2$$

From Table 8.6 (also *NCEES Handbook:* Dry Atmospheric Air Composition), there are 3.773 volumes of nitrogen for every volume of oxygen.

$$(6)(3.773) = 22.6$$

$$C_3H_8 + 6O_2 + 22.6N_2 \rightarrow 3CO_2 + 4H_2O + O_2 + 22.6N_2$$

The percentage of carbon dioxide by weight in flue gas is

$$\begin{aligned} G_{CO_2} &= \frac{m_{CO_2}}{m_{total}} \\ &= \frac{B_{CO_2}(MW_{CO_2})}{\sum B_i(MW_i)} \\ &= \frac{(3)(44.011)}{\begin{array}{r}(3)(44.011) + (4)(18.016) \\ +32 + (22.6)(28.016)\end{array}} \\ &= 0.152 \quad (15\%) \end{aligned}$$

The answer is (C).

6. *Customary U.S. Solution*

For every 1 lbm of coal burned, 0.8 lbm of coal is emitted as carbon. The carbon is balanced between CO and CO_2. Since 12/13 of the emitted carbon goes to CO_2, 0.74 lbm of carbon goes to forming CO_2 for every 1 lbm of coal burned. As in *NCEES Handbook:* Physical Properties of Liquids and Gases—Temperature-Independent Properties, the molecular weight of carbon is 12 lbm/lbmol. The number of moles of CO_2 emitted per 1 lbm of coal is

$$\begin{aligned} n_{CO_2} &= \frac{m_C}{M_C} = \frac{0.74 \text{ lbm}}{12 \, \frac{\text{lbm}}{\text{lbmol}}} \\ &= 0.062 \text{ lbmol } CO_2/\text{lbm coal} \end{aligned}$$

Since the nitrogen does not react, the nitrogen from the flow is from air. From the flue gas sampled, 80 parts of N_2 are released for every 12 parts of CO_2 released. The number of moles of N_2 emitted per 1 lbm of coal is

$$\begin{aligned} n_{N_2} &= \frac{80}{12} n_{CO_2} = \left(\frac{80}{12}\right)\left(0.062 \; \frac{\text{lbmol}}{\text{lbm}}\right) \\ &= 0.41 \text{ lbmol } N_2/\text{lbm coal} \end{aligned}$$

The mass of N_2 is

$$\begin{aligned} m_{N_2} &= n_{N_2} M_{N_2} = \left(0.41 \; \frac{\text{lbmol } N_2}{\text{lbm coal}}\right)\left(28 \; \frac{\text{lbm}}{\text{lbmol}}\right) \\ &= 11.5 \text{ lbm } N_2/\text{lbm coal} \end{aligned}$$

As in *NCEES Handbook:* Dry Atmospheric Air Composition, the mass of oxygen in the air is the product of the molar flow of nitrogen, the ratio of O_2 and N_2 in the air, and the molecular weight of oxygen.

$$\begin{aligned} m_{O_2} &= \left(0.41 \; \frac{\text{lbmol } N_2}{\text{lbm coal}}\right)(0.2315)\left(32 \; \frac{\text{lbm}}{\text{lbmol}}\right) \\ &= 3 \text{ lbm } O_2/\text{lbm coal} \end{aligned}$$

The total air flow per 1 lbm of coal is

$$\begin{aligned} M_{\text{total}} &= m_{N_2} + m_{O_2} = 11.5 \; \frac{\text{lbm } N_2}{\text{lbm coal}} + 3 \; \frac{\text{lbm } O_2}{\text{lbm coal}} \\ &= 14.5 \text{ lbm air/lbm coal} \quad (15 \text{ lbm}) \end{aligned}$$

SI Solution

For every 1 kg of coal burned, 0.8 kg of coal is emitted as carbon. The carbon is balanced between CO and CO_2. Since 12/13 of the emitted carbon goes to CO_2, 0.74 kg of carbon goes to forming CO_2 for every 1 kg of coal burned. As in *NCEES Handbook:* Physical Properties of Liquids and Gases—Temperature-Independent Properties, the molecular weight of carbon is 12 kg/kmol. The number of moles of CO_2 emitted per 1 kg of coal is

$$\begin{aligned} n_{CO_2} &= \frac{m_C}{M_C} = \frac{0.74 \text{ kg}}{12 \; \frac{\text{kg}}{\text{kmol}}} \\ &= 0.062 \text{ kmol } CO_2/\text{kg coal} \end{aligned}$$

Since the nitrogen does not react, the nitrogen from the flow is from air. From the flue gas sampled, 80 parts of N_2 are released for every 12 parts released of CO_2. The number of moles of N_2 is

$$\begin{aligned} n_{N_2} &= \frac{80}{12} n_{CO_2} = \left(\frac{80}{12}\right)\left(0.062 \; \frac{\text{kmol } CO_2}{\text{kg coal}}\right) \\ &= 0.41 \text{ kmol } N_2/\text{kg coal} \end{aligned}$$

The mass of N_2 is

$$\begin{aligned} m_{N_2} &= n_{N_2} M_{N_2} = \left(0.41 \; \frac{\text{kmol } N_2}{\text{kg coal}}\right)\left(28 \; \frac{\text{kg}}{\text{kmol}}\right) \\ &= 11.5 \text{ kg } N_2/\text{kg coal} \end{aligned}$$

As in *NCEES Handbook:* Dry Atmospheric Air Composition, the mass of oxygen in the air is the product of the molar flow of nitrogen, the ratio of O_2 and N_2 in the air, and the molecular weight of oxygen.

$$\begin{aligned} m_{O_2} &= \left(0.41 \; \frac{\text{kmol } N_2}{\text{kg coal}}\right)(0.2315)\left(32 \; \frac{\text{kg}}{\text{kmol}}\right) \\ &= 3 \text{ kg } O_2/\text{kg coal} \end{aligned}$$

The total air flow per kilogram of coal is

$$\begin{aligned} M_{\text{total}} &= m_{N_2} + m_{O_2} = 11.5 \; \frac{\text{kg } N_2}{\text{kg coal}} + 3 \; \frac{\text{kg } O_2}{\text{kg coal}} \\ &= 14.5 \text{ kg air/kg coal} \quad (15 \text{ kg}) \end{aligned}$$

The answer is (B).

7. *Customary U.S. Solution*

step 1: Note the LHV for this fuel.

$$\text{LHV} = 12{,}957 \text{ Btu/lbm}$$

step 2: The gravimetric analysis of 1 lbm of coal is

$$\begin{aligned} &\text{carbon:} \quad 0.75 \text{ lbm} \\ &\text{free hydrogen: } G_{H,\text{total}} - \frac{G_O}{8} = 0.05 - \frac{0.03}{8} \\ &\qquad = 0.0463 \end{aligned}$$

The ratio of the molecular weight of water (18) to the molecular weight of hydrogen (2) is 9.

$$\begin{aligned} &\text{water:} \quad (9)(0.05 - 0.0463) = 0.0333 \\ &\text{nitrogen: } 0.02 \end{aligned}$$

step 3: Using stoichiometry and molecular weights from Table 8.8, the theoretical stack gases per lbm coal for 0.75 lbm of carbon are

$$\begin{aligned} CO_2 &= (0.75)(3.667 \text{ lbm}) = 2.750 \text{ lbm} \\ N_2 &= (0.75)(8.883 \text{ lbm}) = 6.662 \text{ lbm} \end{aligned}$$

All products are calculated similarly (as the table summarizes). All values are per pound of fuel.

	CO_2	N_2	H_2O
from C:	2.750 lbm	6.662 lbm	
from H_2:		1.217 lbm	0.414 lbm
from H_2O:			0.0333 lbm
from O_2:	shows up in	CO_2 and H_2	
from N_2:		0.02 lbm	
total:	2.750 lbm	7.899 lbm	0.4473 lbm

step 4: Assume the combustion gases leave at 1000°F.

As in *NCEES Handbook:* Temperature,

$$T_{\text{ave}} = \left(\frac{1}{2}\right)(60°\text{F} + 1000°\text{F}) = 530°\text{F}$$
$$= 530°\text{F} + 460° = 990°\text{R}$$

The specific heat values are given in Table 8.1 (also *NCEES Handbook* table "Temperature-Dependent Physical Properties of Gases at 14.7 psia (U.S. Units)").

$$c_{p,\text{CO}_2} = 0.251 \text{ Btu/lbm-°F}$$
$$c_{p,\text{N}_2} = 0.255 \text{ Btu/lbm-°F}$$
$$c_{p,\text{H}_2\text{O}} = 0.475 \text{ Btu/lbm-°F}$$

The energy required to raise the combustion products (from 1 lbm of coal) 1°F is

$$m_{\text{CO}_2}c_{p,\text{CO}_2} + m_{\text{N}_2}c_{p,\text{N}_2} + m_{\text{H}_2\text{O}}c_{p,\text{H}_2\text{O}}$$
$$= (2.750 \text{ lbm})\left(0.251 \frac{\text{Btu}}{\text{lbm-°F}}\right)$$
$$+(7.899 \text{ lbm})\left(0.255 \frac{\text{Btu}}{\text{lbm-°F}}\right)$$
$$+(0.4473 \text{ lbm})\left(0.475 \frac{\text{Btu}}{\text{lbm-°F}}\right)$$
$$= 2.92 \text{ Btu/°F}$$

step 5: Assuming all combustion heat goes into the stack gases, the temperature is given by Eq. 8.19.

$$T_{\text{max}} = T_i + \frac{\text{lower heat of combustion}}{\text{energy required}}$$
$$= 60°\text{F} + \frac{12{,}957 \frac{\text{Btu}}{\text{lbm}}}{2.92 \frac{\text{Btu}}{\text{lbm-°F}}}$$
$$= 4497°\text{F} \quad (4500°\text{F})$$

SI Solution

step 1: Note the heating value of this fuel.

$$\text{LHV} = 30.14 \text{ MJ/kg}$$

Steps 2 and 3 are the same as for the customary U.S. solution except that all masses are in kg.

step 4: Assume the combustion gases leave at 550°C.

$$T_{\text{ave}} = \left(\frac{1}{2}\right)(16°\text{C} + 550°\text{C}) + 273° = 556\text{K}$$

Specific heat values are given in Table 8.1 (also *NCEES Handbook* table "Temperature-Dependent Physical Properties of Gases at 0.1 MPa (SI Units)"). Using the footnote for SI units,

$$c_{p,\text{CO}_2} = 1.051 \text{ kJ/kg·K}$$
$$c_{p,\text{N}_2} = 1.068 \text{ kJ/kg·K}$$
$$c_{p,\text{H}_2\text{O}} = 1.989 \text{ kJ/kg·K}$$

The energy required to raise the combustion products (from 1 kg of coal) 1°C is

$$m_{\text{CO}_2}c_{p,\text{CO}_2} + m_{\text{N}_2}c_{p,\text{N}_2} + m_{\text{H}_2\text{O}}c_{p,\text{H}_2\text{O}}$$
$$= (2.750 \text{ kg})\left(1.051 \frac{\text{kJ}}{\text{kg·K}}\right)$$
$$+(7.899 \text{ kg})\left(1.068 \frac{\text{kJ}}{\text{kg·K}}\right)$$
$$+(0.4473 \text{ kg})\left(1.989 \frac{\text{kJ}}{\text{kg·K}}\right)$$
$$= 12.22 \text{ kJ/K}$$

step 5: Assuming all combustion heat goes into the stack gases, the temperature is given by Eq. 8.19.

$$T_{\text{max}} = T_i + \frac{\text{lower heat of combustion}}{\text{energy required}}$$
$$= 16°\text{C} + \frac{\left(30.14 \frac{\text{MJ}}{\text{kg}}\right)\left(1000 \frac{\text{kJ}}{\text{MJ}}\right)}{12.22 \frac{\text{kJ}}{\text{K}}}$$
$$= 2482°\text{C} \quad (2500°\text{C})$$

The answer is (C).

8. Assume the oxygen is in the form of moisture in the fuel. The available hydrogen is

$$G_{H,free} = G_H - \frac{G_O}{8} = 0.1131 - \frac{0.027}{8} = 0.1097$$

SI Solution

step 1: From the PPI manual, find the stoichiometric oxygen required per kilogram of fuel oil.

$$\begin{aligned} C \rightarrow CO_2\text{: } O_2 \text{ required} &= (0.8543)(2.667 \text{ kg}) \\ &= 2.2784 \text{ kg} \\ H_2 \rightarrow H_2O\text{: } O_2 \text{ required} &= (0.1097)(7.936 \text{ kg}) \\ &= 0.8706 \text{ kg} \\ S \rightarrow SO_2\text{: } O_2 \text{ required} &= (0.0034)(0.998 \text{ kg}) \\ &= 0.0034 \text{ kg} \end{aligned}$$

The amount of additional oxygen required per kilogram of fuel oil is

$$2.28 \text{ kg} + 0.871 \text{ kg} + 0.003 \text{ kg} - 0.027 \text{ kg} = 3.154 \text{ kg}$$

step 2: With 60% excess air, the excess oxygen per kilogram of fuel oil is

$$(0.6)(3.154 \text{ kg}) = 1.892 \text{ kg}$$

step 3: The theoretical nitrogen per pound of fuel based on the PPI manual (also the *NCEES Handbook* section titled "Dry Atmospheric Air Composition") is

$$\left(3.154 + 1.892\right) \text{ kg } O_2 * \left(\frac{0.7685 \text{ \% } N_2}{0.2315 \text{ \% } O_2}\right) = 16.75 \text{ kg } N_2$$

Since mass of the fuel is conserved, the wet mass of flue gas produced is

$$\begin{aligned} &1 \text{ kg of fuel} + 5.046 \text{ kg } O_2 + 16.75 \text{ kg } N_2 = \\ &\qquad 21.79 \text{ kg flue gas} \quad (22 \text{ kg}) \end{aligned}$$

The mass of wet flue gas produced per kg fuel burned is most near answer choice (B), 23 kg.

The answer is (B).

9. *Customary U.S. Solution*

step 1: Based on 15,395 lbm of coal burned producing 2816 lbm of ash containing 20.9% carbon by weight, the usable percentage of carbon per pound of fuel is

$$0.5145 - \frac{(2816 \text{ lbm})(0.209)}{15{,}395 \text{ lbm}} = 0.4763$$

step 2: Since moisture is reported separately, assume all of the oxygen and hydrogen are free. (This is not ordinarily the case.) From Table 8.8, find the stoichiometric oxygen required per pound of fuel.

$$\begin{aligned} C \rightarrow CO_2\text{: } O_2 \text{ required} &= (0.4763)(2.667 \text{ lbm}) \\ &= 1.2703 \text{ lbm} \\ H_2 \rightarrow H_2O\text{: } O_2 \text{ required} &= (0.0402)(7.936 \text{ lbm}) \\ &= 0.3190 \text{ lbm} \\ S \rightarrow SO_2\text{: } O_2 \text{ required} &= (0.0392)(0.998 \text{ lbm}) \\ &= 0.0391 \text{ lbm} \end{aligned}$$

The total amount of O_2 required per pound of fuel is

$$\begin{aligned} &1.2703 \text{ lbm} + 0.3190 \text{ lbm} \\ &\quad + 0.0391 \text{ lbm} - 0.0728 \text{ lbm} = 1.5556 \text{ lbm} \end{aligned}$$

step 3: The theoretical air per pound of fuel based on Table 8.6 (also *NCEES Handbook:* Dry Atmospheric Air Composition) is

$$\frac{1.5556 \text{ lbm}}{0.2315} = 6.720 \text{ lbm air}$$

step 4: Ignoring fly ash, the theoretical dry products per pound of fuel are given from Table 8.8.

$$\begin{aligned} CO_2\text{: } &(0.4763)(3.667 \text{ lbm}) &&= 1.7466 \text{ lbm} \\ SO_2\text{: } &(0.0392)(1.998 \text{ lbm}) &&= 0.0783 \text{ lbm} \\ N_2\text{: } &0.0093 \text{ lbm} + (6.720 \text{ lbm})(0.7685) &&= \underline{5.1736 \text{ lbm}} \\ & &\text{total} &= 6.999 \text{ lbm} \end{aligned}$$

step 5: The excess air per pound of fuel is

$$13.3 \text{ lbm} - 6.999 \text{ lbm} = 6.301 \text{ lbm}$$

step 6: The total air supplied per pound of fuel is

$$6.301 \text{ lbm} + 6.720 \text{ lbm} = 13.02 \text{ lbm} \quad (13 \text{ lbm})$$

SI Solution

step 1: Based on 6923 kg of coal burned producing 1267 kg of ash containing 20.9% carbon by weight, the usable percentage of carbon per kilogram of fuel is

$$0.5145 - \frac{(1267 \text{ kg})(0.209)}{6923 \text{ kg}} = 0.4763$$

step 2: From Table 8.8, find the stoichiometric oxygen required per kilogram of fuel.

$$\begin{aligned} C \rightarrow CO_2\text{: } O_2 \text{ required} &= (0.4763)(2.667 \text{ kg}) \\ &= 1.2703 \text{ kg} \\ H_2 \rightarrow H_2O\text{: } O_2 \text{ required} &= (0.0402)(7.936 \text{ kg}) \\ &= 0.3190 \text{ kg} \\ S \rightarrow SO_2\text{: } O_2 \text{ required} &= (0.0392)(0.998 \text{ kg}) \\ &= 0.0391 \text{ kg} \end{aligned}$$

The total amount of O_2 required per kilogram of fuel is

$$\begin{aligned} &1.2703 \text{ kg} + 0.3190 \text{ kg} \\ &\quad +0.0391 \text{ kg} - 0.0728 \text{ kg} = 1.5556 \text{ kg} \end{aligned}$$

step 3: The theoretical air per kilogram of fuel based on Table 8.6 (also *NCEES Handbook:* Dry Atmospheric Air Composition) is

$$\frac{1.5556 \text{ kg}}{0.2315} = 6.720 \text{ kg air}$$

step 4: Ignoring fly ash, the theoretical dry products per kilogram of fuel are given from Table 8.8.

$$\begin{aligned} CO_2\text{: } (0.4763)(3.667 \text{ kg}) &= 1.7466 \text{ kg} \\ SO_2\text{: } (0.0392)(1.998 \text{ kg}) &= 0.0783 \text{ kg} \\ N_2\text{: } 0.0093 \text{ kg} + (6.720 \text{ kg})(0.7685) &= \underline{5.1736 \text{ kg}} \\ \text{total} &= 6.999 \text{ kg} \end{aligned}$$

step 5: The excess air per kilogram of fuel is

$$13.3 \text{ kg} - 6.999 \text{ kg} = 6.301 \text{ kg}$$

step 6: The total air supplied per kilogram of fuel is

$$6.301 \text{ kg} + 6.720 \text{ kg} = 13.02 \text{ kg} \quad (13 \text{ kg})$$

The answer is (A).

10. Each 1 lbm of coal has 0.65 lbm of carbon. If 3% is lost, then 0.63 lbm of carbon make it to the effluent. The number of moles of carbon in the effluent is

$$n_C = \frac{m}{M} = \frac{0.63 \text{ lbm}}{12 \frac{\text{lbm}}{\text{lbmol}}} = 0.0525 \text{ lbmol C}$$

The carbon making it to the effluent appears as CO_2. Using the ratio of N_2 to CO_2 in the effluent, the number of moles of N_2 is

$$n_{N_2} = \left(\frac{81.5}{9.5}\right)(0.0525 \text{ lbmol C}) = 0.45 \text{ lbmol } N_2$$

Using *NCEES Handbook:* Dry Atmospheric Air Composition, the number of moles of air is

$$n_{O_2} = \frac{0.45 \text{ lbmol } N_2}{3.76} = 0.12 \text{ lbmol } O_2$$

Using *NCEES Handbook:* Physical Properties of Liquids and Gases—Temperature-Independent Properties, the total mass of air into the system is

$$\begin{aligned} m_{\text{total}} &= n_{N_2}M_{N_2} + n_{O_2}M_{O_2} \\ &= (0.45 \text{ lbmol } N_2)\left(28 \frac{\text{lbm}}{\text{lbmol}}\right) \\ &\quad +(0.12 \text{ lbmol } O_2)\left(32 \frac{\text{lbm}}{\text{lbmol}}\right) \\ &= 16.44 \text{ lbm air} \end{aligned}$$

As in *NCEES Handbook:* Material Balances With Reaction, the percentage of excess air is

$$\begin{aligned} \% \text{ of excess air} &= \left(\frac{16.44 \text{ lbm} - 9.87 \text{ lbm}}{9.87 \text{ lbm}}\right) \times 100\% \\ &= 67\% \quad (70\%) \end{aligned}$$

The answer is (C).

11. The amount of carbon leaving the flue per 1 lbm of coal is

$$m_C = (1 \text{ lbm})(0.97)(0.6734) = 0.653 \text{ lbm C}$$

The molar percentage of C leaving as CO can be found from the percentages of CO and CO_2 in the flue gas is

$$\%\text{C as CO} = \frac{y_{CO}}{y_{CO} + y_{CO_2}} = \frac{1.6}{1.6 + 15.5} \times 100\% = 9.4\%$$

Since CO and CO_2 have the same amount of carbon, their mass ratios are the same leading to the same mass percent of carbon leaving as CO. The mass of carbon leaving as CO is

$$m_{\text{CasCO}} = (0.094)(0.653 \text{ lbm}) = 0.061 \text{ lbm C as CO}$$

The amount of CO leaving the flue per 1 lbm of coal is

$$n_{CO} = \frac{0.061 \text{ lbm}}{12 \frac{\text{lbm}}{\text{lbmol}}} = 0.00509 \text{ lbmol CO}$$

As in *NCEES Handbook:* Standard Heat of Formation and Combustion, the heat of combustion of CO is 121,700 Btu/lbmol, so the energy lost is

$$\begin{aligned} Q_{\text{lost}} &= (0.00509 \text{ lbmol CO})\left(121{,}700 \ \frac{\text{Btu}}{\text{lbmol}}\right) \\ &= 619.4 \text{ Btu} \quad (600 \text{ Btu}) \end{aligned}$$

The answer is (A).

12. The reactions for the oxidation of the components of the natural gas are

$$CH_4 + 2O_2 \rightarrow CO_2 + 2H_2O$$

$$N_2 \rightarrow \text{no reaction}$$

$$H_2 + 0.5O_2 \rightarrow H_2O$$

$$2CO + O_2 \rightarrow 2CO_2$$

$$O_2 \rightarrow \text{no reaction}$$

$$C_2H_4 + 3O_2 \rightarrow 2CO_2 + 2H_2O$$

$$CO_2 \rightarrow \text{no reaction}$$

$$H_2S + 1.5O_2 \rightarrow SO_2 + H_2O$$

Assume 100 moles of natural gas. To find the total moles of oxygen required, the composition can be multiplied by the stoichiometric coefficient for oxygen in the above reactions. Subtract the amount of oxygen in the natural gas.

$$\begin{aligned} n_{O_2} &= \big((93 \text{ mol})(2) + (1.82 \text{ mol})(0.5) + (0.45 \text{ mol})(0.5) \\ &\quad + (0.25 \text{ mol})(3) + (0.18 \text{ mol})(1.5)\big) - 0.35 \text{ mol} \\ &= 187.81 \text{ mol } O_2 \end{aligned}$$

Using *NCEES Handbook:* Dry Atmospheric Air Composition, the theoretical air requirements are most nearly

$$\begin{aligned} n_{\text{air}} &= \frac{\left(\dfrac{1 \text{ mol air}}{0.209390 \text{ mol } O_2}\right)(187.81 \text{ mol } O_2)}{100 \text{ mol gas}} \\ &= 8.969 \ \frac{\text{mol air}}{\text{mol gas}} \\ &= 8.969 \text{ ft}^3 \text{ air/ft}^3 \text{ fuel} \quad \left(9 \text{ ft}^3 \text{ air/ft}^3 \text{ fuel}\right) \end{aligned}$$

The answer is (C).

13. *Customary U.S. Solution*

step 1: The incoming reactants on a per-pound basis are

0.07 lbm ash
0.05 lbm hydrogen
0.05 lbm oxygen
0.83 lbm carbon

This is an ultimate analysis. Assume that only the hydrogen that is not locked up with oxygen in the form of water is combustible. From Eq. 8.15, the available hydrogen is

$$\begin{aligned} G_{\text{H,available}} &= G_{\text{H,total}} - \frac{G_{\text{O}}}{8} = 0.05 \text{ lbm} - \frac{0.05 \text{ lbm}}{8} \\ &= 0.04375 \text{ lbm} \end{aligned}$$

The mass of water produced is the hydrogen mass plus eight times as much oxygen. The locked hydrogen is

$$0.05 \text{ lbm} - 0.04375 \text{ lbm} = 0.00625 \text{ lbm}$$

$$\begin{aligned} \text{lbm of moisture} &= G_{\text{H}} + G_{\text{O}} = G_{\text{H}} + 8G_{\text{H}} \\ &= 0.00625 \text{ lbm} + (8)(0.00625 \text{ lbm}) \\ &= 0.05625 \text{ lbm} \end{aligned}$$

The air is 23.15% oxygen by weight (see Table 8.6 and *NCEES Handbook:* Dry Atmospheric Air Composition), so other reactants for 26 lbm of air are

$$(0.2315)(26 \text{ lbm}) = 6.019 \text{ lbm } O_2$$
$$(0.7685)(26 \text{ lbm}) = 19.981 \text{ lbm } N_2$$

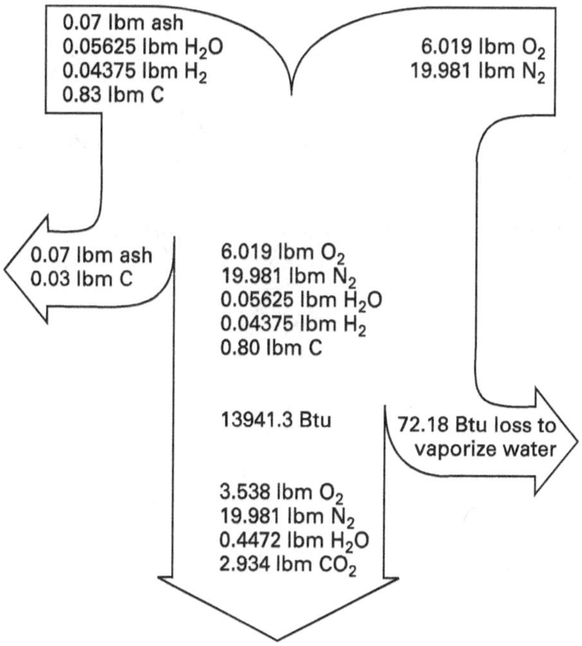

step 2: Ash pit material losses are 10%, or 0.1 lbm, which includes all of the ash.

0.07 lbm ash (noncombustible matter)
0.03 lbm unburned carbon

step 3: Determine what remains.

6.019 lbm oxygen
19.981 lbm nitrogen
0.05625 lbm water
0.04375 lbm hydrogen
0.80 lbm carbon

step 4: Determine the energy loss in vaporizing the moisture.

$$q = (\text{moisture})(h_g - h_f)$$

From App. 10C (also *NCEES Handbook:* Superheated Steam (U.S. Customary Units)), h_g at 550°F and 14.7 psia is 1311.4 Btu/lbm.

From App. 10A (also *NCEES Handbook:* Properties of Saturated Steam (U.S. Customary Units)), h_f at 60°F is 28.08 Btu/lbm.

$$\begin{aligned} q &= (0.05625\ \text{lbm})\left(1311.4\ \frac{\text{Btu}}{\text{lbm}} - 28.08\ \frac{\text{Btu}}{\text{lbm}}\right) \\ &= 72.19\ \text{Btu} \end{aligned}$$

step 5: Calculate the heating value of the remaining fuel components using App. 8A (also *NCEES Handbook:* Standard Heat of Formation and Combustion).

$$\text{HV}_\text{C} = (0.80\ \text{lbm})\left(14{,}093\ \frac{\text{Btu}}{\text{lbm}}\right) = 11{,}274.4\ \text{Btu}$$

$$\text{HV}_\text{H} = (0.04375\ \text{lbm})\left(60{,}958\ \frac{\text{Btu}}{\text{lbm}}\right) = 2666.9\ \text{Btu}$$

The heating value after the coal moisture is evaporated (i.e., the as-delivered heating value) is

$$11{,}274.4\ \text{Btu} + 2666.9\ \text{Btu} - 72.19\ \text{Btu} = 13{,}869\ \text{Btu}$$

step 6: Using Table 8.8, determine the combustion products.

$$\begin{aligned} \text{oxygen required by carbon} &= (0.80)(2.667\ \text{lbm}) \\ &= 2.134\ \text{lbm} \\ \text{oxygen required by hydrogen} &= (0.04375)(7.936\ \text{lbm}) \\ &= 0.3472\ \text{lbm} \\ \begin{array}{c}\text{carbon dioxide} \\ \text{produced by carbon}\end{array} &= (0.8)(3.667\ \text{lbm}) \\ &= 2.934\ \text{lbm} \\ \text{water produced by hydrogen} &= (0.04375)(8.936\ \text{lbm}) \\ &= 0.3910\ \text{lbm} \end{aligned}$$

The remaining oxygen is

$$6.019\ \text{lbm} - 2.134\ \text{lbm} - 0.3472\ \text{lbm} = 3.538\ \text{lbm}$$

step 7: The gaseous products must be heated from 70°F to 550°F. The average temperature is

$$\left(\frac{1}{2}\right)(70°\text{F} + 550°\text{F}) + 460° = 770°\text{R}$$

From Table 8.1 (also *NCEES Handbook* table "Temperature-Dependent Physical Properties of Gases at 14.7 psia (U.S. Units)"), the specific heat of gaseous products, omitting water, is

gas	$\frac{\text{Btu}}{\text{lbm-°R}}$
oxygen	0.228
nitrogen	0.252
carbon dioxide	0.225

$$\begin{aligned} Q_\text{heating} &= \left(\begin{array}{l} (3.538\ \text{lbm})\left(0.228\ \frac{\text{Btu}}{\text{lbm-°R}}\right) \\ \quad +(19.981\ \text{lbm})\left(0.252\ \frac{\text{Btu}}{\text{lbm-°R}}\right) \\ \quad +(2.934\ \text{lbm})\left(0.225\ \frac{\text{Btu}}{\text{lbm-°R}}\right) \end{array}\right) \\ &\quad \times(550°\text{F} - 70°\text{F}) \\ &= 3121\ \text{Btu} \end{aligned}$$

Find the enthalpy of steam at 550°F and 1 atm using *NCEES Handbook* table "Superheated Steam (U.S.

Units)." Find the enthalpy of steam at 70°F and 1 atm using *NCEES Handbook* table "Saturated Steam (U.S. Units)—Temperature Table." So,

$$\begin{aligned} Q_{\text{steam}} &= \dot{m}(h_{550°\text{F}} - h_{70°\text{F}}) \\ &= (0.3910 \text{ lbm})\left(1312 \ \frac{\text{Btu}}{\text{lbm}} - 1092.1 \ \frac{\text{Btu}}{\text{lbm}}\right) \\ &= 86 \text{ Btu} \end{aligned}$$

The total heat is

$$\begin{aligned} Q_{\text{total}} &= Q_{\text{heating}} + Q_{\text{steam}} = 3121 \text{ Btu} + 86 \text{ Btu} \\ &= 3207 \text{ Btu} \end{aligned}$$

step 8: The percentage loss based on the coal's as-delivered heating value is

$$\frac{3207.3 \text{ Btu} + 72.19 \text{ Btu}}{13{,}869 \text{ Btu}} = 0.236 \quad (24\%)$$

SI Solution

Steps 1 through 3 are the same as for the customary U.S. solution except that everything is based on kg.

step 4: Determine the energy loss in the vaporizing moisture.

$$q = (\text{moisture})(h_g - h_f)$$

From App. 10K (also *NCEES Handbook:* Superheated Steam (SI Units)), h_g at 290°C and 101.3 kPa is 3054.5 kJ/kg.

h_f at 15.6°C from App. 10I (also *NCEES Handbook:* Saturated Steam (SI Units)) is 65.49 kJ/kg.

$$\begin{aligned} q &= (0.05625 \text{ kg})\left(3054.5 \ \frac{\text{kJ}}{\text{kg}} - 65.49 \ \frac{\text{kJ}}{\text{kg}}\right) \\ &= 168.1 \text{ kJ} \end{aligned}$$

step 5: Calculate the heating value of the remaining fuel components using App. 8A (also *NCEES Handbook:* Standard Heat of Formation and Combustion) and the table footnote.

$$\begin{aligned} \text{HV}_\text{C} &= (0.80 \text{ kg})\left(14\,093 \ \frac{\text{Btu}}{\text{lbm}}\right)\left(2.326 \ \frac{\frac{\text{kJ}}{\text{kg}}}{\frac{\text{Btu}}{\text{lbm}}}\right) \\ &= 26\,224 \text{ kJ} \end{aligned}$$

$$\begin{aligned} \text{HV}_\text{H} &= (0.04375 \text{ kg})\left(60\,958 \ \frac{\text{Btu}}{\text{lbm}}\right)\left(2.326 \ \frac{\frac{\text{kJ}}{\text{kg}}}{\frac{\text{Btu}}{\text{lbm}}}\right) \\ &= 6203 \text{ kJ} \end{aligned}$$

The heating value after the coal moisture is evaporated is

$$26\,224 \text{ kJ} + 6203 \text{ kJ} - 168.1 \text{ kJ} = 32\,259 \text{ kJ}$$

step 6: This step is the same as for the customary U.S. solution except that all quantities are in kg.

step 7: The gaseous products must be heated from 21°C to 290°C. The average temperature is

$$\left(\frac{1}{2}\right)(21°\text{C} + 290°\text{C}) + 273° = 428.5\text{K}$$

$$(428.5\text{K})\left(1.8 \ \frac{°\text{R}}{\text{K}}\right) = 771°\text{R}$$

From Table 8.1 (also *NCEES Handbook* table "Temperature-Dependent Physical Properties of Gases at 14.7 psia (U.S. Units)"), the specific heat of gaseous products is calculated using the table footnote, and omitting water.

gas	$\frac{\text{kJ}}{\text{kg·K}}$
oxygen	0.955
nitrogen	1.055
carbon dioxide	0.942

$$\begin{aligned} Q_{\text{heating}} &= \left(\begin{array}{l} (3.538 \text{ kg})\left(0.955 \ \frac{\text{kJ}}{\text{kg·K}}\right) \\ \quad +(19.981 \text{ kg})\left(1.055 \ \frac{\text{kJ}}{\text{kg·K}}\right) \\ \quad +(2.934 \text{ kg})\left(0.942 \ \frac{\text{kJ}}{\text{kg·K}}\right) \end{array}\right) \\ &\quad \times (290°\text{C} - 21°\text{C}) \\ &= 7322.8 \text{ kJ} \end{aligned}$$

Find the enthalpy of steam at 290°C and 1 atm using *NCEES Handbook* table "Superheated Steam (U.S. Units)." Find the enthalpy of steam at 21°C and 1 atm using *NCEES Handbook* table "Saturated Steam (SI Units)." So,

$$\begin{aligned} Q_{\text{steam}} &= m(h_{550°\text{F}} - h_{70°\text{F}}) \\ &= (0.3910 \text{ kg})\left(3054.3 \ \frac{\text{kJ}}{\text{kg}} - 2539.9 \ \frac{\text{kJ}}{\text{kg}}\right) \\ &= 201.1 \text{ kJ} \end{aligned}$$

The total heat is

$$\begin{aligned} Q_{\text{total}} &= Q_{\text{heating}} + Q_{\text{steam}} = 7322.8 \text{ kJ} + 201.1 \text{ kJ} \\ &= 7523.9 \text{ kJ} \end{aligned}$$

step 8: The percentage loss is

$$\frac{7523.9\ \text{kJ} + 168.1\ \text{kJ}}{32\,259\ \text{kJ}} = 0.238 \quad (24\%)$$

The answer is (B).

14. As in *NCEES Handbook* table "Temperature-Independent Properties of Liquids and Gases (U.S. Units)," silicon in ash is SiO_2 with a molecular weight of

$$28.09\ \frac{\text{lbm}}{\text{lbmol}} + (2)\left(16\ \frac{\text{lbm}}{\text{lbmol}}\right) = 60.09\ \text{lbm/lbmol}$$

The oxygen used with 6.1% by mass silicon is

$$\left(\frac{(2)(16\ \text{lbm})}{28.09\ \text{lbm}}\right)\left(0.061\ \frac{\text{lbm}}{\text{lbm coal}}\right) = 0.0695\ \text{lbm/lbm coal}$$

Silicon ash produced per lbm of coal is

$$0.061\ \frac{\text{lbm}}{\text{lbm coal}} + 0.0695\ \frac{\text{lbm}}{\text{lbm coal}} = 0.1305\ \text{lbm/lbm coal}$$

Silicon ash produced per hour is

$$\left(0.1305\ \frac{\text{lbm}}{\text{lbm coal}}\right)\left(15{,}300\ \frac{\text{lbm coal}}{\text{hr}}\right) = 1996.7\ \text{lbm/hr}$$

The silicon in 410 lbm/hr refuse is

$$\left(410\ \frac{\text{lbm}}{\text{hr}}\right)(1 - 0.3) = 287\ \text{lbm/hr}$$

The emission rate is

$$1996.7\ \frac{\text{lbm}}{\text{hr}} - 287\ \frac{\text{lbm}}{\text{hr}} = 1709.7\ \text{lbm/hr} \quad (1700\ \text{lbm/hr})$$

The answer is (D).

15. *Customary U.S. Solution*

The stoichiometric reaction for propane is given in Table 8.7 (also *NCEES Handbook* table "Temperature-Independent Properties of Liquids and Gases (U.S. Units)").

$$\begin{array}{lccccccc} & C_3H_8 + & 5O_2 & \rightarrow & 3CO_2 & + & 4H_2O \\ \text{MW} & 44.096 & (5)(31.999) & & (3)(44.010) & & (4)(18.015) \end{array}$$

With 40% excess O_2 by weight,

$$\begin{array}{lcc} & C_3H_8 & (1.4)(5)O_2 \\ \text{MW} & 44.096 + (7)(31.999) & \\ & 44.096 & 224 \end{array}$$

$$\begin{array}{lccccc} & 3CO_2 & & 4H_2O & & 2O_2 \\ \rightarrow & (3)(44.010) & + & (4)(18.015) & + & (2)(31.999) \\ & 132.033 & & 72.064 & & 64 \end{array}$$

The excess oxygen is

$$(2)(32) = 64\ \text{lbm/lbmol}\ C_3H_8$$

The mass ratio of nitrogen to oxygen is

$$\begin{aligned} \frac{G_N}{G_O} &= \left(\frac{B_N}{M_N}\right)\left(\frac{M_O}{B_O}\right) \\ &= \left(\frac{0.40}{28\ \frac{\text{lbm}}{\text{lbmol}}}\right)\left(\frac{32\ \frac{\text{lbm}}{\text{lbmol}}}{0.60}\right) \\ &= 0.762 \end{aligned}$$

The nitrogen accompanying the oxygen is

$$(7)(32)(0.762) = 170.7\ \text{lbm}$$

The mass balance per mole of propane is

$$\begin{array}{ll} & C_3H_8 + O_2 + N_2 \rightarrow CO_2 + H_2O + O_2 + N_2 \\ \text{mass per mole} & 44.096 + 224 + 170.7 \rightarrow 132.033 + 72.064 + 64 + 170.7 \end{array}$$

Standard conditions are 60°F and 1 atm. The propane density is given by Eq. 10.50 (also *NCEES Handbook:* Ideal Gas Law).

$$\rho = \frac{pM}{RT}$$

As in *NCEES Handbook:* Temperature, the absolute temperature, T, is

$$60°\text{F} + 460° = 520°\text{R}$$

$$\rho = \frac{\left(14.7\ \frac{\text{lbf}}{\text{in}^2}\right)\left(12\ \frac{\text{in}}{\text{ft}}\right)^2\left(44.097\ \frac{\text{lbm}}{\text{lbmol}}\right)}{\left(1545.35\ \frac{\text{ft-lbf}}{\text{lbmol-°R}}\right)(520°\text{R})} = 0.1162\ \text{lbm/ft}^3$$

The mass flow rate of propane is

$$\left(250\ \frac{\text{ft}^3}{\text{min}}\right)\left(0.1162\ \frac{\text{lbm}}{\text{ft}^3}\right) = 29.05\ \text{lbm/min}$$

Scaling the other mass balance factors down by

$$\frac{29.05\ \frac{\text{lbm}}{\text{min}}}{44.097\ \frac{\text{lbm}}{\text{lbmol}}} = 0.6588\ \text{lbmol/min}$$

	$C_3H_8 + O_2 + N_2 \rightarrow CO_2 + H_2O + O_2 + N_2$
$\frac{\text{lbm}}{\text{min}}$	$29.05 + 147.57 + 112.5 \rightarrow 86.98 + 47.48 + 42.16 + 112.5$

The oxygen flow rate is 147.57 lbm/min.

From Table 8.7 (also *NCEES Handbook:* Ideal Gas Law), the specific volumes of the reactants are given by

$$\begin{aligned} \upsilon_{C_3H_8} &= \frac{RT}{pM} = \frac{\left(1545.35\ \frac{\text{ft-lbf}}{\text{lbmol-°R}}\right)(540°\text{R})}{\left(14.7\ \frac{\text{lbf}}{\text{in}^2}\right)\left(12\ \frac{\text{in}}{\text{ft}}\right)^2\left(44.097\ \frac{\text{lbm}}{\text{lbmol}}\right)} \\ &= 8.939\ \text{ft}^3/\text{lbm} \end{aligned}$$

$$\begin{aligned} \upsilon_{O_2} &= \frac{RT}{pM} = \frac{\left(1545.35\ \frac{\text{ft-lbf}}{\text{lbmol-°R}}\right)(540°\text{R})}{\left(14.7\ \frac{\text{lbf}}{\text{in}^2}\right)\left(12\ \frac{\text{in}}{\text{ft}}\right)^2\left(32\ \frac{\text{lbm}}{\text{lbmol}}\right)} \\ &= 12.319\ \text{ft}^3/\text{lbm} \end{aligned}$$

$$\begin{aligned} \upsilon_{N_2} &= \frac{RT}{pM} = \frac{\left(1545.35\ \frac{\text{ft-lbf}}{\text{lbmol-°R}}\right)(540°\text{R})}{\left(14.7\ \frac{\text{lbf}}{\text{in}^2}\right)\left(12\ \frac{\text{in}}{\text{ft}}\right)^2\left(28\ \frac{\text{lbm}}{\text{lbmol}}\right)} \\ &= 14.071\ \text{ft}^3/\text{lbm} \end{aligned}$$

The total incoming volume is

$$\begin{aligned} \dot{V} &= \left(29.05\ \frac{\text{lbm}}{\text{min}}\right)\left(8.939\ \frac{\text{ft}^3}{\text{lbm}}\right) \\ &\quad + \left(147.57\ \frac{\text{lbm}}{\text{min}}\right)\left(12.319\ \frac{\text{ft}^3}{\text{lbm}}\right) \\ &\quad + \left(112.5\ \frac{\text{lbm}}{\text{min}}\right)\left(14.071\ \frac{\text{ft}^3}{\text{lbm}}\right) \\ &= 3660\ \text{ft}^3/\text{min} \end{aligned}$$

Since the velocity must be kept below 400 ft/min, the minimum area of inlet pipe is

SI Solution

Following the procedure for the customary U.S. solution, the mass balance per mole of propane is

	$C_3H_8 + O_2 + N_2 \rightarrow CO_2 + H_2O + O_2 + N_2$
kg per mole	$44.096 + 224 + 170.7 \rightarrow 132.033 + 72.064 + 64 + 170.7$

Standard conditions are 16°C and 101.3 kPa. The propane density is given by Eq. 10.50 (also *NCEES Handbook:* Ideal Gas Law).

$$\rho = \frac{pM}{RT}$$

As in *NCEES Handbook:* Temperature, the absolute temperature is

$$T = 16°\text{C} + 273° = 289\text{K}$$

$$\rho = \frac{(101\,300\ \text{Pa})\left(44.097\ \frac{\text{g}}{\text{mol}}\right)}{\left(8314.5\ \frac{\text{J}}{\text{kmol·K}}\right)(289\text{K})} = 1.86\ \text{kg/m}^3$$

The mass flow rate of propane is

$$\frac{\left(118\ \frac{\text{L}}{\text{s}}\right)\left(1.86\ \frac{\text{kg}}{\text{m}^3}\right)}{1000\ \frac{\text{L}}{\text{m}^3}} = 0.2195\ \text{kg/s}$$

Scale the other mass balance factors down.

$$\frac{0.2195\ \frac{\text{kg}}{\text{s}}}{44.097\ \frac{\text{kg}}{\text{kmol}}} = 0.004978\ \text{kmol/s}$$

$$C_3H_8 + O_2 + N_2 \rightarrow CO_2 + H_2O + O_2 + N_2$$
$$\frac{\text{kg}}{\text{s}} \quad 0.2195 + 1.115 + 0.85 \rightarrow 0.6572 + 0.3587 + 0.3186 + 0.85$$

The oxygen flow rate is 1.115 kg/s.

From Table 10.7 (also *NCEES Handbook:* Ideal Gas Law), the specific volumes of the reactants are given by Eq. 10.50.

$$\upsilon_{C_3H_8} = \frac{RT}{pM} = \frac{\left(8314.5\ \frac{\text{J}}{\text{kmol·K}}\right)(300\text{K})}{(101\,300\ \text{Pa})\left(44.097\ \frac{\text{g}}{\text{mol}}\right)} = 0.56\ \text{m}^3/\text{kg}$$

$$\upsilon_{O_2} = \frac{RT}{pM} = \frac{\left(8314.5\ \frac{\text{J}}{\text{kmol·K}}\right)(300\text{K})}{(101300\ \text{Pa})\left(32\ \frac{\text{g}}{\text{mol}}\right)} = 0.7717\ \text{m}^3/\text{kg}$$

$$\upsilon_{N_2} = \frac{RT}{pM} = \frac{\left(8314.5\ \frac{\text{J}}{\text{kmol·K}}\right)(300\text{K})}{(101\,300\ \text{Pa})\left(28\ \frac{\text{g}}{\text{mol}}\right)} = 0.8815\ \text{m}^3/\text{kg}$$

The total incoming volume, $\dot{V}$, is

$$\dot{V} = \left(0.2195\ \frac{\text{kg}}{\text{s}}\right)\left(0.5600\ \frac{\text{m}^3}{\text{kg}}\right) + \left(1.115\ \frac{\text{kg}}{\text{s}}\right)\left(0.7717\ \frac{\text{m}^3}{\text{kg}}\right) + \left(0.85\ \frac{\text{kg}}{\text{s}}\right)\left(0.8815\ \frac{\text{m}^3}{\text{kg}}\right) = 1.73\ \text{m}^3/\text{s}$$

Since the velocity for the reactants must be kept below 2 m/s, the minimum area of inlet pipe is

$$A_{\text{inlet}} = \frac{\dot{V}}{\text{v}} = \frac{1.73\ \frac{\text{m}^3}{\text{s}}}{2\ \frac{\text{m}}{\text{s}}} = 0.866\ \text{m}^2 \quad (0.9\ \text{m}^2)$$

The answer is (A).

16. *Customary U.S. Solution*

Since atmospheric air is not used, the nitrogen and oxygen can be varied independently. Furthermore, since enthalpy increase information is not given for oxygen, a 0% excess of oxygen can be assumed.

The stoichiometric equation for the ideal combustion reaction of propane with only stoichiometric oxygen and forming only carbon dioxide and water is

$$\begin{array}{lcccc} & C_3H_8 + & 5O_2 \rightarrow & 3CO_2 + & 4H_2O \\ \text{moles} & (1) & (5) & (3) & (4) \end{array}$$

Using the equation for standard heats of formations found in the *NCEES Handbook:* section titled "Heat of Reaction," subtract the reactants' standard heat of formation enthalpies from the products' standard heat of formation enthalpies to calculate the heat of reaction. The enthalpy of formation of oxygen is zero since it is an element in its natural state. The heat of reaction (heat of combustion) is

$$\begin{aligned} & n_{CO_2}(\Delta H_f)_{CO_2} + n_{H_2O}(\Delta H_f)_{H_2O} \\ & \quad - n_{C_3H_8}(\Delta H_f)_{C_3H_8} - n_{O_2}(\Delta H_f)_{O_2} \\ & = \left(3\ \frac{\text{lbmol}}{\text{lbmol}}\right)\left(-169{,}300\ \frac{\text{Btu}}{\text{lbmol}}\right) \\ & \quad + \left(4\ \frac{\text{lbmol}}{\text{lbmol}}\right)\left(-104{,}040\ \frac{\text{Btu}}{\text{lbmol}}\right) \\ & \quad - \left(1\ \frac{\text{lbmol}}{\text{lbmol}}\right)\left(28{,}800\ \frac{\text{Btu}}{\text{lbmol}}\right) \\ & \quad - \left(5\ \frac{\text{lbmol}}{\text{lbmol}}\right)\left(0\ \frac{\text{Btu}}{\text{lbmol}}\right) \\ & = -952{,}860\ \text{Btu/lbmol of propane} \end{aligned}$$

The negative sign indicates an exothermic reaction.

Let x be the number of moles of nitrogen per mole of propane. Use the nitrogen to cool the combustion. The heat of reaction will increase the enthalpy of products from the standard reference temperature to 3600°R.

$$\begin{aligned} 952{,}860\ \frac{\text{Btu}}{\text{lbmol}} &= \left(3\ \frac{\text{lbmol}}{\text{lbmol}}\right)\left(39{,}791\ \frac{\text{Btu}}{\text{lbmol}}\right) \\ & \quad + \left(4\ \frac{\text{lbmol}}{\text{lbmol}}\right)\left(31{,}658\ \frac{\text{Btu}}{\text{lbmol}}\right) \\ & \quad + x\left(24{,}471\ \frac{\text{Btu}}{\text{lbmol}}\right) \\ x &= 28.89\ \text{lbmol/lbmol propane} \end{aligned}$$

The mass of nitrogen per pound-mole of propane is

$$m_{N_2} = \left(28.89\ \frac{\text{lbmol}}{\text{lbmol propane}}\right)\left(28.016\ \frac{\text{lbm}}{\text{lbmol}}\right) = 809.4\ \text{lbm/lbmol} \quad (810\ \text{lbm/lbmol})$$

The mass of oxygen per pound-mole of propane is

$$m_{O_2} = \left(5\ \frac{\text{lbmol}}{\text{lbmol}}\right)\left(32\ \frac{\text{lbm}}{\text{lbmol}}\right) = 160\ \text{lbm/lbmol}$$

SI Solution

From the customary U.S. solution, the heat of reaction is

$$\begin{aligned} & n_{CO_2}(\Delta H_f)_{CO_2} + n_{H_2O}(\Delta H_f)_{H_2O} - n_{C_3H_8}(\Delta H_f)_{C_3H_8} - n_{O_2}(\Delta H_f)_{O_2} \\ &= \left(3\ \frac{\text{kmol}}{\text{kmol}}\right)\left(-393.8\ \frac{\text{GJ}}{\text{kmol}}\right) + \left(4\ \frac{\text{kmol}}{\text{kmol}}\right)\left(-242\ \frac{\text{GJ}}{\text{kmol}}\right) \\ &\quad + \left(1\ \frac{\text{kmol}}{\text{kmol}}\right)\left(-67.0\ \frac{\text{GJ}}{\text{kmol}}\right) - \left(5\ \frac{\text{kmol}}{\text{kmol}}\right)\left(0\ \frac{\text{GJ}}{\text{kmol}}\right) \\ &= -2216.4\ \text{GJ/kmol propane} \end{aligned}$$

The negative sign indicates an exothermic reaction.

Let x be the number of moles of nitrogen per mole of propane. Use the nitrogen to cool the combustion. The heat of reaction will increase the enthalpy of products from the standard reference temperature to 1980°C.

$$\begin{aligned} 2216.4\ \frac{\text{GJ}}{\text{mol}} &= \left(3\ \frac{\text{kmol}}{\text{kmol}}\right)\left(92.6\ \frac{\text{GJ}}{\text{kmol}}\right) + \left(4\ \frac{\text{kmol}}{\text{kmol}}\right)\left(73.6\ \frac{\text{GJ}}{\text{kmol}}\right) + x\left(56.9\ \frac{\text{GJ}}{\text{kmol}}\right) \\ x &= 28.90\ \text{kmol/kmol propane} \end{aligned}$$

The mass of nitrogen per mole of propane is

$$m_{N_2} = \left(28.90\ \frac{\text{kmol}}{\text{kmol}}\right)\left(28.016\ \frac{\text{kg}}{\text{kmol}}\right) = 809.7\ \text{kg/kmol} \quad (810\ \text{kg/kmol})$$

The mass of oxygen per kmol of propane is

$$m_{O_2} = \left(5\ \frac{\text{kmol}}{\text{kmol}}\right)\left(32\ \frac{\text{kg}}{\text{kmol}}\right) = 160\ \text{kg/kmol}$$

The answer is (B).

17. The heat rate, Q, required to generate the steam is found from the generation rate of steam, $\dot{m}$, the enthalpy of the saturated vapor, h_g, and the enthalpy of the incoming saturated feedwater, h_f. From App. 10B (also *NCEES Handbook:* Properties of Saturated Steam (U.S. Customary Units)), for an absolute pressure of 200 lbf/in^2, h_g is 1198.8 Btu/lbm. From App. 10A, for a temperature of 160°F, h_f is 128.0 Btu/lbm.

$$\begin{aligned} Q &= \dot{m}(h_g - h_f) \\ &= \left(20{,}070\ \frac{\text{lbm}}{\text{hr}}\right)\left(1198.8\ \frac{\text{Btu}}{\text{lbm}} - 128.0\ \frac{\text{Btu}}{\text{lbm}}\right) \\ &= 21.5 \times 10^6\ \text{Btu/hr} \end{aligned}$$

The lower heating value of the fuel, LHV, is

$$\text{LHV} = \frac{Q}{\dot{m}_{\text{fuel}}} = \frac{21.5 \times 10^6\ \frac{\text{Btu}}{\text{hr}}}{5000\ \frac{\text{lbm}}{\text{hr}}} = 4300\ \text{Btu/lbm}$$

The higher heating value, HHV, includes the heat of vaporization (same as the heat of condensation) of the water in the fuel and generated by the combustion of the fuel. The mass of water in the combustion products produced per point of fuel is

$$m_{\text{water}} = \frac{2000\ \frac{\text{lbm water}}{\text{hr}}}{5000\ \frac{\text{lbm fuel}}{\text{hr}}} = 0.4\ \text{lbm water/lbm fuel}$$

From App. 10B (also *NCEES Handbook:* Properties of Saturated Steam (U.S. Customary Units)), for a partial pressure of 4 lbf/in^2, the enthalpy of vaporization, h_{fg}, is 1006.0 Btu/lbm. From Eq. 8.14 (also *NCEES Handbook:* Combustion Reactions), the higher heating value is

$$\begin{aligned} \text{HHV} &= \text{LHV} + m_{\text{water}} h_{fg} \\ &= 4300\ \frac{\text{Btu}}{\text{lbm}} + \left(0.4\ \frac{\text{lbm}}{\text{lbm}}\right)\left(1006.0\ \frac{\text{Btu}}{\text{lbm}}\right) \\ &= 4702.4\ \text{Btu/lbm} \quad (4700\ \text{Btu/lbm}) \end{aligned}$$

The answer is (B).

9 Energy, Work, and Power

PRACTICE PROBLEMS

1. A solid, cast-iron sphere (density of 0.256 lbm/in^3 (7090 kg/m^3)) of 10 in (25 cm) diameter travels without friction at 30 ft/sec (9 m/s) horizontally. Its kinetic energy is most nearly

(A) 900 ft-lbf (1.2 kJ)

(B) 1200 ft-lbf (1.6 kJ)

(C) 1600 ft-lbf (2.0 kJ)

(D) 1900 ft-lbf (2.3 kJ)

2. The work done when a balloon carries a 12 lbm (5.2 kg) load to 40,000 ft (12 000 m) height is most nearly

(A) 2.4×10^5 ft-lbf (300 kJ)

(B) 4.8×10^5 ft-lbf (610 kJ)

(C) 7.7×10^5 ft-lbf (980 kJ)

(D) 9.9×10^5 ft-lbf (1.3 MJ)

3. The power in horsepower (kW) that is required to lift a 3300 lbm (1500 kg) mass 250 ft (80 m) vertically in 14 sec is most nearly

(A) 40 hp (30 kW)

(B) 70 hp (53 kW)

(C) 90 hp (68 kW)

(D) 110 hp (84 kW)

4. Approximately what volume in ft^3 (m^3) of water can be pumped to a 130 ft (40 m) height in 1 hr by a 7 hp (5 kW) pump? Assume 85% efficiency.

(A) 1500 ft^3 (40 m^3)

(B) 1800 ft^3 (49 m^3)

(C) 2000 ft^3 (54 m^3)

(D) 2400 ft^3 (65 m^3)

SOLUTIONS

1. *Customary U.S. Solution*

Since there is no friction, there is no rotation. The sphere slides. The kinetic energy can be found from the first law of thermodynamics (also *NCEES Handbook:* Open Thermodynamic Systems) or Bernoulli's equation (also *NCEES Handbook:* The Bernoulli Equation).

$$\begin{aligned}
E_{\text{kinetic}} &= \tfrac{1}{2}\left(\frac{m}{g_c}\right)\text{v}^2 = \tfrac{1}{2}\left(\frac{V\rho}{g_c}\right)\text{v}^2 \\
&= \left(\tfrac{1}{2}\right)\left(\tfrac{4}{3}\pi r^3\right)\left(\frac{\rho}{g_c}\right)\text{v}^2 \\
&= \tfrac{2}{3}\pi r^3\left(\frac{\rho}{g_c}\right)\text{v}^2 \\
&= \frac{2}{3}\pi\left(\frac{10\ \text{in}}{2}\right)^3\left(\frac{0.256\ \frac{\text{lbm}}{\text{in}^3}}{32.2\ \frac{\text{lbm-ft}}{\text{lbf-sec}^2}}\right)\left(30\ \frac{\text{ft}}{\text{sec}}\right)^2 \\
&= 1873\ \text{ft-lbf} \quad (1900\ \text{ft-lbf})
\end{aligned}$$

SI Solution

Since there is no friction, there is no rotation. The sphere slides. The kinetic energy can be found from the first law of thermodynamics (also *NCEES Handbook:* Open Thermodynamic Systems) or Bernoulli's equation (also *NCEES Handbook:* The Bernoulli Equation).

$$\begin{aligned}
E_{\text{kinetic}} &= \tfrac{1}{2}m\text{v}^2 = \tfrac{1}{2}(\rho V)\text{v}^2 \\
&= \tfrac{1}{2}\rho\left(\tfrac{4}{3}\pi r^3\right)\text{v}^2 \\
&= \tfrac{2}{3}\pi r^3\rho\text{v}^2 \\
&= \frac{2}{3}\pi\left(\frac{0.25\ \text{m}}{2}\right)^3\left(7090\ \frac{\text{kg}}{\text{m}^3}\right)\left(9\ \frac{\text{m}}{\text{s}}\right)^2 \\
&= 2349\ \text{J} \quad (2.3\ \text{kJ})
\end{aligned}$$

The answer is (D).

2. *Customary U.S. Solution*

From Eq. 9.11(b) (also *NCEES Handbook:* Closed Thermodynamic Systems) and Eq. 9.12 (also *NCEES Handbook:* Open Thermodynamic Systems), the work done by the balloon is

$$\begin{aligned} W = \Delta E_{\text{potential}} &= \frac{mg\Delta h}{g_c} \\ &= \frac{(12\text{ lbm})\left(32.2\ \frac{\text{ft}}{\text{sec}^2}\right)(40{,}000\text{ ft})}{32.2\ \frac{\text{lbm-ft}}{\text{lbf-sec}^2}} \\ &= 4.8\times 10^5\text{ ft-lbf} \end{aligned}$$

SI Solution

From Eq. 9.11(a) (also *NCEES Handbook:* Closed Thermodynamic Systems) and Eq. 9.12 (also *NCEES Handbook:* Open Thermodynamic Systems), the work done by the balloon is

$$\begin{aligned} W = \Delta E_{\text{potential}} &= mg\Delta h \\ &= \frac{(5.2\text{ kg})\left(9.81\ \frac{\text{m}}{\text{s}^2}\right)(12\,000\text{ m})}{1000\ \frac{\text{J}}{\text{kJ}}} \\ &= 612.1\text{ kJ}\quad (610\text{ kJ}) \end{aligned}$$

The answer is (B).

3. *Customary U.S. Solution*

Apply the first law of thermodynamics (also *NCEES Handbook:* Closed Thermodynamic Systems or *NCEES Handbook:* Open Thermodynamic Systems). The work required to lift the mass is

$$W = P\Delta t = \frac{mg\Delta h}{g_c}$$

$$\begin{aligned} P &= \frac{mg\Delta h}{g_c\Delta t} \\ &= \frac{(3300\text{ lbm})\left(32.2\ \frac{\text{ft}}{\text{sec}^2}\right)(250\text{ ft})}{\left(32.2\ \frac{\text{lbm-ft}}{\text{sec}^2\text{-lbf}}\right)(14\text{ sec})\left(550\ \frac{\text{ft-lbf}}{\text{hp-sec}}\right)} \\ &= 107\text{ hp}\quad (110\text{ hp}) \end{aligned}$$

SI Solution

Apply the first law of thermodynamics (also *NCEES Handbook:* Closed Thermodynamic Systems or *NCEES Handbook:* Open Thermodynamic Systems). The work required to lift the mass is

$$W = P\Delta t = mg\Delta h$$

$$\begin{aligned} P = \frac{mg\Delta h}{\Delta t} &= \frac{(1500\text{ kg})\left(9.81\ \frac{\text{m}}{\text{s}^2}\right)(80\text{ m})}{(14\text{ s})\left(1000\ \frac{\text{W}}{\text{kW}}\right)} \\ &= 84.1\text{ kW}\quad (84\text{ kW}) \end{aligned}$$

The answer is (D).

4. *Customary U.S. Solution*

Apply the first law of thermodynamics (also *NCEES Handbook:* Closed Thermodynamic Systems or *NCEES Handbook:* Open Thermodynamic Systems). The volume of water is found from the work performed.

$$\begin{aligned} P_{\text{actual}}\Delta t &= W_{\text{done by pump}} \\ \eta P_{\text{ideal}}\Delta t &= \Delta E_{\text{potential}} \\ &= \frac{mg\Delta h}{g_c} \\ &= \frac{(\rho V)g\Delta h}{g_c} \end{aligned}$$

$$\begin{aligned} V &= \frac{\eta P_{\text{ideal}}\Delta t}{\frac{\rho g\Delta h}{g_c}} \\ &= \frac{(0.85)(7\text{ hp})\left(550\ \frac{\text{ft-lbf}}{\text{hp-sec}}\right)(1\text{ hr})\left(3600\ \frac{\text{sec}}{\text{hr}}\right)}{\frac{\left(62.4\ \frac{\text{lbm}}{\text{ft}^3}\right)\left(32.2\ \frac{\text{ft}}{\text{sec}^2}\right)(130\text{ ft})}{32.2\ \frac{\text{lbm-ft}}{\text{lbf-sec}^2}}} \\ &= 1452\text{ ft}^3\quad (1500\text{ ft}^3) \end{aligned}$$

SI Solution

Apply the first law of thermodynamics (also *NCEES Handbook:* Closed Thermodynamic Systems or Open Thermodynamic Systems). The volume of water is found from the work performed.

$$\begin{aligned} P_{\text{actual}}\Delta t &= W_{\text{done by pump}} \\ \eta P_{\text{ideal}}\Delta t &= \Delta E_{\text{potential}} \\ &= mg\Delta h \\ &= (\rho V)g\Delta h \\ V &= \frac{\eta P_{\text{ideal}}\Delta t}{\rho g\Delta h} \\ &= \frac{(0.85)(5\ \text{kW})\left(1000\ \frac{\text{W}}{\text{kW}}\right)(1\ \text{h})\left(3600\ \frac{\text{s}}{\text{h}}\right)}{\left(1000\ \frac{\text{kg}}{\text{m}^3}\right)\left(9.81\ \frac{\text{m}}{\text{s}^2}\right)(40\ \text{m})} \\ &= 39.0\ \text{m}^3 \quad (40\ \text{m}^3) \end{aligned}$$

The answer is (A).

Thermodynamics

10 Thermodynamic Properties of Substances

PRACTICE PROBLEMS

1. The molar enthalpy of 250°F (120°C) steam with a quality of 92% is most nearly

(A) 16,000 Btu/lbmol (37 MJ/kmol)

(B) 18,000 Btu/lbmol (41 MJ/kmol)

(C) 20,000 Btu/lbmol (46 MJ/kmol)

(D) 22,000 Btu/lbmol (51 MJ/kmol)

2. The ratio of specific heats for air at 600°F (300°C) is most nearly

(A) 1.33

(B) 1.38

(C) 1.41

(D) 1.67

3. The density of helium at 600°F (300°C) and one standard atmosphere is most nearly

(A) 0.0052 lbm/ft^3 (0.085 kg/m^3)

(B) 0.0061 lbm/ft^3 (0.098 kg/m^3)

(C) 0.0076 lbm/ft^3 (0.12 kg/m^3)

(D) 0.0095 lbm/ft^3 (0.15 kg/m^3)

4. What is the thermodynamic state of water at 600°F and 300 psia?

(A) subcooled liquid

(B) saturated vapor

(C) superheated vapor

(D) real or ideal gas

5. 106°F water flows in a closed feedwater heater. At the pressure in the feedwater heater, water has a saturation temperature of 293°F and a saturation enthalpy of 262 Btu/lbm. What is most nearly the enthalpy of the water?

(A) 75 Btu/lbm

(B) 110 Btu/lbm

(C) 120 Btu/lbm

(D) 130 Btu/lbm

6. A 25% mixture (by volume) of ethylene glycol and water is used in a solar heating application. The mixture is intended to operate at standard atmospheric pressure and an average temperature of 150°F. At 150°F, ethylene glycol has a specific gravity of 1.11 (based on 60°F water) and a specific heat of 0.63 Btu/lbm-°F. What is most nearly the specific heat of the mixture at 150°F?

(A) 0.892 Btu/lbm-°F

(B) 0.899 Btu/lbm-°F

(C) 0.908 Btu/lbm-°F

(D) 0.913 Btu/lbm-°F

SOLUTIONS

1. *Customary U.S. Solution*

From App. 10A (also *NCEES Handbook:* Properties of Saturated Steam (U.S. Customary Units)), for 250°F steam, the enthalpy of saturated liquid, h_f, is 218.6 Btu/lbm. The heat of vaporization, h_{fg}, is 945.4 Btu/lbm. The enthalpy is given by Eq. 10.40 (also *NCEES Handbook:* Properties for Two-Phase (Vapor-Liquid) Systems).

$$\begin{aligned} h &= h_f + xh_{fg} = 218.6\ \frac{\text{Btu}}{\text{lbm}} + (0.92)\left(945.4\ \frac{\text{Btu}}{\text{lbm}}\right) \\ &= 1088.4\ \text{Btu/lbm} \end{aligned}$$

As in *NCEES Handbook* table "Temperature-Independent Properties of Liquids and Gases (U.S. Units)," the molecular weight of water is 18 lbm/lbmol. The molar enthalpy is given by Eq. 10.14.

$$\begin{aligned} H &= (\text{MW})h = \left(18\ \frac{\text{lbm}}{\text{lbmol}}\right)\left(1088.4\ \frac{\text{Btu}}{\text{lbm}}\right) \\ &= 19{,}591\ \text{Btu/lbmol} \quad (20{,}000\ \text{Btu/lbmol}) \end{aligned}$$

SI Solution

From App. 10I (also *NCEES Handbook:* Saturated Steam (SI Units), for 120°C steam, the enthalpy of saturated liquid, h_f, is 503.81 kJ/kg. The heat of vaporization, h_{fg}, is 2202.1 kJ/kg. The enthalpy is given by Eq. 10.40 (also *NCEES Handbook:* Properties for Two-Phase (Vapor-Liquid) Systems).

$$\begin{aligned} h &= h_f + xh_{fg} = 503.81\ \frac{\text{kJ}}{\text{kg}} + (0.92)\left(2202.1\ \frac{\text{kJ}}{\text{kg}}\right) \\ &= 2529.7\ \text{kJ/kg} \end{aligned}$$

As in *NCEES Handbook* table "Temperature-Independent Properties of Liquids and Gases (SI Units)," the molecular weight of water is 18 kg/kmol. Molar enthalpy is given by Eq. 10.14.

$$\begin{aligned} H &= \text{MW} \times h = \left(18\ \frac{\text{kg}}{\text{kmol}}\right)\left(2529.7\ \frac{\text{kJ}}{\text{kg}}\right) \\ &= 45\,535\ \text{kJ/kmol} \quad (46\ \text{MJ/kmol}) \end{aligned}$$

The answer is (C).

2. *Customary U.S. Solution*

As in *NCEES Handbook:* Temperature, the absolute temperature is

$$600°\text{F} + 460° = 1060°\text{R}$$

From Table 8.1 (also *NCEES Handbook:* Temperature-Dependent Properties of Air (U.S. Customary Units)), the specific heat at constant pressure for air at 1060°R is

$$c_p = 0.250\ \frac{\text{Btu}}{\text{lbm-°R}}$$

From Eq. 10.95(b) (also *NCEES Handbook:* Ideal Gas Law),

$$\begin{aligned} c_v &= c_p - \frac{R}{J} \\ &= 0.250\ \frac{\text{Btu}}{\text{lbm-°R}} - \frac{1545.35\ \frac{\text{ft-lbf}}{\text{lbmol-°R}}}{\left(778\ \frac{\text{ft-lbf}}{\text{Btu}}\right)\left(28.967\ \frac{\text{lbm}}{\text{lbmol}}\right)} \\ &= 0.1814\ \text{Btu/lbm-°R} \end{aligned}$$

The ratio of specific heats is given by Eq. 10.28.

$$\begin{aligned} k &= \frac{c_p}{c_v} = \frac{0.250\ \frac{\text{Btu}}{\text{lbm-°R}}}{0.1814\ \frac{\text{Btu}}{\text{lbm-°R}}} \\ &= 1.378 \quad (1.38) \end{aligned}$$

SI Solution

From Table 8.1 (also *NCEES Handbook:* Temperature-Dependent Properties of Air (SI Units)), the specific heat at constant pressure for air is

$$c_p = 1.047\ \text{kJ/kg·K}$$

From Eq. 10.95(a) (also *NCEES Handbook:* Ideal Gas Law),

$$\begin{aligned} c_v &= c_p - \frac{R}{J} \\ &= \left(1.047\ \frac{\text{kJ}}{\text{kg·K}}\right)\left(1000\ \frac{\text{J}}{\text{kJ}}\right) - \frac{8.3145\ \frac{\text{J}}{\text{mol·K}}}{0.0289\ \frac{\text{kg}}{\text{mol}}} \\ &= 760\ \text{J/kg·K} \end{aligned}$$

The ratio of specific heats is given by Eq. 10.28.

$$k = \frac{c_p}{c_v} = \frac{\left(1.047 \ \frac{\text{kJ}}{\text{kg·K}}\right)\left(1000 \ \frac{\text{J}}{\text{kJ}}\right)}{760 \ \frac{\text{J}}{\text{kg·K}}}$$

$$= 1.377 \quad (1.38)$$

The answer is (B).

3. *Customary U.S. Solution*

As in *NCEES Handbook:* Temperature, the absolute temperature is

$$600°\text{F} + 460° = 1060°\text{R}$$

From Eq. 10.50 (also *NCEES Handbook:* Ideal Gas Law), the density of helium is

$$\rho = \frac{p}{RT} = \frac{\left(14.7 \ \frac{\text{lbf}}{\text{in}^2}\right)\left(12 \ \frac{\text{in}}{\text{ft}}\right)^2\left(4 \ \frac{\text{lbm}}{\text{lbmol}}\right)}{\left(1545.35 \ \frac{\text{ft-lbf}}{\text{lbmol-°R}}\right)(1060°\text{R})}$$

$$= 0.00517 \text{ lbm/ft}^3 \quad \left(0.0052 \text{ lbm/ft}^3\right)$$

SI Solution

As in *NCEES Handbook:* Temperature, the absolute temperature is

$$300°\text{C} + 273° = 573\text{K}$$

From Eq. 10.50 (also *NCEES Handbook:* Ideal Gas Law), the density of helium is

$$\rho = \frac{p}{RT} = \frac{(101\,300 \text{ Pa})\left(4 \ \frac{\text{g}}{\text{mol}}\right)}{\left(8314.5 \ \frac{\text{J}}{\text{kmol·K}}\right)(573\text{K})}$$

$$= 0.085 \text{ kg/m}^3$$

The answer is (A).

4. As in *NCEES Handbook* table "Superheated Steam (U.S. Units)" the saturation temperature for 300 psia steam is 417°F, so since the water's temperature is higher than this, the water is either a superheated vapor or a gas. The critical temperature for water is 705°F, so the water can't be considered a gas. Therefore, it is a superheated vapor.

The answer is (C).

5. Since the pressure of the water in the feedwater heater is not given, a subcooled liquid table can't be used directly. (The saturation temperature could be used to find the pressure, however, if this approach was taken.) As in *NCEES Handbook* table "Physical Properties of Liquid Water (U.S. Units)" the specific heat of liquid water is approximately 1 Btu/lbm-°F, which is the reason that saturated liquid enthalpy and saturation temperature have essentially the same numerical values. Calculate the subcooled enthalpy by subtracting the sensible heat from 106°F to 293°F.

$$h_{106°\text{F}} = h_{\text{saturation}} - c_p(T_{\text{saturation}} - T)$$

$$= 262 \ \frac{\text{Btu}}{\text{lbm}} - \left(1 \ \frac{\text{Btu}}{\text{lbm-°F}}\right)(293°\text{F} - 106°\text{F})$$

$$= 75 \text{ Btu/lbm}$$

The answer is (A).

6. Use a saturated steam table, such as App. 10A (also *NCEES Handbook* table "Properties of Saturated Steam (U.S. Units)—Temperature Table") to get the properties of 150°F water. The specific volume is the reciprocal of the density. Since $\Delta h = c_p \Delta T$, the specific heat can be found from the change in enthalpy over a known temperature range. Use the saturation enthalpies at 140°F and 160°F.

$$\rho = \frac{1}{v_f}$$

$$= \frac{1}{0.01634 \ \frac{\text{ft}^3}{\text{lbm}}}$$

$$= 61.20 \text{ lbm/ft}^3$$

$$c_{p,\text{water}} = \frac{h_{\text{sat},T_2} - h_{\text{sat},T_1}}{T_2 - T_1}$$

$$= \frac{128.00 \ \frac{\text{Btu}}{\text{lbm}} - 107.99 \ \frac{\text{Btu}}{\text{lbm}}}{160°\text{F} - 140°\text{F}}$$

$$= 1.00 \text{ Btu/lbm-°F}$$

Specific heats of liquid mixtures are gravimetrically weighted.

Thermodynamics

Consider 1 ft^3 of mixture, containing 0.25 ft^3 of ethylene glycol and $1 - 0.25\ ft^3 = 0.75\ ft^3$ water. The gravimetric fraction of ethylene glycol, G_{glycol}, in the mixture is

$$\begin{aligned}
G_{glycol} &= \frac{m_{glycol}}{m_{glycol} + m_{water}} \\
&= \frac{SG_{glycol}\rho_{water,60°F}V_{glycol}}{SG_{glycol}\rho_{water,60°F}V_{glycol} + \rho_{water,150°F}V_{water}} \\
&= \frac{(1.11)\left(62.4\ \frac{lbm}{ft^3}\right)(0.25\ ft^3)}{(1.11)\left(62.4\ \frac{lbm}{ft^3}\right)(0.25\ ft^3) + \left(61.2\ \frac{lbm}{ft^3}\right)(0.75\ ft^3)} \\
&= 0.274
\end{aligned}$$

The gravimetric fraction of water in the mixture is

$$G_{water} = 1 - G_{glycol} = 1 - 0.274 = 0.726$$

The specific heat of the mixture is

$$\begin{aligned}
c_{p,mixture} &= G_{glycol}c_{p,glycol} + G_{water}c_{p,water} \\
&= (0.274)\left(0.63\ \frac{Btu}{lbm\text{-}°F}\right) + (0.726)\left(1.00\ \frac{Btu}{lbm\text{-}°F}\right) \\
&= 0.899\ Btu/lbm\text{-}°F
\end{aligned}$$

The answer is (B).

11 Changes in Thermodynamic Properties

PRACTICE PROBLEMS

1. Which statement is FALSE?

(A) The availability of a system depends on the location of the system.

(B) A heat engine cycle can have a thermal efficiency of 100% given that the transfer of heat to the working fluid is at a constant temperature in a reversible process over a reasonably finite time.

(C) The gas temperature always decreases when the gas expands isentropically.

(D) Water in a liquid state cannot exist at any temperature if the pressure is less than the triple point pressure.

2. Which statement is true?

(A) Entropy does not change in an adiabatic process.

(B) The entropy of a closed system cannot decrease.

(C) Entropy increases when a refrigerant passes through a throttling valve.

(D) The entropy of air inside a closed room with a running, electrically driven fan will always increase over time.

3. Which process CANNOT be modeled as an isenthalpic process?

(A) viscous drag on an object moving through air

(B) an ideal gas accelerating to supersonic speed through a converging-diverging nozzle

(C) refrigerant passing through a pressure-reducing throttling valve

(D) high-pressure steam escaping through a spring-loaded pressure-relief (safety) valve

4. Air expands isentropically in a steady-flow process from 700°F and 400 psia to 50 psia. Based on the specific heat capacity at room temperature, what is most nearly the change in enthalpy per lbm assuming air as an ideal gas and the ratio of heat capacities is 1.4?

(A) −680 Btu/lbm

(B) −120 Btu/lbm

(C) −90 Btu/lbm

(D) −14 Btu/lbm

5. A perfectly insulated, rigid enclosure contains 0.60 kg of air ($c_p = 1.005$ kJ/kg·K; $c_v = 0.718$ kJ/kg·K). The ambient conditions outside the enclosure are 95 kPa and 20°C. The air inside the enclosure is initially at 200 kPa and 20°C. Subsequently, an internal impeller within the enclosure raises the air's pressure to 230 kPa through a shaft from an external motor with a motor efficiency of 65%. Approximately how much shaft work is required to raise the pressure from 200 kPa to 230 kPa?

(A) 19 kJ

(B) 25 kJ

(C) 32 kJ

(D) 140 kJ

6. Cast iron is heated from 80°F to 780°F (27°C to 416°C). The heat required per unit mass is most nearly

(A) 70 Btu/lbm (160 kJ/kg)

(B) 120 Btu/lbm (280 kJ/kg)

(C) 170 Btu/lbm (390 kJ/kg)

(D) 320 Btu/lbm (740 kJ/kg)

7. The ventilation rate in a building is 3×10^5 ft^3/hr (2.4 m^3/s). The air is heated from 35°F to 75°F (2°C to 24°C) by water whose temperature decreases from 180°F to 150°F (82°C to 66°C). The water flow rate in gal/min (L/s) is most nearly

(A) 9 gal/min (0.58 L/s)

(B) 13 gal/min (0.83 L/s)

(C) 15 gal/min (0.96 L/s)

(D) 22 gal/min (1.4 L/s)

8. 8.0 ft^3 (0.25 m^3) of 180°F, 14.7 psia (82°C, 101.3 kPa) air is cooled to 100°F (38°C) in a constant-pressure process. The amount of work done is most nearly

(A) −2100 ft-lbf (−3.1 kJ)

(B) −1500 ft-lbf (−2.3 kJ)

(C) −1100 ft-lbf (−1.5 kJ)

(D) −900 ft-lbf (−1.3 kJ)

9. Most nearly, the enthalpy change of an isentropic process using steam with an initial quality of 95% and operating between 300 psia and 50 psia (2 MPa and 0.35 MPa) is

(A) 100 Btu/lbm (230 kJ/kg)

(B) 130 Btu/lbm (300 kJ/kg)

(C) 210 Btu/lbm (480 kJ/kg)

(D) 340 Btu/lbm (780 kJ/kg)

10. A closed air heater receives 540°F, 100 psia (280°C, 700 kPa) air and heats it to 1540°F (840°C). The outside temperature is 100°F (40°C). The pressure of the air drops 20 psi (150 kPa) as it passes through the heater. The heat transfer can be approximated as the outside temperature times the change in entropy. The percentage loss in available energy due to the pressure drop is most nearly

(A) 6%

(B) 12%

(C) 18%

(D) 34%

11. Carbon dioxide gas at 20 psia and 70°F (150 kPa and 21°C) is compressed to 3800 psia and 70°F (25 MPa and 21°C) by a compressor/heat exchanger combination. The compressed gas is stored at 70°F (21°C) in a 100 ft^3 (3 m^3) rigid tank initially charged with carbon dioxide gas at 20 psia (150 kPa). The average mass flow rate of carbon dioxide gas into the tank if the compressor fills the tank in exactly one hour is most nearly

(A) 6500 lbm/hr (0.8 kg/s)

(B) 9700 lbm/hr (1.3 kg/s)

(C) 12,000 lbm/hr (1.6 kg/s)

(D) 14,000 lbm/hr (1.9 kg/s)

12. The mass of an insulated 20 ft^3 (0.6 m^3) steel tank is 40 lbm (20 kg). The steel has a specific heat of 0.11 Btu/lbm-°R (0.46 kJ/kg·K). The tank is placed in a room where the surrounding air is 70°F and 14.7 psia (21°C and 101.3 kPa). After the tank is evacuated to 1 psia and 70°F (7 kPa and 21°C), a valve is suddenly opened, allowing the tank to fill with room air. The air enters the tank in a well-mixed, turbulent condition and increases the system's final temperature to 300°F (149° C). The work performed by the room air entering the tank is most nearly

(A) −42 Btu (−49 000 J)

(B) −34 Btu (−38 000 J)

(C) −8 Btu (−9500 J)

(D) −4 Btu (−4700 J)

SOLUTIONS

1. Availability depends on the temperature of the environment, T_L. Option A is true.

A heat engine's maximum efficiency is that of a Carnot engine cycle, which is always less than 100%. Option B is false.

An expansion includes a drop in enthalpy, and since $h = u + pv$, both u (manifested as temperature) and p decrease. Option C is true.

Liquid water cannot exist below the triple point pressure. Option D is true.

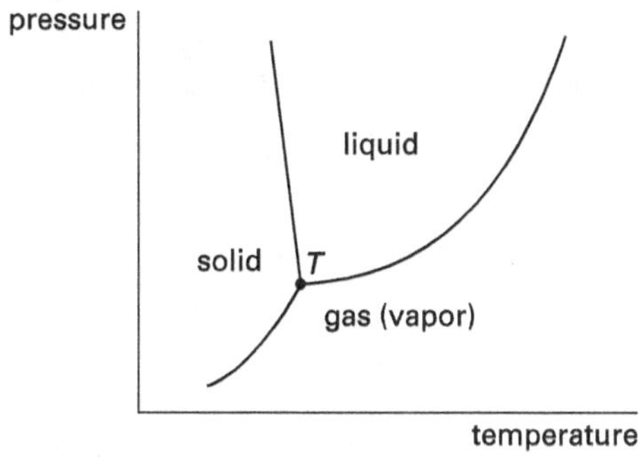

The answer is (B).

2. Entropy does not change in an isentropic (reversible adiabatic) process. However, not all adiabatic processes are reversible. Option A is false.

The entropy of a closed system can be decreased by decreasing the temperature. Option B is false.

When a refrigerant is throttled, the enthalpy remains constant, the pressure drops, and the entropy increases. Option C is true.

The fan and motor will certainly increase the entropy of the air inside the room. However, if the air temperature is decreased by heat loss to the outside, the entropy will decrease. Option D is false.

The answer is (C).

3. A throttling process (including flow through a control valve, safety relief valve, throttling valve, nozzle, or orifice) is modeled as an isenthalpic process. Consider steam expanding through a safety relief valve. To squeeze through the narrow restriction between the disk and the valve seat, steam has to accelerate to a high speed. It does so by converting enthalpy into kinetic energy. The process is not frictionless, but passage past the valve seat occurs so quickly as to be essentially so, and the process is considered to be isentropic. Once past the narrow restriction, the steam expands into the lower pressure region in the valve outlet. The steam decelerates as the flow area increases from the valve passageway to the downstream pipe. This decrease in velocity (kinetic energy) is manifested as an increase in temperature and enthalpy. The enthalpy drop associated with the initial increase in kinetic energy is reclaimed (except for a small portion lost due to the effects of friction). Therefore, the final process is essentially isenthalpic.

Viscous drag of an object in any fluid is an adiabatic process without work being performed on or by the fluid. The duration of the contact between fluid and object is short. Therefore, viscous drag is isenthalpic.

The increase in velocity in a supersonic nozzle comes entirely at the expense of enthalpy.

The answer is (B).

4. From Eq. 11.97 (also the *NCEES Handbook:* section titled "Closed Thermodynamic Systems") (applicable to both closed and open systems),

$$T_2 = T_1\left(\frac{p_2}{p_1}\right)^{\frac{k-1}{k}} = (1160°\text{R})\left(\frac{50\ \frac{\text{lbf}}{\text{in}^2}}{400\ \frac{\text{lbf}}{\text{in}^2}}\right)^{\frac{1.4-1}{1.4}} = 640.4°\text{R}$$

As in the *NCEES Handbook:* section titled "Temperature-Dependent Properties of Air (U.S. Customary Units)," the value for c_p at room temperature is 0.240 Btu/lbm-°F.

From Eq. 11.125 (also the *NCEES Handbook:* section titled "Turbines"), for turbines (i.e., gas expansion), the change in enthalpy is

$$\begin{aligned}\Delta h &= c_p\Delta T = \left(0.240\ \frac{\text{Btu}}{\text{lbm-°R}}\right)(640.4°\text{R} - 1160°\text{R}) \\ &= -124.7\ \text{Btu/lbm} \quad (-120\ \text{Btu/lbm})\end{aligned}$$

The answer is (B).

5. Using the ideal gas law (see *NCEES Handbook:* Ideal Gas Law),

$$T_2 = \frac{T_1 p_2}{p_1} = \frac{(20°\text{C} + 273°)(230\ \text{kPa})}{200\ \text{kPa}} = 337\text{K}$$

$$T_{2,°\text{C}} = 337\ \text{K} - 273° = 64°\text{C}$$

This is a closed system. The first law of thermodynamics is applicable.

$$Q = \Delta U + W$$

Since the system is insulated, it is adiabatic, and $Q = 0$.

$$W = -\Delta U$$

Work on a unit mass (not molar) basis. Use Eq. 11.51 (also *NCEES Handbook:* Closed Thermodynamic Systems), valid for any process.

$$\begin{aligned} W &= -\Delta U = -(m)(c_v)(T_2 - T_1) \\ &= (0.6 \text{ kg})\left(0.718 \ \frac{\text{kJ}}{\text{kg·K}}\right)(64°\text{C} - 20°\text{C}) \\ &= 19 \text{ kJ} \end{aligned}$$

The answer is (A).

6. *Customary U.S. Solution*

From Table 10.2 (also *NCEES Handbook* table "Physical Properties of Metals at 68°F (U.S. Units)"), the approximate value of specific heat for cast iron is $c_p = 0.10$ Btu/lbm-°F.

The heat required per unit mass is

$$\begin{aligned} q &= c_p(T_2 - T_1) = \left(0.10 \ \frac{\text{Btu}}{\text{lbm-°F}}\right)(780°\text{F} - 80°\text{F}) \\ &= 70 \text{ Btu/lbm} \end{aligned}$$

SI Solution

From Table 10.2 (also *NCEES Handbook* table "Physical Properties of Metals at 20°C (SI Units)"), the approximate value of specific heat of cast iron is $c_p = 0.42$ kJ/kg·K.

The heat required per unit mass is

$$\begin{aligned} q &= c_p(T_2 - T_1) \\ &= \left(0.42 \ \frac{\text{kJ}}{\text{kg·K}}\right)(416°\text{C} - 27°\text{C}) \\ &= 163.4 \text{ kJ/kg} \quad (160 \text{ kJ/kg}) \end{aligned}$$

The answer is (A).

7. *Customary U.S. Solution*

First calculate the mass flow rate of air to be heated by using the ideal gas law. (Usually the air mass would be evaluated at the entering conditions. This problem is ambiguous. The building conditions are used because a building ventilation rate was specified.)

$$\dot{m}_{\text{air}} = \frac{p\dot{V}}{RT}$$

As in *NCEES Handbook:* Temperature, the absolute temperature is

$$\begin{aligned} T &= 75°\text{F} + 460° = 535°\text{R} \\ \dot{m}_{air} &= \frac{\left(14.7 \ \frac{\text{lbf}}{\text{in}^2}\right)\left(12 \ \frac{\text{in}}{\text{ft}}\right)^2\left(3 \times 10^5 \ \frac{\text{ft}^3}{\text{hr}}\right)\left(29 \ \frac{\text{lbm}}{\text{lbmol}}\right)}{\left(1545.35 \ \frac{\text{ft-lbf}}{\text{lbmol-°R}}\right)(535°\text{R})} \\ &= 2.227 \times 10^4 \text{ lbm/hr} \end{aligned}$$

As in *NCEES Handbook:* Ideal Gas Law, the heat lost by the water is equal to the heat gained by the air.

$$\begin{aligned} &\dot{m}_w c_{p,w}(T_{1,w} - T_{2,w}) = \dot{m}_{\text{air}} c_{p,\text{air}}(T_{2,\text{air}} - T_{1,\text{air}}) \\ &\dot{m}_w\left(1 \ \frac{\text{Btu}}{\text{lbm-°F}}\right)(180°\text{F} - 150°\text{F}) \\ &\quad = \left(2.227 \times 10^4 \ \frac{\text{lbm}}{\text{hr}}\right)\left(0.241 \ \frac{\text{Btu}}{\text{lbm-°F}}\right)(75°\text{F} - 35°\text{F}) \\ &\dot{m}_w = 7156.1 \text{ lbm/hr} \end{aligned}$$

From App. 18A (also *NCEES Handbook* table "Physical Properties of Liquid Water (U.S. Units)"), the density of water at 165°F is approximately 61 lbm/ft^3. The water volume flow rate is

$$\begin{aligned} \dot{V}_w &= \frac{\dot{m}}{\rho} = \frac{\left(7156.1 \ \frac{\text{lbm}}{\text{hr}}\right)\left(7.481 \ \frac{\text{gal}}{\text{ft}^3}\right)}{\left(61 \ \frac{\text{lbm}}{\text{ft}^3}\right)\left(60 \ \frac{\text{min}}{\text{hr}}\right)} \\ &= 14.63 \text{ gal/min} \quad (15 \text{ gal/min}) \end{aligned}$$

SI Solution

First calculate the mass flow rate of air to be heated by using the ideal gas law. (Usually the air mass would be evaluated at the entering conditions. This problem is ambiguous. The building conditions are used because a building ventilation rate was specified.)

$$\dot{m}_{\text{air}} = \frac{p\dot{V}}{RT}$$

As in *NCEES Handbook* table "Conversion Table for Temperature Units," the absolute temperature is

$$\begin{aligned} T &= 24°\text{C} + 273° = 297\text{K} \\ \dot{m}_{air} &= \frac{(101\,300 \text{ Pa})\left(2.4 \ \frac{\text{m}^3}{\text{s}}\right)\left(29 \ \frac{\text{g}}{\text{mol}}\right)}{\left(8314.5 \ \frac{\text{J}}{\text{kmol·K}}\right)(297\text{K})} \\ &= 2.85 \text{ kg/s} \end{aligned}$$

As in *NCEES Handbook:* Ideal Gas Law, the heat lost by the water is equal to the heat gained by the air.

$$\dot{m}_w c_{p,w}(T_{1,w} - T_{2,w}) = \dot{m}_{\text{air}} c_{p,\text{air}}(T_{2,\text{air}} - T_{1,\text{air}})$$

$$\dot{m}_w\left(4.190\ \frac{\text{kJ}}{\text{kg}\cdot\text{K}}\right)(82°\text{C} - 66°\text{C}) = \left(2.85\ \frac{\text{kg}}{\text{s}}\right)\left(1.005\ \frac{\text{kJ}}{\text{kg}\cdot\text{K}}\right)(24°\text{C} - 2°\text{C})$$

$$\dot{m}_w = 0.940\ \text{kg/s}$$

From App. 18B (also *NCEES Handbook* table "Physical Properties of Liquid Water (SI Units)"), the density of water at 74°C is approximately 976 kg/m^3. The water volume flow rate is

$$\dot{V}_w = \frac{\dot{m}}{\rho} = \frac{\left(0.940\ \frac{\text{kg}}{\text{s}}\right)\left(1000\ \frac{\text{L}}{\text{m}^3}\right)}{976\ \frac{\text{kg}}{\text{m}^3}} = 0.963\ \text{L/s}\quad(0.96\ \text{L/s})$$

The answer is (C).

8. *Customary U.S. Solution*

As in *NCEES Handbook:* Ideal Gas Law, the mass of air is

$$m = \frac{pV}{RT} = \frac{\left(14.7\ \frac{\text{lbf}}{\text{in}^2}\right)\left(12\ \frac{\text{in}}{\text{ft}}\right)^2(8\ \text{ft}^3)\left(29\ \frac{\text{lbm}}{\text{lbmol}}\right)}{\left(1545.35\ \frac{\text{ft-lbf}}{\text{lbmol-°R}}\right)(640°\text{R})} = 0.4964\ \text{lbm}$$

For a constant pressure process from Eq. 11.55 (also *NCEES Handbook:* Closed Thermodynamic Systems), on a per unit mass basis,

$$W = p(v_2 - v_1) = R(T_2 - T_1)$$

Since $\Delta T_{°\text{R}} = \Delta T_{°\text{F}}$, the total work for m in lbm is

$$W = mR(T_2 - T_1) = \frac{(0.4964\ \text{lbm})\left(1545.35\ \frac{\text{ft-lbf}}{\text{lbmol-°R}}\right)(100°\text{F} - 180°\text{F})}{29\ \frac{\text{lbm}}{\text{lbmol}}} = -2116.6\ \text{ft-lbf}\quad(-2100\ \text{ft-lbf})$$

This is negative because work is done on the system.

SI Solution

As in *NCEES Handbook:* Ideal Gas Law, the mass of air is

$$m = \frac{pV}{RT} = \frac{(101\,300\ \text{Pa})(0.25\ \text{m}^3)\left(29\ \frac{\text{g}}{\text{mol}}\right)}{\left(8314.5\ \frac{\text{J}}{\text{kmol}\cdot\text{K}}\right)(355\text{K})} = 0.2486\ \text{kg}$$

For a constant pressure process from Eq. 11.55 (also *NCEES Handbook:* Closed Thermodynamic Systems), on a per unit mass basis,

$$W = p(v_2 - v_1) = R(T_2 - T_1)$$

The total work for m in kg is

$$W = mR(T_2 - T_1) = \frac{(0.2486\ \text{kg})\left(8314.5\ \frac{\text{J}}{\text{kmol}\cdot\text{K}}\right)(38°\text{C} - 82°\text{C})}{29\ \frac{\text{g}}{\text{mol}}} = -3136.1\ \text{J}\quad(-3.1\ \text{kJ})$$

The answer is (A).

9. *Customary U.S. Solution*

From App. 10B (also *NCEES Handbook* table "Saturated Steam (U.S. Units)—Temperature Table"), for 300 psia, the enthalpy of saturated liquid, h_f, is 394.0 Btu/lbm. The heat of vaporization, h_{fg}, is 809.4 Btu/lbm. The enthalpy is given by Eq. 10.40 (also *NCEES Handbook:* Properties for Two-Phase (Vapor-Liquid) Systems).

$$h_1 = h_f + xh_{fg} = 394.0\ \frac{\text{Btu}}{\text{lbm}} + (0.95)\left(809.4\ \frac{\text{Btu}}{\text{lbm}}\right) = 1162.9\ \text{Btu/lbm}$$

From the Mollier diagram (also *NCEES Handbook* diagram "Temperature-Entropy (T-S) Diagram (U.S. Customary Units)"), for an isentropic process from 300 psia to 50 psia, $h_2 = 1031$ Btu/lbm.

The change in enthalpy for the isentropic process is

$$h_1 - h_2 = 1162.9\ \frac{\text{Btu}}{\text{lbm}} - 1031\ \frac{\text{Btu}}{\text{lbm}} = 131.9\ \text{Btu/lbm}\quad(130\ \text{Btu/lbm})$$

SI Solution

From App. 10J (also *NCEES Handbook* table "Saturated Steam (SI Units)—Temperature Table"), for 2 MPa, the enthalpy of saturated liquid, h_f, is 908.50 kJ/kg. The heat of vaporization, h_{fg}, is 1889.8 kJ/kg. The enthalpy is given by Eq. 10.40 (also *NCEES Handbook:* Properties for Two-Phase (Vapor-Liquid) Systems).

$$\begin{aligned} h_1 = h_f + x h_{fg} &= 908.50\ \frac{\text{kJ}}{\text{kg}} + (0.95)\left(1889.8\ \frac{\text{kJ}}{\text{kg}}\right) \\ &= 2703.8\ \text{kJ/kg} \end{aligned}$$

From the Mollier diagram (also *NCEES Handbook* diagram "Temperature-Entropy (T-S) Diagram (U.S. Customary Units)," for an isentropic process from 2 MPa to 0.35 MPa, $h_2 = 2405$ kJ/kg.

The change in enthalpy for the isentropic process is

$$\begin{aligned} h_1 - h_2 &= 2703.8\ \frac{\text{kJ}}{\text{kg}} - 2405\ \frac{\text{kJ}}{\text{kg}} \\ &= 298.8\ \text{kJ/kg} \quad (300\ \text{kJ/kg}) \end{aligned}$$

The answer is (B).

10. *Customary U.S. Solution*

Using the first law of thermodynamics, and substituting with the expressions for the change in enthalpy and the change in entropy from *NCEES Handbook*: Ideal Gas Law, work can be calculated as

$$\begin{aligned} w &= q - \Delta h = T_s \Delta s - c_p \Delta T \\ &= T_s\left(c_p \ln\left(\frac{T_2}{T_1}\right) - \frac{R}{MW}\ln\left(\frac{P_2}{P_1}\right)\right) - c_p(T_2 - T_1) \end{aligned}$$

where T_s is the temperature of the surroundings.

The absolute temperature at the inlet of the air heater is

$$T_1 = 540°\text{F} + 460° = 1000°\text{R}$$

The absolute temperature at the outlet of the air heater is

$$T_2 = 1540°\text{F} + 460° = 2000°\text{R}$$

As in *NCEES Handbook:* Temperature-Dependent Properties of Air (U.S. Customary Units), the heat capacity of air is $c_p = 0.24$ Btu/lbm°F. Since 1 degree Fahrenheit is equal to 1 degree Rankine, the heat capacity is the same when put in terms of Rankine. The work without the pressure term is

$$\begin{aligned} w &= T_s c_p \ln\left(\frac{T_2}{T_1}\right) - c_p(T_2 - T_1) \\ &= c_p\left[T_s \ln\left(\frac{T_2}{T_1}\right) - (T_2 - T_1)\right] \\ &= \left(0.24\ \frac{\text{Btu}}{\text{lbm}\times°\text{R}}\right) \\ &\quad \times\left[(100 + 460°\text{R})\ln\left(\frac{2000°\text{R}}{1000°\text{R}}\right) - (2000°\text{R} - 1000°\text{R})\right] \\ &= -146.8\ \text{Btu/lbm} \end{aligned}$$

The work with the pressure term is

$$\begin{aligned} w &= -146.8\ \frac{\text{Btu}}{\text{lbm}} + T_s\left(-\frac{R}{MW}\ln\left(\frac{P_2}{P_1}\right)\right) \\ &= -146.8\ \frac{\text{Btu}}{\text{lbm}} - T_s\left(\frac{R}{MW}\ln\left(\frac{P_2}{P_1}\right)\right) \\ &= -146.8\ \frac{\text{Btu}}{\text{lbm}} - (100 + 460°\text{R}) \\ &\quad \times\left(\frac{1.987\ \frac{\text{Btu}}{\text{lbmol}\times°\text{R}}}{29\ \frac{\text{lbm}}{\text{lbmol}}}\ln\left(\frac{80\ \text{psia}}{100\ \text{psia}}\right)\right) \\ &= -146.8\ \frac{\text{Btu}}{\text{lbm}} - \left(-8.6\ \frac{\text{Btu}}{\text{lbm}}\right) \\ &= -138.2\ \frac{\text{Btu}}{\text{lbm}} \end{aligned}$$

The percentage loss in available energy due to pressure drop is

$$\begin{aligned} \frac{W_{\text{max}} - W_{\text{max,p loss}}}{W_{\text{max}}} \times 100\% &= \left(\frac{-146.8\ \frac{\text{Btu}}{\text{lbm}} - \left(-138.2\ \frac{\text{Btu}}{\text{lbm}}\right)}{-146.8\ \frac{\text{Btu}}{\text{lbm}}}\right) \\ &\quad \times 100\% \\ &= 5.8\% \quad (6\%) \end{aligned}$$

SI Solution

Using the first law of thermodynamics, and substituting with the expressions for the change in enthalpy and the change in entropy from *NCEES Handbook*: Ideal Gas Law, work can be calculated as

$$\begin{aligned} w &= q - \Delta h = T_s \Delta s - c_p \Delta T \\ &= T_s\left(c_p \ln\left(\frac{T_2}{T_1}\right) - \frac{R}{MW}\ln\left(\frac{P_2}{P_1}\right)\right) - c_p(T_2 - T_1) \end{aligned}$$

where T_s is the temperature of the surroundings.

The absolute temperature at the inlet of the air heater is

$$T_1 = 280°\text{C} + 273° = 553\text{K}$$

The absolute temperature at the outlet of the air heater is

$$T_2 = 840°\text{C} + 273° = 1113\text{K}$$

As in *NCEES Handbook:* Temperature-Dependent Properties of Air (U.S. Customary Units), the heat capacity is $c_p = 1.01$ kJ/kg·K. The work without the pressure change is

$$\begin{aligned} w &= T_s c_p \ln\left(\frac{T_2}{T_1}\right) - c_p(T_2 - T_1) \\ &= c_p\left[T_s \ln\left(\frac{T_2}{T_1}\right) - (T_2 - T_1)\right] \\ &= \left(1.01\ \frac{\text{kJ}}{\text{kg}\times\text{K}}\right) \\ &\quad \times\left[(40 + 273\ \text{K})\ln\left(\frac{1100\ \text{K}}{550\ \text{K}}\right) - (1100\ \text{K} - 550\ \text{K})\right] \\ &= -336.4\ \text{kJ/kg} \end{aligned}$$

The work with the pressure change is

$$\begin{aligned} w &= -336.4\ \frac{\text{kJ}}{\text{kg}} + T_s\left(-\frac{R}{MW}\ln\left(\frac{P_2}{P_1}\right)\right) \\ &= -336.4\ \frac{\text{kJ}}{\text{kg}} - T_s\left(\frac{R}{MW}\ln\left(\frac{P_2}{P_1}\right)\right) \\ &= -336.4\ \frac{\text{kJ}}{\text{kg}} - (40 + 273\text{K}) \\ &\quad \times\left(\frac{8.314\ \frac{\text{kJ}}{\text{kmol}\times\text{K}}}{29\ \frac{\text{kg}}{\text{kmol}}}\ln\left(\frac{550\ \text{kPa}}{700\ \text{kPa}}\right)\right) \\ &= -336.4\ \frac{\text{kJ}}{\text{kg}} - \left(-21.6\ \frac{\text{kJ}}{\text{kg}}\right) \\ &= -314.8\ \frac{\text{kJ}}{\text{kg}} \end{aligned}$$

The percentage loss in available energy due to pressure drop is

$$\begin{aligned} \frac{W_{\text{max}} - W_{\text{max, }p\text{ loss}}}{W_{\text{max}}} \times 100\% &= \left(\frac{-336.4\ \frac{\text{kJ}}{\text{kg}} - \left(-314.8\ \frac{\text{kJ}}{\text{kg}}\right)}{-336.4\ \frac{\text{kJ}}{\text{kg}}}\right) \\ &\quad \times 100\% \\ &= 6.42\% \quad (6\%) \end{aligned}$$

The answer is (A).

11. *Customary U.S. Solution*

As in *NCEES Handbook* table "Conversion Table for Temperature Units," the absolute temperature is

$$T = 70°\text{F} + 460° = 530°\text{R}$$

The mass is

$$\begin{aligned} m = \frac{pV}{RT} &= \frac{\left(20\ \frac{\text{lbf}}{\text{in}^2}\right)\left(12\ \frac{\text{in}}{\text{ft}}\right)^2(100\ \text{ft}^3)\left(44\ \frac{\text{lbm}}{\text{lbmol}}\right)}{\left(1545.35\ \frac{\text{ft-lbf}}{\text{lbmol-°R}}\right)(530°\text{R})} \\ &= 15.5\ \text{lbm} \end{aligned}$$

From *NCEES Handbook:* Physical Properties of Liquids and Gases—Temperature-Independent Properties, the critical temperature and pressure of carbon dioxide are

$$\begin{aligned} T_c &= 87.8°\text{F} + 460° = 547.8°\text{R} \\ p_c &= 1070\ \text{lbf/in}^2 \end{aligned}$$

Using *NCEES Handbook:* Compressibility and the Theorem of Corresponding States, the reduced temperature and pressure of carbon dioxide are

$$T_r = \frac{T}{T_c} = \frac{530°\text{R}}{547.8°\text{R}} = 0.97$$

$$p_r = \frac{p}{p_c} = \frac{3800\ \frac{\text{lbf}}{\text{in}^2}}{1070\ \frac{\text{lbf}}{\text{in}^2}} = 3.55$$

From the compressibility chart, $Z = 0.45$.

The final mass is

$$m = \frac{pV}{ZRT} = \frac{\left(3800\ \frac{\text{lbf}}{\text{in}^2}\right)\left(12\ \frac{\text{in}}{\text{ft}}\right)^2(100\ \text{ft}^3)\left(44\ \frac{\text{lbm}}{\text{lbmol}}\right)}{(0.45)\left(1545.35\ \frac{\text{ft-lbf}}{\text{lbmol-°R}}\right)(530\text{°R})}$$
$$= 6532.5\ \text{lbm} \quad (6530\ \text{lbm})$$

The average mass flow rate of carbon dioxide is

$$\dot{m} = \frac{6530\ \text{lbm} - 15.5\ \text{lbm}}{1\ \text{hr}} = 6517\ \text{lbm/hr} \quad (6500\ \text{lbm/hr})$$

SI Solution

As in *NCEES Handbook* table "Conversion Table for Temperature Units," the absolute temperature is

$$T = 21\text{°C} + 273° = 294\text{K}$$

The mass is

$$m = \frac{pV}{RT} = \frac{(150\ \text{kPa})(3\ \text{m}^3)\left(44\ \frac{\text{g}}{\text{mol}}\right)\left(\frac{10^3\ \text{Pa/kPa}}{\text{kPa}}\right)}{\left(8314.5\ \frac{\text{J}}{\text{kmol·K}}\right)(294\text{K})}$$
$$= 8.1\ \text{kg}$$

As in *NCEES Handbook:* Physical Properties of Liquids and Gases—Temperature-Independent Properties, the critical temperature and pressure of carbon dioxide are

$$T_c = 31\text{°C} + 273° = 304\text{K}$$

$$p_c = 7.38\ \text{MPa}$$

Using *NCEES Handbook:* Compressibility and the Theorem of Corresponding States, the reduced temperature and pressure of carbon dioxide are

$$T_r = \frac{T}{T_c} = \frac{294\text{K}}{304\text{K}} = 0.97$$

$$p_r = \frac{p}{p_c} = \frac{25\ \text{MPa}}{7.38\ \text{MPa}} = 3.39$$

From the compressibility chart, $Z = 0.45$.

The final mass is

$$m = \frac{pV}{ZRT} = \frac{(25\ \text{MPa})(3\ \text{m}^3)\left(44\ \frac{\text{g}}{\text{mol}}\right)\left(10^6\ \text{Pa/kPa}\right)}{(0.45)\left(8314.5\ \frac{\text{J}}{\text{kmol·K}}\right)(294\text{K})}$$
$$= 3000\ \text{kg}$$

The average mass flow rate of carbon dioxide is

$$\dot{m} = \frac{3000\ \frac{\text{kg}}{\text{hr}} - 8.1\ \frac{\text{kg}}{\text{hr}}}{3600\ \frac{\text{sec}}{\text{hr}}} = 0.8\ \text{kg/s}$$

The answer is (A).

12. Choose the control volume to include the air outside the tank that is pushed into the tank (subscript "*e*" for "entering"), as well as the tank volume.

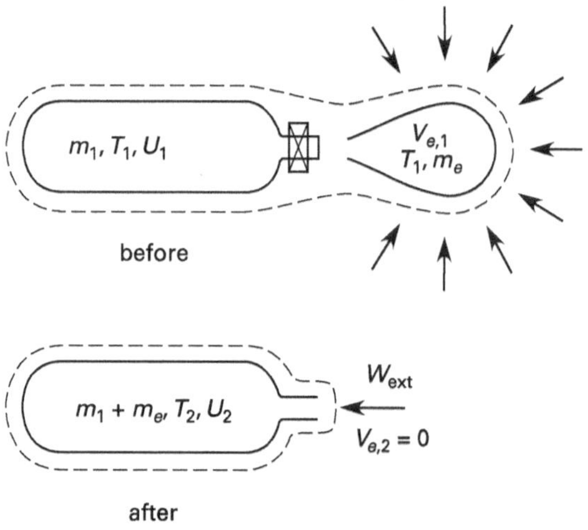

Customary U.S. Solution

As in the conversion table for temperature units, the absolute temperature of the air in the tank when evacuated is

$$T_1 = 70\text{°F} + 460° = 530\text{°R}$$

Using ideal gas law, the mass, m_1, is

$$m_1 = \frac{P_1 V(\text{MW})}{RT_1}$$
$$= \frac{\left(1\ \frac{\text{lbf}}{\text{in}^2}\right)\left(12\ \frac{\text{in}}{\text{ft}}\right)^2(20\ \text{ft}^3)\left(29\ \frac{\text{lbm}}{\text{lbmol}}\right)}{\left(1545\ \frac{\text{ft-lbf}}{\text{lbmol-°R}}\right)(530\text{°R})}$$
$$= 0.102\ \text{lbm}$$

The final system temperature, T_2, 300°F. The absolute temperature is

$$T_2 = 300\text{°F} + 460° = 760\text{°R}$$

The final system mass of air, m_2, is

$$m_2 = m_1 + m_e = \frac{P_2 V(\text{MW})}{RT_2}$$

$$= \frac{\left(14.7\ \frac{\text{lbf}}{\text{in}^2}\right)\left(12\ \frac{\text{in}}{\text{ft}}\right)^2 (20\ \text{ft}^3)\left(29\ \frac{\text{lbm}}{\text{lbmol}}\right)}{\left(1545\ \frac{\text{ft-lbf}}{\text{lbmol-°R}}\right)(760\text{°R})}$$

$$= 1.046\ \text{lbm}$$

$$m_1 + m_e \approx 1.046\ \text{lbm}$$

$$m_e \approx 1.046\ \text{lbm} - m_1 = 1.046\ \text{lbm} - 0.102\ \text{lbm} = 0.944\ \text{lbm}$$

Using the ideal gas law, the initial volume of external air entering the system, $V_{e,1}$, is

$$V_{e,1} = \frac{m_2 R T_1}{MWP_{\text{air}}} = \frac{(0.944\ \text{lbm})\left(1545\ \frac{\text{ft-lbf}}{\text{lbmol-°R}}\right)(530\text{°R})}{\left(29\ \frac{\text{lbm}}{\text{lbmol}}\right)\left(14.7\ \frac{\text{lbf}}{\text{in}^2}\right)\left(12\ \frac{\text{in}}{\text{ft}}\right)^2}$$

$$= 12.59\ \text{ft}^3$$

From *NCEES Handbook*: Closed System Changes, the isobaric (constant pressure), closed system total work is

$$W_{\text{ext}} = P(V_{e,2} - V_{e,1})$$

$$= \frac{\left(14.7\ \frac{\text{lbf}}{\text{in}^2}\right)\left(12\ \frac{\text{in}}{\text{ft}}\right)^2 (0\ \text{ft} - 12.59\ \text{ft}^3)}{778.21\ \frac{\text{ft-lbf}}{\text{Btu}}}$$

$$= -34.25\ \text{Btu} \quad (-34\ \text{Btu}) \quad \left[\begin{array}{c}\text{surroundings do}\\ \text{work on the system}\end{array}\right]$$

The answer is (B).

SI Solution

As in *NCEES Handbook:* Temperature, the absolute temperature of the air in the tank when evacuated is

$$T_1 = 21\text{°C} + 273° = 294\text{K}$$

As in *NCEES Handbook:* Ideal Gas Law,

$$m_1 = \frac{P_1 V(\text{MW})}{RT_1} = \frac{(7000\ \text{Pa})(0.60\ \text{m}^3)\left(29\ \frac{\text{g}}{\text{mol}}\right)}{\left(8.314\ \frac{\text{m}^3\cdot\text{Pa}}{\text{mol}\cdot\text{K}}\right)(294\ \text{K})}$$

$$= 49.8\ g = 0.0498\ \text{kg}$$

The final system temperature $T_2 = 149\text{°C}$. The absolute temperature is

$$T_2 = 149\text{°C} + 273° = 422\ \text{K}$$

The final system mass of air, m_2, is

$$m_2 = m_1 + m_e = \frac{P_2 V(\text{MW})}{RT_2}$$

$$= \frac{(101.3\ \text{kPa})\left(1000\ \frac{\text{Pa}}{\text{kPa}}\right)(0.60\ \text{m}^3)\left(29\ \frac{\text{g}}{\text{mol}}\right)}{\left(8.314\ \frac{\text{m}^3\cdot\text{Pa}}{\text{mol}\cdot\text{K}}\right)(422\ \text{K})}$$

$$= 502.39\ \text{g}$$

$$m_1 + m_e \approx 0.5024\ \text{kg}$$

$$m_e \approx 0.5024\ \text{kg} - m_1 = 0.5024\ \text{kg} - 0.0498\ \text{kg} = 0.4526\ \text{kg}$$

Using the ideal gas law, the initial volume of external air entering the system, $V_{e,1}$, is

$$V_{e,1} = \frac{m_2 R T_1}{(\text{MW})P_{\text{air}}}$$

$$= \frac{(0.4526\ \text{kg})\left(1000\ \frac{\text{g}}{\text{kg}}\right)\left(8.314\ \frac{\text{m}^3\cdot\text{Pa}}{\text{mol}\cdot\text{K}}\right)(294\ \text{K})}{\left(29\ \frac{\text{g}}{\text{mol}}\right)(101.3\ \text{kPa})\left(1000\ \frac{\text{Pa}}{\text{kPa}}\right)}$$

$$= 0.3766\ \text{m}^3$$

From *NCEES Handbook*: Closed System Changes, the isobaric (constant pressure), closed system total work is

$$W_{\text{ext}} = P(V_{e,2} - V_{e,1})$$

$$= (101.3\ \text{kPa})\left(1000\ \frac{\text{Pa}}{\text{kPa}}\right)(0\ \text{m}^3 - 0.3766\ \text{m}^3)$$

$$= -38\,150\ \text{J} \quad (-38\,000\ \text{J}) \quad \left[\begin{array}{c}\text{surroundings do}\\ \text{work on the system}\end{array}\right]$$

The answer is (B).

Thermodynamics

12 Vapor Power Equipment

PRACTICE PROBLEMS

1. "BLEVE" is an acronym that is used to describe

(A) a boiler demand leveling valve

(B) a catastrophic boiler explosion

(C) the vapor from a bleed valve

(D) a burnout caused by insufficient feedwater

2. A small power generating plant has two furnace boilers and two turbine generators. During long periods of low electrical demand, capacity is achieved with a single boiler. To achieve a functional cross-coupled configuration, which valves should be opened and which should be closed?

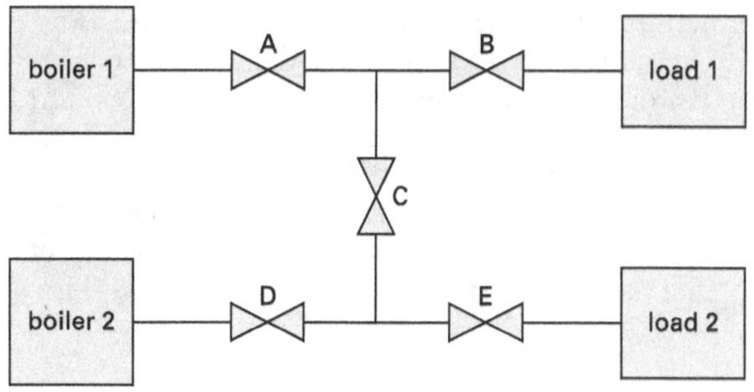

(A) open: A, B, D, and E
closed: C

(B) open: A, B, C, and D
closed: E

(C) open: A, B, C, and E
closed: D

(D) open: A, C, and E
closed: B, D

3. A steam generator receives liquid water at 80°F and produces saturated steam at 300 psia. Most nearly, what is the ratio of boiler sensible to boiler latent heat supplied to the water and steam?

(A) 0.06

(B) 0.12

(C) 0.27

(D) 0.43

4. What is most nearly the isentropic efficiency of a process that expands dry steam from 100 psia to 3 psia (700 kPa to 20 kPa) and 90% quality?

(A) 40%

(B) 50%

(C) 60%

(D) 70%

5. A 5000 kW steam turbine uses 200 psia (1.4 MPa) steam with 100°F (40°C) of superheat. The condenser is at 1 in Hg (3.4 kPa) absolute. Assuming an ideal turbine, the water rate per kilowatt of generated power at full load is most nearly

(A) 4.7 lbm/kW-hr (0.0006 kg/kW·s)

(B) 8.8 lbm/kW-hr (0.0012 kg/kW·s)

(C) 25 lbm/kW-hr (0.0031 kg/kW·s)

(D) 37 lbm/kW-hr (0.0046 kg/kW·s)

6. A 10,000 kW steam turbine operates on 400 psia (3.0 MPa), 750°F (420°C) dry steam, expanding to 2 in Hg (6.8 kPa) absolute. The maximum adiabatic heat drop available for power production is most nearly

(A) 450 Btu/lbm (1100 kJ/kg)

(B) 600 Btu/lbm (1400 kJ/kg)

(C) 750 Btu/lbm (1800 kJ/kg)

(D) 900 Btu/lbm (2200 kJ/kg)

7. A 750 kW steam turbine has a water rate of 20 lbm/kW-hr (2.5×10^{-3} kg/kW·s). Steam with 50°F (30°C) of superheat is expanded from 165 psia (1.0 MPa absolute) to 26 in Hg (90 kPa) absolute. 65°F (18°C) cooling water is available. The terminal temperature difference is zero. The quantity of cooling water required is most nearly

(A) 45,000 lbm/hr (5.9 kg/s)

(B) 60,000 lbm/hr (7.8 kg/s)

(C) 80,000 lbm/hr (10 kg/s)

(D) 95,000 lbm/hr (12 kg/s)

8. 332,000 lbm/hr (41.8 kg/s) of 81°F (27°C) water enters a two-pass, counterflow heat exchanger. The heat exchanger is constructed of 1850 ft² (172 m²) of ⅝ in (15.9 mm) copper tubing. Saturated steam is bled from a turbine at 4.45 psia (30.6 kPa) and condenses to saturated liquid. The heated water leaves at 150°F (65.6° C) with an enthalpy of 1100 Btu/lbm (2.56 MJ/kg). The overall heat transfer coefficient is most nearly

(A) 1900 Btu/hr-ft²-°F (11 kW/m²·°C)

(B) 4400 Btu/hr-ft²-°F (25 kW/m²·°C)

(C) 6500 Btu/hr-ft²-°F (37 kW/m²·°C)

(D) 7700 Btu/hr-ft²-°F (44 kW/m²·°C)

9. A two-pass surface condenser constructed of 1 in (25.4 mm) BWG tubing receives 82,000 lbm/hr (10.3 kg/s) of steam from a turbine. Steam enters the condenser with an enthalpy of 980 Btu/lbm (2.280 MJ/kg) The condenser operates at a pressure of 1 in Hg (3.4 kPa) absolute. Water is circulated at 8 ft/sec (2.4 m/s) through an equivalent length of 120 ft (36 m) of extra strong 30 in (76.2 cm) steel pipe. An additional head loss of 6 in wg (1.5 kPa) is incurred in the intake screens. If the water temperature increases 10°F (5.6°C) across the condenser, what is most nearly the circulation rate of cooling water?

(A) 9000 gal/min (34 kL/min)

(B) 11,000 gal/min (42 kL/min)

(C) 13,000 gal/min (49 kL/min)

(D) 15,000 gal/min (57 kL/min)

10. 100 lbm/hr (0.013 kg/s) of 60°F (16°C) water is turned into 14.7 psia (101.3 kPa) saturated steam in an electric boiler. Radiation losses are 35% of the supplied energy. If electricity is \$0.04 per kW-hr, the cost is most nearly

(A) \$2/hr

(B) \$3/hr

(C) \$4/hr

(D) \$5/hr

11. A gas burner produces 250 lbm/hr (0.032 kg/s) of 98% dry steam at 40 psia (300 kPa) from 60°F (16°C) feedwater. The fuel gas enters at 80°F (26°C) and 4 in Hg (13.6 kPa) and has a heating value of 550 Btu/ft³ (20.5 MJ/m³) at standard industrial conditions. The barometric pressure is 30.2 in Hg (102.4 kPa). 13.5 ft³/min (6.4 L/s) of fuel gas is consumed. The efficiency of the boiler is most nearly

(A) 37%

(B) 43%

(C) 57%

(D) 66%

12. A boiler evaporates 8.23 lbm (8.23 kg) of 120°F (50°C) water per pound (per kilogram) of coal fired, producing 100 psia (700 kPa) saturated steam. The coal is 2% moisture by weight as fired, and dry coal is 5% ash. 1% of the coal is removed from the ash pit. (The ash pit loss has the same composition as unfired, dry coal.) Coal is initially at 60°F (16°C), and combustion occurs at 14.7 psia (101.3 kPa). The combustion products leave at 600°F (315°C). The heating value of the dry coal is 12,800 Btu/lbm (29.80 MJ/kg). The efficiency of the boiler is most nearly

(A) 53%

(B) 68%

(C) 73%

(D) 82%

13. 500 psia (3.5 MPa) steam is superheated to 1000°F (500°C) before expanding through a 75% efficient turbine to 5 psia (30 kPa). No subcooling occurs. The pump work is negligible compared to the 200 MW generated. The quantity of steam required is most nearly

(A) 8.6×10^5 lbm/hr (120 kg/s)

(B) 1.4×10^6 lbm/hr (190 kg/s)

(C) 2.0×10^6 lbm/hr (270 kg/s)

(D) 4.2×10^6 lbm/hr (310 kg/s)

14. 191,000 lbm/hr (24 kg/s) of 635°F (335°C) combustion gases flow through a 20 ft (6 m) wide boiler stack whose front and back plates are 5 ft 10 in (1.78 m) apart. An integral crossflow economizer is being designed to heat water from 212°F to 285°F (100°C to 140°C) by dropping the stack gas temperature to 470°F (240°C). Layers of 24 tubes with dimensions 0.957 in (24.3 mm) ID, 1.315 in (33.4 mm) OD, and 20 ft (6 m) length will be placed on a 2.315 in (58.8 mm) pitch in horizontal banks. The overall coefficient of heat transfer for the tubes is 10 Btu/hr-ft^2-°F (57 W/m^2·°C). Assuming the heat capacity of the combustion gas can be approximated by that of Nitrogen, how many 24-tube layers are required?

(A) 5

(B) 9

(C) 12

(D) 17

15. Water is used in an adiabatic steam desuperheater. 1000 lbm/hr (0.13 kg/s) of 200 psia (1.5 MPa), 600°F (300°C) steam enters with negligible velocity. 50 lbm/hr (0.0063 kg/s) of 82°F (28°C) water enters with negligible velocity. 100 psia (700 kPa) steam leaves the desuperheater at 2000 ft/sec (600 m/s). The temperature of the leaving steam is most nearly

(A) 330°F (165°C)

(B) 360°F (180°C)

(C) 400°F (200°C)

(D) 470°F (240°C)

16. Waste steam at 400°F (200°C) and 100 psia (700 kPa) was originally used only for heating cold water. Cold water entered at 70°F (21°C) and 60 psia (400 kPa). 2000 lbm/hr (0.25 kg/s) of hot water at 180°F (80°C) and 20 psia (150 kPa) were produced. Now, the same quantity of steam is to be expanded through a low-pressure turbine. The low-pressure turbine has an isentropic efficiency of 60% and a mechanical efficiency of 96%. The steam will then flow through a mixing heater (see diagram). A pressure drop of 5 psi (30 kPa) occurs through the heater. The heater output must remain at 180°F (80°C) and 20 psia (150 kPa), but the output at point F may decrease.

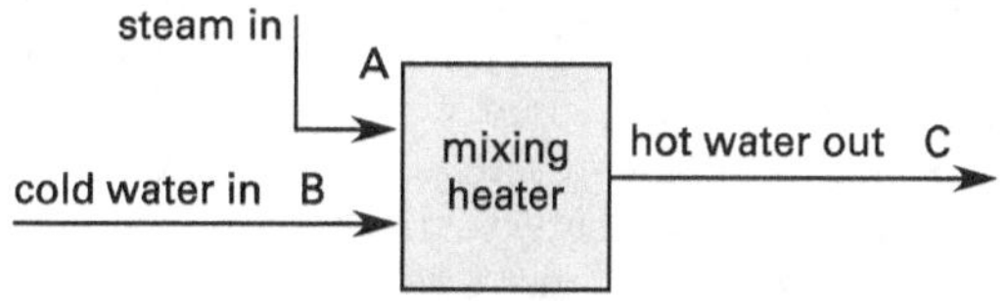

original use

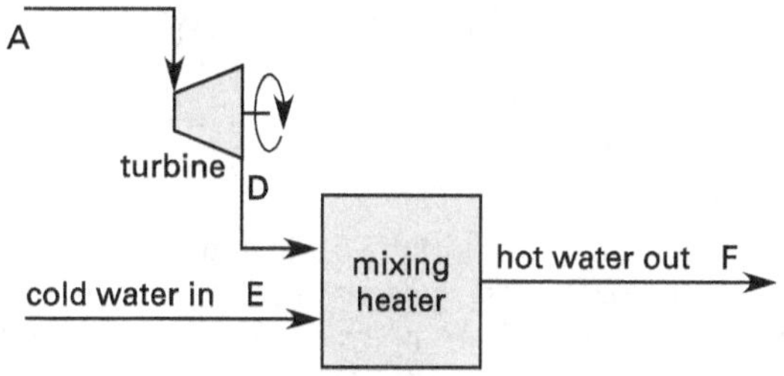

proposed use of waste steam

The steam flow is most nearly

(A) 160 lbm/hr (0.019 kg/s)

(B) 180 lbm/hr (0.022 kg/s)

(C) 210 lbm/hr (0.024 kg/s)

(D) 250 lbm/hr (0.029 kg/s)

SOLUTIONS

1. "BLEVE" is the acronym for "boiling liquid-expanding vapor explosion," essentially the catastrophic failure of a container whose walls cannot withstand the pressure of vaporization. BLEVEs are caused when a leak allows the pressurized liquid contents to flash into a vapor. The term is most frequently used to describe explosions of tanks containing flammable liquids, but it is equally applicable to boilers. Common elements of BLEVE events are weakening of tank walls (usually by fire) and inadequate or nonfunctional pressure-relief valves.

The answer is (B).

2. Cross-coupling provides flexibility integrating redundant, back-up, excess-capacity, and auxiliary units. Cross-coupling permits any available source to be connected to any load, facilitating operation when units are idle or down for maintenance. In addition, when both all loads are online, cross-coupling provides smoother transitions and faster response to fluctuating demand. With a two-two system (i.e., two sources and two loads) in cross-coupled mode, both sources are connected to both loads during periods of high demand, but only one source is connected to either one load or the other during periods of low demand.

It is unlikely (unnecessary) that both sources would be connected to a single load. Also, it is unlikely that a single source would be sufficient for two loads. Operation with all cross-control valves closed is not a cross-controlled configuration.

The answer is (D).

3. Sensible heat changes temperature, while latent heat changes phase. The pressure throughout the steam generator is 300 psia. Water enters subcooled at 80°F and 300 psia. The water remains in liquid form as it is heated (sensible heat) to the saturation temperature corresponding to 300 psia, which (from App. 10B and *NCEES Handbook:* Properties of Saturated Steam (U.S. Customary Units)) is 417.35°F. Additional energy (latent heat) converts the saturated water to saturated steam at 300 psia.

The enthalpy of 80°F liquid water is essentially a function of temperature only. From App. 10A (also *NCEES Handbook:* Properties of Saturated Steam (U.S. Customary Units)), $h_{80°F} = 48.07$ Btu/lbm. From App. 10B (also *NCEES Handbook:* Properties of Saturated Steam (U.S. Customary Units)), $h_{f,300\text{ psia}} = 394.0$ Btu/lbm, and $h_{fg,300\text{ psia}} = 809.4$ Btu/lbm.

The ratio of sensible heat to latent heat is

$$\frac{q_{\text{sensible}}}{q_{\text{latent}}} = \frac{h_{f,300\text{ psia}} - h_{80°\text{F}}}{h_{fg,300\text{ psia}}} = \frac{394.0\ \frac{\text{Btu}}{\text{lbm}} - 48.07\ \frac{\text{Btu}}{\text{lbm}}}{809.4\ \frac{\text{Btu}}{\text{lbm}}}$$
$$= 0.427 \quad (0.43)$$

The answer is (D).

4. *Customary U.S. Solution*

From App. 10B (also *NCEES Handbook:* Properties of Saturated Steam (U.S. Customary Units)), for 100 psia, the enthalpy of dry steam is $h_1 = 1187.2$ Btu/lbm.

From *NCEES Handbook* figure "Temperature-Entropy (T-S) Diagram (U.S. Customary Units)," for an isentropic process from 100 psia to 3 psia, $h_2 = 950$ Btu/lbm.

From App. 10B (also *NCEES Handbook:* Properties of Saturated Steam (U.S. Customary Units)), for 3 psia,

$$h_f = 109.4 \text{ Btu/lbm}$$
$$h_g = 1122.6 \text{ Btu/lbm}$$
$$h_{fg} = 1122.6\ \frac{\text{Btu}}{\text{lbm}} - 109.4\ \frac{\text{Btu}}{\text{lbm}} = 1013.2 \text{ Btu/lbm}$$
$$h_2' = h_f + xh_{fg}$$
$$= 109.4\ \frac{\text{Btu}}{\text{lbm}} + (0.9)\left(1013.2\ \frac{\text{Btu}}{\text{lbm}}\right)$$
$$= 1021.3 \text{ Btu/lbm}$$

From Eq. 12.17 (also *NCEES Handbook:* Open Thermodynamic Systems), the isentropic efficiency is

$$\eta_s = \frac{h_1 - h_2'}{h_1 - h_2} = \frac{1187.2\ \frac{\text{Btu}}{\text{lbm}} - 1021.3\ \frac{\text{Btu}}{\text{lbm}}}{1187.2\ \frac{\text{Btu}}{\text{lbm}} - 950\ \frac{\text{Btu}}{\text{lbm}}}$$
$$= 0.699 \quad (70\%)$$

SI Solution

From App. 10J (also *NCEES Handbook:* Saturated Steam (SI Units)), for 700 kPa, the enthalpy of dry steam is $h_1 = 2763.5$ kJ/kg.

From *NCEES Handbook* figure "Temperature-Entropy (T-S) Diagram (SI Units)," for an isentropic process from 700 kPa to 20 kPa, $h_2 = 2200$ kJ/kg.

From App. 10J (also *NCEES Handbook:* Saturated Steam (SI Units)), for 20 kPa,

$$h_f = 251.4 \text{ kJ/kg}$$
$$h_g = 2609.7 \text{ kJ/kg}$$
$$h_{fg} = 2609.7 \ \frac{\text{kJ}}{\text{kg}} - 251.4 \ \frac{\text{kJ}}{\text{kg}} = 2358.3 \text{ kJ/kg}$$
$$h_2' = h_f + xh_{fg}$$
$$= 251.4 \ \frac{\text{kJ}}{\text{kg}} + (0.9)\left(2358.3 \ \frac{\text{kJ}}{\text{kg}}\right)$$
$$= 2373.9 \text{ kJ/kg}$$

From Eq. 12.17 (also *NCEES Handbook:* Open Thermodynamic Systems), the isentropic efficiency is

$$\eta_s = \frac{h_1 - h_2'}{h_1 - h_2} = \frac{2763.5 \ \frac{\text{kJ}}{\text{kg}} - 2373.9 \ \frac{\text{kJ}}{\text{kg}}}{2763.5 \ \frac{\text{kJ}}{\text{kg}} - 2200 \ \frac{\text{kJ}}{\text{kg}}}$$
$$= 0.691 \quad (70\%)$$

The answer is (D).

5. *Customary U.S. Solution*

From App. 10A (also *NCEES Handbook:* Properties of Saturated Steam (U.S. Customary Units)), T_{sat} for 200 psia is 381.80°F.

The steam temperature is

$$381.80°\text{F} + 100°\text{F} = 481.80°\text{F}$$

Interpolate from App. 10C (also *NCEES Handbook:* Superheated Steam (U.S. Customary Units)), $h_1 =$ 1257.8 Btu/lbm.

1 in Hg is approximately 0.5 psia. From *NCEES Handbook* figure "Temperature-Entropy (T-S) Diagram (U.S. Customary Units)," assuming isentropic expansion and dropping straight down to the 0.5 psia line, $h_2 \approx 870$ Btu/lbm.

For isentropic expansion, $\eta_{\text{turbine}} = 1$, and the steam mass flow rate through the turbine is given by Eq. 12.22 (also *NCEES Handbook:* Turbines).

$$\dot{m} = \frac{P_{\text{turbine}}}{h_1 - h_2} = \frac{(5000 \text{ kW})\left(3412 \ \frac{\text{Btu}}{\text{kW-hr}}\right)}{1257.8 \ \frac{\text{Btu}}{\text{lbm}} - 870 \ \frac{\text{Btu}}{\text{lbm}}}$$
$$= 4.399 \times 10^4 \text{ lbm/hr}$$

The water rate per kilowatt of generated power at full load is

$$\text{WR} = \frac{\dot{m}}{P_{\text{turbine}}} = \frac{4.399 \times 10^4 \ \frac{\text{lbm}}{\text{hr}}}{5000 \text{ kW}}$$
$$= 8.798 \text{ lbm/kW-hr} \quad (8.8 \text{ lbm/kW-hr})$$

SI Solution

From App. 10J (also *NCEES Handbook:* Saturated Steam (SI Units)—Temperature Table), T_{sat} for 1.4 MPa is 195.1°C.

The steam temperature is

$$195.1°\text{C} + 40°\text{C} = 235.1°\text{C}$$

Interpolate from App. 10K (also *NCEES Handbook:* Superheated Steam (SI Units)), $h_1 = 2890$ kJ/kg, and $s_1 = 6.6724$ kJ/kg·K.

Interpolate from App. 10I (also *NCEES Handbook:* Saturated Steam (SI Units)—Temperature Table), for 3.4 kPa, the entropy of saturated liquid, s_f, the entropy of saturated vapor, s_g, the enthalpy of saturated liquid, h_f, and the enthalpy of vaporization, h_g, are

$$s_f = 0.3817 \text{ kJ/kg·K}$$
$$s_g = 8.5364 \text{ kJ/kg·K}$$
$$h_f = 109.48 \text{ kJ/kg}$$
$$h_g = 2549.06 \text{ kJ/kg}$$

For isentropic expansion, $s_1 = s_2 = 6.6724$ kJ/kg·K.

Since $s_2 < s_g$, the expanded steam is in the liquid-vapor region. The quality of the mixture is given by Eq. 10.41 (also *NCEES Handbook:* Properties for Two-Phase (Vapor-Liquid) Systems).

$$x = \frac{s - s_f}{s_{fg}} = \frac{s - s_f}{s_g - s_f}$$
$$= \frac{6.6724 \ \frac{\text{kJ}}{\text{kg·K}} - 0.3817 \ \frac{\text{kJ}}{\text{kg·K}}}{8.5364 \ \frac{\text{kJ}}{\text{kg·K}} - 0.3817 \ \frac{\text{kJ}}{\text{kg·K}}}$$
$$= 0.7714$$

The final enthalpy is given by Eq. 10.40 (also *NCEES Handbook:* Properties for Two-Phase (Vapor-Liquid) Systems).

$$h = h_f + xh_{fg} = 109.48 \ \frac{\text{kJ}}{\text{kg}} + (0.7714)\left(2549.06 \ \frac{\text{kJ}}{\text{kg}}\right)$$
$$= 2075.8 \text{ kJ/kg}$$

For isentropic expansion, $\eta_{turbine} = 1$, and the steam mass flow rate through a turbine is given by Eq. 12.22 (also *NCEES Handbook:* Turbines).

$$\dot{m} = \frac{P_{turbine}}{h_1 - h_2} = \frac{5000 \text{ kW}}{2890 \ \frac{\text{kJ}}{\text{kg}} - 2075.8 \ \frac{\text{kJ}}{\text{kg}}}$$
$$= 6.141 \text{ kg/s}$$

The water rate per kilowatt of generated power at full load is

$$\text{WR} = \frac{\dot{m}}{P_{turbine}} = \frac{6.141 \ \frac{\text{kg}}{\text{s}}}{5000 \text{ kW}}$$
$$= 1.23 \times 10^{-3} \text{ kg/kW·s} \quad (0.0012 \text{ kg/kW·s})$$

The answer is (B).

6. *Customary U.S. Solution*

From App. 10C (also *NCEES Handbook:* Superheated Steam (U.S. Customary Units)), the enthalpy, h_1, of dry steam at 400 psia and 750°F is $h_1 = 1389.9$ Btu/lbm.

From *NCEES Handbook* figure "Temperature-Entropy (T-S) Diagram (U.S. Customary Units)," assuming isentropic expansion to 2 in Hg (about 1 psia), $h_2 = 935$ Btu/lbm.

The maximum adiabatic heat drop is

$$h_1 - h_2 = 1389.9 \ \frac{\text{Btu}}{\text{lbm}} - 935 \ \frac{\text{Btu}}{\text{lbm}}$$
$$= 454.9 \text{ Btu/lbm} \quad (450 \text{ Btu/lbm})$$

SI Solution

From App. 10K (also *NCEES Handbook:* Superheated Steam (SI Units)), the enthalpy, h_1, of dry steam at 3.0 MPa and 420°C is $h_1 = 3277.1$ kJ/kg.

From *NCEES Handbook* figure "Temperature-Entropy (T-S) Diagram (SI Units)," assuming isentropic expansion, $h_2 \approx 2210$ kJ/kg.

The maximum adiabatic heat drop is

$$h_1 - h_2 = 3277.1 \ \frac{\text{kJ}}{\text{kg}} - 2210 \ \frac{\text{kJ}}{\text{kg}}$$
$$= 1067.1 \text{ kJ/kg} \quad (1100 \text{ kJ/kg})$$

The answer is (A).

7. *Customary U.S. Solution*

The water rate is

$$\text{WR} = \frac{\dot{m}}{P_{turbine}}$$

The steam flow rate is

$$\dot{m}_{steam} = (\text{WR})P_{turbine}$$
$$= \left(20 \ \frac{\text{lbm}}{\text{kW-hr}}\right)(750 \text{ kW})$$
$$= 15{,}000 \text{ lbm/hr}$$

From App. 10B, (also *NCEES Handbook:* Properties of Saturated Steam (U.S. Customary Units)), for 165 psia, $T_{sat} = 366$°F.

The steam temperature is

$$366°\text{F} + 50°\text{F} = 416°\text{F}$$

From App. 10C (also *NCEES Handbook:* Superheated Steam (U.S. Customary Units)), $h_1 = 1226.0$ Btu/lbm.

From Eq. 12.22 (also *NCEES Handbook:* Turbines),

$$P_{turbine} = \dot{m}(h_1 - h_2)$$

$$h_1 - h_2 = \frac{P_{turbine}}{\dot{m}} = \frac{(750 \text{ kW})\left(3412 \ \frac{\text{Btu}}{\text{kW-hr}}\right)}{15{,}000 \ \frac{\text{lbm}}{\text{hr}}}$$
$$= 170.6 \text{ Btu/lbm}$$
$$h_2 = h_1 - 170.6 \ \frac{\text{Btu}}{\text{lbm}} = 1226.0 \ \frac{\text{Btu}}{\text{lbm}} - 170.6 \ \frac{\text{Btu}}{\text{lbm}}$$
$$= 1055.4 \text{ Btu/lbm}$$
$$p_2 = (26 \text{ in Hg})\left(0.491 \ \frac{\text{lbf}}{\text{in}^2}\right)$$
$$= 12.77 \text{ psia}$$

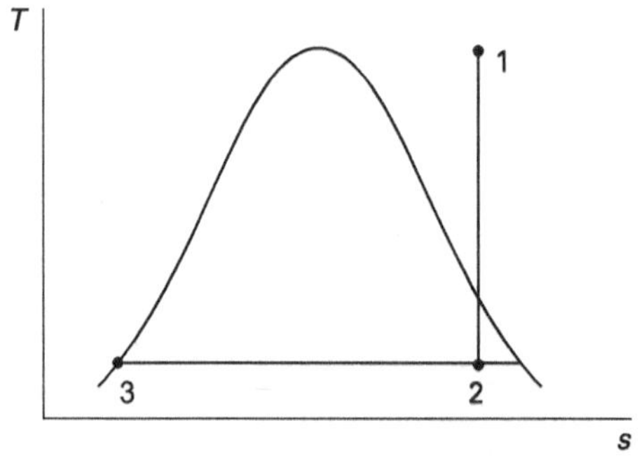

From App. 10B (also *NCEES Handbook:* Properties of Saturated Steam (U.S. Customary Units)), for $p_2 = 12.77$ psia,

$$T_{\text{sat}} = T_2 = T_3 \approx 204°\text{F}$$
$$h_{f,3} \approx 172.3 \text{ Btu/lbm}$$

From App. 18A (also *NCEES Handbook:* Physical Properties of Water), the specific heat of water at 65°F is $c_{p,\text{water}} = 0.999$ Btu.

Assuming the water and steam leave in thermal equilibrium, the heat lost by steam is equal to the heat gained by water.

$$\dot{m}_{\text{water}} c_{p,\text{water}}(T_{\text{water,out}} - T_{\text{water,in}}) = \dot{m}_{\text{steam}}(h_2 - h_{f,3})$$

Since the terminal temperature difference is zero, $T_{\text{water,out}} = T_3$.

$$\dot{m}_{\text{water}}\left(0.999 \frac{\text{Btu}}{\text{lbm-°F}}\right)(204°\text{F} - 65°\text{F}) = \left(15{,}000 \frac{\text{lbm}}{\text{hr}}\right)\left(1055.3 \frac{\text{Btu}}{\text{lbm}} - 172.3 \frac{\text{Btu}}{\text{lbm}}\right)$$

$$\dot{m}_{\text{water}} = 9.54 \times 10^4 \text{ lbm/hr} \quad (95{,}000 \text{ lbm/hr})$$

SI Solution

The steam flow rate is

$$\begin{aligned} \dot{m}_{\text{water}} &= (\text{WR})P_{\text{turbine}} \\ &= \left(2.5 \times 10^{-3} \frac{\text{kg}}{\text{kW·s}}\right)(750 \text{ kW}) \\ &= 1.875 \text{ kg/s} \end{aligned}$$

From App. 10J (also *NCEES Handbook:* Saturated Steam (SI Units)), for 1 MPa, $T_{\text{sat}} = 179.9°\text{C}$.

The steam temperature is

$$179.9°\text{C} + 30°\text{C} = 209.9°\text{C}$$

From App. 10K (also *NCEES Handbook:* Superheated Steam (SI Units)), $h_1 = 2851.0$ kJ/kg.

From Eq. 12.22 (also *NCEES Handbook:* Turbines),

$$\begin{aligned} P_{\text{turbine}} &= \dot{m}(h_1 - h_2) \\ h_1 - h_2 &= \frac{P_{\text{turbine}}}{\dot{m}} = \frac{750 \text{ kW}}{1.875 \frac{\text{kg}}{\text{s}}} \\ &= 400 \text{ kJ/kg} \\ h_2 &= h_1 - 400 \frac{\text{kJ}}{\text{kg}} = 2851.0 \frac{\text{kJ}}{\text{kg}} - 400 \frac{\text{kJ}}{\text{kg}} \\ &= 2451.0 \text{ kJ/kg} \end{aligned}$$

From App. 10J (also *NCEES Handbook:* Saturated Steam (SI Units)), for $p_2 = 90$ kPa,

$$T_{\text{sat}} = T_2 = T_3 = 96.69°\text{C}$$
$$h_{f,3} = 405.20 \text{ kJ/kg}$$

From App. 18B (also *NCEES Handbook:* Physical Properties of Water), the specific heat of water at 18°C is $c_p = 4.186$ kJ/kg·K.

Assuming water and steam leave in thermal equilibrium, the heat lost by steam is equal to the heat gained by water.

$$\dot{m}_{\text{water}} c_{p,\text{water}}(T_{\text{water,out}} - T_{\text{water,in}}) = \dot{m}_{\text{steam}}(h_2 - h_{f,3})$$

Since the terminal difference is zero, $T_{\text{water,out}} = T_3$.

$$\dot{m}_{\text{water}}\left(4.186 \frac{\text{kJ}}{\text{kg·K}}\right)(96.71°\text{C} - 18°\text{C}) = \left(1.875 \frac{\text{kg}}{\text{s}}\right)\left(2451.0 \frac{\text{kJ}}{\text{kg}} - 405.20 \frac{\text{kJ}}{\text{kg}}\right)$$

$$\dot{m}_{\text{water}} = 11.64 \text{ kg/s} \quad (12 \text{ kg/s})$$

The answer is (D).

8. *Customary U.S. Solution*

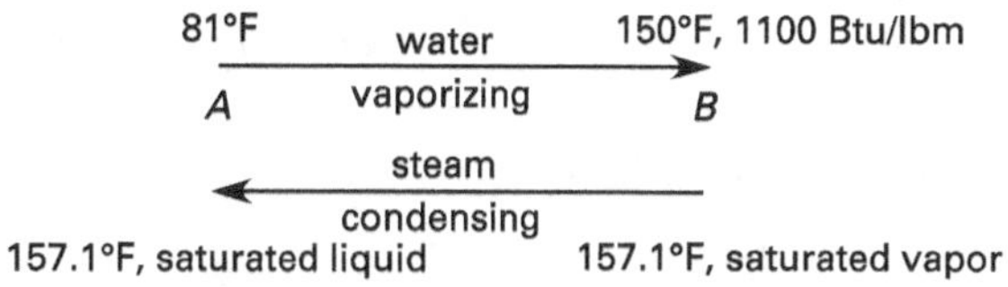

From App. 10A (also *NCEES Handbook:* Saturated Steam (U.S. Units)—Temperature Table), at 81°F, $h_{\text{water},1} \approx 49.06$ Btu/lbm.

From App. 10B (also *NCEES Handbook:* Saturated Steam (U.S. Units)—Temperature Table), for 4.45 psia, $T_{\text{sat}} = 157.1°\text{F}$.

The heat transferred to the water is

$$\begin{aligned} Q &= \dot{m}_{\text{water}}(h_{\text{water},2} - h_{\text{water},1}) \\ &= \left(332{,}000 \frac{\text{lbm}}{\text{hr}}\right)\left(1100 \frac{\text{Btu}}{\text{lbm}} - 49.06 \frac{\text{Btu}}{\text{lbm}}\right) \\ &= 3.489 \times 10^8 \text{ Btu/hr} \end{aligned}$$

The two end temperature differences are

$$\Delta T_A = 157.1°\text{F} - 81°\text{F} = 76.1°\text{F}$$
$$\Delta T_B = 157.1°\text{F} - 150°\text{F} = 7.1°\text{F}$$

Thermodynamics

The logarithmic mean temperature difference, T_{lm}, from Eq. 19.68 (also *NCEES Handbook:* Log-Mean Temperature Difference), is

$$\Delta T_{\text{lm}} = \frac{\Delta T_A - \Delta T_B}{\ln \dfrac{\Delta T_A}{\Delta T_B}} = \frac{76.1°\text{F} - 7.1°\text{F}}{\ln \dfrac{76.1°\text{F}}{7.1°\text{F}}}$$
$$= 29.09°\text{F}$$

Since $T_{\text{steam},A} = T_{\text{steam},B}$, the correction factor, F_c, for ΔT_{lm} is 1.

The overall heat transfer coefficient is calculated from Eq. 19.69 (also *NCEES Handbook:* Combination of Heat-Transfer Mechanisms).

$$Q = U_o A_o F_c \Delta T_{\text{lm}}$$

$$U_o = \frac{Q}{A_o F_c \Delta T_{\text{lm}}} = \frac{3.489 \times 10^8 \ \dfrac{\text{Btu}}{\text{hr}}}{(1850 \text{ ft}^2)(1)(29.09°\text{F})}$$
$$= 6483 \text{ Btu/hr-ft}^2\text{-°F} \quad (6500 \text{ Btu/hr-ft}^2\text{-°F})$$

SI Solution

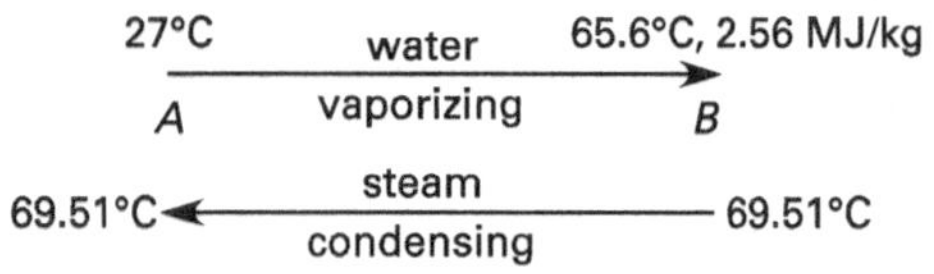

From App. 10I (also *NCEES Handbook:* Saturated Steam (SI Units)), at 27°C, $h_{\text{water},1} = 113.19$ kJ/kg.

From App. 10J, for 30.6 kPa (0.306 bar), $T_{\text{sat}} = 69.51$°C.

The heat transferred to the water is

$$Q = \dot{m}_{\text{water}}(h_{\text{water},2} - h_{\text{water},1})$$
$$= \left(41.8 \ \frac{\text{kg}}{\text{s}}\right)\left(\left(2.56 \ \frac{\text{MJ}}{\text{kg}}\right)\left(1000 \ \frac{\text{kJ}}{\text{MJ}}\right) - 113.19 \ \frac{\text{kJ}}{\text{kg}}\right)$$
$$= 102\,277 \text{ kJ/s}$$

The two end temperature differences are

$$\Delta T_A = 69.51°\text{C} - 27°\text{C} = 42.5°\text{C}$$
$$\Delta T_B = 69.51°\text{C} - 65.6°\text{C} = 3.9°\text{C}$$

The logarithmic mean temperature difference, T_{lm}, from Eq. 19.68 (also *NCEES Handbook:* Log-Mean Temperature Difference), is

$$\Delta T_{\text{lm}} = \frac{\Delta T_A - \Delta T_B}{\ln \dfrac{\Delta T_A}{\Delta T_B}} = \frac{42.5°\text{C} - 3.9°\text{C}}{\ln \dfrac{42.5°\text{C}}{3.9°\text{C}}}$$
$$= 16.16°\text{C}$$

Since $T_{\text{steam},A} = T_{\text{steam},B}$, the correction factor, F_c, for ΔT_{lm} is 1.

The overall heat transfer coefficient is calculated from Eq. 19.69 (also *NCEES Handbook:* Combination of Heat-Transfer Mechanisms).

$$Q = U_o A_o F_c \Delta T_{\text{lm}}$$

$$U_o = \frac{Q}{A_o F_c \Delta T_{\text{lm}}} = \frac{\left(102\,277 \ \dfrac{\text{kJ}}{\text{s}}\right)\left(1000 \ \dfrac{\text{W}}{\text{kW}}\right)}{(172 \text{ m}^2)(1)(16.16°\text{C})}$$
$$= 36\,797 \text{ W/m}^2\text{·°C} \quad (37 \text{ kW/m}^2\text{·°C})$$

The answer is (C).

9.

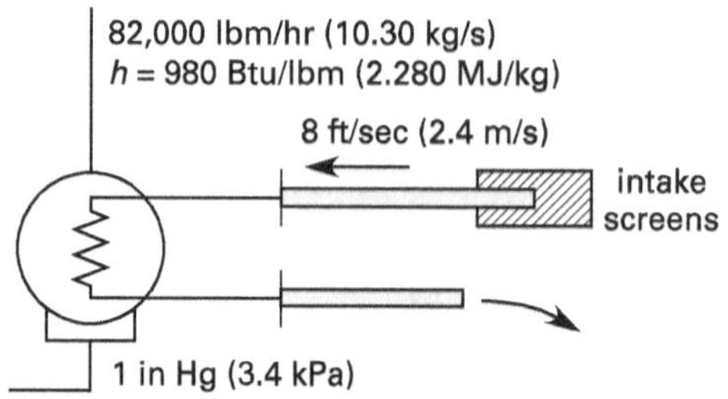

Customary U.S. Solution

The condenser pressure is

$$(1 \text{ in Hg})\left(0.491 \ \frac{\frac{\text{lbf}}{\text{in}^2}}{\text{in Hg}}\right) = 0.5 \text{ psia}$$

From App. 10B (also *NCEES Handbook:* Properties of Saturated Steam (U.S. Customary Units)), the enthalpy of the saturated liquid is $h_f = 47.08$ Btu/lbm.

As in *NCEES Handbook:* Physical Properties of Water, the specific heat of water is $c_{p,\text{water}} = 1$ Btu/lbm-°F.

The heat lost by the steam is equal to the heat gained by the water.

$$\dot{m}_{\text{water}} c_{p,\text{water}} \Delta T = \dot{m}_{\text{steam}}(h_2 - h_f)$$

$$\dot{m}_{\text{water}}\left(1 \ \frac{\text{Btu}}{\text{lbm-°F}}\right)(10°\text{F}) = \left(82{,}000 \ \frac{\text{lbm}}{\text{hr}}\right) \times \left(980 \ \frac{\text{Btu}}{\text{lbm}} - 47.08 \ \frac{\text{Btu}}{\text{lbm}}\right)$$

$$\dot{m}_{\text{water}} = 7.6499 \times 10^6 \text{ lbm/hr}$$

The density of water, ρ, is 62.4 lbm/ft^3.

$$Q = \frac{\dot{m}_{\text{water}}}{\rho} = \frac{\left(7.6499\times 10^6\ \frac{\text{lbm}}{\text{hr}}\right)\left(7.48\ \frac{\text{gal}}{\text{ft}^3}\right)}{\left(62.4\ \frac{\text{lbm}}{\text{ft}^3}\right)\left(60\ \frac{\text{min}}{\text{hr}}\right)}$$
$$= 1.528\times 10^4\ \text{gal/min}\quad (15{,}000\ \text{gal/min})$$

(The flow rate can also be determined from the velocity and pipe area. However, this does not use the 10° data or perform an energy balance.)

SI Solution

The condenser pressure is 3.4 kPa.

From App. 10I (also *NCEES Handbook:* Saturated Steam (SI Units)), the enthalpy of saturated liquid is $h_f = 109.8$ kJ/kg.

As in *NCEES Handbook:* Physical Properties of Water, the specific heat of water is $c_{p,\text{water}} = 4.187$ kJ/kg·K.

The heat lost by the steam is equal to the heat gained by the water.

$$\dot{m}_{\text{water}} c_{p,\text{water}} \Delta T = \dot{m}_{\text{steam}}(h_2 - h_f)$$
$$\dot{m}_{\text{water}}\left(4.187\ \frac{\text{kJ}}{\text{kg·K}}\right)(5.6^\circ\text{C}) = \left(10.3\ \frac{\text{kg}}{\text{s}}\right)\left(\left(2.280\ \frac{\text{MJ}}{\text{kg}}\right)\times\left(1000\ \frac{\text{kJ}}{\text{MJ}}\right) - 109.8\ \frac{\text{kJ}}{\text{kg}}\right)$$
$$\dot{m}_{\text{water}} = 953.3\ \text{kg/s}$$

The density of water, ρ, is 1000 kg/m^3.

$$Q = \frac{\dot{m}_{\text{water}}}{\rho} = \left(\frac{953.3\ \frac{\text{kg}}{\text{s}}}{1000\ \frac{\text{kg}}{\text{m}^3}}\right)\left(1000\ \frac{\text{L}}{\text{m}^3}\right)\left(60\ \frac{\text{s}}{\text{min}}\right)$$
$$= 57\,200\ \text{L/min}\quad (57\ \text{kL/min})$$

The answer is (D).

10. *Customary U.S. Solution*

From App. 10A (also *NCEES Handbook:* Properties of Saturated Steam (U.S. Customary Units)), the enthalpy of saturated liquid at 60°F is $h_1 = 28.08$ Btu/lbm.

From App. 10B, the enthalpy of saturated steam at 14.7 psia is $h_2 = 1150.3$ Btu/lbm.

The heat transfer rate to the water is

$$Q = \dot{m}(h_2 - h_1)$$
$$= \left(100\ \frac{\text{lbm}}{\text{hr}}\right)\left(1150.3\ \frac{\text{Btu}}{\text{lbm}} - 28.08\ \frac{\text{Btu}}{\text{lbm}}\right)$$
$$= 1.122\times 10^5\ \text{Btu/hr}$$
$$= \frac{1.122\times 10^5\ \frac{\text{Btu}}{\text{hr}}}{\left(1000\ \frac{\text{W}}{\text{kW}}\right)\left(3.412\ \frac{\text{Btu}}{\text{W-hr}}\right)}$$
$$= 32.89\ \text{kW}$$
$$\text{cost} = \frac{(32.89\ \text{kW})\left(\frac{\$0.04}{\text{kW-hr}}\right)}{1-0.35}$$
$$= \$2.02/\text{hr}\quad (\$2/\text{hr})$$

SI Solution

From App. 10I (also *NCEES Handbook:* Saturated Steam (SI Units)—Temperature Table), for saturated liquid at 16°C, $h_1 = 67.17$ kJ/kg.

From App. 10J, the enthalpy of saturated steam at 101.3 kPa is $h_2 = 2675.4$ kJ/kg.

The heat transfer rate to the water is

$$Q = \dot{m}(h_2 - h_1)$$
$$= \left(0.013\ \frac{\text{kg}}{\text{s}}\right)\left(2675.4\ \frac{\text{kJ}}{\text{kg}} - 67.17\ \frac{\text{kJ}}{\text{kg}}\right)$$
$$= 33.907\ \text{kW}$$
$$\text{cost} = \frac{(33.907\ \text{kW})\left(\frac{\$0.04}{\text{kW·h}}\right)}{1-0.35}$$
$$= \$2.09/\text{h}\quad (\$2/\text{h})$$

The answer is (A).

11. *Customary U.S. Solution*

From App. 10A (also *NCEES Handbook:* Properties of Saturated Steam (U.S. Customary Units)), the enthalpy of saturated liquid at 60°F is $h_1 = 28.08$ Btu/lbm.

From App. 10B, for 40 psia steam, the enthalpy of saturated liquid, h_f, is 236.1 Btu/lbm. The heat of vaporization, h_{fg}, is 933.7 Btu/lbm. The enthalpy is given by Eq. 10.40 (also *NCEES Handbook:* Properties for Two-Phase (Vapor-Liquid) Systems).

$$h_2 = h_f + x h_{fg}$$
$$= 236.1\ \frac{\text{Btu}}{\text{lbm}} + (0.98)\left(933.7\ \frac{\text{Btu}}{\text{lbm}}\right)$$
$$= 1151.13\ \text{Btu/lbm}$$

The heat transfer rate is

$$Q = \dot{m}(h_2 - h_1)$$
$$= \frac{\left(250\ \frac{\text{lbm}}{\text{hr}}\right)\left(1151.13\ \frac{\text{Btu}}{\text{lbm}} - 28.08\ \frac{\text{Btu}}{\text{lbm}}\right)}{60\ \frac{\text{min}}{\text{hr}}}$$
$$= 4679.4\ \text{Btu/min}$$

Find the volume of gas used at standard conditions for a heating gas (60°F).

$$\dot{V}_{\text{std}} = \dot{V}\left(\frac{T_0}{T}\right)\left(\frac{p}{p_0}\right)$$
$$= \left(13.5\ \frac{\text{ft}^3}{\text{min}}\right)\left(\frac{60°\text{F} + 460°}{80°\text{F} + 460°}\right) \times \left(\frac{(4\ \text{in Hg} + 30.2\ \text{in Hg})\left(0.491\ \frac{\text{lbf}}{\text{in}^3}\right)}{14.7\ \frac{\text{lbf}}{\text{in}^2}}\right)$$
$$= 14.85\ \text{ft}^3/\text{min}$$

The efficiency of the boiler is

$$\eta = \frac{Q}{\text{heat input}} = \frac{4679.4\ \frac{\text{Btu}}{\text{min}}}{\left(14.85\ \frac{\text{ft}^3}{\text{min}}\right)\left(550\ \frac{\text{Btu}}{\text{ft}^3}\right)}$$
$$= 0.573 \quad (57\%)$$

SI Solution

From App. 10I (also *NCEES Handbook:* Saturated Steam (SI Units)), the enthalpy of saturated liquid at 16°C is $h_1 = 67.17$ kJ/kg.

From App. 10J, for 300 kPa steam, the enthalpy of saturated liquid, h_f, is 561.43 kJ/kg. The enthalpy of vaporization, h_{fg}, is 2163.5 kJ/kg. The enthalpy is given by Eq. 10.40 (also *NCEES Handbook:* Properties for Two-Phase (Vapor-Liquid) Systems).

$$h_2 = h_f + xh_{fg} = 561.43\ \frac{\text{kJ}}{\text{kg}} + (0.98)\left(2163.5\ \frac{\text{kJ}}{\text{kg}}\right)$$
$$= 2681.7\ \text{kJ/kg}$$

The heat transfer rate is

$$Q = \dot{m}(h_2 - h_1)$$
$$= \left(0.032\ \frac{\text{kg}}{\text{s}}\right)\left(2681.7\ \frac{\text{kJ}}{\text{kg}} - 67.17\ \frac{\text{kJ}}{\text{kg}}\right)$$
$$= 83.66\ \text{kJ/s}$$

Find the volume of the gas used at standard conditions for a heating gas (16°C).

$$\dot{V}_{\text{std}} = \dot{V}\left(\frac{T_0}{T}\right)\left(\frac{p}{p_0}\right)$$
$$= \frac{\left(6.4\ \frac{\text{L}}{\text{s}}\right)\left(\frac{16°\text{C} + 273°}{26°\text{C} + 273°}\right)\left(\frac{13.6\ \text{kPa} + 102.4\ \text{kPa}}{101.3\ \text{kPa}}\right)}{1000\ \frac{\text{L}}{\text{m}^3}}$$
$$= 0.00708\ \text{m}^3/\text{s}$$

The efficiency of the boiler is

$$\eta = \frac{Q}{\text{heat input}}$$
$$= \frac{83.66\ \frac{\text{kJ}}{\text{s}}}{\left(0.00708\ \frac{\text{m}^3}{\text{s}}\right)\left(20.5\ \frac{\text{MJ}}{\text{m}^3}\right)\left(1000\ \frac{\text{kJ}}{\text{MJ}}\right)}$$
$$= 0.576 \quad (57\%)$$

The answer is (C).

12. *Customary U.S. Solution*

step 1: Determine the actual gravimetric analysis of the coal as fired. 1 lbm of coal contains 0.02 lbm moisture, leaving 0.98 lbm dry coal. 1% is lost to the ash pit. The remainder is 0.98 lbm − 0.01 lbm = 0.97 lbm.

step 2: Calculate the heating value of the remaining coal.

$$\text{HV} = (0.97\ \text{lbm})\left(12{,}800\ \frac{\text{Btu}}{\text{lbm}}\right) = 12{,}416\ \text{Btu}$$

step 3: Find the energy, Q, required to produce steam. From App. 10A (also *NCEES Handbook:* Properties of Saturated Steam (U.S. Customary Units)), the enthalpy of water at 120°F is $h_1 = 88.00$ Btu/lbm.

From App. 10B, the enthalpy of saturated steam at 100 psia is $h_2 = 1187.5$ Btu/lbm.

$$\begin{aligned} Q &= m(h_2 - h_1) \\ &= (8.23\ \text{lbm})\left(1187.5\ \frac{\text{Btu}}{\text{lbm}} - 88.00\ \frac{\text{Btu}}{\text{lbm}}\right) \\ &= 9048.9\ \text{Btu} \end{aligned}$$

step 4: The combustion efficiency is

$$\eta = \frac{Q}{\text{HV}} = \frac{9048.9\ \text{Btu}}{12{,}416\ \text{Btu}} = 0.729 \quad (73\%)$$

SI Solution

Since boiler data are based on 1 unit mass of coal fired, step 1 will be the same as for the customary U.S. solution. Repeat the rest of the steps as follows.

step 2: Calculate the heating value of the remaining coal.

$$\text{HV} = (0.97\ \text{kg})\left(29.80\ \frac{\text{MJ}}{\text{kg}}\right)\left(1000\ \frac{\text{kJ}}{\text{MJ}}\right) = 28\,906\ \text{kJ}$$

step 3: Find the energy, Q, required to produce steam. From App. 10I (also *NCEES Handbook:* Saturated Steam (SI Units)), the enthalpy of water at 50°C is $h_1 = 209.34$ kJ/kg.

From App. 10J, the enthalpy of saturated steam at 700 kPa is $h_2 = 2762.8$ kJ/kg.

$$\begin{aligned} Q &= m(h_2 - h_1) \\ &= (8.23\ \text{kg})\left(2762.8\ \frac{\text{kJ}}{\text{kg}} - 209.34\ \frac{\text{kJ}}{\text{kg}}\right) \\ &= 21\,015\ \text{kJ} \end{aligned}$$

step 4: The combustion efficiency is

$$\eta = \frac{Q}{\text{HV}} = \frac{21\,015\ \text{kJ}}{28\,906\ \text{kJ}} = 0.727 \quad (73\%)$$

The answer is (C).

13. *Customary U.S. Solution*

Refer to Fig. 13.1. Leaving the superheater and entering the turbine, the enthalpy, h_D, and entropy, s_D, can be obtained from App. 10C (also *NCEES Handbook:* Superheated Steam (U.S. Customary Units)).

$$\begin{aligned} h_D &= 1521.0\ \text{Btu/lbm} \\ s_D &= 1.7376\ \text{Btu/lbm-°F} \end{aligned}$$

For isentropic expansion through the turbine, $s_E = s_D$. From App. 10B (also *NCEES Handbook:* Properties of Saturated Steam (U.S. Customary Units)), the enthalpy of saturated liquid, h_f, the enthalpy of evaporation, h_{fg}, the entropy of saturated liquid, s_f, and the entropy of evaporation, s_{fg}, are

$$\begin{aligned} h_f &= 130.2\ \text{Btu/lbm} \\ h_{fg} &= 1000.5\ \text{Btu/lbm} \\ s_f &= 0.2349\ \text{Btu/lbm-°F} \\ s_{fg} &= 1.6089\ \text{Btu/lbm-°F} \end{aligned}$$

The quality of the mixture for an isentropic process (at point E) is given by Eq. 10.41 (also *NCEES Handbook:* Properties for Two-Phase (Vapor-Liquid) Systems).

$$\begin{aligned} x_E &= \frac{s_E - s_f}{s_{fg}} = \frac{1.7376\ \frac{\text{Btu}}{\text{lbm-°F}} - 0.2349\ \frac{\text{Btu}}{\text{lbm-°F}}}{1.6089\ \frac{\text{Btu}}{\text{lbm-°F}}} \\ &= 0.9340 \end{aligned}$$

The isentropic enthalpy (at point E) is given by Eq. 10.40 (also *NCEES Handbook:* Properties for Two-Phase (Vapor-Liquid) Systems).

$$\begin{aligned} h_E &= h_f + x_E h_{fg} \\ &= 130.2\ \frac{\text{Btu}}{\text{lbm}} + (0.9340)\left(1000.5\ \frac{\text{Btu}}{\text{lbm}}\right) \\ &= 1064.7\ \text{Btu/lbm} \end{aligned}$$

From Eq. 12.19 (also *NCEES Handbook:* Open Thermodynamic Systems), the actual enthalpy of steam at point E is

$$\begin{aligned} h'_E &= h_D - \eta_s(h_D - h_E) \\ &= 1521\ \frac{\text{Btu}}{\text{lbm}} - (0.75)\left(1521\ \frac{\text{Btu}}{\text{lbm}} - 1064.7\ \frac{\text{Btu}}{\text{lbm}}\right) \\ &= 1178.8\ \text{Btu/lbm} \end{aligned}$$

Since the pump work is negligible, the mass flow rate of steam is

$$\begin{aligned} \dot{m} &= \frac{P}{W_{\text{turbine}}} = \frac{P}{h_D - h'_E} \\ &= \frac{(200\ \text{MW})\left(10^6\ \frac{\text{W}}{\text{MW}}\right)\left(3.412\ \frac{\text{Btu}}{\text{W-hr}}\right)}{1521\ \frac{\text{Btu}}{\text{lbm}} - 1178.8\ \frac{\text{Btu}}{\text{lbm}}} \\ &= 1.994 \times 10^6\ \text{lbm/hr} \quad (2.0 \times 10^6\ \text{lbm/hr}) \end{aligned}$$

SI Solution

Refer to Fig. 13.1. Leaving the superheater and entering the turbine, the enthalpy, h_D, and entropy, s_D, can be obtained from the Mollier figure (also *NCEES Handbook* figure "Temperature-Entropy (T-S) Diagram (SI Units)").

$$h_D \approx 3430 \text{ kJ/kg}$$
$$s_D \approx 7.25 \text{ kJ/kg·K}$$

For isentropic expansion through the turbine, $s_E = s_D$. From App. 10J (also *NCEES Handbook:* Saturated Steam (SI Units)), the enthalpy of saturated liquid, h_f, the enthalpy of evaporation, h_{fg}, the entropy of saturated liquid, s_f, and the entropy of saturated vapor, s_g, are

$$h_f = 289.27 \text{ kJ/kg}$$
$$h_{fg} = 2335.3 \text{ kJ/kg}$$
$$s_f = 0.9441 \text{ kJ/kg·K}$$
$$s_g = 7.7675 \text{ kJ/kg·K}$$

The quality of the mixture for an isentropic process (at point E) is given by Eq. 10.41 (also *NCEES Handbook:* Properties for Two-Phase (Vapor-Liquid) Systems) as

$$\begin{aligned} x_E &= \frac{s_E - s_f}{s_{fg}} = \frac{s_E - s_f}{s_g - s_f} \\ &= \frac{7.25 \ \frac{\text{kJ}}{\text{kg·K}} - 0.9441 \ \frac{\text{kJ}}{\text{kg·K}}}{7.7675 \ \frac{\text{kJ}}{\text{kg·K}} - 0.9441 \ \frac{\text{kJ}}{\text{kg·K}}} \\ &= 0.9242 \end{aligned}$$

The isentropic enthalpy (at point E) is given by Eq. 10.40 (also *NCEES Handbook:* Properties for Two-Phase (Vapor-Liquid) Systems) as

$$\begin{aligned} h_E &= h_f + x_E h_{fg} \\ &= 289.27 \ \frac{\text{kJ}}{\text{kg}} + (0.9242)\left(2335.3 \ \frac{\text{kJ}}{\text{kg}}\right) \\ &= 2447.6 \text{ kJ/kg} \end{aligned}$$

From Eq. 12.19 (also *NCEES Handbook:* Open Thermodynamic Systems), the actual enthalpy of steam at point E is

$$\begin{aligned} h'_E &= h_D - \eta_s(h_D - h_E) \\ &= 3430 \ \frac{\text{kJ}}{\text{kg}} - (0.75)\left(3430 \ \frac{\text{kJ}}{\text{kg}} - 2447.6 \ \frac{\text{kJ}}{\text{kg}}\right) \\ &= 2693.2 \text{ kJ/kg} \end{aligned}$$

Since the pump work is negligible, the mass flow rate of steam is

$$\begin{aligned} \dot{m} &= \frac{P}{W_{\text{turbine}}} = \frac{P}{h_D - h'_E} = \frac{(200 \text{ MW})\left(10^3 \ \frac{\text{kW}}{\text{MW}}\right)}{3430 \ \frac{\text{kJ}}{\text{kg}} - 2693.2 \ \frac{\text{kJ}}{\text{kg}}} \\ &= 271.4 \text{ kg/s} \quad (270 \text{ kg/s}) \end{aligned}$$

The answer is (C).

14. *Customary U.S. Solution*

The illustration shows one of N layers. Each layer consists of $n = 24$ tubes, only 3 of which are shown.

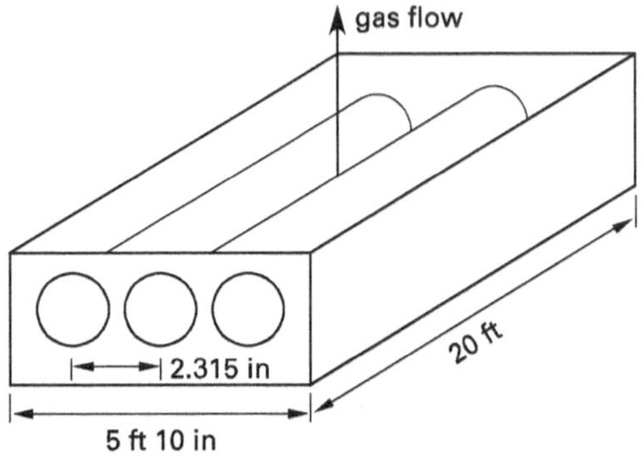

Assume stack gases consist primarily of nitrogen. The average gas temperature is

$$T_{\text{ave}} = \left(\frac{1}{2}\right)(635°\text{F} + 470°\text{F}) + 460° = 1012.5°\text{R}$$

From Eq. 10.94 (also *NCEES Handbook* table "Temperature-Dependent Physical Properties of Gases at 14.7 psia (U.S. Units)"), the specific heat of nitrogen is 0.2554 Btu/lbm-°R using constants given in Table 10.9.

For the purpose of calculating the logarithmic temperature difference, assume counterflow operation. The two end temperature differences are

$$\Delta T_A = 635°\text{F} - 285°\text{F} = 350°\text{F}$$
$$\Delta T_B = 470°\text{F} - 212°\text{F} = 258°\text{F}$$

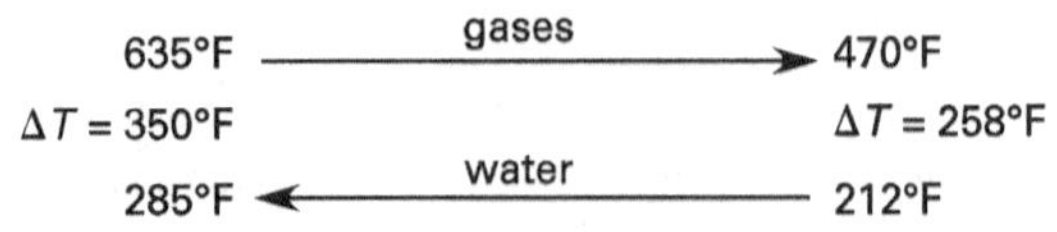

The logarithmic temperature difference (see Eq. 19.68 and *NCEES Handbook:* Log-Mean Temperature Difference) is

$$\Delta T_{\text{lm}} = \frac{\Delta T_A - \Delta T_B}{\ln \frac{\Delta T_A}{\Delta T_B}} = \frac{350°\text{F} - 258°\text{F}}{\ln \frac{350°\text{F}}{258°\text{F}}} = 301.7°\text{F}$$

Determine F_c for crossflow.

$$R = \frac{635°\text{F} - 470°\text{F}}{285°\text{F} - 212°\text{F}} = 2.26$$

$$x = \frac{285°\text{F} - 212°\text{F}}{635°\text{F} - 212°\text{F}} = 0.17$$

From App. 19D (also *NCEES Handbook:* F-Factor Charts), F_c for all heat exchanger configurations is close to 1.0 for this set of parameters.

The heat transfer from the temperature gain of water is equal to the heat transfer based on the logarithmic mean temperature difference. From Eq. 19.69 (also *NCEES Handbook:* Overall Heat-Transfer Coefficient),

$$Q = \dot{m}c_p\Delta T = U_oA_oF_c\Delta T_{\text{lm}}$$

$$\left(191{,}000\ \frac{\text{lbm}}{\text{hr}}\right)\left(0.2554\ \frac{\text{Btu}}{\text{lbm-°R}}\right)(635°\text{F} - 470°\text{F}) = \left(10\ \frac{\text{Btu}}{\text{hr-ft}^2\text{-°F}}\right)A_o(1.0)(301.7°\text{F})$$

$$A_o = 2667.9\ \text{ft}^2$$

The tube area per bank is

$$A_{\text{bank}} = n\pi D_oL = \frac{24\pi(1.315\ \text{in})(20\ \text{ft})}{12\ \frac{\text{in}}{\text{ft}}} = 165.2\ \text{ft}^2$$

$$N = \frac{A_o}{A_{\text{bank}}} = \frac{2667.9\ \text{ft}^2}{165.2\ \text{ft}^2} = 16.1 \quad (17)$$

SI Solution

The illustration shows one of N layers. Each layer consists of $n = 24$ tubes, only 3 of which are shown.

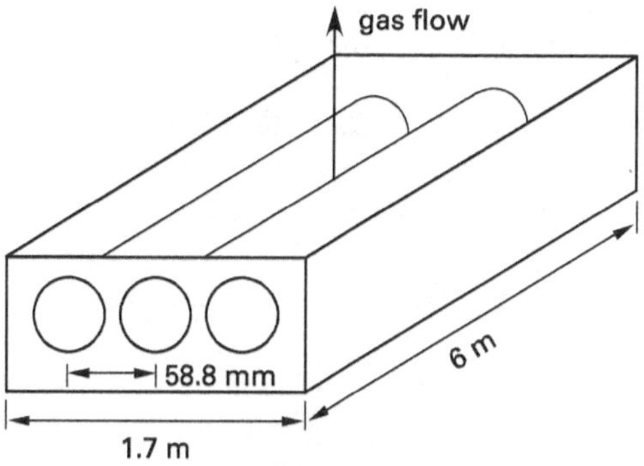

Assume stack gas consists primarily of nitrogen. The average gas temperature is

$$T_{\text{ave}} = \left(\frac{1}{2}\right)(335°\text{C} + 240°\text{C}) + 273° = 560.5\text{K} \quad (1009°\text{R})$$

From Table 10.9 (also *NCEES Handbook* table "Temperature-Dependent Physical Properties of Gases at 0.1 MPa (SI Units)"), the specific heat of nitrogen is 1069 J/kg·°C.

For the purpose of calculating the logarithmic temperature difference, assume counterflow operation. The two end temperature differences are

$$\Delta T_A = 335°\text{C} - 140°\text{C} = 195°\text{C}$$
$$\Delta T_B = 240°\text{C} - 100°\text{C} = 140°\text{C}$$

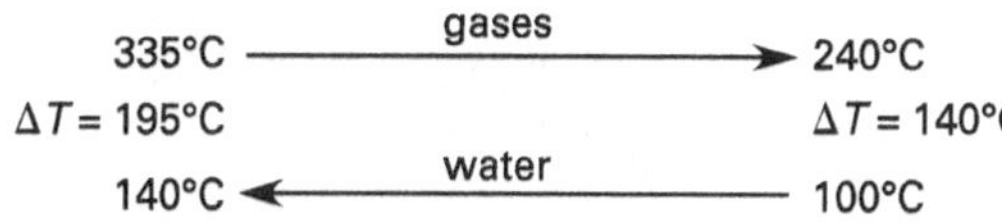

The logarithmic mean temperature difference is (see Eq. 19.68 and *NCEES Handbook:* Log-Mean Temperature Difference)

$$\Delta T_{\text{lm}} = \frac{\Delta T_A - \Delta T_B}{\ln \frac{\Delta T_A}{\Delta T_B}} = \frac{195°\text{C} - 140°\text{C}}{\ln \frac{195°\text{C}}{140°\text{C}}} = 166.0°\text{C}$$

$$R = \frac{335°\text{C} - 240°\text{C}}{140°\text{C} - 100°\text{C}} = 2.38$$

$$x = \frac{140°\text{C} - 100°\text{C}}{335°\text{C} - 100°\text{C}} = 0.17$$

From App. 19D (also *NCEES Handbook:* F-Factor Charts), F_c for all heat exchanger configurations is close to 1.0 for this set of conditions.

The heat transfer from the temperature gain of water is equal to the heat transfer based on the logarithmic mean temperature difference. From Eq. 19.69 (also *NCEES Handbook:* Log-Mean Temperature Difference),

$$Q = \dot{m}c_p\Delta T = U_oA_oF_c\Delta T_{lm}$$

$$\left(24\ \frac{\text{kg}}{\text{s}}\right)\left(1069\ \frac{\text{J}}{\text{kg·°C}}\right)(335°\text{C} - 240°\text{C}) = \left(57\ \frac{\text{W}}{\text{m}^2\text{·°C}}\right)A_o(1.0)(166.0°\text{C})$$

$$A_o = 257.6\ \text{m}^2$$

The tube area per bank is

$$A_{\text{bank}} = n\pi D_oL = \frac{24\pi(33.4\ \text{mm})(6\ \text{m})}{1000\ \frac{\text{mm}}{\text{m}}} = 15.1\ \text{m}^2$$

$$N = \frac{A_o}{A_{\text{bank}}} = \frac{257.6\ \text{m}^2}{15.1\ \text{m}^2} = 17.1\quad (17)$$

The answer is (D).

15.

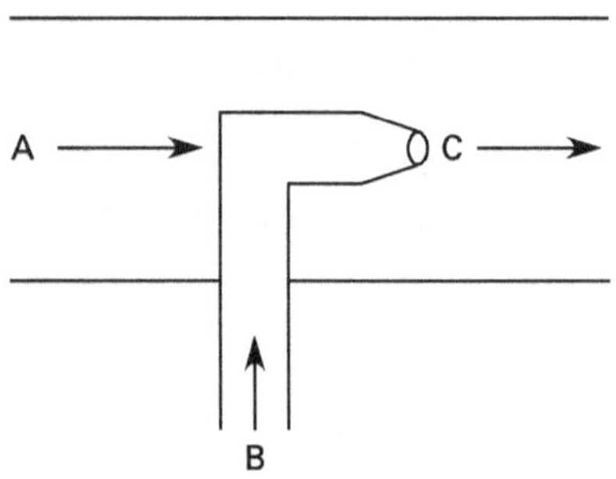

Customary U.S. Solution

The enthalpy, h_A, and entropy, s_A, for steam at 600°F and 200 psia can be obtained from App. 10C (also *NCEES Handbook:* Superheated Steam (U.S. Customary Units)).

$$h_A = 1322.3\ \text{Btu/lbm}$$
$$s_A = 1.6771\ \text{Btu/lbm-°R}$$

For water at 82°F, enthalpy can be obtained for saturated water from App. 10A (also *NCEES Handbook:* Properties of Saturated Steam (U.S. Customary Units)).

$$h_B = 50.06\ \frac{\text{Btu}}{\text{lbm}}$$

From App. 10A, $s_B = 0.09697$ Btu/lbm-°F. From an energy balance equation (with $Q = 0$, $v_1 = 0$, $\Delta z = 0$, and $W = 0$), Eq. 11.33(b) can be written as

$$\begin{aligned}0 &= h_C(m_A + m_B) - (h_Am_A + h_Bm_B) + \left(\frac{v_C^2}{2g_cJ}\right)(m_A + m_B)\\ &= h_C\left(1000\ \frac{\text{lbm}}{\text{hr}} + 50\ \frac{\text{lbm}}{\text{hr}}\right) - \left(\begin{array}{l}\left(1322.3\ \frac{\text{Btu}}{\text{lbm}}\right)\left(1000\ \frac{\text{lbm}}{\text{hr}}\right)\\ +\left(50.06\ \frac{\text{Btu}}{\text{lbm}}\right)\left(50\ \frac{\text{lbm}}{\text{hr}}\right)\end{array}\right)\\ &\quad + \frac{\left(2000\ \frac{\text{ft}}{\text{sec}}\right)^2\left(1000\ \frac{\text{lbm}}{\text{hr}} + 50\ \frac{\text{lbm}}{\text{hr}}\right)}{(2)\left(32.2\ \frac{\text{lbm-ft}}{\text{lbf-sec}^2}\right)\left(778\ \frac{\text{ft-lbf}}{\text{Btu}}\right)}\end{aligned}$$

$$h_C = 1181.9\ \text{Btu/lbm}$$
$$p_C = 100\ \text{psia}$$

Since the enthalpy of saturated steam at 100 psia, h_g, from App. 10B (also *NCEES Handbook:* Properties of Saturated Steam (U.S. Customary Units)), is greater than h_C, steam leaving the desuperheater is not 100% saturated. From App. 10B, the temperature of the saturated steam mix is 327.81°F (330°F).

SI Solution

(a) The enthalpy, h_A, and entropy, s_A, for steam at 300°C and 1.5 MPa can be obtained from App. 10K (also *NCEES Handbook:* Superheated Steam (SI Units)).

$$h_A = 3038.2\ \text{kJ/kg}$$
$$s_A = 6.9198\ \text{kJ/kg·K}$$

From App. 10I (also *NCEES Handbook:* Saturated Steam (SI Units)), for water at 28°C, $h_B = 117.37$ kJ/kg.

Similarly, from App. 10I, $s_B = 0.4091$ kJ/kg·K.

From an energy balance equation (with $Q=0$, $v_1=0$, $\Delta z=0$, and $W=0$), Eq. 11.33(a) (also *NCEES Handbook:* Open Thermodynamic Systems) can be written as

$$\begin{aligned}0 &= h_C(m_A+m_B)-(m_Ah_A+m_Bh_B)\\ &\quad+\left(\frac{v_C^2}{2}\right)(m_A+m_B)\\ &= h_C\left(0.13\ \frac{kg}{s}+0.0063\ \frac{kg}{s}\right)\\ &\quad-\left(\begin{array}{l}\left(0.13\ \frac{kg}{s}\right)\left(3038.2\ \frac{kJ}{kg}\right)\\ \quad+\left(0.0063\ \frac{kg}{s}\right)\left(117.37\ \frac{kJ}{kg}\right)\end{array}\right)\\ &\quad+\left(\frac{\left(600\ \frac{m}{s}\right)^2}{2}\right)\left(0.13\ \frac{kg}{s}+0.0063\ \frac{kg}{s}\right)\\ &\quad\times\left(\frac{1\ kJ}{1000\ J}\right)\\ h_C &= 2723.2\ kJ/kg\\ p_C &= 700\ kPa\end{aligned}$$

Since the enthalpy of saturated steam at 700 kPa, from App. 10J (also *NCEES Handbook:* Saturated Steam (SI Units)), is greater than h_C, steam leaving the desuperheater is not 100% saturated. From App. 10J (also *NCEES Handbook:* Saturated Steam (SI Units)), the temperature of the steam mix is 164.9°C (165°C).

The answer is (A).

16. *Customary U.S. Solution*

Work with the original system to find the steam flow.

From App. 10C (also *NCEES Handbook:* Superheated Steam (U.S. Customary Units)), at 100 psia and 400°F, $h_A = 1227.7$ Btu/lbm.

From App. 10A (also *NCEES Handbook:* Properties of Saturated Steam (U.S. Customary Units)) at 70°F, $h_B = 38.08$ Btu/lbm.

From App. 10A (also *NCEES Handbook:* Properties of Saturated Steam (U.S. Customary Units)) at 180°F, $h_C = 148.04$ Btu/lbm.

Let $x =$ fraction of steam in mixture.

From the energy balance equation,

$$m_Ah_A+m_Bh_B=m_Ch_C$$

$$\begin{aligned}&x\left(1227.7\ \frac{Btu}{lbm}\right)\\ &\quad+(1-x)\left(38.08\ \frac{Btu}{lbm}\right)=(1)\left(148.04\ \frac{Btu}{lbm}\right)\\ &x=0.0924\end{aligned}$$

The steam flow is

$$\begin{aligned}\dot{m} &= x\left(2000\ \frac{lbm}{hr}\right)=(0.0924)\left(2000\ \frac{lbm}{hr}\right)\\ &= 184.8\ lbm/hr\quad(180\ lbm/hr)\end{aligned}$$

SI Solution

Work with the original system to find the steam flow.

From App. 10K (also *NCEES Handbook:* Superheated Steam (SI Units)), at 700 kPa and 200°C, $h_A = 2845.3$ kJ/kg.

From App. 10I (also *NCEES Handbook:* Saturated Steam (SI Units)), at 21°C, $h_B = 88.10$ kJ/kg.

From App. 10I (also *NCEES Handbook:* Saturated Steam (SI Units)), at 80°C, $h_C = 335.01$ kJ/kg.

Let $x =$ fraction of steam in mixture.

From the energy balance equation,

$$m_Ah_A+m_Bh_B=m_Ch_C$$

$$\begin{aligned}&x\left(2845.3\ \frac{kJ}{kg}\right)\\ &\quad+(1-x)\left(88.10\ \frac{kJ}{kg}\right)=(1)\left(335.01\ \frac{kJ}{kg}\right)\\ &x=0.0896\end{aligned}$$

The steam flow is

$$\begin{aligned}\dot{m} &= x\left(0.25\ \frac{kg}{s}\right)=(0.0896)\left(0.25\ \frac{kg}{s}\right)\\ &= 0.0224\ kg/s\quad(0.022\ kg/s)\end{aligned}$$

The answer is (B).

13 Vapor Power Cycles

PRACTICE PROBLEMS

1. An air combustion cycle operates between 650°F and 100°F (340°C and 38°C). The maximum possible thermal efficiency is most nearly

(A) 42%

(B) 49%

(C) 54%

(D) 58%

2. A steam Carnot cycle operates between 650°F and 100°F (340°C and 38°C). The turbine and compressor (pump) isentropic efficiencies are 90% and 80%, respectively. The thermal efficiency is most nearly

(A) 32%

(B) 37%

(C) 42%

(D) 48%

3. A steam turbine cycle produces 600 MWe (megawatts of electrical power). The condenser load is 3.07×10^9 Btu/hr (900 MW). The thermal efficiency is most nearly

(A) 32%

(B) 37%

(C) 40%

(D) 44%

4. A steam Rankine cycle operates with 100 psia (700 kPa) saturated steam that is reduced to 1 atm through expansion in a turbine with an isentropic efficiency of 80%. There is significant subcooling, and water is at 80°F (27°C) and 1 atm when it enters the boilerfeed pump. The pump's isentropic efficiency is 60%. The cycle's thermal efficiency is most nearly

(A) 10%

(B) 14%

(C) 21%

(D) 26%

5. A turbine and the boilerfeed pumps in a reheat cycle have isentropic efficiencies of 88% and 96%, respectively. The cycle starts with water at 60°F (16°C) at the entrance to the boilerfeed pump and produces 600°F (300°C), 600 psia (4 MPa) steam. The steam is reheated when its pressure drops during the first expansion to 20 psia (150 kPa).

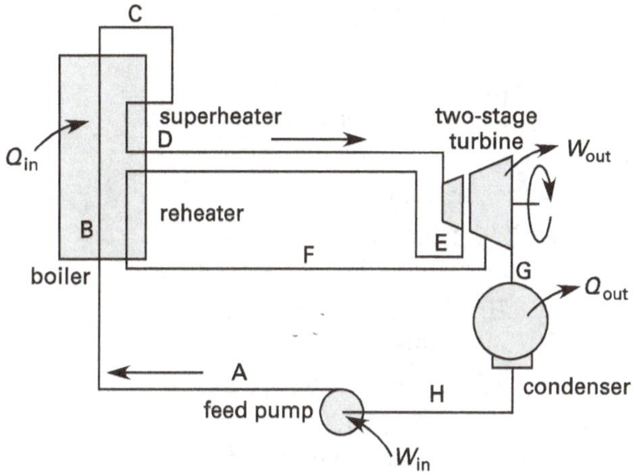

The thermal efficiency of the cycle is most nearly

(A) 28%

(B) 34%

(C) 39%

(D) 42%

6. A reheat steam cycle operates as shown. The pump work between points F and G is 0.15 Btu/lbm (0.3 kJ/kg).

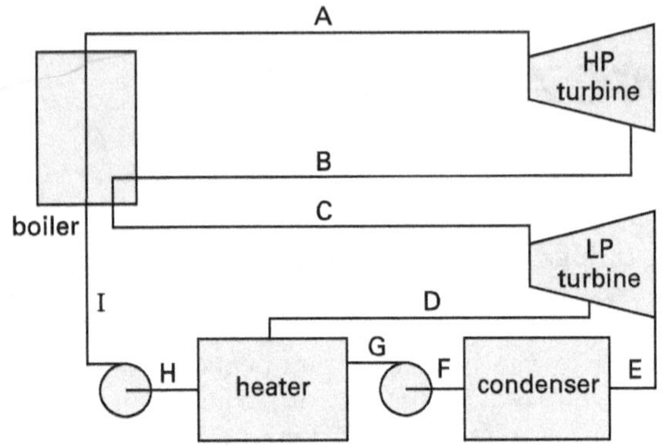

At A: 900 psia (6.2 MPa)
800°F (420°C)

At B: 200 psia (1.5 MPa)
1270 Btu/lbm (2960 kJ/kg)

At C: 190 psia (1.4 MPa)
800°F (420°C)

At D: 50 psia (350 kPa)
1280 Btu/lbm (2980 kJ/kg)

At E: 2 in Hg absolute (6.8 kPa)
1075 Btu/lbm (2500 kJ/kg)

At F: 69.73 Btu/lbm (162.5 kJ/kg)

At H: 250.2 Btu/lbm (583.0 kJ/kg)
0.0173 ft^3/lbm (0.0011 m^3/kg)

At I: 253.1 Btu/lbm (589.7 kJ/kg)

The thermal efficiency of the cycle is most nearly

(A) 22%

(B) 29%

(C) 34%

(D) 41%

SOLUTIONS

1. *Customary U.S. Solution*

As in the *NCEES Handbook:* section titled "Temperature," the absolute temperatures are

$$T_{high} = 650°F + 460° = 1110°R$$
$$T_{low} = 100°F + 460° = 560°R$$

The maximum possible thermal efficiency is given by the Carnot cycle. From Eq. 13.8 (also the *NCEES Handbook:* section titled "Gas Power Cycles"),

$$\eta_{th} = \frac{T_{high} - T_{low}}{T_{high}} = \frac{1110°R - 560°R}{1110°R} = 0.495 \quad (49\%)$$

SI Solution

As in the *NCEES Handbook:* section titled "Temperature"), the absolute temperatures are

$$T_{high} = 340°C + 273° = 613K$$
$$T_{low} = 38°C + 273° = 311K$$

The maximum possible thermal efficiency is given by the Carnot cycle. From Eq. 13.8 (also the *NCEES Handbook*: section titled "Gas Power Cycles"),

$$\eta_{th} = \frac{T_{high} - T_{low}}{T_{high}} = \frac{613K - 311K}{613K} = 0.493 \quad (49\%)$$

The answer is (B).

2. *Customary U.S. Solution*

Refer to the Carnot cycle (see Fig. 13.2 and *NCEES Handbook:* Power Cycles).

At point A, from App. 10A (also *NCEES Handbook* table "Saturated Steam (U.S. Units)—Temperature Table"), for saturated liquid at $T_A = 650°F$,

$$h_A = 696.0 \text{ Btu/lbm}$$
$$s_A = 0.8833 \text{ Btu/lbm-°F}$$

At point B, from App. 10A, for saturated vapor at $T_B = 650°F$,

$$h_B = 1119.7 \text{ Btu/lbm}$$
$$s_B = 1.2651 \text{ Btu/lbm-°F}$$

At point C,

$$T_C = 100°F$$
$$s_C = s_B = 1.2651 \text{ Btu/lbm-°F}$$

From App. 10A,

$$s_f = 0.1296 \text{ Btu/lbm-°F}$$
$$s_g = 1.9819 \text{ Btu/lbm-°F}$$
$$h_f = 68.03 \text{ Btu/lbm}$$
$$h_{fg} = 1036.7 \text{ Btu/lbm}$$

As in *NCEES Handbook:* Properties for Two-Phase (Vapor-Liquid) Systems, the quality of the system at point C is

$$x_C = \frac{s_C - s_f}{s_g - s_f} = \frac{1.2651 \frac{\text{Btu}}{\text{lbm-°F}} - 0.1296 \frac{\text{Btu}}{\text{lbm-°F}}}{1.9819 \frac{\text{Btu}}{\text{lbm-°F}} - 0.1296 \frac{\text{Btu}}{\text{lbm-°F}}} = 0.613$$

The specific enthalpy of the system at point C is

$$h_C = h_f + x_C h_{fg} = 68.03 \frac{\text{Btu}}{\text{lbm}} + (0.613)\left(1036.7 \frac{\text{Btu}}{\text{lbm}}\right) = 703.5 \text{ Btu/lbm}$$

At point D,

$$T_D = 100°F$$
$$s_D = s_A = 0.8833 \text{ Btu/lbm-°F}$$

As in *NCEES Handbook:* Properties for Two-Phase (Vapor-Liquid) Systems, the quality of the system at point D is

$$x_D = \frac{s_D - s_f}{s_g - s_f} = \frac{0.8833 \frac{\text{Btu}}{\text{lbm-°F}} - 0.1296 \frac{\text{Btu}}{\text{lbm-°F}}}{1.9819 \frac{\text{Btu}}{\text{lbm-°F}} - 0.1296 \frac{\text{Btu}}{\text{lbm-°F}}} = 0.407$$

The specific enthalpy of the system at point D is

$$h_D = h_f + x_D h_{fg} = 68.03 \frac{\text{Btu}}{\text{lbm}} + (0.407)\left(1036.7 \frac{\text{Btu}}{\text{lbm}}\right) = 489.8 \text{ Btu/lbm}$$

From Eq. 13.9 (also *NCEES Handbook:* Open Thermodynamic Systems), due to the inefficiency of the turbine,

$$h'_C = h_B - \eta_{s,\text{turbine}}(h_B - h_C) = 1119.7 \frac{\text{Btu}}{\text{lbm}} - (0.9)\left(1119.7 \frac{\text{Btu}}{\text{lbm}} - 703.5 \frac{\text{Btu}}{\text{lbm}}\right) = 745.1 \text{ Btu/lbm}$$

From Eq. 13.10 (also *NCEES Handbook:* Open Thermodynamic Systems), due to the inefficiency of the pump,

$$h'_A = h_D + \frac{h_A - h_D}{\eta_{s,\text{pump}}} = 489.8 \frac{\text{Btu}}{\text{lbm}} + \frac{696.0 \frac{\text{Btu}}{\text{lbm}} - 489.8 \frac{\text{Btu}}{\text{lbm}}}{0.8} = 747.6 \text{ Btu/lbm}$$

From Eq. 13.8 (also *NCEES Handbook:* Power Cycles), the thermal efficiency of the entire cycle is

$$\eta_{th} = \frac{(h_B - h'_C) - (h'_A - h_D)}{h_B - h'_A} = \frac{\left(1119.7 \frac{\text{Btu}}{\text{lbm}} - 745.1 \frac{\text{Btu}}{\text{lbm}}\right) - \left(747.6 \frac{\text{Btu}}{\text{lbm}} - 489.8 \frac{\text{Btu}}{\text{lbm}}\right)}{1119.7 \frac{\text{Btu}}{\text{lbm}} - 747.6 \frac{\text{Btu}}{\text{lbm}}} = 0.314 \quad (32\%)$$

SI Solution

At point A, $T_A = 340°C$.

From App. 10I (also *NCEES Handbook* table "Saturated Steam (U.S. Units)—Temperature Table"), for saturated liquid,

$$h_A = 1594.5 \text{ kJ/kg}$$
$$s_A = 3.6601 \text{ kJ/kg·K}$$

At point B, $T_B = 340°C$.

From App. 10I, for saturated vapor,

$$h_B = 2621.9 \text{ kJ/kg}$$
$$s_B = 5.3356 \text{ kJ/kg·K}$$

At point C,

$$T_C = 38°\text{C}$$
$$s_C = s_B = 5.3356 \text{ kJ/kg·K}$$

From App. 10I,

$$s_f = 0.5456 \text{ kJ/kg·K}$$
$$s_g = 8.2935 \text{ kJ/kg·K}$$
$$h_f = 159.17 \text{ kJ/kg}$$
$$h_{fg} = 2410.7 \text{ kJ/kg}$$

As in *NCEES Handbook:* Properties for Two-Phase (Vapor-Liquid) Systems, the quality of the system at point C is

$$x_C = \frac{s_C - s_f}{s_g - s_f} = \frac{5.3356 \frac{\text{kJ}}{\text{kg·K}} - 0.5456 \frac{\text{kJ}}{\text{kg·K}}}{8.2935 \frac{\text{kJ}}{\text{kg·K}} - 0.5456 \frac{\text{kJ}}{\text{kg·K}}} = 0.618$$

The specific enthalpy of the system at point C is

$$h_C = h_f + x_C h_{fg} = 159.17 \frac{\text{kJ}}{\text{kg}} + (0.618)\left(2410.7 \frac{\text{kJ}}{\text{kg}}\right) = 1649.0 \text{ kJ/kg}$$

At point D,

$$T_D = 38°\text{C}$$
$$s_D = s_A = 3.6601 \text{ kJ/kg·K}$$

As in *NCEES Handbook:* Properties for Two-Phase (Vapor-Liquid) Systems, the quality of the system at point D is

$$x_D = \frac{s_D - s_f}{s_g - s_f} = \frac{3.6601 \frac{\text{kJ}}{\text{kg·K}} - 0.5456 \frac{\text{kJ}}{\text{kg·K}}}{8.2935 \frac{\text{kJ}}{\text{kg·K}} - 0.5456 \frac{\text{kJ}}{\text{kg·K}}} = 0.402$$

The specific enthalpy of the system at point D is

$$h_D = h_f + x_D h_{fg} = 159.17 \frac{\text{kJ}}{\text{kg}} + (0.402)\left(2410.7 \frac{\text{kJ}}{\text{kg}}\right) = 1128.3 \text{ kJ/kg}$$

From Eq. 13.9 (also *NCEES Handbook:* Open Thermodynamic Systems), due to the inefficiency of the turbine,

$$h_C' = h_B - \eta_{s,\text{turbine}}(h_B - h_C) = 2621.9 \frac{\text{kJ}}{\text{kg}} - (0.9)\left(2621.9 \frac{\text{kJ}}{\text{kg}} - 1649.0 \frac{\text{kJ}}{\text{kg}}\right) = 1746.3 \text{ kJ/kg}$$

From Eq. 13.10 (also *NCEES Handbook:* Open Thermodynamic Systems), due to the inefficiency of the pump,

$$h_A' = h_D + \frac{h_A - h_D}{\eta_{s,\text{pump}}} = 1128.3 \frac{\text{kJ}}{\text{kg}} + \frac{1594.5 \frac{\text{kJ}}{\text{kg}} - 1128.3 \frac{\text{kJ}}{\text{kg}}}{0.8} = 1711.1 \text{ kJ/kg}$$

From Eq. 13.8 (also *NCEES Handbook:* Power Cycles), the thermal efficiency of the entire cycle is

$$\eta_{th} = \frac{(h_B - h_C') - (h_A' - h_D)}{h_B - h_A'} = \frac{\left(2621.9 \frac{\text{kJ}}{\text{kg}} - 1746.3 \frac{\text{kJ}}{\text{kg}}\right) - \left(1711.1 \frac{\text{kJ}}{\text{kg}} - 1128.3 \frac{\text{kJ}}{\text{kg}}\right)}{2621.9 \frac{\text{kJ}}{\text{kg}} - 1711.1 \frac{\text{kJ}}{\text{kg}}} = 0.321 \quad (32\%)$$

The answer is (A).

3. *Customary U.S. Solution*

The condenser load is $Q_{out} = 3.07 \times 10^9$ Btu/hr.

The net work is

$$W_{net} = Q_{in} - Q_{out}$$

The boiler load is

$$\begin{aligned} Q_{in} &= W_{net} + Q_{out} \\ &= (600 \text{ MW})\left(1000 \ \frac{\text{kW}}{\text{MW}}\right)\left(3412 \ \frac{\text{Btu}}{\text{kW-hr}}\right) \\ &\quad + 3.07 \times 10^9 \ \frac{\text{Btu}}{\text{hr}} \\ &= 5.12 \times 10^9 \text{ Btu/hr} \end{aligned}$$

From Eq. 13.1 (also *NCEES Handbook:* Power Cycles), the thermal efficiency is

$$\eta_{th} = \frac{Q_{in} - Q_{out}}{Q_{in}} = \frac{5.12 \times 10^9 \ \frac{\text{Btu}}{\text{hr}} - 3.07 \times 10^9 \ \frac{\text{Btu}}{\text{hr}}}{5.12 \times 10^9 \ \frac{\text{Btu}}{\text{hr}}}$$
$$= 0.400 \quad (40\%)$$

SI Solution

The condenser load is $Q_{out} = 900$ MW.

The boiler load is

$$\begin{aligned} Q_{in} &= W_{net} + Q_{out} = 600 \text{ MW} + 900 \text{ MW} \\ &= 1500 \text{ MW} \end{aligned}$$

From Eq. 13.1 (also *NCEES Handbook:* Power Cycles), the thermal efficiency is

$$\begin{aligned} \eta_{th} &= \frac{Q_{in} - Q_{out}}{Q_{in}} = \frac{1500 \text{ MW} - 900 \text{ MW}}{1500 \text{ MW}} \\ &= 0.400 \quad (40\%) \end{aligned}$$

The answer is (C).

4. *Customary U.S. Solution*

At point A, $p_A = 100$ psia.

From App. 10B (also *NCEES Handbook* table "Saturated Steam (U.S. Units)—Temperature Table"), the enthalpy of saturated liquid is $h_A = 298.5$ Btu/lbm.

From App. 10B, the enthalpy and entropy of saturated vapor are

$$\begin{aligned} h_B &= 1187.5 \text{ Btu/lbm} \\ s_B &= 1.6032 \text{ Btu/lbm-°R} \end{aligned}$$

At point C,

$$\begin{aligned} p_C &= 1 \text{ atm} \\ s_C &= s_B = 1.6032 \text{ Btu/lbm-°R} \end{aligned}$$

From App. 10B, the entropy and enthalpy of saturated liquid and entropy and enthalpy of vaporization are

$$\begin{aligned} s_f &= 0.3122 \text{ Btu/lbm-°R} \\ h_f &= 180.2 \text{ Btu/lbm} \\ s_{fg} &= 1.4445 \text{ Btu/lbm-°R} \\ h_{fg} &= 970.1 \text{ Btu/lbm} \end{aligned}$$

As in *NCEES Handbook:* Properties for Two-Phase (Vapor-Liquid) Systems, the quality of the system at point C is

$$\begin{aligned} x_C &= \frac{s_C - s_f}{s_{fg}} \\ &= \frac{1.6032 \ \frac{\text{Btu}}{\text{lbm-°R}} - 0.3122 \ \frac{\text{Btu}}{\text{lbm-°R}}}{1.4445 \ \frac{\text{Btu}}{\text{lbm-°R}}} \\ &= 0.894 \end{aligned}$$

The specific enthalpy of the system at point C is

$$\begin{aligned} h_C &= h_f + x_C h_{fg} = 180.2 \ \frac{\text{Btu}}{\text{lbm}} + (0.894)\left(970.1 \ \frac{\text{Btu}}{\text{lbm}}\right) \\ &= 1047.5 \text{ Btu/lbm} \end{aligned}$$

At point D, $T = 80°F$ and $p_D = 1$ atm (subcooled). h and v are essentially independent of pressure.

From App. 10A, the enthalpy and specific volume of saturated liquid are

$$\begin{aligned} h_D &= 48.07 \text{ Btu/lbm} \\ v_D &= 0.01607 \text{ ft}^3\text{/lbm} \end{aligned}$$

At point E, $p_E = p_A = 100$ psia.

From Eq. 13.14 (also *NCEES Handbook:* Open Thermodynamic Systems),

$$\begin{aligned} h_E &\approx h_D + v_D(p_E - p_D) \\ &= 48.07 \ \frac{\text{Btu}}{\text{lbm}} + \left(0.01607 \ \frac{\text{ft}^3}{\text{lbm}}\right) \\ &\quad \times \frac{\left(100 \ \frac{\text{lbf}}{\text{in}^2} - 14.7 \ \frac{\text{lbf}}{\text{in}^2}\right)\left(12 \ \frac{\text{in}}{\text{ft}}\right)^2}{778 \ \frac{\text{ft-lbf}}{\text{Btu}}} \\ &= 48.32 \text{ Btu/lbm} \end{aligned}$$

From Eq. 13.18 (also *NCEES Handbook:* Open Thermodynamic Systems), due to the inefficiency of the turbine,

$$\begin{aligned} h'_C &= h_B - \eta_{s,\text{turbine}}(h_B - h_C) \\ &= 1187.5\ \frac{\text{Btu}}{\text{lbm}} - (0.80)\left(1187.5\ \frac{\text{Btu}}{\text{lbm}} - 1047.5\ \frac{\text{Btu}}{\text{lbm}}\right) \\ &= 1075.5\ \text{Btu/lbm} \end{aligned}$$

From Eq. 13.19 (also *NCEES Handbook:* Open Thermodynamic Systems), due to the inefficiency of the pump,

$$\begin{aligned} h'_E &= h_D + \frac{h_E - h_D}{\eta_{s,\text{pump}}} \\ &= 48.07\ \frac{\text{Btu}}{\text{lbm}} + \frac{48.32\ \frac{\text{Btu}}{\text{lbm}} - 48.07\ \frac{\text{Btu}}{\text{lbm}}}{0.6} \\ &= 48.49\ \text{Btu/lbm} \end{aligned}$$

From Eq. 13.17 (also *NCEES Handbook:* Power Cycles), the thermal efficiency of the cycle is

$$\begin{aligned} \eta_{th} &= \frac{(h_B - h'_C) - (h'_E - h_D)}{h_B - h'_E} \\ &= \frac{\left(1187.5\ \frac{\text{Btu}}{\text{lbm}} - 1075.5\ \frac{\text{Btu}}{\text{lbm}}\right) - \left(48.49\ \frac{\text{Btu}}{\text{lbm}} - 48.07\ \frac{\text{Btu}}{\text{lbm}}\right)}{1187.5\ \frac{\text{Btu}}{\text{lbm}} - 48.49\ \frac{\text{Btu}}{\text{lbm}}} \\ &= 0.098 \quad (10\%) \end{aligned}$$

SI Solution

At point A, $p_A = 700$ kPa (7 bars).

From App. 10J (also *NCEES Handbook* table "Saturated Steam (SI Units)—Temperature Table"), the enthalpy of saturated liquid is $h_A = 697.00$ kJ/kg.

From App. 10J, the enthalpy and entropy of saturated vapor are

$$\begin{aligned} h_B &= 2762.8\ \text{kJ/kg} \\ s_B &= 6.7071\ \text{kJ/kg·K} \end{aligned}$$

At point C,

$$\begin{aligned} p_C &= 1\ \text{atm} \\ s_C &= s_B = 6.7071\ \text{kJ/kg·K} \end{aligned}$$

From App. 10J, the entropy and enthalpy of saturated liquid, the entropy of saturated vapor, and the enthalpy of vaporization are

$$\begin{aligned} s_f &= 1.3062\ \text{kJ/kg·K} \\ h_f &= 418.79\ \text{kJ/kg} \\ s_g &= 7.3553\ \text{kJ/kg·K} \\ h_{fg} &= 2256.6\ \text{kJ/kg} \end{aligned}$$

As in *NCEES Handbook:* Properties for Two-Phase (Vapor-Liquid) Systems,

$$\begin{aligned} x_C &= \frac{s_C - s_f}{s_g - s_f} = \frac{6.7071\ \frac{\text{kJ}}{\text{kg·K}} - 1.3062\ \frac{\text{kJ}}{\text{kg·K}}}{7.3553\ \frac{\text{kJ}}{\text{kg·K}} - 1.3062\ \frac{\text{kJ}}{\text{kg·K}}} \\ &= 0.893 \\ h_C &= h_f + x_C h_{fg} = 418.79\ \frac{\text{kJ}}{\text{kg}} + (0.893)\left(2256.6\ \frac{\text{kJ}}{\text{kg}}\right) \\ &= 2433.9\ \text{kJ/kg} \end{aligned}$$

At point D, $T = 27°\text{C}$ and $p_D = 1$ atm (subcooled). h and υ are essentially independent of pressure.

From App. 10I, the enthalpy and specific volume of saturated liquid are

$$\begin{aligned} h_D &= 113.19\ \text{kJ/kg} \\ \upsilon_D &= 1.0035\ \text{cm}^3/\text{g} \end{aligned}$$

At point E, $p_E = p_A = 700$ kPa.

From Eq. 13.14,

$$\begin{aligned} h_E &\approx h_D + \upsilon_D(p_E - p_D) \\ &= 113.19\ \frac{\text{kJ}}{\text{kg}} + \frac{\left(1.0035\ \frac{\text{cm}^3}{\text{g}}\right)\left(1000\ \frac{\text{g}}{\text{kg}}\right) \times (700\ \text{kPa} - 101\ \text{kPa})}{10^6\ \frac{\text{cm}^3}{\text{m}^3}} \\ &= 113.79\ \text{kJ/kg} \end{aligned}$$

From Eq. 13.18 (also *NCEES Handbook:* Open Thermodynamic Systems), due to the inefficiency of the turbine,

$$\begin{aligned} h'_C &= h_B - \eta_{s,\text{turbine}}(h_B - h_C) \\ &= 2762.8\ \frac{\text{kJ}}{\text{kg}} - (0.80)\left(2762.8\ \frac{\text{kJ}}{\text{kg}} - 2433.9\ \frac{\text{kJ}}{\text{kg}}\right) \\ &= 2499.7\ \text{kJ/kg} \end{aligned}$$

From Eq. 13.19 (also *NCEES Handbook:* Open Thermodynamic Systems), due to the inefficiency of the pump,

$$\begin{aligned} h'_{\text{E}} &= h_{\text{D}} + \frac{h_{\text{E}} - h_{\text{D}}}{\eta_{s,\text{pump}}} \\ &= 113.19\ \frac{\text{kJ}}{\text{kg}} + \frac{113.79\ \frac{\text{kJ}}{\text{kg}} - 113.19\ \frac{\text{kJ}}{\text{kg}}}{0.6} \\ &= 114.19\ \text{kJ/kg} \end{aligned}$$

From Eq. 13.17 (also *NCEES Handbook:* Power Cycles), the thermal efficiency of the cycle is

$$\begin{aligned} \eta_{\text{th}} &= \frac{(h_{\text{B}} - h'_{\text{C}}) - (h'_{\text{E}} - h_{\text{D}})}{h_{\text{B}} - h'_{\text{E}}} \\ &= \frac{\left(2762.8\ \frac{\text{kJ}}{\text{kg}} - 2499.7\ \frac{\text{kJ}}{\text{kg}}\right) - \left(114.19\ \frac{\text{kJ}}{\text{kg}} - 113.19\ \frac{\text{kJ}}{\text{kg}}\right)}{2762.8\ \frac{\text{kJ}}{\text{kg}} - 114.19\ \frac{\text{kJ}}{\text{kg}}} \\ &= 0.10 \quad (10\%) \end{aligned}$$

The answer is (A).

5. *Customary U.S. Solution*

At point B, $p_{\text{B}} = 600$ psia.

From App. 10B (also *NCEES Handbook* table "Saturated Steam (U.S. Units)—Temperature Table"), the enthalpy of the saturated liquid is $h_{\text{B}} = 471.7$ Btu/lbm.

At point C, $p_{\text{C}} = 600$ psia.

From App. 10B, the enthalpy of the saturated vapor is $h_{\text{C}} = 1203.8$ Btu/lbm.

At point D,

$$\begin{aligned} T_{\text{D}} &= 600°\text{F} \\ p_{\text{D}} &= 600\ \text{psia} \end{aligned}$$

From App. 10C (also *NCEES Handbook:* Superheated Steam (U.S. Customary Units)), the enthalpy and entropy of superheated vapor are

$$\begin{aligned} h_{\text{D}} &= 1289.9\ \text{Btu/lbm} \\ s_{\text{D}} &= 1.5326\ \text{Btu/lbm-°R} \end{aligned}$$

At point E,

$$\begin{aligned} p_{\text{E}} &= 20\ \text{psia} \\ s_{\text{E}} &= s_{\text{D}} = 1.5326\ \text{Btu/lbm-°R} \end{aligned}$$

From App. 10B (also *NCEES Handbook* table "Saturated Steam (U.S. Units)—Temperature Table"), the various saturation properties are

$$\begin{aligned} s_f &= 0.3358\ \text{Btu/lbm-°R} \\ s_{fg} &= 1.3961\ \text{Btu/lbm-°R} \\ h_f &= 196.3\ \text{Btu/lbm} \\ h_{fg} &= 959.9\ \text{Btu/lbm} \end{aligned}$$

As in *NCEES Handbook:* Properties for Two-Phase (Vapor-Liquid) Systems,

$$\begin{aligned} x_{\text{E}} &= \frac{s_{\text{E}} - s_f}{s_g - s_f} = \frac{1.3961\ \frac{\text{Btu}}{\text{lbm-°R}} - 0.3358\ \frac{\text{Btu}}{\text{lbm-°R}}}{1.5236\ \frac{\text{Btu}}{\text{lbm-°R}}} \\ &= 0.89 \\ h_{\text{E}} &= h_f + x_{\text{E}} h_{fg} \\ &= 196.3\ \frac{\text{Btu}}{\text{lbm}} + (0.89)\left(959.9\ \frac{\text{Btu}}{\text{lbm}}\right) \\ &= 1050.6\ \text{Btu/lbm} \end{aligned}$$

From Eq. 13.38 (also *NCEES Handbook:* Open Thermodynamic Systems), due to the inefficiency of the turbine,

$$\begin{aligned} h'_{\text{E}} &= h_{\text{D}} - \eta_{s,\text{turbine}}(h_{\text{D}} - h_{\text{E}}) \\ &= 1289.9\ \frac{\text{Btu}}{\text{lbm}} - (0.88) \times \left(1289.9\ \frac{\text{Btu}}{\text{lbm}} - 1050.6\ \frac{\text{Btu}}{\text{lbm}}\right) \\ &= 1079.3\ \text{Btu/lbm} \end{aligned}$$

At point F, the temperature has been returned to 600°F, but the pressure stays at the expansion pressure, p_{E}.

$$\begin{aligned} p_{\text{F}} &= 20\ \text{psia} \\ T_{\text{F}} &= 600°\text{F} \end{aligned}$$

From App. 10C (also *NCEES Handbook:* Superheated Steam (U.S. Customary Units)), the enthalpy and entropy of the superheated vapor are

$$\begin{aligned} h_{\text{F}} &= 1334.9\ \text{Btu/lbm} \\ s_{\text{F}} &= 1.9399\ \text{Btu/lbm-°R} \end{aligned}$$

Thermodynamics

At point G,

$$T_G = 60°\text{F}$$
$$s_G = s_F = 1.9399\ \text{Btu/lbm-°R}$$

From App. 10A (also *NCEES Handbook* "Saturated Steam (U.S. Units)—Temperature Table"), the various saturation properties are

$$s_f = 0.05554\ \text{Btu/lbm-°R}$$
$$s_g = 2.0940\ \text{Btu/lbm-°R}$$
$$h_f = 28.08\ \text{Btu/lbm}$$
$$h_{fg} = 1059.3\ \text{Btu/lbm}$$

From *NCEES Handbook:* Properties for Two-Phase (Vapor-Liquid) Systems,

$$x_G = \frac{s_G - s_f}{s_g - s_f} = \frac{1.9399\ \frac{\text{Btu}}{\text{lbm-°R}} - 0.05554\ \frac{\text{Btu}}{\text{lbm-°R}}}{2.0940\ \frac{\text{Btu}}{\text{lbm-°R}} - 0.05554\ \frac{\text{Btu}}{\text{lbm-°R}}} = 0.924$$

$$h_G = h_f + x_G h_{fg} = 28.08\ \frac{\text{Btu}}{\text{lbm}} + (0.924)\left(1059.3\ \frac{\text{Btu}}{\text{lbm}}\right) = 1006.9\ \text{Btu/lbm}$$

From Eq. 13.39 (also *NCEES Handbook:* Open Thermodynamic Systems, due to the inefficiency of the turbine,

$$h'_G = h_F - \eta_{s,\text{turbine}}(h_F - h_G) = 1334.9\ \frac{\text{Btu}}{\text{lbm}} - (0.88)\left(1334.9\ \frac{\text{Btu}}{\text{lbm}} - 1006.9\ \frac{\text{Btu}}{\text{lbm}}\right) = 1046.3\ \text{Btu/lbm}$$

At point H, $T_H = 60°\text{F}$.

From App. 10A (also *NCEES Handbook* table "Saturated Steam (U.S. Units)—Temperature Table"), the saturation pressure, enthalpy, and specific volume of the saturated liquid are

$$p_H = 0.2564\ \text{psia}$$
$$h_H = 28.08\ \text{Btu/lbm}$$
$$\upsilon_H = 0.01604\ \text{ft}^3/\text{lbm}$$

At point A, $p_A = 600$ psia.

From Eq. 13.14 (also *NCEES Handbook:* Open Thermodynamic Systems),

$$h_A \approx h_H + \upsilon_H(p_A - p_H) = 28.08\ \frac{\text{Btu}}{\text{lbm}} + \frac{\left(0.01604\ \frac{\text{ft}^3}{\text{lbm}}\right) \times \left(600\ \frac{\text{lbf}}{\text{in}^2} - 0.2564\ \frac{\text{lbf}}{\text{in}^2}\right)\left(12\ \frac{\text{in}}{\text{ft}}\right)^2}{778\ \frac{\text{ft-lbf}}{\text{Btu}}} = 29.86\ \text{Btu/lbm}$$

From Eq. 13.40 (also *NCEES Handbook:* Open Thermodynamic Systems), due to the inefficiency of the pump,

$$h'_A = h_H + \frac{h_A - h_H}{\eta_{s,\text{pump}}} = 28.08\ \frac{\text{Btu}}{\text{lbm}} + \frac{29.86\ \frac{\text{Btu}}{\text{lbm}} - 28.08\ \frac{\text{Btu}}{\text{lbm}}}{0.96} = 29.93\ \text{Btu/lbm}$$

From Eq. 13.37 (also *NCEES Handbook:* Power Cycles), the thermal efficiency of the cycle for a non-isentropic process for the turbine and the pump is

$$\eta_{th} = \frac{(h_D - h'_A) + (h_F - h'_E) - (h'_G - h_H)}{(h_D - h'_A) + (h_F - h'_E)}$$
$$= \frac{\left(1289.9\ \frac{\text{Btu}}{\text{lbm}} - 29.93\ \frac{\text{Btu}}{\text{lbm}}\right) + \left(1334.9\ \frac{\text{Btu}}{\text{lbm}} - 1079.3\ \frac{\text{Btu}}{\text{lbm}}\right) - \left(1046.3\ \frac{\text{Btu}}{\text{lbm}} - 28.08\ \frac{\text{Btu}}{\text{lbm}}\right)}{\left(1289.9\ \frac{\text{Btu}}{\text{lbm}} - 29.93\ \frac{\text{Btu}}{\text{lbm}}\right) + \left(1334.9\ \frac{\text{Btu}}{\text{lbm}} - 1051.4\ \frac{\text{Btu}}{\text{lbm}}\right)}$$
$$= 0.33\quad (33\%)$$

SI Solution

At point B, $p_B = 4$ MPa.

From App. 10J (also *NCEES Handbook* "Saturated Steam (SI Units)—Temperature Table"), the enthalpy of saturated liquid is $h_B = 1082.5$ kJ/kg.

At point C, $p_C = 4$ MPa.

From App. 10J, the enthalpy of saturated vapor is $h_C = 2800.8$ kJ/kg.

At point D, $p_D = 4$ MPa and $T_D = 300°C$.

As in *NCEES Handbook* figure "Temperature-Entropy (T-S) Diagram (SI Units)," the enthalpy of superheated vapor is $h_D = 2980$ kJ/kg.

At point E, in the diagram, assuming isentropic expansion, $h_E = 2395$ kJ/kg.

From Eq. 13.38 (also *NCEES Handbook:* Open Thermodynamic Systems), due to the inefficiency of the turbine,

$$\begin{aligned} h'_E &= h_D - \eta_{s,\text{turbine}}(h_D - h_E) \\ &= 2980\ \frac{\text{kJ}}{\text{kg}} - (0.88)\left(2980\ \frac{\text{kJ}}{\text{kg}} - 2395\ \frac{\text{kJ}}{\text{kg}}\right) \\ &= 2465.2\ \text{kJ/kg} \end{aligned}$$

At point F,

$$p_F = 150\ \text{kPa}$$
$$T_F = 300°C$$

From App. 10K (also *NCEES Handbook* table "Saturated Steam (SI Units)—Temperature Table"), the enthalpy and entropy of the superheated vapor are

$$h_F = 3073.3\ \text{kJ/kg}$$
$$s_F = 8.0284\ \text{kJ/kg·K}$$

At point G,

$$T_G = 16°C$$
$$s_G = s_F = 8.0284\ \text{kJ/kg·K}$$

From App. 10I (also *NCEES Handbook* table "Saturated Steam (SI Units)—Temperature Table"), the various saturation properties are

$$s_f = 0.2390\ \text{kJ/kg·K}$$
$$s_g = 8.7570\ \text{kJ/kg·K}$$
$$h_f = 67.17\ \text{kJ/kg}$$
$$h_{fg} = 2463.0\ \text{kJ/kg}$$

$$\begin{aligned} x_G &= \frac{s_G - s_f}{s_g - s_f} \\ &= \frac{8.0284\ \frac{\text{kJ}}{\text{kg·K}} - 0.2390\ \frac{\text{kJ}}{\text{kg·K}}}{8.7570\ \frac{\text{kJ}}{\text{kg·K}} - 0.2390\ \frac{\text{kJ}}{\text{kg·K}}} \\ &= 0.914 \end{aligned}$$

$$\begin{aligned} h_G &= h_f + x_G h_{fg} \\ &= 67.17\ \frac{\text{kJ}}{\text{kg}} + (0.914)\left(2463.0\ \frac{\text{kJ}}{\text{kg}}\right) \\ &= 2318.4\ \text{kJ/kg} \end{aligned}$$

From Eq. 13.39, due to the inefficiency of the turbine,

$$\begin{aligned} h'_G &= h_F - \eta_{s,\text{turbine}}(h_F - h_G) \\ &= 3073.3\ \frac{\text{kJ}}{\text{kg}} - (0.88)\left(3073.3\ \frac{\text{kJ}}{\text{kg}} - 2318.4\ \frac{\text{kJ}}{\text{kg}}\right) \\ &= 2409.0\ \text{kJ/kg} \end{aligned}$$

At point H, $T_H = 16°C$.

From App. 10I (also *NCEES Handbook* "Saturated Steam (SI Units)—Temperature Table"), the saturation pressure, enthalpy, and specific volume of the saturated liquid are

$$p_H = (0.01819\ \text{bar})\left(100\ \frac{\text{kPa}}{\text{bar}}\right) = 1.819\ \text{kPa}$$
$$h_H = 67.17\ \text{kJ/kg}$$
$$\begin{aligned} v_H &= \frac{\left(1.0011\ \frac{\text{cm}^3}{\text{g}}\right)\left(1000\ \frac{\text{g}}{\text{kg}}\right)}{\left(100\ \frac{\text{cm}}{\text{m}}\right)^3} \\ &= 1.0011 \times 10^{-3}\ \text{m}^3/\text{kg} \end{aligned}$$

At point A,

$$p_A = (4\ \text{MPa})\left(1000\ \frac{\text{kPa}}{\text{MPa}}\right) = 4000\ \text{kPa}$$

From Eq. 13.14 (also *NCEES Handbook:* Open Thermodynamic Systems),

$$\begin{aligned} h_A &\approx h_H + v_H(p_A - p_H) \\ &= 67.17\ \frac{\text{kJ}}{\text{kg}} + \left(1.0011 \times 10^{-3}\ \frac{\text{m}^3}{\text{kg}}\right) \\ &\quad \times (4000\ \text{kPa} - 1.819\ \text{kPa}) \\ &= 71.17\ \text{kJ/kg} \end{aligned}$$

From Eq. 13.40 (also *NCEES Handbook:* Open Thermodynamic Systems), due to the inefficiency of the pump,

$$h'_A = h_H + \frac{h_A - h_H}{\eta_{s,\text{pump}}}$$
$$= 67.17\ \frac{\text{kJ}}{\text{kg}} + \frac{71.17\ \frac{\text{kJ}}{\text{kg}} - 67.17\ \frac{\text{kJ}}{\text{kg}}}{0.96}$$
$$= 71.34\ \text{kJ/kg}$$

From Eq. 13.37 (also *NCEES Handbook:* Power Cycles), the thermal efficiency of the cycle for a non-isentropic process for the turbine and the pump is

$$\eta_{\text{th}} = \frac{(h_D - h'_A) + (h_F - h'_E) - (h'_G - h_H)}{(h_D - h'_A) + (h_F - h'_E)}$$
$$= \frac{\left(2980\ \frac{\text{kJ}}{\text{kg}} - 71.34\ \frac{\text{kJ}}{\text{kg}}\right) + \left(3073.3\ \frac{\text{kJ}}{\text{kg}} - 2465.2\ \frac{\text{kJ}}{\text{kg}}\right) - \left(2409.0\ \frac{\text{kJ}}{\text{kg}} - 67.17\ \frac{\text{kJ}}{\text{kg}}\right)}{\left(2980\ \frac{\text{kJ}}{\text{kg}} - 71.34\ \frac{\text{kJ}}{\text{kg}}\right) + \left(3073.3\ \frac{\text{kJ}}{\text{kg}} - 2465.2\ \frac{\text{kJ}}{\text{kg}}\right)}$$
$$= 0.334 \quad (34\%)$$

The answer is (B).

6. *Customary U.S. Solution*

Refer to the given illustration in the problem statement and to the following diagram.

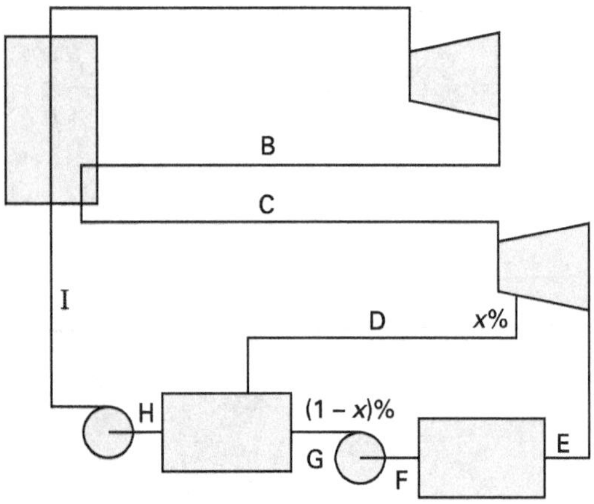

At point A,

$$p_A = 900\ \text{psia}$$
$$T_A = 800°\text{F}$$

Using *NCEES Handbook* table "Superheated Steam (U.S. Units),"

$$h_A = 1393.9\ \text{Btu/lbm}$$
$$s_A = 1.5816\ \text{Btu/lbm-°R}$$

From the Mollier diagram, assuming isentropic expansion to 200 psia, $h_B = 1230$ Btu/lbm.

As in *NCEES Handbook:* Open Thermodynamic Systems, isentropic efficiency of the high pressure turbine is

$$\eta_{s,\text{turbine}} = \frac{h_A - h'_B}{h_A - h_B}$$
$$= \left(\frac{1393.9\ \frac{\text{Btu}}{\text{lbm}} - 1270\ \frac{\text{Btu}}{\text{lbm}}}{1393.9\ \frac{\text{Btu}}{\text{lbm}} - 1230\ \frac{\text{Btu}}{\text{lbm}}}\right) \times 100\%$$
$$= 75.59\%$$

At point C,

$$p_C = 190\ \text{psia}$$
$$T_C = 800°\text{F}$$

Using App. 10C (also *NCEES Handbook* table "Superheated Steam (U.S. Units)"), $h_C = 1426.0$ Btu/lbm.

At D: $h'_D = 1280\ \text{Btu/lbm}$

At E: $h'_E = 1075\ \text{Btu/lbm}$

At F: $h_F = 69.73\ \text{Btu/lbm}$

At G: $W_{\text{pump}} = 0.15\ \text{Btu/lbm}$

$$W_{\text{pump}} = h'_G - h_F$$
$$h'_G = W_{\text{pump}} + h_F$$
$$= 0.15\ \frac{\text{Btu}}{\text{lbm}} + 69.73\ \frac{\text{Btu}}{\text{lbm}}$$
$$= 69.88\ \text{Btu/lbm}$$

At H: $h_H = 250.2\ \text{Btu/lbm}$

At I: $h'_I = 253.1\ \text{Btu/lbm}$

From an energy balance in the heater,

$$xh'_D + (1-x)h'_G = h_H$$
$$x(h'_D - h'_G) = h_H - h'_G$$
$$x = \frac{h_H - h'_G}{h'_D - h'_G} = \frac{250.2\ \frac{\text{Btu}}{\text{lbm}} - 69.88\ \frac{\text{Btu}}{\text{lbm}}}{1280\ \frac{\text{Btu}}{\text{lbm}} - 69.88\ \frac{\text{Btu}}{\text{lbm}}} = 0.149$$

From Eq. 13.8 (also *NCEES Handbook:* Power Cycles), the thermal efficiency of the cycle is

$$\eta_{th} = \frac{W_{out} - W_{in}}{Q_{in}} = \frac{(h_A - h'_B) + (h_C - h'_D) + (1-x)(h'_D - h'_E) - (h'_I - h_H) - (1-x)(h'_G - h_F)}{(h_A - h'_I) + (h_C - h'_B)}$$

$$= \frac{\left(1393.9\ \frac{\text{Btu}}{\text{lbm}} - 1270\ \frac{\text{Btu}}{\text{lbm}}\right) + \left(1426.0\ \frac{\text{Btu}}{\text{lbm}} - 1280\ \frac{\text{Btu}}{\text{lbm}}\right) + (1-0.149)\left(1280\ \frac{\text{Btu}}{\text{lbm}} - 1075\ \frac{\text{Btu}}{\text{lbm}}\right) - \left(253.1\ \frac{\text{Btu}}{\text{lbm}} - 250.2\ \frac{\text{Btu}}{\text{lbm}}\right) - (1-0.149)\left(69.88\ \frac{\text{Btu}}{\text{lbm}} - 69.73\ \frac{\text{Btu}}{\text{lbm}}\right)}{\left(1393.9\ \frac{\text{Btu}}{\text{lbm}} - 253.1\ \frac{\text{Btu}}{\text{lbm}}\right) + \left(1426.0\ \frac{\text{Btu}}{\text{lbm}} - 1270\ \frac{\text{Btu}}{\text{lbm}}\right)}$$

$$= 0.340 \quad (34\%)$$

SI Solution

Refer to the illustration in the problem statement and to the diagram in the customary U.S. solution.

At point A,

$$p_A = 6.2\ \text{MPa}$$
$$T_A = 420°\text{C}$$

From (also *NCEES Handbook* figure "Temperature-Entropy (T-S) Diagram (SI Units)"),

$$h_A = 3235.0\ \text{kJ/kg}$$
$$s_A = 6.65\ \text{kJ/kg·K}$$

Use *NCEES Handbook* table "Pressure-Enthalpy (p-H) Diagram (SI Units)." Assuming isentropic expansion to 1.5 MPa, $h_B = 2860$ kJ/kg.

As in *NCEES Handbook:* Open Thermodynamic Systems, isentropic efficiency of the high pressure turbine is

$$\eta_{s,\text{turbine}} = \frac{h_A - h'_B}{h_A - h_B} = \frac{3235.0\ \frac{\text{kJ}}{\text{kg}} - 2960\ \frac{\text{kJ}}{\text{kg}}}{3235.0\ \frac{\text{kJ}}{\text{kg}} - 2860\ \frac{\text{kJ}}{\text{kg}}} = 0.733 \quad (73\%)$$

At point C,

$$p_C = 1.4\ \text{MPa}$$
$$T_C = 420°\text{C}$$

Using App. 10K (also *NCEES Handbook* table "Superheated Steam (SI Units)"), $h_C = 3301.3$ kJ/kg.

At point D, $h'_D = 2980$ kJ/kg.

At point E, $h'_E = 2500$ kJ/kg.

At point F, $h'_F = 162.5$ kJ/kg.

At point G,

$$W_{pump} = 0.3\ \text{kJ/kg} = h'_G - h_F$$
$$h'_G = W_{pump} + h_F = 0.3\ \frac{\text{kJ}}{\text{kg}} + 162.5\ \frac{\text{kJ}}{\text{kg}} = 162.8\ \text{kJ/kg}$$

At point H, $h_H = 583.0$ kJ/kg.

At point I, $h'_I = 589.7$ kJ/kg.

From an energy balance in the heater,

$$xh'_D + (1-x)h'_G = h_H$$
$$x(h'_D - h'_G) = h_H - h'_G$$
$$x = \frac{h_H - h_G}{h'_D - h'_G} = \frac{583.0\ \frac{\text{kJ}}{\text{kg}} - 162.8\ \frac{\text{kJ}}{\text{kg}}}{2980\ \frac{\text{kJ}}{\text{kg}} - 162.8\ \frac{\text{kJ}}{\text{kg}}} = 0.149$$

From Eq. 13.8 (also *NCEES Handbook:* Power Cycles), the thermal efficiency of the cycle is

$$\eta_{\text{th}} = \frac{W_{\text{out}} - W_{\text{in}}}{Q_{\text{in}}}$$

$$= \frac{\begin{array}{c}(h_A - h_B') + (h_C - h_D') + (1-x)(h_D' - h_E') \\ -(h_I' - h_H) - (1-x)(h_G' - h_F)\end{array}}{(h_A - h_I') + (h_C - h_B')}$$

$$= \frac{\begin{array}{l}\left(3235.0\ \frac{\text{kJ}}{\text{kg}} - 2960\ \frac{\text{kJ}}{\text{kg}}\right) \\ \quad + \left(3301.3\ \frac{\text{kJ}}{\text{kg}} - 2980\ \frac{\text{kJ}}{\text{kg}}\right) \\ \quad + (1-0.149)\left(2980\ \frac{\text{kJ}}{\text{kg}} - 2500\ \frac{\text{kJ}}{\text{kg}}\right) \\ \quad - \left(598.7\ \frac{\text{kJ}}{\text{kg}} - 583.0\ \frac{\text{kJ}}{\text{kg}}\right) \\ \quad - (1-0.149)\left(162.8\ \frac{\text{kJ}}{\text{kg}} - 162.5\ \frac{\text{kJ}}{\text{kg}}\right)\end{array}}{\begin{array}{l}\left(3235.0\ \frac{\text{kJ}}{\text{kg}} - 598.7\ \frac{\text{kJ}}{\text{kg}}\right) \\ \quad + \left(3301.3\ \frac{\text{kJ}}{\text{kg}} - 2960\ \frac{\text{kJ}}{\text{kg}}\right)\end{array}}$$

$$= 0.332 \quad (34\%)$$

The answer is (C).

14 Combustion Power Cycles

PRACTICE PROBLEMS

1. Air expands isentropically at the rate of 10 ft^3/sec (280 L/s) from 200 psia and 1500°F (1.4 MPa and 820°C) to 50 psia (350 kPa). Using air tables, the air's enthalpy change is most nearly

(A) −350 Btu/lbm (−770 kJ/kg)

(B) −230 Btu/lbm (−530 kJ/kg)

(C) −160 Btu/lbm (−380 kJ/kg)

(D) −110 Btu/lbm (−250 kJ/kg)

2. In an air-standard gas turbine, air at 14.7 psia and 60°F (101.3 kPa and 16°C) enters a compressor and is compressed through a volume ratio of 5:1. The compressor efficiency is 83%. Air enters the turbine at 1500°F (820°C) and expands to 14.7 psia (101.3 kPa). The turbine efficiency is 92%.

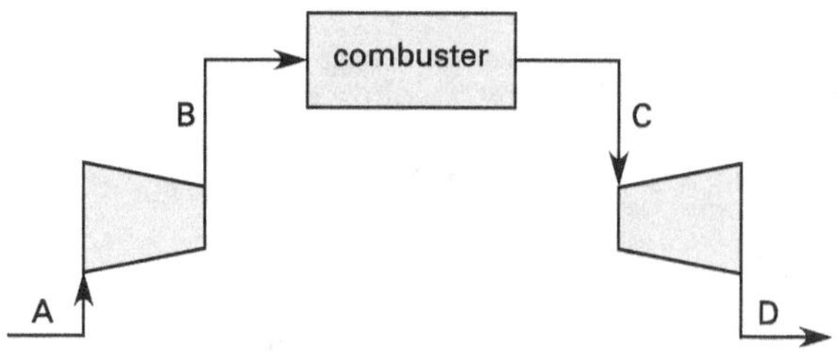

The thermal efficiency of the cycle is most nearly

(A) 33%

(B) 39%

(C) 44%

(D) 51%

3. In an air-standard gas turbine, air at 14.7 psia and 60°F (101.3 kPa and 16°C) enters a compressor and is compressed through a volume ratio of 5:1. The compressor efficiency is 83%. Air enters the turbine at 1500°F (820°C) and expands to 14.7 psia (101.3 kPa). The turbine efficiency is 92%. A 65% efficient regenerator is added to the gas turbine.

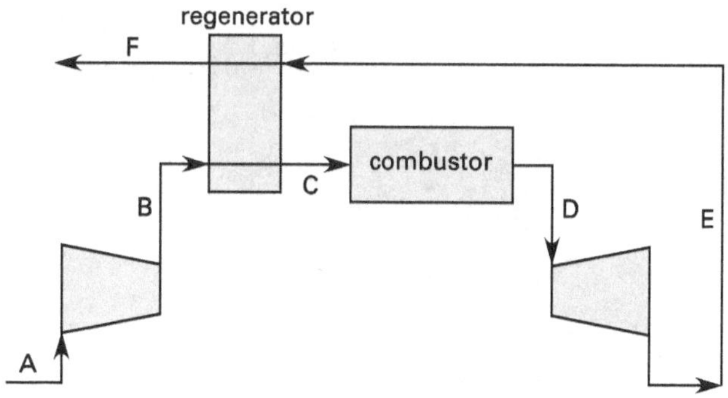

Assume the specific heat remains constant. The new thermal efficiency is most nearly

(A) 24%

(B) 28%

(C) 34%

(D) 41%

4. A gas turbine operating on the Brayton cycle with an 8:1 pressure ratio is located at 7000 ft (2100 m) altitude. The conditions at that altitude are 12 psia and 35°F (82 kPa and 2°C). While consuming 0.609 lbm/hp-hr (100 kg/GJ) of fuel and 50,000 cfm (23 500 L/s) of air, the turbine develops 6000 bhp (4.5 MW). The turbine efficiency is 80%, and the compressor efficiency is 85%. The fuel has a lower heating value of 19,000 Btu/lbm (44 MJ/kg). The fuel mass is small compared to the air mass. There is no pressure loss in the combustor. The turbine receives combustor gases at 1800°F (980°C). The air inlet filter area is 254 ft^2 (22.9 m^2). The combustion efficiency and combustor temperature remain the same. At 7000 ft, the ideal fuel rate is most nearly

(A) 3000 lbm/hr (0.37 kg/s)

(B) 3200 lbm/hr (0.40 kg/s)

(C) 3400 lbm/hr (0.42 kg/s)

(D) 3600 lbm/hr (0.44 kg/s)

5. A precision air turbine is used to drive a small dentist's drill. 140°F (60°C) air enters the turbine at the rate of 15 lbm/hr (1.9 g/s). The output of the turbine is 0.25 hp (0.19 kW). The turbine exhausts to 15 psia (103.5 kPa). The flow is steady. The isentropic efficiency of the expansion process is 60%. The change in entropy through the turbine is most nearly

(A) 0.08 Btu/lbm-°R (0.33 kJ/kg·K)

(B) 0.10 Btu/lbm-°R (0.46 kJ/kg·K)

(C) 0.18 Btu/lbm-°R (0.82 kJ/kg·K)

(D) 0.32 Btu/lbm-°R (1.5 kJ/kg·K)

SOLUTIONS

1. *Customary U.S. Solution*

As in *NCEES Handbook* table "Conversion Table for Temperature Units," the absolute temperature is

$$T_1 = 1500°\text{F} + 460° = 1960°\text{R}$$

From *NCEES Handbook:* Closed Thermodynamic Systems, the new temperature after isentropic expansion is

$$\begin{aligned} T_2 &= T_1\left(\frac{p_2}{p_1}\right)^{(k-1)/k} = (1960°\text{R})\left(\frac{50\ \frac{\text{lbf}}{\text{in}^2}}{200\ \frac{\text{lbf}}{\text{in}^2}}\right)^{(1.4-1)/1.4} \\ &= 1319°\text{R} \end{aligned}$$

As in *NCEES Handbook:* Temperature-Dependent Properties of Air (U.S. Customary Units), the heat capacity is $c_p = 0.2567$ Btu/lbm-°F. From *NCEES Handbook:* Ideal Gas Law, the air's enthalpy change is

$$\begin{aligned} \Delta h &= c_p\Delta T = \left(0.2567\ \frac{\text{Btu}}{\text{lbm-°F}}\right)(1319°\text{R} - 1960°\text{R}) \\ &= -164.5\ \text{Btu/lbm} \quad (-160\ \text{Btu/lbm}) \end{aligned}$$

SI Solution

As in *NCEES Handbook* table "Conversion Table for Temperature Units," the absolute temperature is

$$\begin{aligned} T_1 &= 820°\text{C} + 273° \\ &= 1093\text{K} \end{aligned}$$

As in *NCEES Handbook:* Closed Thermodynamic Systems, the new temperature after isentropic expansion is

$$\begin{aligned} T_2 &= T_1\left(\frac{p_2}{p_1}\right)^{(k-1)/k} = (1093\text{K})\left(\frac{350\ \text{kPa}}{1400\ \text{kPa}}\right)^{(1.4-1)/1.4} \\ &= 736\text{K} \end{aligned}$$

From *NCEES Handbook:* Temperature-Dependent Properties of Air (SI Units), the heat capacity is $c_p =$ 1.068 kJ/kg·K. From *NCEES Handbook:* Ideal Gas Law, the air's enthalpy change is

$$\begin{aligned} \Delta h &= c_p\Delta T \\ &= \left(1.068\ \frac{\text{kJ}}{\text{kg·K}}\right)(736\text{K} - 1093\text{K}) \\ &= -381.3\ \text{kJ/kg} \quad (-380\ \text{kJ/kg}) \end{aligned}$$

The answer is (C).

2. *Customary U.S. Solution*

As in *NCEES Handbook* table "Conversion Table for Temperature Units," the absolute temperature is

$$T_A = 60°F + 460° = 520°R$$

As in *NCEES Handbook:* Closed Thermodynamic Systems, the temperature at point B after an isentropic compression with a volume ratio of 5 is

$$T_B = T_A\left(\frac{V_A}{V_B}\right)^{k-1} = (520°R)(5)^{1.4-1} = 990°R$$

The pressure at point B is

$$p_B = p_A\left(\frac{V_A}{V_B}\right)^k = \left(14.7\ \frac{\text{lbf}}{\text{in}^2}\right)(5)^{1.4} = 139.9\ \text{lbf/in}^2$$

At point C,

$$T_C = 1500°F + 460° = 1960°R$$
$$p_C = p_B = 139.9\ \text{lbf/in}^2$$

At point D, the pressure is 14.7 psi. Since C-D is an isentropic process, the temperature at point D is

$$\begin{aligned} T_D &= T_C\left(\frac{p_D}{p_C}\right)^{(k-1)/k} \\ &= (1960°R)\left(\frac{14.7\ \frac{\text{lbf}}{\text{in}^2}}{139.9\ \frac{\text{lbf}}{\text{in}^2}}\right)^{(1.4-1)/1.4} \\ &= 1030°R \end{aligned}$$

From *NCEES Handbook:* Temperature-Dependent Properties of Air (U.S. Customary Units), $c_p = 0.24$ Btu/lbm-°F for both compressor and turbine. Using *NCEES Handbook:* Open Thermodynamic Systems, the actual work for the compressor is

$$\begin{aligned} W_{\text{actual}} &= \frac{W_{\text{rev}}}{\eta_{\text{isen}}} = \frac{c_p(T_A - T_B)}{\eta_{\text{isen}}} \\ &= \frac{\left(0.24\ \frac{\text{Btu}}{\text{lbm-°F}}\right)(520°R - 990°R)}{0.83} \\ &= -135.9\ \text{Btu/lbm} \end{aligned}$$

The actual work for the turbine is

$$\begin{aligned} W_{\text{actual}} &= \eta_{\text{isen}} W_{\text{rev}} = \eta_{\text{isen}} c_p(T_C - T_D) \\ &= (0.92)\left(0.24\ \frac{\text{Btu}}{\text{lbm-°F}}\right)(1960°R - 1030°R) \\ &= 205.3\ \text{Btu/lbm} \end{aligned}$$

Rearrange the above equation to find the actual T_B.

$$\begin{aligned} c_p(T_A - T_B') &= \frac{c_p(T_A - T_B)}{\eta_{\text{isen}}} \\ T_B' &= T_A - \frac{(T_A - T_B)}{\eta_{\text{isen}}} \\ &= 520°R - \frac{520°R - 990°R}{0.83} \\ &= 1086°R \end{aligned}$$

As in *NCEES Handbook:* Power Cycles, the cycle efficiency is

$$\begin{aligned} \eta_{cycle} &= \frac{W_{AB} + W_{CD}}{q_{BC}} = \frac{W_{AB} + W_{CD}}{c_p(T_C - T_B)} \\ &= \frac{-135.9\ \frac{\text{Btu}}{\text{lbm}} + 205.3\ \frac{\text{Btu}}{\text{lbm}}}{0.24\ \frac{\text{Btu}}{\text{lbm°F}}(1960 - 1086)°F} \\ &= 0.33\quad (33\%) \end{aligned}$$

SI Solution

As in *NCEES Handbook* table "Conversion Table for Temperature Units," the absolute temperature is

$$T_A = 16°C + 273° = 289K$$

As in *NCEES Handbook:* Closed Thermodynamic Systems, the temperature at point B after an isentropic compression with a volume ratio of 5 is

$$\begin{aligned} T_B &= T_A\left(\frac{V_A}{V_B}\right)^{k-1} \\ &= (289K)(5)^{1.4-1} \\ &= 550.2K \end{aligned}$$

The pressure at point B is

$$\begin{aligned} p_B &= p_A\left(\frac{V_A}{V_B}\right)^k \\ &= (101.3\ \text{kPa})(5)^{1.4} \\ &= 964.2\ \text{kPa} \end{aligned}$$

At point C,

$$T_A = 820°C + 273° = 1093K$$

$$p_c = p_a = 964.2 \text{ kPa}$$

At point D, the pressure is 101.3 kPa. Since C-D is an isentropic process, the temperature at point D is

$$\begin{aligned} T_D &= T_C\left(\frac{p_D}{p_C}\right)^{(k-1)/k} = (1093K)\left(\frac{101.3 \text{ kPa}}{964.2 \text{ kPa}}\right)^{(1.4-1)/1.4} \\ &= 574K \end{aligned}$$

From *NCEES Handbook:* Temperature-Dependent Properties of Air (SI Units), $c_p = 1.009$ kJ/kg·K for both compressor and turbine. Using *NCEES Handbook:* Open Thermodynamic Systems, the actual work for the compressor is

$$\begin{aligned} W_{actual} &= \frac{W_{rev}}{\eta_{isen}} = \frac{c_p(T_A - T_B)}{\eta_{isen}} \\ &= \frac{\left(1.009 \frac{\text{kJ}}{\text{kg·K}}\right)(289K - 550.2K)}{0.83} \\ &= -317.59 \text{ kJ/kg} \end{aligned}$$

The actual work for the turbine is

$$\begin{aligned} W_{actual} &= \eta_{isen} W_{rev} = \eta_{isen} c_p (T_C - T_D) \\ &= (0.92)\left(1.009 \frac{\text{kJ}}{\text{kg·K}}\right)(1093K - 574K) \\ &= 481.8 \text{ kJ/kg} \end{aligned}$$

Rearrange the expression to find the actual T'_B.

$$\begin{aligned} c_p(T_A - T'_B) &= \frac{c_p(T_A - T_B)}{\eta_{isen}} \\ T'_B &= T_A - \frac{T_A - T_B}{\eta_{isen}} \\ &= 289K - \frac{289K - 550.2K}{0.83} \\ &= 603.7K \end{aligned}$$

As in *NCEES Handbook:* Power Cycles, the cycle efficiency is

$$\begin{aligned} \eta_{cycle} &= \frac{W_{AB} + W_{CD}}{q_{BC}} = \frac{W_{AB} + W_{CD}}{c_p(T_C - T_B)} \\ &= \frac{-317.59 \frac{\text{kJ}}{\text{kg}} + 481.8 \frac{\text{kJ}}{\text{kg}}}{1.009 \frac{\text{kJ}}{\text{kg·K}}(1093 - 603.7)K} \\ &= 0.33 \quad (33\%) \end{aligned}$$

The answer is (A).

3. *Customary U.S. Solution*

Since specific heats are constant, use ideal gas equations rather than air tables. To convert temperatures, use *NCEES Handbook* table "Conversion Table for Temperature Units."

At A:

$$T_A = 60°F + 460° = 520°R \quad \text{[given]}$$
$$p_A = 14.7 \text{ psia} \quad \text{[given]}$$

From *NCEES Handbook:* Closed Thermodynamic Systems, at B:

$$T_B = T_A\left(\frac{v_A}{v_B}\right)^{k-1} = (520°R)(5)^{1.4-1} = 989.9°R$$

$$\begin{aligned} p_B &= p_A\left(\frac{v_A}{v_B}\right)^k = (14.7 \text{ psia})(5)^{1.4} \\ &= 139.9 \text{ psia} \end{aligned}$$

To convert temperatures, use *NCEES Handbook* table "Conversion Table for Temperature Units" for isentropic processes, at D:

$$T_D = 1500°F + 460° = 1960°R \quad \text{[given]}$$
$$p_D = p_B = 139.9 \text{ psia}$$

To convert temperatures, use *NCEES Handbook* table "Conversion Table for Temperature Units" for isentropic processes, at E:

$$p_E = 14.7 \text{ psia} \quad \text{[given]}$$

$$\begin{aligned} T_E &= T_D\left(\frac{p_E}{p_D}\right)^{(k-1)/k} = (1960°R)\left(\frac{14.7 \text{ psia}}{139.9 \text{ psia}}\right)^{(1.4-1)/1.4} \\ &= 1029.6°R \end{aligned}$$

From Eq. 14.102 (also *NCEES Handbook:* Open Thermodynamic Systems),

$$\begin{aligned} T'_B &= T_A + \frac{T_B - T_A}{\eta_{s,\text{compressor}}} \\ &= 520°R + \frac{989.9°R - 520°R}{0.83} \\ &= 1086.1°R \end{aligned}$$

From Eq. 14.104 (also *NCEES Handbook:* Open Thermodynamic Systems),

$$\begin{aligned} T'_E &= T_D - \eta_{s,\text{turbine}}(T_D - T_E) \\ &= 1960°R - (0.92)(1960°R - 1029.6°R) \\ &= 1104.0°R \end{aligned}$$

From the problem statement,

$$\eta_{\text{regenerator}} = \frac{h_C - h'_B}{h'_E - h'_B}$$

For constant c_p,

$$\begin{aligned} \eta_{\text{regenerator}} &= \frac{T_C - T'_B}{T'_E - T'_B} \\ 0.65 &= \frac{T_C - 1086.1°R}{1104.0°R - 1086.1°R} \\ T_C &= 1097.7°R \end{aligned}$$

From Eq. 14.106 (also *NCEES Handbook:* Power Cycles), with constant specific heats,

$$\begin{aligned} \eta_{\text{th}} &= \frac{(T_D - T'_E) - (T'_B - T_A)}{T_D - T_C} \\ &= \frac{(1960°R - 1104.0°R) - (1086.1°R - 520°R)}{1960°R - 1097.7°R} \\ &= 0.336 \quad (34\%) \end{aligned}$$

SI Solution

From Prob. 2,

$$\begin{aligned} T_A &= 289K \\ T'_B &= 603.7K \\ T_D &= 1093K \\ T'_E &= T'_D = 615.7K \end{aligned}$$

For constant specific heats,

$$\begin{aligned} \eta_{\text{regenerator}} &= \frac{T_C - T'_B}{T'_E - T'_B} \\ 0.65 &= \frac{T_C - 603.7K}{615.7\,K - 603.7K} \\ T_C &= 611.5K \end{aligned}$$

From Eq. 14.106 (also *NCEES Handbook:* Power Cycles with constant specific heats),

$$\begin{aligned} \eta_{\text{th}} &= \frac{(T_D - T'_E) - (T'_B - T_A)}{T_D - T_C} \\ &= \frac{(1093\,K - 615.7K) - (603.7K - 289K)}{1093\,K - 611.5K} \\ &= 0.338 \quad (34\%) \end{aligned}$$

The answer is (C).

4. Use the illustration shown for both the customary U.S. and SI solutions.

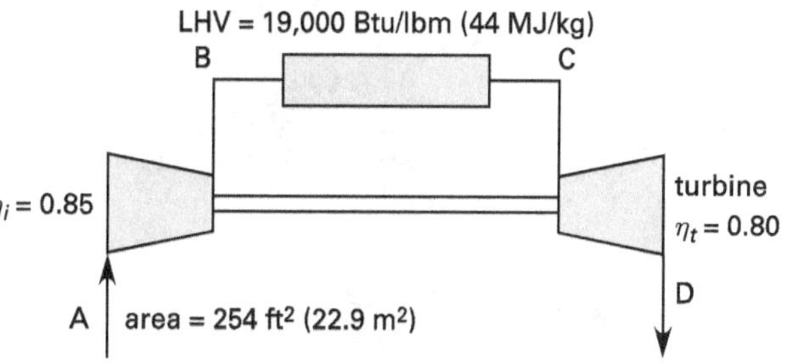

Customary U.S. Solution

At 7000 ft altitude:

$$\begin{aligned} \dot{V}_{a1} &= 50{,}000 \text{ ft}^3/\text{min} \\ \text{BHP}_1 &= 6000 \text{ hp} \\ \text{BSCF}_1 &= 0.609 \text{ lbm/hp-hr} \end{aligned}$$

At A:

$$\begin{aligned} p_A &= 12 \text{ psia} \\ T_A &= 35°F \end{aligned}$$

As in *NCEES Handbook* table "Conversion Table for Temperature Units," the absolute temperature at A is

$$T_A = 35°F + 460° = 495°R$$

As in *NCEES Handbook:* Ideal Gas Law, the air density is

$$\rho_{a1} = \frac{p_A}{RT_A} = \frac{\left(12\ \frac{\text{lbf}}{\text{in}^2}\right)\left(12\ \frac{\text{in}}{\text{ft}}\right)^2}{\left(53.35\ \frac{\text{ft-lbf}}{\text{lbm-°R}}\right)(495\text{°R})}$$
$$= 0.0654\ \text{lbm/ft}^3 \quad (0.065\ \text{lbm/ft}^3)$$

From Eq. 14.79, the air mass flow rate is

$$\begin{aligned} \dot{m}_{a1} &= \dot{V}_{a1}\rho_{a1} \\ &= \left(50{,}000\ \frac{\text{ft}^3}{\text{min}}\right)\left(0.0654\ \frac{\text{lbm}}{\text{ft}^3}\right) \\ &= 3270\ \text{lbm/min} \end{aligned}$$

At B:

$$\begin{aligned} T_B &= T_A\left(\frac{p_B}{p_A}\right)^{(k-1)/k} \\ &= (495\text{°R})(8)^{(1.4-1)/1.4} \\ &= 896.7\text{°R} \end{aligned}$$

From Eq. 14.102 (also *NCEES Handbook:* Open Thermodynamic Systems), due to the inefficiency of the compressor,

$$\begin{aligned} T'_B &= T_A - \frac{T_A - T_B}{\eta_{\text{isen}}} \\ &= 495\text{°R} - \frac{495\text{°R} - 896.7\text{°R}}{0.85} \\ &= 967.6\text{°R} \end{aligned}$$

The work of the compressor is

$$\begin{aligned} W_{\text{compression}} &= c_p(T_A - T'_B) \\ &= \left(0.24\ \frac{\text{Btu}}{\text{lbm-°F}}\right)(495\text{°R} - 967.6\text{°R}) \\ &= -113.4\ \text{Btu/lbm} \end{aligned}$$

At C:

The absolute temperature is

$$T_C = 1800\text{°F} + 460° = 2260\text{°R} \quad \text{[no change if moved]}$$

Since there is no pressure drop across the combustor, $p_C = 96$ psia.

The energy requirement from the fuel is

$$\begin{aligned} \dot{m}\Delta h &= \dot{m}_{a1}(h_C - h'_B) = \dot{m}_{a1}c_p(T_C - T'_B) \\ &= \left(3270\ \frac{\text{lbm}}{\text{min}}\right)\left(0.24\ \frac{\text{Btu}}{\text{lbm-°F}}\right)(2260\text{°R} - 967.6\text{°R}) \\ &= 1.014 \times 10^6\ \text{Btu/min} \end{aligned}$$

The ideal fuel rate is

$$\begin{aligned} \frac{\dot{m}\Delta h}{\text{LHV}} &= \frac{\left(1.014 \times 10^6\ \frac{\text{Btu}}{\text{min}}\right)\left(60\ \frac{\text{min}}{\text{hr}}\right)}{19{,}000\ \frac{\text{Btu}}{\text{lbm}}} \\ &= 3202\ \text{lbm/hr} \quad (3200\ \text{lbm/hr}) \end{aligned}$$

SI Solution

At 2100 m altitude:

At A:

$$p_A = 82\ \text{kPa} \quad \text{[given]}$$

As in *NCEES Handbook* table "Conversion Table for Temperature Units," the absolute temperature is

$$T_A = 2\text{°C} + 273° = 275\text{K} \quad \text{[given]}$$

As in *NCEES Handbook:* Ideal Gas Law, the air density is

$$\rho_{a1} = \frac{p_A}{RT_A} = \frac{(82\ \text{kPa})\left(1000\ \frac{\text{Pa}}{\text{kPa}}\right)}{\left(287.03\ \frac{\text{J}}{\text{kg·K}}\right)(275\text{K})}$$
$$= 1.0389\ \text{kg/m}^3 \quad (1.0\ \text{kg/m}^3)$$

From Eq. 14.79, the air mass flow rate is

$$\begin{aligned} \dot{m}_{a1} &= \dot{V}_{a1}\rho_{a1} \\ &= \frac{\left(23\,500\ \frac{\text{L}}{\text{s}}\right)\left(1.0389\ \frac{\text{kg}}{\text{m}^3}\right)}{1000\ \frac{\text{L}}{\text{m}^3}} \\ &= 24.41\ \text{kg/s} \quad (24\ \text{kg/s}) \end{aligned}$$

At B:

$$p_C = 8p_A = (8)(82\ \text{kPa}) = 656\ \text{kPa}$$

Assuming isentropic compression,

$$T_B = T_A\left(\frac{p_B}{p_A}\right)^{(k-1)/k} = (275\text{K})(8)^{(1.4-1)/1.4} = 498.1\text{K}$$

Due to the inefficiency of the compressor, from Eq. 14.102 (also *NCEES Handbook:* Open Thermodynamic Systems),

$$\begin{aligned} T'_B &= T_A - \frac{(T_A - T_B)}{\eta_{\text{isen}}} = 275\text{K} - \frac{275\text{K} - 498.1\text{K}}{0.85} \\ &= 537.5\text{K} \end{aligned}$$

The work of the compressor is

$$\begin{aligned} W_{\text{compression}} &= c_p(T_A - T'_B) \\ &= \left(1.009\ \frac{\text{kJ}}{\text{kg·K}}\right)(275\text{K} - 537.5\text{K}) \\ &= -264.9\ \text{kJ/kg} \end{aligned}$$

At C:

The absolute temperature is

$$T_C = 980°\text{C} + 273° = 1253\text{K}$$

Since there is no pressure drop across the combustor, $p_C = 656$ kPa.

The power provided by the fuel is

$$\begin{aligned} \dot{m}\Delta h &= \dot{m}_{a1}(h_C - h'_B) = \dot{m}_{a1}c_p(T_C - T'_B) \\ &= \left(24.41\ \frac{\text{kg}}{\text{s}}\right)\left(1.009\ \frac{\text{kJ}}{\text{kg·K}}\right)(1253\text{K} - 537.5\text{K}) \\ &= 17\,622\ \text{kJ/s} \end{aligned}$$

The ideal fuel rate is

$$\begin{aligned} \frac{\dot{m}\Delta h}{\text{LHV}} &= \frac{17\,622\ \frac{\text{kJ}}{\text{s}}}{\left(44\ \frac{\text{MJ}}{\text{kg}}\right)\left(1000\ \frac{\text{kJ}}{\text{MJ}}\right)} \\ &= 0.40\ \text{kg/s} \end{aligned}$$

The answer is (B).

5. Use the illustration shown for both the customary U.S. and SI solutions.

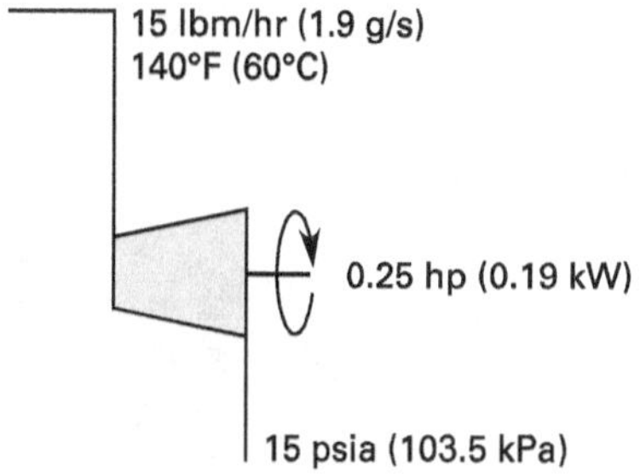

Customary U.S. Solution

The drill power is

$$\begin{aligned} P &= 0.25\ \text{hp}\quad [\text{given}] \\ &= (0.25\ \text{hp})\left(2545\ \frac{\text{Btu}}{\text{hp-hr}}\right) \\ &= 636.25\ \text{Btu/hr} \end{aligned}$$

The absolute inlet temperature is

$$T_1 = 140°\text{F} + 460° = 600°\text{R}$$

From Eq. 12.18,

$$P = \dot{m}(h_1 - h'_2)$$

$$\eta_{s,\text{turbine}} = \frac{h_1 - h'_2}{h_1 - h_2} = \frac{c_p(T_1 - T'_2)}{c_p(T_1 - T_2)}$$

Combining equations and rearranging,

$$\begin{aligned} P &= \dot{m}\eta_{s,\text{turbine}}(h_1 - h_2) \\ &= \dot{m}\eta_{s,\text{turbine}}c_p(T_1 - T_2) \\ T_2 &= T_1 - \frac{P}{\dot{m}\eta_{s,\text{turbine}}c_p} \\ &= 600°\text{R} - \frac{636.25\ \frac{\text{Btu}}{\text{hr}}}{\left(15\ \frac{\text{lbm}}{\text{hr}}\right)(0.6)\left(0.24\ \frac{\text{Btu}}{\text{lbm-°F}}\right)} \\ &= 305.4°\text{R} \end{aligned}$$

The actual temperature is

$$\begin{aligned} T'_2 &= T_1 - \eta_{s,\text{turbine}}(T_1 - T_2) \\ &= 600°\text{R} - (0.6)(600°\text{R} - 305.4°\text{R}) \\ &= 423.2°\text{R} \end{aligned}$$

Using an equation for isentropic processes from *NCEES Handbook:* Closed Thermodynamic Systems, the exit pressure is

$$\begin{aligned} P_1 &= P_2\left(\frac{T_1}{T_2}\right)^{k/(k-1)} \\ &= \left(15\ \frac{\text{lbf}}{\text{in}^2}\right)\left(\frac{600^\circ\text{R}}{305.4^\circ\text{R}}\right)^{1.4/(1.4-1)} \\ &= 159.4\ \text{psi} \end{aligned}$$

Using the molecular weight from *NCEES Handbook:* Combustion Reactions and *NCEES Handbook:* Ideal Gas Law, the change in entropy is

$$\begin{aligned} \Delta s &= c_p \ln\frac{T_2}{T_1} - \frac{R\ln\left(\frac{p_2}{p_1}\right)}{MW} \\ &= \left(0.24\ \frac{\text{Btu}}{\text{lbm-}^\circ\text{F}}\right)\left(\ln\frac{423.2^\circ\text{R}}{600^\circ\text{R}}\right) \\ &\quad - \left(\frac{1545\ \dfrac{\text{ft-lbf}}{\text{lbmol-}^\circ\text{R}}}{\left(29\ \dfrac{\text{lbm}}{\text{lbmol}}\right)\left(778\ \dfrac{\text{ft-lbf}}{\text{Btu}}\right)}\right)\left(\ln\frac{15\ \dfrac{\text{lbf}}{\text{in}^2}}{159.4\ \dfrac{\text{lbf}}{\text{in}^2}}\right) \\ &= 0.0773\ \text{Btu/lbm-}^\circ\text{R} \quad (0.08\ \text{Btu/lbm-}^\circ\text{R}) \end{aligned}$$

SI Solution

The absolute inlet temperature is

$$T_1 = 60^\circ\text{C} + 273^\circ = 333\text{K}$$

From Eq. 12.18,

$$P = \dot{m}(h_1 - h_2')$$

$$\eta_{s,\text{turbine}} = \frac{h_1 - h_2'}{h_1 - h_2} = \frac{c_p(T_1 - T_2')}{c_p(T_1 - T_2)}$$

Combining equations and rearranging,

$$\begin{aligned} P &= \dot{m}\eta_{s,\text{turbine}}(h_1 - h_2) = \dot{m}\eta_{s,\text{turbine}}c_p(T_1 - T_2) \\ T_2 &= T_1 - \frac{P}{\dot{m}\eta_{s,\text{turbine}}c_p} \\ &= 333\text{K} - \frac{(0.19\ \text{kW})\left(1000\dfrac{\text{g}}{\text{kg}}\right)}{\left(1.9\ \dfrac{\text{g}}{\text{s}}\right)(0.6)\left(1.009\ \dfrac{\text{kJ}}{\text{kg·K}}\right)} \\ &= 167.8\text{K} \end{aligned}$$

The actual temperature is

$$\begin{aligned} T_2' &= T_1 - \eta_{s,\text{turbine}}(T_1 - T_2) \\ &= 333\text{K} - (0.6)(333\text{K} - 167.8\text{K}) \\ &= 233.9\text{K} \end{aligned}$$

Using an equation for isentropic processes from *NCEES Handbook:* Closed Thermodynamic Systems, the exit pressure is

$$\begin{aligned} P_1 &= P_2\left(\frac{T_1}{T_2}\right)^{k/(k-1)} \\ &= (103.5\ \text{kPa})\left(\frac{333\text{K}}{167.8\text{K}}\right)^{1.4/(1.4-1)} \\ &= 1140\ \text{kPa} \end{aligned}$$

Using the molecular weight from *NCEES Handbook:* Combustion Reactions and *NCEES Handbook:* Ideal Gas Law, the change in entropy is

$$\begin{aligned} \Delta s &= c_p \ln\frac{T_2}{T_1} - \frac{R\ln\left(\frac{p_2}{p_1}\right)}{MW} \\ &= \left(1.009\ \frac{\text{kJ}}{\text{kg·K}}\right)\left(\ln\frac{233.9\text{K}}{333\text{K}}\right) \\ &\quad - \left(\frac{8.314\ \dfrac{\text{kJ}}{\text{kmol·K}}}{29\ \dfrac{\text{kg}}{\text{kmol}}}\right)\left(\ln\frac{103.5\ \text{kPa}}{1140\ \text{kPa}}\right) \\ &= 0.3314\ \text{kJ/kg·K} \quad (0.33\ \text{kJ/kg·K}) \end{aligned}$$

The answer is (A).

15 Gas Compression Cycles

PRACTICE PROBLEMS

1. The isentropic efficiency of a compressor that takes in air at 8 psia and −10°F (55 kPa and −20°C) and discharges it at 40 psia and 315°F (275 kPa and 160°C) is most nearly

(A) 61%

(B) 66%

(C) 74%

(D) 81%

2. Air at 14.7 psia and 500°F (101.3 kPa and 260°C) is compressed in a centrifugal compressor to 6 atm. The isentropic efficiency of the compression process is 65%. The increase in entropy is most nearly

(A) 0.05 Btu/lbm-°R (0.20 kJ/kg·K)

(B) 0.099 Btu/lbm-°R (0.47 kJ/kg·K)

(C) 0.23 Btu/lbm-°R (1.0 kJ/kg·K)

(D) 0.44 Btu/lbm-°R (2.1 kJ/kg·K)

3. Compressors A and B both discharge into a common tank. The storage tank contains 100 psia, 90°F air (700 kPa, 32°C). Both compressors receive air at 14.7 psia and 80°F (101.3 kPa and 27°C). The flow rate through compressor A is 600 cfm (300 L/s). Process C uses 100 cfm of 80 psia, 85°F (50 L/s of 550 kPa, 29°C) air. Process D uses 120 cfm of 85 psia, 80°F (60 L/s of 590 kPa, 27°C) air. Process E uses 8 lbm/min (0.06 kg/s) of 85°F (29°C) air. Air is an ideal gas, and compressibility effects are to be disregarded.

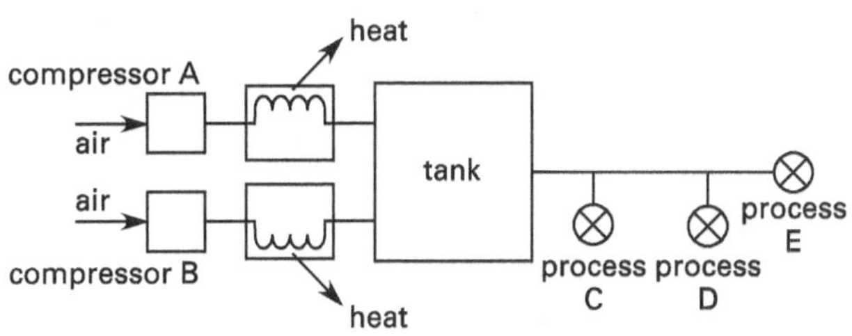

The volumetric flow rate through compressor B is most nearly

(A) 620 ft^3/min (320 L/s)

(B) 740 ft^3/min (370 L/s)

(C) 900 ft^3/min (450 L/s)

(D) 1100 ft^3/min (550 L/s)

4. 300 cfm (150 L/s) of air at 14.7 psia and 90°F (101.3 kPa and 32°C) enters a compressor. The compressor discharges air into a water-cooled heat exchanger. The compressed air is stored at 300 psig and 90°F (2.07 MPa and 32°C) in a 1000 ft^3 (27 m^3) tank. The tank feeds three air-driven tools with the flow rates and properties listed. The pressures to the air-driven tools are regulated and remain constant at the minimum required operating pressure as the tank pressure changes. Air is an ideal gas, and compressibility effects are to be disregarded.

	tool 1	tool 2	tool 3
flow rate			
(cfm)	40	15	unknown
(L/s)	19	7	unknown
flow rate			
(lbm/min)	unknown	unknown	6
(kg/s)	unknown	unknown	0.045
minimum pressure			
(psig)	90	50	80
(kPa)	620	350	550
temperature			
(°F)	90	85	80
(°C)	32	29	27

Approximately how long can the system run?

(A) 0.6 hr (1.2 h)

(B) 1.8 hr (3.6 h)

(C) 3.4 hr (6.8 h)

(D) 6.6 hr (13 h)

SOLUTIONS

1.

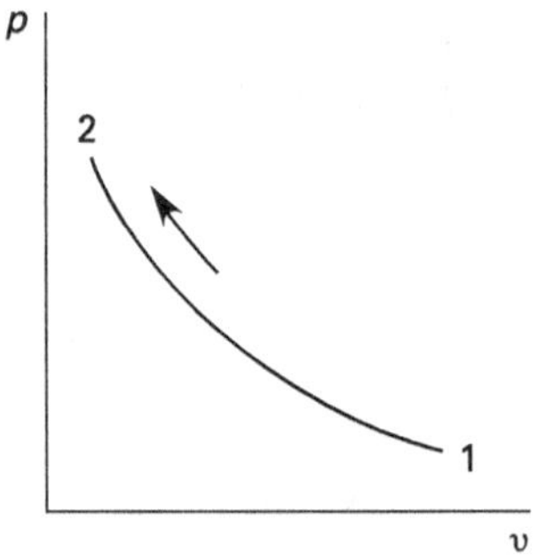

Customary U.S. Solution

Assume air is an ideal gas.

From Eq. 14.21 (also *NCEES Handbook:* Closed Thermodynamic Systems), the isentropic temperature at point 2 is

$$T_2 = T_1\left(\frac{p_2}{p_1}\right)^{(k-1)/k}$$

As in *NCEES Handbook* table "Conversion Table for Temperature Units," the absolute temperature at point 1 is

$$T_1 = -10°\text{F} + 460° = 450°\text{R}$$

$$\begin{aligned} T_2 &= (450°\text{R})\left(\frac{40 \text{ psia}}{8 \text{ psia}}\right)^{(1.4-1)/1.4} \\ &= 712.7°\text{R} \\ T_2 &= 712.7°\text{R} - 460° = 252.7°\text{F} \end{aligned}$$

The efficiency of the compressor is given by Eq. 14.102 (also *NCEES Handbook:* Open Thermodynamic Systems).

$$\begin{aligned} \eta_{s,\text{compressor}} &= \frac{T_2 - T_1}{T_2' - T_1} \\ &= \frac{252.7°\text{F} - (-10°\text{F})}{315°\text{F} - (-10°\text{F})} \\ &= 0.808 \quad (81\%) \end{aligned}$$

SI Solution

Assume air is an ideal gas.

From Eq. 14.21 (also *NCEES Handbook:* Closed Thermodynamic Systems), the isentropic temperature at point 2 is

$$T_2 = T_1\left(\frac{p_2}{p}\right)^{(k-1)/k}$$

As in *NCEES Handbook* table "Conversion Table for Temperature Units," the absolute temperature at point 1 is

$$\begin{aligned} T_1 &= -20°\text{C} + 273° = 253\text{K} \\ T_2 &= (253\text{K})\left(\frac{275 \text{ kPa}}{55 \text{ kPa}}\right)^{(1.4-1)/1.4} \\ &= 400.7\text{K} \\ T_2 &= 400.7\text{K} - 273° = 127.7°\text{C} \end{aligned}$$

The efficiency of the compressor is given by Eq. 14.102 (also *NCEES Handbook:* Open Thermodynamic Systems).

$$\begin{aligned} \eta_{s,\text{compressor}} &= \frac{T_2 - T_1}{T_2' - T_1} \\ &= \frac{127.7°\text{C} - (-20°\text{C})}{160°\text{C} - (-20°\text{C})} \\ &= 0.821 \quad (81\%) \end{aligned}$$

The answer is (D).

2. *Customary U.S. Solution*

As in *NCEES Handbook* table "Conversion Table for Temperature Units," the absolute temperature is

$$T_1 = 500°\text{F} + 460° = 960°\text{R}$$

From Eq. 14.21 (also *NCEES Handbook:* Closed Thermodynamic Systems), the isentropic temperature after expansion is

$$\begin{aligned} T_2 &= T_1\left(\frac{p_2}{p_1}\right)^{(k-1)/k} \\ &= (960°\text{R})\left(\frac{6 \text{ atm}}{1 \text{ atm}}\right)^{(1.4-1)/1.4} \\ &= 1602°\text{R} \end{aligned}$$

As in *NCEES Handbook:* Open Thermodynamic Systems, the actual temperature, taking into account isentropic efficiency, is

$$\begin{aligned} T_2' &= T_1 - \frac{T_1 - T_2}{\eta_{\text{isen}}} \\ &= 960^\circ\text{R} - \frac{960^\circ\text{R} - 1602^\circ\text{R}}{0.65} \\ &= 1948^\circ\text{R} \end{aligned}$$

Using Eq. 10.39 (also the molecular weight as explained in *NCEES Handbook:* Combustion Reactions and *NCEES Handbook:* Ideal Gas Law), the increase in entropy is

$$\begin{aligned} \Delta s &= c_p \ln \frac{T_2}{T_1} - R \ln \frac{p_2}{p_1} \\ &= \left(0.24\ \frac{\text{Btu}}{\text{lbm-}^\circ\text{F}}\right)\left(\ln \frac{1948^\circ\text{R}}{960^\circ\text{R}}\right) \\ &\quad - \left(\frac{1545\ \frac{\text{ft-lbf}}{\text{lbmol-}^\circ\text{R}}}{\left(29\ \frac{\text{lbm}}{\text{lbmol}}\right)\left(778\ \frac{\text{ft-lbf}}{\text{Btu}}\right)}\right)\left(\ln \frac{6\ \text{atm}}{1\ \text{atm}}\right) \\ &= 0.047\ \text{Btu/lbm-}^\circ\text{R} \quad (0.05\ \text{Btu/lbm-}^\circ\text{R}) \end{aligned}$$

SI Solution

As in *NCEES Handbook* table "Conversion Table for Temperature Units," the absolute inlet temperature is

$$T_1 = 260^\circ\text{C} + 273^\circ = 533\text{K}$$

From Eq. 14.21 (also *NCEES Handbook:* Closed Thermodynamic Systems), the isentropic temperature after expansion is

$$T_2 = T_1 \left(\frac{p_2}{p_1}\right)^{(k-1)/k} = (533\text{K})\left(\frac{6\ \text{atm}}{1\ \text{atm}}\right)^{(1.4-1)/1.4} = 889\text{K}$$

As in *NCEES Handbook:* Open Thermodynamic Systems, the actual temperature, taking into account isentropic efficiency, is

$$\begin{aligned} T_2' &= T_1 - \frac{(T_1 - T_2)}{\eta_{\text{isen}}} = 533\text{K} - \frac{533\text{K} - 889\text{K}}{0.65} \\ &= 1081\text{K} \end{aligned}$$

Using Eq. 10.39 (also the molecular weight as explained in *NCEES Handbook:* Combustion Reactions and *NCEES Handbook:* Ideal Gas Law), the increase in entropy is

$$\begin{aligned} \Delta s &= c_p \ln \frac{T_2}{T_1} - R \ln \frac{p_2}{p_1} \\ &= \left(1.009\ \frac{\text{kJ}}{\text{kg}\cdot\text{K}}\right)\left(\ln \frac{1081\text{K}}{533\text{K}}\right) \\ &\quad - \left(\frac{8.314\ \frac{\text{kJ}}{\text{kmol}\cdot\text{K}}}{29\ \frac{\text{kg}}{\text{kmol}}}\right)\left(\ln \frac{6\ \text{atm}}{1\ \text{atm}}\right) \\ &= (0.1998\ \text{kJ/kg}\cdot\text{K}) \quad (0.20\ \text{kJ/kg}\cdot\text{K}) \end{aligned}$$

The answer is (A).

3. *Customary U.S. Solution*

For process C, the absolute temperature is 85°F + 460°= 545°R. Using ideal gas laws, the mass flow rate for process C is

$$\begin{aligned} \dot{m}_\text{C} &= \frac{p_\text{C} \dot{V}_\text{C} M}{R T_\text{C}} = \frac{\left(80\ \frac{\text{lbf}}{\text{in}^2}\right)\left(12\ \frac{\text{in}}{\text{ft}}\right)^2\left(100\ \frac{\text{ft}^3}{\text{min}}\right)\left(29\ \frac{\text{lbm}}{\text{lbmol}}\right)}{\left(1545\ \frac{\text{ft-lbf}}{\text{lbmol-}^\circ\text{R}}\right)(545^\circ\text{R})} \\ &= 39.66\ \text{lbm/min} \end{aligned}$$

Similarly, the mass flow rate for process D with an absolute temperature of 80°F + 460° = 540°R is

$$\begin{aligned} \dot{m}_\text{D} &= \frac{p_\text{D} \dot{V}_\text{D} M}{R T_\text{D}} \\ &= \frac{\left(85\ \frac{\text{lbf}}{\text{in}^2}\right)\left(12\ \frac{\text{in}}{\text{ft}}\right)^2\left(120\ \frac{\text{ft}^3}{\text{min}}\right)\left(29\ \frac{\text{lbm}}{\text{lbmol}}\right)}{\left(1545\ \frac{\text{ft-lbf}}{\text{lbmol-}^\circ\text{R}}\right)(540^\circ\text{R})} \\ &= 51.03\ \text{lbm/min} \end{aligned}$$

For process E, the mass flow rate is given as $\dot{m}_\text{E} = 8$ lbm/min.

The total mass flow rate for all three processes is

$$\begin{aligned} \dot{m}_\text{total} &= \dot{m}_\text{C} + \dot{m}_\text{D} + \dot{m}_\text{E} \\ &= 39.66\ \frac{\text{lbm}}{\text{min}} + 51.03\ \frac{\text{lbm}}{\text{min}} + 8\ \frac{\text{lbm}}{\text{min}} \\ &= 98.69\ \text{lbm/min} \end{aligned}$$

Thermodynamics

The mass flow rate into compressor A is with an absolute temperature of 80°F + 460° = 540°R.

$$\dot{m}_A = \frac{p_A \dot{V}_A M}{RT_A} = \frac{\left(14.7\ \frac{\text{lbf}}{\text{in}^2}\right)\left(12\ \frac{\text{in}}{\text{ft}}\right)^2\left(600\ \frac{\text{ft}^3}{\text{min}}\right)\left(29\ \frac{\text{lbm}}{\text{lbmol}}\right)}{\left(1545\ \frac{\text{ft-lbf}}{\text{lbmol-°R}}\right)(540°\text{R})} = 44.13\ \text{lbm/min}$$

The required input for compressor B is

$$\begin{aligned}\dot{m}_B &= \dot{m}_{\text{total}} - \dot{m}_A \\ &= 98.69\ \frac{\text{lbm}}{\text{min}} - 44.13\ \frac{\text{lbm}}{\text{min}} \\ &= 54.56\ \text{lbm/min}\end{aligned}$$

The volumetric flow rate for compressor B is

$$\dot{V}_B = \frac{\dot{m}_B R T_B}{p_A M} = \frac{\left(54.56\ \frac{\text{lbm}}{\text{min}}\right)\left(1545\ \frac{\text{ft-lbf}}{\text{lbmol-°R}}\right)(540°\text{R})}{\left(14.7\ \frac{\text{lbf}}{\text{in}^2}\right)\left(12\ \frac{\text{in}}{\text{ft}}\right)^2\left(29\ \frac{\text{lbm}}{\text{lbmol}}\right)} = 742\ \text{ft}^3/\text{min} \quad (740\ \text{ft}^3/\text{min})$$

SI Solution

The absolute temperature for process C is 29°C + 273° = 302K. Using ideal gas laws, the mass flow rate for process C is

$$\dot{m}_C = \frac{p_C \dot{V}_C M}{RT_C} = \frac{(550\ \text{kPa})\left(50\ \frac{\text{L}}{\text{s}}\right)\left(29\ \frac{\text{kg}}{\text{kmol}}\right)}{\left(8.314\ \frac{\text{kJ}}{\text{kmol·K}}\right)(302\text{K})\left(1000\ \frac{\text{L}}{\text{m}^3}\right)} = 0.3173\ \text{kg/s}$$

The absolute temperature for process D is 27°C + 273° = 300K. Using ideal gas laws, the mass flow rate for process D is

$$\dot{m}_D = \frac{p_D \dot{V}_D M}{RT_D} = \frac{(590\ \text{kPa})\left(60\ \frac{\text{L}}{\text{s}}\right)\left(29\ \frac{\text{kg}}{\text{kmol}}\right)}{\left(8.314\ \frac{\text{kJ}}{\text{kmol·K}}\right)(300\text{K})\left(1000\ \frac{\text{L}}{\text{m}^3}\right)} = 0.4111\ \text{kg/s}$$

For process E, the mass flow rate is given as $\dot{m}_E = 0.06$ kg/s.

The total mass flow rate for all three processes is

$$\begin{aligned}\dot{m}_{\text{total}} &= \dot{m}_C + \dot{m}_D + \dot{m}_E \\ &= 0.3173\ \frac{\text{kg}}{\text{s}} + 0.4111\ \frac{\text{kg}}{\text{s}} + 0.06\ \frac{\text{kg}}{\text{s}} \\ &= 0.7884\ \text{kg/s}\end{aligned}$$

The absolute temperature for air into compressors A and B is 27°C + 273° = 300K.

The mass flow rate into compressor A is

$$\dot{m}_A = \frac{p_A \dot{V}_A M}{RT_A} = \frac{(101.3\ \text{kPa})\left(300\frac{\text{L}}{\text{s}}\right)\left(29\ \frac{\text{kg}}{\text{kmol}}\right)}{\left(8.314\ \frac{\text{kJ}}{\text{kmol·K}}\right)(300\text{K})\left(1000\ \frac{\text{L}}{\text{m}^3}\right)} = 0.3530\ \text{kg/s}$$

The required input for compressor B is

$$\begin{aligned}\dot{m}_B &= \dot{m}_{\text{total}} - \dot{m}_A \\ &= 0.7884\ \frac{\text{kg}}{\text{s}} - 0.3530\ \frac{\text{kg}}{\text{s}} \\ &= 0.4354\ \text{kg/s}\end{aligned}$$

The volumetric flow rate for compressor B is

$$\dot{V}_B = \frac{\dot{m}_B R T_B}{p_A M} = \frac{\left(0.4354\ \frac{\text{kg}}{\text{s}}\right)\left(8.314\ \frac{\text{kJ}}{\text{kmol·K}}\right)(300\text{K})}{(101.3\ \text{kPa})\left(29\ \frac{\text{kg}}{\text{kmol}}\right)} = 370\ \text{L/s}$$

The answer is (B).

4.

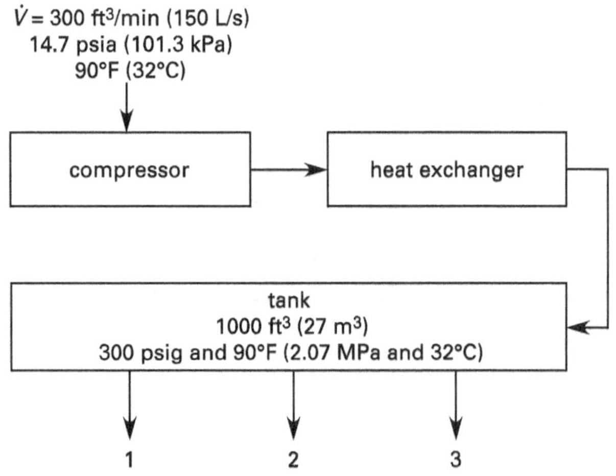

Customary U.S. Solution

Assume steady flow and constant properties.

As in *NCEES Handbook* table "Conversion Table for Temperature Units," the absolute temperature for compressor air is $90°F + 460° = 550°R$. Using ideal gas laws, the mass flow rate of air into the compressor is

$$\begin{aligned}\dot{m} &= \frac{p\dot{V}M}{RT}\\ &= \frac{\left(14.7\ \frac{\text{lbf}}{\text{in}^2}\right)\left(12\ \frac{\text{in}}{\text{ft}}\right)^2\left(300\ \frac{\text{ft}^3}{\text{min}}\right)\left(29\ \frac{\text{lbm}}{\text{lbmol}}\right)}{\left(1545\ \frac{\text{ft-lbf}}{\text{lbmol-°R}}\right)(550°\text{R})}\\ &= 21.7\ \text{lbm/min}\end{aligned}$$

As in *NCEES Handbook* table "Conversion Table for Temperature Units," the absolute temperature and absolute pressure of stored compressed air are

$$\begin{aligned}T &= 90°\text{F} + 460°\\ &= 550°\text{R}\\ p &= 300\ \text{psig} + 14.7\ \text{psia}\\ &= 314.7\ \text{psia}\end{aligned}$$

The mass of stored compressed air in a 1000 ft^3 tank is

$$\begin{aligned}m &= \frac{pVM}{RT}\\ &= \frac{\left(314.7\ \frac{\text{lbf}}{\text{in}^2}\right)\left(12\ \frac{\text{in}}{\text{ft}}\right)^2(1000\ \text{ft}^3)\left(29\ \frac{\text{lbm}}{\text{lbmol}}\right)}{\left(1545\ \frac{\text{ft-lbf}}{\text{lbmol-°R}}\right)(550°\text{R})}\\ &= 1545.9\ \text{lbm}\end{aligned}$$

Assuming that each tool operates at its minimum pressure, the mass leaving the system can be calculated as follows.

Tool 1:

The absolute pressure is

$$90\ \text{psig} + 14.7\ \text{psia} = 104.7\ \text{psia}$$

As in *NCEES Handbook* table "Conversion Table for Temperature Units," the absolute temperature is

$$90°\text{F} + 460° = 550°\text{R}$$

$$\begin{aligned}\dot{m}_{\text{tool 1}} &= \frac{p\dot{V}M}{RT}\\ &= \frac{\left(104.7\ \frac{\text{lbf}}{\text{in}^2}\right)\left(12\ \frac{\text{in}}{\text{ft}}\right)^2\left(40\ \frac{\text{ft}^3}{\text{min}}\right)\left(29\ \frac{\text{lbm}}{\text{lbmol}}\right)}{\left(1545\ \frac{\text{ft-lbf}}{\text{lbmol-°R}}\right)(550°\text{R})}\\ &= 20.57\ \text{lbm/min}\end{aligned}$$

Tool 2:

The absolute pressure is

$$50\ \text{psig} + 14.7\ \text{psia} = 64.7\ \text{psia}$$

As in *NCEES Handbook* table "Conversion Table for Temperature Units," the absolute temperature is

$$85°\text{F} + 460° = 545°\text{R}$$

$$\begin{aligned}\dot{m}_{\text{tool 2}} &= \frac{p\dot{V}M}{RT}\\ &= \frac{\left(64.7\ \frac{\text{lbf}}{\text{in}^2}\right)\left(12\ \frac{\text{in}}{\text{ft}}\right)^2\left(15\ \frac{\text{ft}^3}{\text{min}}\right)\left(29\ \frac{\text{lbm}}{\text{lbmol}}\right)}{\left(1545\ \frac{\text{ft-lbf}}{\text{lbmol-°R}}\right)(545°\text{R})}\\ &= 4.81\ \text{lbm/min}\end{aligned}$$

Tool 3:

$$\dot{m}_{\text{tool 3}} = 6\ \text{lbm/min}\quad\text{[given]}$$

The total mass flow leaving the system is

$$\begin{aligned}\dot{m}_{\text{total}} &= \dot{m}_{\text{tool 1}} + \dot{m}_{\text{tool 2}} + \dot{m}_{\text{tool 3}}\\ &= 20.57\ \frac{\text{lbm}}{\text{min}} + 4.81\ \frac{\text{lbm}}{\text{min}} + 6\ \frac{\text{lbm}}{\text{min}}\\ &= 31.38\ \text{lbm/min}\end{aligned}$$

The critical pressure of 104.7 psia is required for tool 1 operation. The mass in the tank when critical pressure is achieved is

$$\begin{aligned}m_{\text{tank critical}} &= \frac{pVM}{RT}\\ &= \frac{\left(104.7\ \frac{\text{lbf}}{\text{in}^2}\right)\left(12\ \frac{\text{in}}{\text{ft}}\right)^2(1000\ \text{ft}^3)\left(29\ \frac{\text{lbm}}{\text{lbmol}}\right)}{\left(1545\ \frac{\text{ft-lbf}}{\text{lbmol-°R}}\right)(550°\text{R})}\\ &= 514.3\ \text{lbm}\end{aligned}$$

The amount in the tank to be depleted is

$$\begin{aligned} m_{\text{depleted}} &= m_{\text{tank}} - m_{\text{tank,critical}} = 1545.9 \text{ lbm} - 514.3 \text{ lbm} \\ &= 1031.6 \text{ lbm} \end{aligned}$$

The net flow rate of air to the tank is

$$\begin{aligned} \dot{m}_{\text{net}} &= \dot{m}_{\text{in}} - \dot{m}_{\text{out}} \\ &= 21.7 \ \frac{\text{lbm}}{\text{min}} - 31.38 \ \frac{\text{lbm}}{\text{min}} \\ &= -9.68 \text{ lbm/min} \end{aligned}$$

The time the system can run is

$$\begin{aligned} \frac{m_{\text{depleted}}}{\dot{m}_{\text{net}}} &= \frac{1031.6 \text{ lbm}}{9.68 \ \frac{\text{lbm}}{\text{min}}} \\ &= 106.6 \text{ min} \quad (1.8 \text{ hr}) \end{aligned}$$

SI Solution

As in *NCEES Handbook* table "Conversion Table for Temperature Units," the absolute temperature of the compressor air is

$$32°\text{C} + 273° = 305\text{K}$$

Using ideal gas laws, the mass flow rate of air into the compressor is

$$\begin{aligned} \dot{m} &= \frac{p\dot{V}M}{RT} = \frac{(101.3 \text{ kPa})\left(150 \ \frac{\text{L}}{\text{s}}\right)\left(29 \ \frac{\text{kg}}{\text{kmol}}\right)}{\left(8.314 \ \frac{\text{kJ}}{\text{kmol·K}}\right)(305\text{K})\left(1000 \ \frac{\text{L}}{\text{m}^3}\right)} \\ &= 0.1736 \text{ kg/s} \end{aligned}$$

As in *NCEES Handbook* table "Conversion Table for Temperature Units," the absolute temperature and pressure of stored compressed air are

$$\begin{aligned} T &= 32°\text{C} + 273° = 305\text{K} \\ p &= (2.07 \text{ MPa})\left(10^6 \ \frac{\text{Pa}}{\text{MPa}}\right) \\ &= 2.07 \times 10^6 \text{ Pa} \end{aligned}$$

The mass of stored compressed air in a 1000 ft^3 tank is

$$\begin{aligned} m_{\text{tank}} &= \frac{pVM}{RT} \\ &= \frac{(2.07 \times 10^6 \text{ Pa})(27 \text{ m}^3)\left(29 \ \frac{\text{kg}}{\text{kmol}}\right)}{\left(8.314 \ \frac{\text{kJ}}{\text{kmol·K}}\right)(305\text{K})} \\ &= 639.1 \text{ kg} \end{aligned}$$

Assuming that each tool operates at its minimum pressure, the mass leaving the system can be calculated as follows.

Tool 1:

As in *NCEES Handbook* table "Conversion Table for Temperature Units," the absolute temperature is $32°\text{C} + 273° = 305\text{K}$.

$$\begin{aligned} \dot{m}_{\text{tool 1}} &= \frac{p\dot{V}M}{RT} \\ &= \frac{(620 \text{ kPa})\left(19 \ \frac{\text{L}}{\text{s}}\right)\left(29 \ \frac{\text{kg}}{\text{kmol}}\right)}{\left(8.314 \ \frac{\text{kJ}}{\text{kmol·K}}\right)(305\text{K})\left(1000 \ \frac{\text{L}}{\text{m}^3}\right)} \\ &= 0.1346 \text{ kg/s} \end{aligned}$$

Tool 2:

As in *NCEES Handbook* table "Conversion Table for Temperature Units," the absolute temperature is $29°\text{C} + 273° = 302\text{K}$.

$$\begin{aligned} \dot{m}_{\text{tool 2}} &= \frac{p\dot{V}}{RT} \\ &= \frac{(350 \text{ kPa})\left(1000 \ \frac{\text{Pa}}{\text{kPa}}\right)\left(7 \ \frac{\text{L}}{\text{s}}\right)}{\left(287 \ \frac{\text{J}}{\text{kg·K}}\right)(302\text{K})\left(1000 \ \frac{\text{L}}{\text{m}^3}\right)} \\ &= 0.02827 \text{ kg/s} \end{aligned}$$

Tool 3:

$$\dot{m}_{\text{tool 3}} = 0.045 \text{ kg/s} \quad \text{[given]}$$

The total mass flow rate leaving the system is

$$\begin{aligned} \dot{m}_{\text{total}} &= \dot{m}_{\text{tool 1}} + \dot{m}_{\text{tool 2}} + \dot{m}_{\text{tool 3}} \\ &= 0.1346 \ \frac{\text{kg}}{\text{s}} + 0.02827 \ \frac{\text{kg}}{\text{s}} + 0.045 \ \frac{\text{kg}}{\text{s}} \\ &= 0.2079 \text{ kg/s} \end{aligned}$$

The critical pressure of 620 kPa is required for tool 1 operation. The mass in the tank when critical pressure is achieved is

$$\begin{aligned} m_{\text{tank critical}} &= \frac{pVM}{RT} = \frac{(620 \text{ kPa})(27 \text{ m}^3)\left(29 \dfrac{\text{kg}}{\text{kmol}}\right)}{\left(8.314 \dfrac{\text{kJ}}{\text{kmol}\cdot\text{K}}\right)(305\text{K})} \\ &= 191.2 \text{ kg} \end{aligned}$$

The amount in the tank to be depleted is

$$\begin{aligned} m_{\text{depleted}} &= m_{\text{tank}} - m_{\text{tank,critical}} = 638.5 \text{ kg} - 191.2 \text{ kg} \\ &= 447.3 \text{ kg} \end{aligned}$$

The net flow rate of air into the tank is

$$\begin{aligned} \dot{m}_{\text{net}} &= \dot{m}_{\text{in}} - \dot{m}_{\text{out}} = 0.1736 \ \frac{\text{kg}}{\text{s}} - 0.2079 \ \frac{\text{kg}}{\text{s}} \\ &= -0.0343 \text{ kg/s} \end{aligned}$$

The time the system can run is

$$\frac{m_{\text{depleted}}}{\dot{m}_{\text{net}}} = \frac{447.3 \text{ kg}}{0.0343 \ \dfrac{\text{kg}}{\text{s}}} = 13\,041 \text{ s} \quad (3.6 \text{ h})$$

The answer is (B).

16 Refrigeration Cycles

PRACTICE PROBLEMS

1. A heat pump operates on the Carnot cycle between 40°F and 700°F (4°C and 370°C). The coefficient of performance is most nearly

(A) 1.5

(B) 1.8

(C) 2.2

(D) 2.7

2. A refrigerator uses refrigerant R-12. The input power is 585 W. Heat absorbed from the cooled space is 450 Btu/hr (0.13 kW). The coefficient of performance is most nearly

(A) 0.2

(B) 0.4

(C) 0.7

(D) 0.9

3. Ammonia is used in a reversed Carnot cycle refrigerator with reservoirs at 110°F (45°C) and 10°F (−10°C). 1000 Btu/hr (1000 kJ/h) are to be removed. The rejected heat is most nearly

(A) 1000 Btu/hr (1000 kJ/h)

(B) 1200 Btu/hr (1200 kJ/h)

(C) 1400 Btu/hr (1400 kJ/h)

(D) 1600 Btu/hr (1600 kJ/h)

4. A refrigerator cools a continuous aqueous solution ($c_p = 1$ Btu/lbm-°F; 4.19 kJ/kg·°C) flow of 100 gal/min (0.4 m^3/min) from 80°F (25°C) to 20°F (−5°C) in an 80°F (25°C) environment. The minimum power requirement is most nearly

(A) 82 hp (63 kW)

(B) 100 hp (74 kW)

(C) 130 hp (90 kW)

(D) 150 hp (94 kW)

5. An air refrigeration cycle compresses air from 70°F (20°C) and 14.7 psia (101 kPa) to 60 psia (400 kPa) in a 70% efficient compressor. The air is cooled to 25°F (−4.0° C) in a constant pressure process before entering a turbine with isentropic efficiency of 0.80. Assume air is an ideal gas. The coefficient of performance of the cycle is most nearly

(A) 0.7

(B) 0.8

(C) 0.9

(D) 1.1

SOLUTIONS

1. *Customary U.S. Solution*

The coefficient of performance for a heat pump operating on the Carnot cycle is given by Eq. 16.9 (also *NCEES Handbook:* Power Cycles).

$$\mathrm{COP}_{\text{heat pump}} = \frac{T_{\text{high}}}{T_{\text{high}} - T_{\text{low}}}$$

The absolute temperatures (see also *NCEES Handbook:* Temperature) are

$$\begin{aligned} T_{\text{high}} &= 700°\text{F} + 460° = 1160°\text{R} \\ T_{\text{low}} &= 40°\text{F} + 460° = 500°\text{R} \\ \mathrm{COP}_{\text{heat pump}} &= \frac{1160°\text{R}}{1160°\text{R} - 500°\text{R}} \\ &= 1.76 \quad (1.8) \end{aligned}$$

SI Solution

The coefficient of performance for a heat pump operating on the Carnot cycle is given by Eq. 16.9 (also *NCEES Handbook:* Power Cycles).

$$\mathrm{COP}_{\text{heat pump}} = \frac{T_{\text{high}}}{T_{\text{high}} - T_{\text{low}}}$$

The absolute temperatures (see also *NCEES Handbook:* Temperature) are

$$\begin{aligned} T_{\text{high}} &= 370°\text{C} + 273° = 643\text{K} \\ T_{\text{low}} &= 4°\text{C} + 273° = 277\text{K} \\ \mathrm{COP}_{\text{heat pump}} &= \frac{643\text{K}}{643\text{K} - 277\text{K}} \\ &= 1.76 \quad (1.8) \end{aligned}$$

The answer is (B).

2. *Customary U.S. Solution*

The coefficient of performance for a refrigerator is given by Eq. 16.1 (also *NCEES Handbook:* Power Cycles).

$$\begin{aligned} \mathrm{COP}_{\text{refrigerator}} &= \frac{Q_{\text{in}}}{W_{\text{in}}} = \frac{450\ \frac{\text{Btu}}{\text{hr}}}{(585\ \text{W})\left(3.4121\ \frac{\frac{\text{Btu}}{\text{hr}}}{\text{W}}\right)} \\ &= 0.225 \quad (0.2) \end{aligned}$$

SI Solution

The coefficient of performance for a refrigerator is given by Eq. 16.1 (also *NCEES Handbook:* Power Cycles).

$$\begin{aligned} \mathrm{COP}_{\text{refrigerator}} &= \frac{Q_{\text{in}}}{W_{\text{in}}} = \frac{(0.13\ \text{kW})\left(1000\ \frac{\text{W}}{\text{kW}}\right)}{585\ \text{W}} \\ &= 0.222 \quad (0.2) \end{aligned}$$

The answer is (A).

3. *Customary U.S. Solution*

Absolute temperature is

$$\begin{aligned} T_H &= 110°\text{F} + 459.67 \\ &= 569.67°\text{R} \\ T_C &= 10°\text{F} + 459.67 \\ &= 469.67°\text{R} \end{aligned}$$

As in the *NCEES* Handbook: section titled "Power Cycles—Refrigeration Cycles," the coefficient of performance is

$$\begin{aligned} \mathrm{COP}_{R,\text{Carnot}} &= \frac{1}{\left(\frac{T_H}{T_C}\right) - 1} \\ &= \frac{1}{\left(\frac{569.67°\text{R}}{469.67°\text{R}}\right) - 1} \\ &= 4.7 \end{aligned}$$

The work input is

$$\begin{aligned} W_{\text{net, in}} &= \frac{Q_C}{\mathrm{COP}_R} = \frac{1000\ \frac{\text{Btu}}{\text{hr}}}{4.7} \\ &= 212.77\ \text{Btu/hr} \end{aligned}$$

The rejected heat is

$$\begin{aligned} Q_{\text{out}} &= Q_C + W_{\text{net, in}} = 1000\ \frac{\text{Btu}}{\text{hr}} + 212.77\ \frac{\text{Btu}}{\text{hr}} \\ &= 1212.77\ \text{Btu/hr} \quad (1200\ \text{Btu/hr}) \end{aligned}$$

SI Solution

Absolute temperature is

$$\begin{aligned} T_H &= 45°\text{C} + 273.15 \\ &= 318.15\text{K} \\ T_C &= -10°\text{C} + 273.15 \\ &= 263.15\text{K} \end{aligned}$$

As in the *NCEES Handbook:* section titled "Power Cycles—Refrigeration Cycles," the coefficient of performance is

$$\begin{aligned} \text{COP}_{R,\text{Carnot}} &= \frac{1}{\left(\frac{T_H}{T_C}\right) - 1} \\ &= \frac{1}{\left(\frac{318.15\text{K}}{263.15\text{K}}\right) - 1} \\ &= 4.785 \end{aligned}$$

The work input is

$$\begin{aligned} W_{\text{net, in}} &= \frac{Q_C}{\text{COP}_R} = \frac{1000\ \frac{\text{kJ}}{\text{h}}}{4.785} \\ &= 208.99\ \text{kJ/h} \end{aligned}$$

The rejected heat is

$$\begin{aligned} Q_{\text{out}} &= Q_C + W_{\text{net, in}} = 1000\ \frac{\text{kJ}}{\text{h}} + 208.99\ \frac{\text{kJ}}{\text{h}} \\ &= 1208.99\ \text{kJ/h} \quad (1200\ \text{kJ/h}) \end{aligned}$$

The answer is (B).

4. *Customary U.S. Solution*

The density of an aqueous solution is essentially the same as water.

$$\begin{aligned} \dot{m} &= \dot{V}\rho = \left(100\ \frac{\text{gal}}{\text{min}}\right)\left(0.1337\ \frac{\text{ft}^3}{\text{gal}}\right)\left(62.4\ \frac{\text{lbm}}{\text{ft}^3}\right) \\ &= 834.3\ \text{lbm/min} \end{aligned}$$

$$\begin{aligned} \dot{Q}_{\text{in}} &= \dot{m}c_p\Delta T \\ &= \left(834.3\ \frac{\text{lbm}}{\text{min}}\right)\left(1\ \frac{\text{Btu}}{\text{lbm-°F}}\right)(80°\text{F} - 20°\text{F}) \\ &= 50{,}058\ \text{Btu/min} \end{aligned}$$

$$\begin{aligned} \dot{Q}_{\text{in,ton refrigeration}} &= \left(50{,}058\ \frac{\text{Btu}}{\text{min}}\right)\left(\frac{60\ \text{min}}{1\ \text{hr}}\right) \\ &\quad \times \left(\frac{1\ \text{ton refrigeration}}{1.2 \times 10^4\ \frac{\text{Btu}}{\text{hr}}}\right) \\ &= 250.29\ \text{ton refrigeration} \end{aligned}$$

As in the *NCEES Handbook* section titled "Power Cycles," the coefficient of performance is

$$\text{COP} = \frac{T_{\text{low}}}{T_{\text{high}} - T_{\text{low}}} = \frac{20°\text{F} + 460°}{80°\text{F} - 20°\text{F}} = 8$$

$$\begin{aligned} W_{\text{in, hp}} &= \frac{\left(\frac{4.7141\ \text{hp}}{1\ \text{ton refrigeration}}\right)\dot{Q}_{\text{in, ton refrigeration}}}{COP} \\ &= \frac{\left(\frac{4.7141\ \text{hp}}{1\ \text{ton refrigeration}}\right)(250.29\ \text{ton refrigeration})}{8} \\ &= 147.5\ \text{hp} \quad (150\ \text{hp}) \end{aligned}$$

SI Solution

The density of an aqueous solution is essentially the same as water.

$$\begin{aligned} \dot{m} &= \dot{V}\rho = \left(0.4\ \frac{\text{m}^3}{\text{min}}\right)\left(1000\ \frac{\text{kg}}{\text{m}^3}\right) \\ &= 400\ \text{kg/min} \\ \dot{Q}_{\text{in}} &= \dot{m}c_p\Delta T \\ &= \left(400\ \frac{\text{kg}}{\text{min}}\right)\left(4.19\ \frac{\text{kJ}}{\text{kg}\cdot°\text{C}}\right)(25°\text{C} - (-5°\text{C})) \\ &= 50\,280\ \text{kJ/min} \end{aligned}$$

As in the *NCEES Handbook:* section titled "Power Cycles," the coefficient of performance is

$$\begin{aligned} \text{COP} &= \frac{T_{\text{low}}}{T_{\text{high}} - T_{\text{low}}} = \frac{-5°\text{C} + 273.15°}{25°\text{C} - (-5°\text{C})} \\ &= 8.938 \\ \dot{W}_{\text{in}} &= \frac{\dot{Q}_{\text{in}}}{\text{COP}} = \frac{50\,280\ \frac{\text{kJ}}{\text{min}}}{(8.938)\left(60\ \frac{\text{s}}{\text{min}}\right)} \\ &= 93.76\ \text{kW} \quad (94\ \text{kW}) \end{aligned}$$

The answer is (D).

5. *Customary U.S. Solution*

Assuming ideal gas, from *NCEES Handbook:* Closed Thermodynamic Systems,

$$\frac{T_{\text{D}}}{T_{\text{C}}} = \left(\frac{p_{\text{high}}}{p_{\text{low}}}\right)^{(k-1)/k}$$

For air, $k = 1.4$.

$$T_D = T_C\left(\frac{60\ \text{psia}}{14.7\ \text{psia}}\right)^{(1.4-1)/1.4} = (70°\text{F} + 460°)(1.495) = 792.1°\text{R}$$

As in *NCEES Handook*: Open Thermodynamic Systems, the temperature leaving the compressor if the process is not isentropic is

$$T_D' = T_C + \frac{T_D - T_C}{\eta_{\text{compressor}}} = 530°\text{R} + \frac{792.1°\text{R} - 530°\text{R}}{0.7} = 904.4°\text{R}$$

From *NCEES Handbook:* Closed Thermodynamic Systems, find the temperature at B.

$$\frac{T_A}{T_B} = \left(\frac{p_{\text{high}}}{p_{\text{low}}}\right)^{(k-1)/k}$$

$$T_B = \frac{T_A}{\left(\frac{p_{\text{high}}}{p_{\text{low}}}\right)^{(k-1)/k}} = \frac{25°\text{F} + 460°}{\left(\frac{60\ \text{psia}}{14.7\ \text{psia}}\right)^{(1.4-1)/1.4}} = 324.5°\text{R}$$

As in *NCEES Handbook:* Open Thermodynamic Systems, the temperature leaving the turbine if the process is not isentropic is

$$T_B' = T_A - \eta_{\text{turbine}}(T_A - T_B) = 485°\text{R} - (0.80)(485°\text{R} - 324.5°\text{R}) = 356.6°\text{R}$$

As in *NCEES Handbook:* Power Cycles, the coefficient of performance of the cycle is

$$\text{COP} = \frac{T_C - T_B'}{(T_D' - T_A) - (T_C - T_B')} = \frac{530°\text{R} - 356.6°\text{R}}{(904.5°\text{R} - 485°\text{R}) - (530°\text{R} - 356.6°\text{R})} = 0.705 \quad (0.7)$$

SI Solution

From *NCEES Handbook:* Closed Thermodynamic Systems, assuming air is an ideal gas with $k = 1.4$,

$$\frac{T_D}{T_C} = \left(\frac{p_{\text{high}}}{p_{\text{low}}}\right)^{(k-1)/k}$$

$$T_C = 20°\text{C} + 273.15° = 293.15\text{K}$$

$$T_D = T_C\left(\frac{p_{\text{high}}}{p_{\text{low}}}\right)^{(k-1)/k} = (293.15\text{K})\left(\frac{400\ \text{kPa}}{101\ \text{kPa}}\right)^{(1.4-1)/1.4} = 434.4\text{K}$$

As in *NCEES Handbook:* Open Thermodynamic Systems, the temperature leaving the compressor if the process is not isentropic is

$$T_D' = T_C + \frac{T_D - T_C}{\eta_{\text{compressor}}} = 293.15\text{K} + \frac{434.4\text{K} - 293.15\text{K}}{0.7} = 494.9\text{K}$$

From *NCEES Handbook:* Closed Thermodynamic Systems, find the temperature at B.

$$T_A = -4°\text{C} + 273.15° = 269.15\text{K}$$

$$T_B = \frac{T_A}{\left(\frac{p_{\text{high}}}{p_{\text{low}}}\right)^{(k-1)/k}} = \frac{269.15\text{K}}{\left(\frac{400\ \text{kPa}}{101\ \text{kPa}}\right)^{(1.4-1)/1.4}} = 181.6\text{K}$$

As in *NCEES Handbook:* Open Thermodynamic Systems, the temperature leaving the turbine if the process is not isentropic is

$$T_B' = T_A - \eta_{\text{turbine}}(T_A - T_B) = 269.15\ \text{K} - (0.80)(269.15\ \text{K} - 181.6\text{K}) = 199.1\text{K}$$

As in *NCEES Handbook:* Power Cycles, the coefficient of performance of the cycle is

$$\text{COP} = \frac{T_C - T_B'}{(T_D' - T_A) - (T_C - T_B')} = \frac{293.15\text{K} - 199.1\text{K}}{(494.9\text{K} - 269.15\text{K}) - (293.15\text{K} - 199.1\text{K})} = 0.714 \quad (0.7)$$

The answer is (A).

17 Heat Transfer: Conduction

PRACTICE PROBLEMS

1. Experiments have shown that the thermal conductivity, k, of a particular material varies with temperature, T, according to the following relationship.

$$k_T = (0.030)(1 + 0.0015T)$$

What is most nearly the value of k that should be used for a transfer of heat through the material if the hot-side temperature is 350° (°F or °C) and the cold-side temperature is 150° (°F or °C)?

(A) 0.04

(B) 0.06

(C) 0.10

(D) 0.20

2. A composite wall is made up of 3.0 in (7.6 cm) of material A exposed to 1000°F (540°C), 5.0 in (13 cm) of material B, and 6.0 in (15 cm) of material C exposed to 200°F (90°C). The mean thermal conductivities for materials A, B, and C are 0.06, 0.5, and 0.8 Btu-ft/hr-ft^2-°F (0.1, 0.9, and 1.4 W/m·K), respectively.

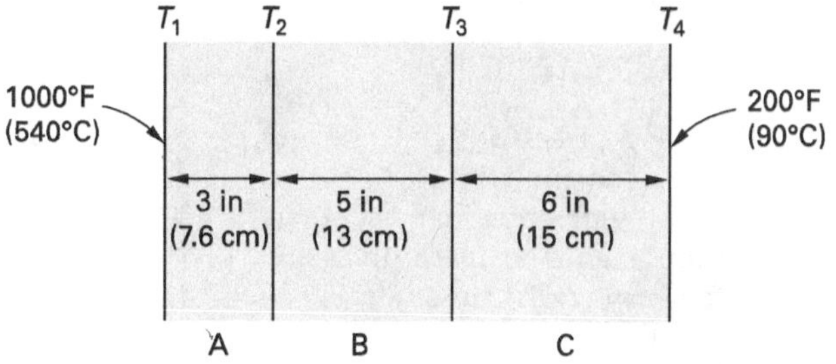

The temperature at the A-B material interface is most nearly

(A) 300°F (100°C)

(B) 350°F (150°C)

(C) 400°F (200°C)

(D) 500°F (250°C)

3. What is most nearly the heat flow through insulating brick 1.0 ft (30 cm) thick in an oven wall with a thermal conductivity of 0.038 Btu-ft/hr-ft^2-°F (0.066 W/m·K)? The thermal gradient is 350°F (177°C).

(A) 7 Btu/hr-ft^2 (22 W/m^2)

(B) 9 Btu/hr-ft^2 (28 W/m^2)

(C) 11 Btu/hr-ft^2 (35 W/m^2)

(D) 13 Btu/hr-ft^2 (40 W/m^2)

4. In an old factory, a hot liquid flows through a 30 ft long copper tube that has a 1 in inner diameter and a ⅓ in wall thickness. The copper tube is covered with a ½ in thick layer of glass wool insulation, and the glass wool is covered with a ½ in thick layer of asbestos. The outside temperatures of the copper tube and asbestos layer are 500°F and 150°F, respectively. The thermal conductivities of copper, glass wool, and asbestos are 220 Btu/hr-ft-°F, 0.0315 Btu/hr-ft-°F, and 0.115 Btu/hr-ft-°F, respectively. The temperature at the interface of the glass wool and asbestos is most nearly

(A) 190°F

(B) 200°F

(C) 260°F

(D) 350°F

5. In an old factory, a hot liquid flows through a 30 ft long copper tube that has a 1 in inner diameter and a ⅓ in wall thickness. The copper tube is covered with a ½ in thick layer of glass wool insulation, and the glass wool is covered with a layer of asbestos of unknown thickness. The outside temperatures of the copper tube and asbestos layer are 500°F and 80°F, respectively. The heat transfer rate per unit length is 110 Btu/hr-ft. The thermal conductivities of copper, glass wool, and asbestos are 220 Btu/hr-ft-°F, 0.0315 Btu/hr-ft-°F, and 0.115 Btu/hr-ft-°F, respectively. The thickness of the asbestos layer is most nearly

(A) 0.2 in

(B) 0.3 in

(C) 1.0 in

(D) 2.5 in

6. A chromium spherical tank with a 4 ft outer diameter contains hot oil. The tank is covered with a 5 in thick spherical layer of magnesia insulation. The magnesia insulation is covered with a 2 in thick spherical layer of Styrofoam. The outside temperatures of the chromium sphere and the Styrofoam layers are 250°F and 70°F, respectively. The thermal conductivities of magnesia and Styrofoam are 0.04 Btu/hr-ft-°F and 0.02 Btu/hr-ft-°F, respectively. The temperature at the interface of the magnesia and Styrofoam layers is most nearly

(A) 100°F

(B) 110°F

(C) 140°F

(D) 220°F

7. A metallic, thin-walled spherical tank is covered with a layer of insulation. The thickness of the insulation is equal to the radius of the spherical tank. To increase the volumetric capacity, the original tank will be replaced with a new, insulated, thin-walled spherical tank that has a new tank radius equal to twice the original tank radius. The insulation layer material and thickness are unchanged. Simultaneously, the process is modified so that the outside temperatures for the sphere and its insulated layer are also doubled. The ratio of the heat transfer of the new tank to the heat transfer of the original tank is most nearly

(A) 1:2

(B) 2:1

(C) 3:1

(D) 6:1

8. The heat supply of a large building is turned off at 5:00 p.m. when the interior temperature is 70°F (21°C). The outdoor temperature is constant at 40°F (4°C). The thermal capacity of the building and its contents is 100,000 Btu/°F (60 MJ/K), and the conductance is 6500 Btu/hr-°F (1.1 kW/K). The interior temperature at 1:00 a.m. is most nearly

(A) 45°F (7°C)

(B) 50°F (10°C)

(C) 60°F (15°C)

(D) 65°F (20°C)

9. Steel ball bearings varying in diameter, d, from $\frac{1}{4}$ in to $1\frac{1}{2}$ in (6.35 mm to 38.1 mm) are quenched from 1800°F (980°C) in an oil bath that remains at 110°F (43°C). The ball bearings are removed when their internal (center) temperature reaches 250°F (120°C). The average film coefficient is 56 Btu/hr-ft^2-°F (320 W/m^2·K).

A linear equation for the time the ball bearings remain in the oil bath as a function of diameter is most nearly

(A) $0.013d\,(1800d)$

(B) $0.023d\,(3000d)$

(C) $0.033d\,(4700d)$

(D) $0.042d\,(5900d)$

10. The walls of a building are made of a brick layer 1 m thick with a conductive coefficient, k, of 0.70 W/m·C, a rock-wool insulation layer 0.080 m thick with $k = 0.50$ W/m·C, and a layer of sheet rock 0.0375 m thick with $k = 0.45$ W/m·C. If the inside and outside temperatures are a constant 27 and −8.0°C, respectively, what is most nearly the heat loss due to conduction through the wall?

(A) 21 W/m^2

(B) 26 W/m^2

(C) 50 W/m^2

(D) 58 W/m^2

11. A 0.4 in (1.0 cm) diameter uranium dioxide fuel rod with a thermal conductivity of 1.1 Btu-ft/hr-ft^2-°F (1.9 W/m·K) is clad with 0.020 in (0.5 mm) of stainless steel. The fuel rod generates 4×10^7 Btu/hr-ft^3 (4.1×10^8 W/m^3) internally. A coolant at 500°F (260°C) circulates around the clad rod. The outside film coefficient is 10,000 Btu/hr-ft^2-°F (57 kW/m^2·K). The temperature at the longitudinal centerline of the rod is most nearly

(A) 2100°F (1200°C)

(B) 2400°F (1300°C)

(C) 2700°F (1500°C)

(D) 3100°F (1700°C)

12. Two long pieces of $\frac{1}{16}$ in (1.6 mm) copper wire are connected end-to-end with a hot soldering iron. The minimum melting temperature of the solder is 450°F (230°C). The surrounding air temperature is 80°F (27°C). The unit film coefficient is 3 Btu/hr-ft^2-°F (17 W/m^2·K). The thermal conductivity of the solder is 215 Btu-ft/hr-ft^2-°F (372.1 W/m·K). Disregard radiation losses. The minimum rate of heat application to keep the solder molten is most nearly

(A) 4.3 Btu/hr (1.3 W)

(B) 11 Btu/hr (3.3 W)

(C) 20 Btu/hr (6.0 W)

(D) 47 Btu/hr (14 W)

SOLUTIONS

1. Use the value of k at an average temperature of $\frac{1}{2}(T_1 + T_2)$.

$$\begin{aligned} T &= \left(\frac{1}{2}\right)(150° + 350°) = 250° \\ k &= (0.030)(1 + 0.0015T) \\ &= (0.030)\big(1 + (0.0015)(250°)\big) \\ &= 0.04125 \quad (0.04) \end{aligned}$$

The answer is (A).

2. *Customary U.S. Solution*

Since the wall temperatures are given, it is not necessary to consider films.

From Eq. 17.28 (also *NCEES Handbook:* Conduction), the heat flow through the composite wall is

$$Q = \frac{A(T_1 - T_4)}{\sum\limits_{i=1}^{n} \frac{L_i}{k_i}}$$

On a per unit area basis,

$$\begin{aligned} \frac{Q}{A} &= \frac{T_1 - T_4}{\sum\limits_{i=1}^{n} \frac{L_i}{k_i}} \\ &= \frac{T_1 - T_2}{\frac{L_A}{k_A}} \\ &= \frac{T_2 - T_3}{\frac{L_B}{k_B}} \\ &= \frac{1000°\text{F} - 200°\text{F}}{\begin{array}{l}\dfrac{3 \text{ in}}{\left(\left(0.06 \dfrac{\text{Btu-ft}}{\text{hr-ft}^2\text{-°F}}\right) \times \left(12 \dfrac{\text{in}}{\text{ft}}\right)\right)} + \dfrac{5 \text{ in}}{\left(\left(0.5 \dfrac{\text{Btu-ft}}{\text{hr-ft}^2\text{-°F}}\right) \times \left(12 \dfrac{\text{in}}{\text{ft}}\right)\right)} \\ \quad + \dfrac{6 \text{ in}}{\left(0.8 \dfrac{\text{Btu-ft}}{\text{hr-ft}^2\text{-°F}}\right)\left(12 \dfrac{\text{in}}{\text{ft}}\right)}\end{array}} \\ &= 142.2 \text{ Btu/hr-ft}^2 \end{aligned}$$

To find the temperature at the A-B interface, T_2, use

$$\begin{aligned} \frac{Q}{A} &= \frac{T_1 - T_2}{\frac{L_A}{k_A}} \\ T_2 &= T_1 - \left(\frac{Q}{A}\right)\left(\frac{L_A}{k_A}\right) \\ &= 1000°\text{F} - \left(142.2 \frac{\text{Btu}}{\text{hr-ft}^2}\right) \\ &\quad \times \left(\frac{3 \text{ in}}{\left(0.06 \frac{\text{Btu-ft}}{\text{hr-ft}^2\text{-°F}}\right)\left(12 \frac{\text{in}}{\text{ft}}\right)}\right) \\ &= 407.5°\text{F} \quad (400°\text{F}) \end{aligned}$$

SI Solution

Since the wall temperatures are given, it is not necessary to consider films.

From Eq. 17.28 (also *NCEES Handbook:* Conduction), the heat flow through the composite wall is

$$Q = \frac{A(T_1 - T_4)}{\sum\limits_{i=1}^{n} \frac{L_i}{k_i}}$$

On a per unit area basis,

$$\begin{aligned} \frac{Q}{A} &= \frac{T_1 - T_4}{\sum\limits_{i=1}^{n} \frac{L_i}{k_i}} \\ &= \frac{T_1 - T_2}{\frac{L_A}{k_A}} \\ &= \frac{T_2 - T_3}{\frac{L_B}{k_B}} \\ &= \frac{540°\text{C} - 90°\text{C}}{\begin{array}{l}\dfrac{7.6 \text{ cm}}{\left(0.1 \dfrac{\text{W}}{\text{m·K}}\right)\left(100 \dfrac{\text{cm}}{\text{m}}\right)} + \dfrac{13 \text{ cm}}{\left(0.9 \dfrac{\text{W}}{\text{m·K}}\right)\left(100 \dfrac{\text{cm}}{\text{m}}\right)} \\ \quad + \dfrac{15 \text{ cm}}{\left(1.4 \dfrac{\text{W}}{\text{m·K}}\right)\left(100 \dfrac{\text{cm}}{\text{m}}\right)}\end{array}} \\ &= 444.8 \text{ W/m}^2 \end{aligned}$$

To find the temperature at the A-B interface, T_2, use

$$\frac{Q}{A} = \frac{T_1 - T_2}{\frac{L_A}{k_A}}$$

$$T_2 = T_1 - \left(\frac{Q}{A}\right)\left(\frac{L_A}{k_A}\right)$$

$$= 540°C - \left(444.8\ \frac{W}{m^2}\right)\left(\frac{7.6\ cm}{\left(0.1\ \frac{W}{m{\cdot}K}\right)\left(100\ \frac{cm}{m}\right)}\right)$$

$$= 202.0°C\quad (200°C)$$

The answer is (C).

3. *Customary U.S. Solution*

From the *NCEES Handbook* section titled "Fundamentals of Heat Transfer Fourier's Law of Conduction," the heat flux per unit area through a plane wall is

$$\frac{Q_{1-2}}{A} = \frac{k\Delta T}{L} = \frac{\left(0.038\ \frac{Btu\text{-}ft}{hr\text{-}ft^2\text{-}°F}\right)(350°F)}{1.0\ ft}$$

$$= 13.3\ Btu/hr\text{-}ft^2\quad (13\ Btu/hr\text{-}ft^2)$$

SI Solution

From the *NCEES Handbook* section titled "Fundamentals of Heat Transfer Fourier's Law of Conduction," the heat flux per unit area through a plane wall is

$$\frac{Q_{1-2}}{A} = \frac{k\Delta T}{L} = \frac{\left(0.066\ \frac{W}{m{\cdot}K}\right)(177\ K)\left(100\ \frac{cm}{m}\right)}{30\ cm}$$

$$= 38.9\ W/m^2\quad (40\ W/m^2)$$

The answer is (D).

4. Because the temperature outside the copper tube is known, the thermal conductivity of the copper and the pipe's inside radius can be disregarded. r_a is the radius at the interface between the copper and the first layer of insulation. r_b is the radius at the interface between the glass wool insulation and the asbestos layer. r_c is the outer radius of the glass wool layer plus the thickness of the asbestos layer.

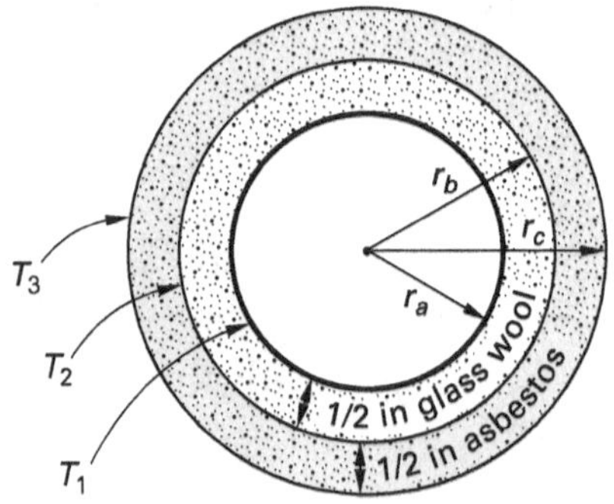

$$r_a = r_{inside} + t_{pipe} = \frac{1}{2}\ in + \frac{1}{3}\ in = 0.833\ in$$

$$r_b = r_a + t_{glass\ wool} = 0.833\ in + \frac{1}{2}\ in = 1.333\ in$$

$$r_c = r_b + t_{asbestos} = 1.333\ in + \frac{1}{2}\ in = 1.833\ in$$

The copper tube's length is not necessary for the solution. Use the equation for conduction through a cylindrical wall with resistances in series from the *NCEES PE Chemical Reference Handbook*, Section 4.2.1.8, and the known temperatures to calculate the heat transfer per unit length from the outside of the copper tube through the asbestos.

$$\frac{Q}{L} = \frac{2\pi(T_1 - T_3)}{\frac{\ln\frac{r_b}{r_a}}{k_{glass\ wool}} + \frac{\ln\frac{r_c}{r_b}}{k_{asbestos}}}$$

$$= \frac{2\pi(500°F - 150°F)}{\frac{\ln\frac{1.333\ in}{0.833\ in}}{0.0315\ \frac{Btu}{hr\text{-}ft\text{-}°F}} + \frac{\ln\frac{1.833\ in}{1.333\ in}}{0.115\ \frac{Btu}{hr\text{-}ft\text{-}°F}}}$$

$$= 124.28\ Btu/hr\text{-}ft$$

Use the equation for conduction through a cylindrical wall from the *NCEES PE Chemical Reference Handbook*, Section 4.2.1.3, again to calculate the temperature at the interface of glass wool and the asbestos from the heat layer through a single layer.

$$\frac{Q}{L} = \frac{2\pi(T_2 - T_3)}{\dfrac{\ln \dfrac{r_c}{r_b}}{k_{\text{asbestos}}}}$$

$$\begin{aligned} T_2 &= \frac{\left(\dfrac{Q}{L}\right)\left(\dfrac{\ln \dfrac{r_c}{r_b}}{k_{\text{asbestos}}}\right)}{2\pi} + T_3 \\ &= \frac{\left(124.28\ \dfrac{\text{Btu}}{\text{hr-ft}}\right)\left(\dfrac{\ln \dfrac{1.833\ \text{in}}{1.333\ \text{in}}}{0.115\ \dfrac{\text{Btu}}{\text{hr-ft-°F}}}\right)}{2\pi} + 150°\text{F} \\ &= 204.79°\text{F} \quad (200°\text{F}) \end{aligned}$$

The answer is (B).

5. r_a is the radius at the interface between the copper pipe and the first layer of insulation. r_b is the radius at the interface between the glass wool insulation and the asbestos layer.

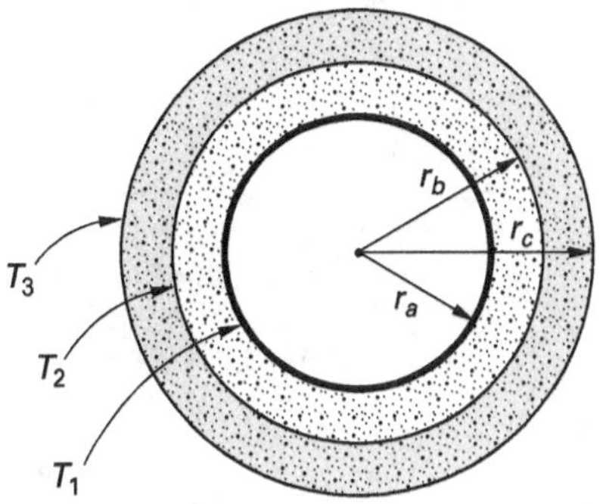

$$r_a = r_{\text{inside}} + t_{\text{pipe}} = \frac{1}{2}\ \text{in} + \frac{1}{3}\ \text{in} = 0.833\ \text{in}$$

$$r_b = r_a + t_{\text{glass wool}} = 0.833\ \text{in} + \frac{1}{2}\ \text{in} = 1.333\ \text{in}$$

The copper tube's length is not necessary for the solution. Use Eq. 17.40 (also *NCEES Handbook:* Conduction) to calculate the unknown radius, r_c, which is the outer radius of the glass wool layer plus the thickness of the asbestos layer.

$$\frac{Q}{L} = \frac{2\pi(T_1 - T_3)}{\dfrac{1}{k_{\text{glass wool}}}\ln \dfrac{r_b}{r_a} + \dfrac{1}{k_{\text{asbestos}}}\ln \dfrac{r_c}{r_b}}$$

$$110\ \frac{\text{Btu}}{\text{hr-ft}} = \frac{2\pi(500°\text{F} - 80°\text{F})}{\left(\dfrac{1}{0.0315\ \dfrac{\text{Btu}}{\text{hr-ft-°F}}}\right)\ln \dfrac{1.333\ \text{in}}{0.833\ \text{in}} + \left(\dfrac{1}{0.115\ \dfrac{\text{Btu}}{\text{hr-ft-°F}}}\right)\ln \dfrac{r_c}{1.333\ \text{in}}}$$

$$\ln \frac{r_c}{1.333\ \text{in}} = 1.0425$$

$$\begin{aligned} r_c &= (1.333\ \text{in})e^{1.0425} \\ &= 3.781\ \text{in} \end{aligned}$$

The thickness of the asbestos is

$$\begin{aligned} t_{\text{asbestos}} &= r_c - r_b = 3.781\ \text{in} - 1.333\ \text{in} \\ &= 2.448\ \text{in} \quad (2.5\ \text{in}) \end{aligned}$$

The answer is (D).

6. r_a is the outer radius of the tank. r_b is the radius at the interface between the magnesia insulation and the Styrofoam layer. r_c is the outer radius of the magnesia insulation layer plus the thickness of the Styrofoam layer.

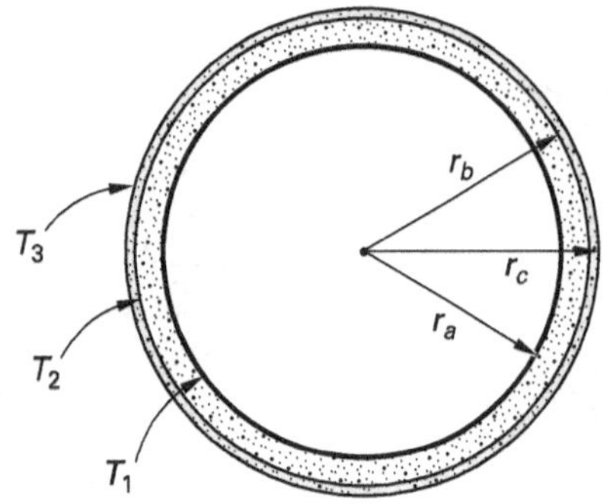

$$r_a = \frac{r_o}{2} = \frac{4\ \text{ft}}{2} = 2\ \text{ft}$$

$$r_b = r_a + t_{\text{magnesia}} = 2\ \text{ft} + \frac{5\ \text{in}}{12\ \frac{\text{in}}{\text{ft}}} = 2.42\ \text{ft}$$

$$r_c = r_b + t_{\text{Styrofoam}} = 2.42\ \text{ft} + \frac{2\ \text{in}}{12\ \frac{\text{in}}{\text{ft}}} = 2.58\ \text{ft}$$

Use Eq. 17.49 (also *NCEES Handbook:* Conduction) to calculate the heat transfer from the chromium sphere through the Styrofoam.

$$\begin{aligned} Q &= \frac{4\pi(T_1 - T_3)}{\dfrac{\dfrac{1}{r_a} - \dfrac{1}{r_b}}{k_{\text{magnesia}}} + \dfrac{\dfrac{1}{r_b} - \dfrac{1}{r_c}}{k_{\text{Styrofoam}}}} \\ &= \frac{4\pi(250^\circ\text{F} - 70^\circ\text{F})}{\dfrac{\dfrac{1}{2\ \text{ft}} - \dfrac{1}{2.42\ \text{ft}}}{0.04\ \dfrac{\text{Btu}}{\text{hr-ft-}^\circ\text{F}}} + \dfrac{\dfrac{1}{2.42\ \text{ft}} - \dfrac{1}{2.58\ \text{ft}}}{0.02\ \dfrac{\text{Btu}}{\text{hr-ft-}^\circ\text{F}}}} \\ &= 655.5\ \text{Btu/hr-ft} \end{aligned}$$

Use Eq. 17.49 (also *NCEES Handbook:* Conduction) again to calculate the temperature at the interface of the magnesia and the Styrofoam from the heat transfer through a single layer.

$$T_1 - T_2 = \frac{Q\left(\dfrac{\dfrac{1}{r_a} - \dfrac{1}{r_b}}{k_{\text{magnesia}}}\right)}{4\pi}$$

$$\begin{aligned} T_2 &= T_1 - \frac{Q\left(\dfrac{\dfrac{1}{r_a} - \dfrac{1}{r_b}}{k_{\text{magnesia}}}\right)}{4\pi} \\ &= 250^\circ\text{F} - \frac{\left(655.5\ \dfrac{\text{Btu}}{\text{hr-ft}}\right)\left(\dfrac{\dfrac{1}{2\ \text{ft}} - \dfrac{1}{2.42\ \text{ft}}}{0.04\ \dfrac{\text{Btu}}{\text{hr-ft-}^\circ\text{F}}}\right)}{4\pi} \\ &= 136.8^\circ\text{F} \quad (140^\circ\text{F}) \end{aligned}$$

The answer is (C).

7. $r_{a,1}$ is the outside radius of the original tank.

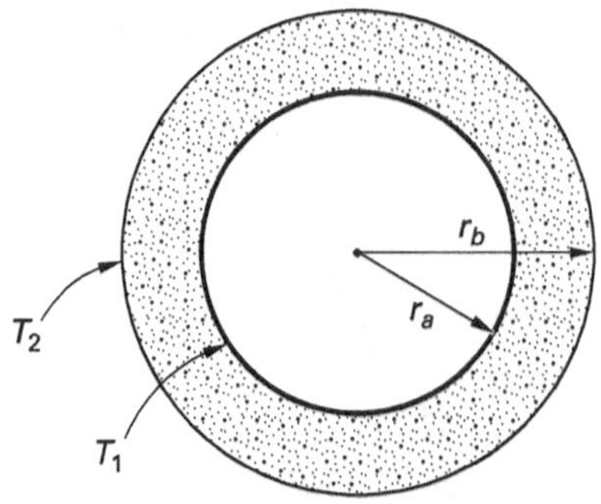

$$r_{a,1} = r$$

$$r_{b,1} = r + r = 2r$$

The heat transfer rate through an insulated, spherical tank is found from Eq. 17.49 (also *NCEES Handbook:* Conduction). For the original tank,

$$\begin{aligned} Q_1 &= \frac{4\pi(T_{a,1} - T_{b,1})}{\dfrac{\dfrac{1}{r_{a,1}} - \dfrac{1}{r_{b,1}}}{k}} = \frac{4\pi(T_{a,1} - T_{b,1})}{\dfrac{\dfrac{1}{r} - \dfrac{1}{2r}}{k}} \\ &= 8\pi kr(T_{a,1} - T_{b,1}) \end{aligned}$$

For the new tank,

$$r_{a,2} = 2r$$
$$r_{b,2} = 2r + r = 3r$$

The outside temperatures of the tank and insulation are doubled. From Eq. 17.49, the heat transfer rate for the new tank, Q_2, can be simplified to

$$\begin{aligned} Q_2 &= \frac{4\pi(T_{a,2} - T_{b,2})}{\dfrac{\dfrac{1}{r_{a,2}} - \dfrac{1}{r_{b,2}}}{k}} \\ &= \frac{4\pi(2T_{a,1} - 2T_{b,1})}{\dfrac{\dfrac{1}{2r} - \dfrac{1}{3r}}{k}} \\ &= 48\pi kr(T_{a,1} - T_{b,1}) \end{aligned}$$

The ratio of the heat transfer of the new tank to the heat transfer of the original tank is

$$\frac{Q_2}{Q_1} = \frac{48\pi kr(T_{a,1} - T_{b,1})}{8\pi kr(T_{a,1} - T_{b,1})} = 6 \quad (6{:}1)$$

The answer is (D).

8. *Customary U.S. Solution*

This is a transient problem. The total time is from 5 p.m. to 1 a.m., which is eight hours.

The thermal capacitance (capacity), C_e, is given as 100,000 Btu/°F.

Resistance and conductance are reciprocals. The thermal resistance is

$$\begin{aligned} R_e &= \frac{1}{\text{thermal conductance}} = \frac{1}{6500\ \dfrac{\text{Btu}}{\text{hr-°F}}} \\ &= 0.0001538\ \text{hr-°F/Btu} \end{aligned}$$

From Eq. 17.59 (also *NCEES Handbook:* Conduction),

$$\begin{aligned} T_t &= T_\infty + (T_0 - T_\infty)e^{-t/R_eC_e} \\ T_{8\ \text{hr}} &= 40°\text{F} + (70°\text{F} - 40°\text{F}) \\ &\quad \times \exp\left(\frac{-8\ \text{hr}}{\left(0.0001538\ \dfrac{\text{hr-°F}}{\text{Btu}}\right)\left(100{,}000\ \dfrac{\text{Btu}}{°\text{F}}\right)}\right) \\ &= 57.8°\text{F} \quad (60°\text{F}) \end{aligned}$$

SI Solution

The thermal capacitance (capacity) is

$$C_e = \left(60\ \frac{\text{MJ}}{\text{K}}\right)\left(1000\ \frac{\text{kJ}}{\text{MJ}}\right) = 60\,000\ \text{kJ/K} \quad \text{[given]}$$

Resistance and conductance are reciprocals. The thermal resistance is

$$R_e = \frac{1}{\text{thermal conductance}} = \frac{1}{1.1\ \dfrac{\text{kW}}{\text{K}}} = 0.909\ \text{K/kW}$$

From Eq. 17.59 (also *NCEES Handbook:* Conduction),

$$\begin{aligned} T_t &= T_\infty + (T_0 - T_\infty)e^{-t/R_eC_e} \\ &= 4°\text{C} + (21°\text{C} - 4°\text{C}) \\ &\quad \times \exp\left(\frac{(-8\ \text{h})\left(3600\ \dfrac{\text{s}}{\text{h}}\right)}{\left(0.909\ \dfrac{\text{K}}{\text{kW}}\right)\left(60\,000\ \dfrac{\text{kJ}}{\text{K}}\right)}\right) \\ &= 14.0°\text{C} \quad (15°\text{C}) \end{aligned}$$

The answer is (C).

9. *Customary U.S. Solution*

This is a transient problem. Check the Biot number to see if the lumped parameter method can be used.

The characteristic length, from Eq. 17.14, is

$$L_c = \frac{V}{A_s} = \frac{\left(\dfrac{\pi}{6}\right)d^3}{\pi d^2} = \frac{d}{6}$$

For the largest ball, $d = 1.5$ in.

$$L_c = \frac{d}{6} = \frac{1.5\ \text{in}}{(6)\left(12\ \dfrac{\text{in}}{\text{ft}}\right)} = 0.0208\ \text{ft}$$

Evaluate the thermal conductivity, k, of steel at

$$\left(\frac{1}{2}\right)(1800°\text{F} + 250°\text{F}) = 1025°\text{F}$$

From App. 17B (also *NCEES Handbook* table "Physical Properties of Metals at 68°F (U.S. Units)"), for steel, $k \approx 22.0$ Btu-ft/hr-ft^2-°F.

Heat Transfer

From Eq. 17.16 (also *NCEES Handbook:* Conduction), the Biot number is

$$\mathrm{Bi} = \frac{hL_c}{k} = \frac{\left(56\ \frac{\mathrm{Btu}}{\mathrm{hr\text{-}ft^2\text{-}^\circ F}}\right)(0.0208\ \mathrm{ft})}{22.0\ \frac{\mathrm{Btu\text{-}ft}}{\mathrm{hr\text{-}ft^2\text{-}^\circ F}}} = 0.053$$

For small balls, Bi will be even smaller.

Since $\mathrm{Bi} < 0.10$, the lumped parameter method can be used.

The assumptions are

- homogeneous body temperature
- minimal radiation losses
- oil bath temperature remains constant
- h remains constant

From Eq. 17.57 and Eq. 17.58 (also *NCEES Handbook:* Conduction), the time constant is

$$C_eR_e = c_p\rho V\left(\frac{1}{hA_s}\right) = \left(\frac{c_p\rho}{h}\right)\left(\frac{V}{A_s}\right) = \left(\frac{c_p\rho}{h}\right)L_c = \left(\frac{c_p\rho}{h}\right)\left(\frac{d}{6}\right)$$

From App. 17B (also *NCEES Handbook* table "Physical Properties of Metals at 68°F (U.S. Units)"), $\rho = 490\ \mathrm{lbm/ft^3}$ and $c_p = 0.11\ \mathrm{Btu/lbm\text{-}^\circ F}$, even though those values are for 32°F.

The time constant in hours is (measuring d in inches)

$$C_eR_e = \frac{\left(\frac{\left(0.11\ \frac{\mathrm{Btu}}{\mathrm{lbm\text{-}^\circ F}}\right)\left(490\ \frac{\mathrm{lbm}}{\mathrm{ft^3}}\right)}{56\ \frac{\mathrm{Btu}}{\mathrm{hr\text{-}ft^2\text{-}^\circ F}}}\right)\left(\frac{d}{6}\right)}{12\ \frac{\mathrm{in}}{\mathrm{ft}}} = 0.01337d$$

Take the natural log of the transient equation.

$$T_t = 250^\circ\mathrm{F}$$
$$T_\infty = 110^\circ\mathrm{F}$$
$$\Delta T = 1800^\circ\mathrm{F} - 110^\circ\mathrm{F} = 1690^\circ\mathrm{F}$$

$$\ln(T_t - T_\infty) = \ln\left(\Delta T e^{\frac{-t}{R_eC_e}}\right) = \ln \Delta T + \ln e^{\frac{-t}{R_eC_e}} = \ln \Delta T - \frac{t}{R_eC_e}$$

$$\ln(250^\circ\mathrm{F} - 110^\circ\mathrm{F}) = \ln(1690^\circ\mathrm{F}) - \frac{t}{0.01337d}$$

$$4.942 = 7.432 - \frac{t}{0.01337d}$$

$$t = 0.0333d \quad (0.033d)$$

SI Solution

For the largest ball, the characteristic length is

$$L_c = \frac{d}{6} = \frac{38.1\ \mathrm{mm}}{(6)\left(1000\ \frac{\mathrm{mm}}{\mathrm{m}}\right)} = 6.35 \times 10^{-3}\ \mathrm{m}$$

Evaluate the thermal conductivity, k, of steel at

$$\left(\frac{1}{2}\right)(980^\circ\mathrm{C} + 120^\circ\mathrm{C}) = 550^\circ\mathrm{C}$$

From App. 17B (also *NCEES Handbook* table "Physical Properties of Metals at 20°C (SI Units)") and its footnote,

$$k \approx \left(22.0\ \frac{\mathrm{Btu\text{-}ft}}{\mathrm{hr\text{-}ft^2\text{-}^\circ F}}\right)\left(\frac{1.7307\ \mathrm{W{\cdot}hr{\cdot}ft^2{\cdot}^\circ F}}{\mathrm{m{\cdot}K{\cdot}Btu{\cdot}ft}}\right) = 38.08\ \mathrm{W/m{\cdot}K}$$

From Eq. 17.16 (also *NCEES Handbook:* Conduction), the Biot number is

$$\mathrm{Bi} = \frac{hL_c}{k} = \frac{\left(320\ \frac{\mathrm{W}}{\mathrm{m^2{\cdot}K}}\right)(6.35 \times 10^{-3}\ \mathrm{m})}{38.08\ \frac{\mathrm{W}}{\mathrm{m{\cdot}K}}} = 0.053$$

For small balls, Bi will be even smaller.

Since Bi < 0.10, the lumped method can be used. The assumptions are given in the customary U.S. solution. From Eq. 17.57 and Eq. 17.58 (also *NCEES Handbook:* Conduction), the time constant in hours is

$$C_eR_e = c_p\rho V\left(\frac{1}{hA_s}\right) = \left(\frac{c_p\rho}{h}\right)\left(\frac{V}{A_s}\right) = \left(\frac{c_p\rho}{h}\right)L_c$$
$$= \left(\frac{c_p\rho}{h}\right)\left(\frac{d}{6}\right)$$

From App. 17B (also *NCEES Handbook* table "Physical Properties of Metals at 20°C (SI Units)") and its footnote,

$$\rho = \left(490\ \frac{\text{lbm}}{\text{ft}^3}\right)\left(16.0185\ \frac{\text{kg}\cdot\text{ft}^3}{\text{m}^3\cdot\text{lbm}}\right)$$
$$= 7849.1\ \text{kg/m}^3$$
$$c_p = \left(0.11\ \frac{\text{Btu}}{\text{lbm-°F}}\right)\left(4186.8\ \frac{\text{J}\cdot\text{lbm}\cdot\text{°F}}{\text{kg}\cdot\text{K}\cdot\text{Btu}}\right)$$
$$= 460.5\ \text{J/kg}\cdot\text{K}$$
$$C_eR_e = \left(\frac{\left(460.5\ \frac{\text{J}}{\text{kg}\cdot\text{K}}\right)\left(7849.1\ \frac{\text{kg}}{\text{m}^3}\right)}{320\ \frac{\text{W}}{\text{m}^2\cdot\text{K}}}\right)\left(\frac{d}{6}\right)$$
$$= 1882.6d$$

Take the natural log of the transient equation.

$$T_t = 120°\text{C}$$
$$T_\infty = 43°\text{C}$$
$$\Delta T = 980°\text{C} - 43°\text{C} = 937°\text{C}$$
$$\ln(T_t - T_\infty) = \ln \Delta T - \frac{t}{1882.6d}$$
$$\ln(120°\text{C} - 43°\text{C}) = \ln(937°\text{C}) - \frac{t}{1882.6d}$$
$$4.344 = 6.843 - \frac{t}{R_eC_e}$$
$$t = 4704.6d \quad (4700d)$$

The answer is (C).

10. The problem can be set up as a thermal circuit.

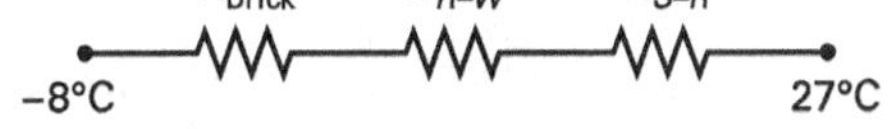

The total thermal resistance is

$$R_{\text{thermal}} = \sum_i \frac{L_i}{k_i}$$
$$= \frac{1\ \text{m}}{0.7\ \frac{\text{W}}{\text{m}\cdot\text{C}}} + \frac{0.080\ \text{m}}{0.50\ \frac{\text{W}}{\text{m}\cdot\text{C}}} + \frac{0.0375\ \text{m}}{0.45\ \frac{\text{W}}{\text{m}\cdot\text{C}}}$$
$$= 1.67\ \text{m}^2\cdot\text{C/W}$$

The heat loss per unit area is

$$q = \frac{\Delta T}{R_{\text{thermal}}} = \frac{27°\text{C} - (-8°\text{C})}{1.67\ \frac{\text{m}^2\cdot\text{C}}{\text{W}}} = 21\ \text{W/m}^2$$

The answer is (A).

11. *Customary U.S. Solution*

The volume of the rod is

$$V = AL = \left(\frac{\pi}{4}\right)d^2L$$

The volume per unit length is

$$\frac{V}{L} = \frac{\pi}{4}d^2 = \frac{\left(\frac{\pi}{4}\right)(0.4\ \text{in})^2}{\left(12\ \frac{\text{in}}{\text{ft}}\right)^2}$$
$$= 8.727 \times 10^{-4}\ \text{ft}^3/\text{ft}$$

The heat output per unit length of rod is

$$\frac{Q}{L} = \left(\frac{V}{L}\right)G = \left(8.727 \times 10^{-4}\ \frac{\text{ft}^3}{\text{ft}}\right)\left(4 \times 10^7\ \frac{\text{Btu}}{\text{hr-ft}^3}\right)$$
$$= 3.491 \times 10^4\ \text{Btu/hr-ft}$$

The diameter of the cladding is

$$d_o = 0.4\ \text{in} + (2)(0.020\ \text{in}) = 0.44\ \text{in}$$

The surface area per unit length of cladding is

$$A = \pi d_o = \frac{\pi(0.44\ \text{in})}{12\ \frac{\text{in}}{\text{ft}}} = 0.1152\ \text{ft}^2/\text{ft}$$

Heat Transfer

From Eq. 17.30 (also *NCEES Handbook:* Conduction), the surface temperature of the cladding is

$$
\begin{aligned}
T_s &= \frac{Q}{hA} + T_\infty \\
&= \frac{3.491 \times 10^4 \ \frac{\text{Btu}}{\text{hr-ft}}}{\left(10{,}000 \ \frac{\text{Btu}}{\text{hr-ft}^2\text{-°F}}\right)\left(0.1152 \ \frac{\text{ft}^2}{\text{ft}}\right)} + 500\text{°F} \\
&= 530.3\text{°F}
\end{aligned}
$$

For the cladding,

$$r_o = \frac{d_o}{2} = \frac{0.44 \text{ in}}{(2)\left(12 \ \frac{\text{in}}{\text{ft}}\right)} = 0.01833 \text{ ft}$$

$$r_i = \frac{d}{2} = \frac{0.4 \text{ in}}{(2)\left(12 \ \frac{\text{in}}{\text{ft}}\right)} = 0.01667 \text{ ft}$$

From App. 17B (also *NCEES Handbook* table "Physical Properties of Metals at 68°F (U.S. Units)"), k for stainless steel is 9.4 Btu-ft/hr-ft^2-°F at 68°F.

From Eq. 17.7 and Eq. 17.26 (also *NCEES Handbook:* Conduction) for a cylinder,

$$
\begin{aligned}
T_{\text{inside}} - T_{\text{outside}} &= \frac{Q \ln \frac{r_o}{r_i}}{2\pi k L} = \frac{\frac{Q}{L} \ln \frac{r_o}{r_i}}{2\pi k} \\
&= \frac{\left(3.491 \times 10^4 \ \frac{\text{Btu}}{\text{hr-ft}}\right) \ln \frac{0.01833 \text{ ft}}{0.01667 \text{ ft}}}{2\pi\left(9.4 \ \frac{\text{Btu-ft}}{\text{hr-ft}^2\text{-°F}}\right)} \\
&= 56.1\text{°F} \\
T_{\text{inside cladding}} &= T_{\text{outside fuel rod}} + 56.1\text{°F} \\
&= 530.3\text{°F} + 56.1\text{°F} \\
&= 586.4\text{°F}
\end{aligned}
$$

From Eq. 17.65 (also *NCEES Handbook:* Conduction), using $r_{o,\text{fuel rod}} = r_{i,\text{cladding}}$, and using k for the uranium dioxide (not the stainless steel),

$$
\begin{aligned}
T_{\text{center}} &= T_o + \frac{G r_o^2}{4k} \\
&= 586.4\text{°F} + \frac{\left(4 \times 10^7 \ \frac{\text{Btu}}{\text{hr-ft}^3}\right)(0.01667 \text{ ft})^2}{(4)\left(1.1 \ \frac{\text{Btu-ft}}{\text{hr-ft}^2\text{-°F}}\right)} \\
&= 3112\text{°F} \quad (3100\text{°F})
\end{aligned}
$$

SI Solution

The rod volume per unit length is

$$
\begin{aligned}
\frac{V}{L} &= \left(\frac{\pi}{4}\right) d^2 = \frac{\left(\frac{\pi}{4}\right)(1.0 \text{ cm})^2}{\left(100 \ \frac{\text{cm}}{\text{m}}\right)^2} \\
&= 7.854 \times 10^{-5} \text{ m}^3/\text{m}
\end{aligned}
$$

The heat output per unit length of rod is

$$
\begin{aligned}
\frac{Q}{L} &= \left(\frac{V}{L}\right) G = \left(7.854 \times 10^{-5} \ \frac{\text{m}^3}{\text{m}}\right)\left(4.1 \times 10^8 \ \frac{\text{W}}{\text{m}^3}\right) \\
&= 32\,201.4 \text{ W/m}
\end{aligned}
$$

The diameter of the cladding is

$$d_o = \frac{1.0 \text{ cm} + \frac{(2)(0.5 \text{ mm})}{10 \ \frac{\text{mm}}{\text{cm}}}}{100 \ \frac{\text{cm}}{\text{m}}} = 0.011 \text{ m}$$

The surface area per unit length of cladding is

$$A = \pi d_o = \pi(0.011 \text{ m}) = 0.0346 \text{ m}^2/\text{m}$$

From Eq. 17.30 (also *NCEES Handbook:* Conduction), the surface temperature of the cladding is

$$
\begin{aligned}
T_s &= \frac{Q}{hA} + T_\infty \\
&= \frac{32\,201.4 \ \frac{\text{W}}{\text{m}}}{\left(57 \ \frac{\text{kW}}{\text{m}^2\cdot\text{K}}\right)\left(1000 \ \frac{\text{W}}{\text{kW}}\right)\left(0.0346 \ \frac{\text{m}^2}{\text{m}}\right)} + 260\text{°C} \\
&= 276.3\text{°C}
\end{aligned}
$$

For the cladding,

$$r_o = \frac{d_o}{2} = \frac{0.011 \text{ m}}{2} = 0.0055 \text{ m}$$

$$r_i = \frac{d_i}{2} = \frac{0.01 \text{ m}}{2} = 0.0050 \text{ m}$$

From App. 17B (also *NCEES Handbook* table "Physical Properties of Metals at 20°C (SI Units)") and the table's footnote, the k for stainless steel is 16.3 W/m·K at 30°C.

Heat Transfer

For a cylinder, from Eq. 17.7 and Eq. 17.26 (also *NCEES Handbook:* Conduction),

$$T_{\text{inside}} - T_{\text{outside}} = \frac{Q \ln \frac{r_o}{r_i}}{2\pi k L} = \left(\frac{Q}{L}\right)\left(\frac{\ln \frac{r_o}{r_i}}{2\pi k}\right)$$

$$= \left(32\,201.4\ \frac{\text{W}}{\text{m}}\right)\left(\frac{\ln \frac{0.0055\ \text{m}}{0.0050\ \text{m}}}{(2\pi)\left(16.3\ \frac{\text{W}}{\text{m}\cdot\text{K}}\right)}\right)$$

$$= 30^\circ\text{C}$$

$$T_{\text{inside cladding}} = T_{\text{outside fuel rod}} + 30^\circ\text{C} = 276.3^\circ\text{C} + 30^\circ\text{C} = 306.3^\circ\text{C}$$

From Eq. 17.65 (also *NCEES Handbook:* Conduction), using $r_{o,\text{fuel rod}} = r_{i,\text{cladding}}$, and using k for the uranium dioxide (not the stainless steel),

$$T_{\text{center}} = T_o + \frac{Gr_o^2}{4k}$$

$$= 306.3^\circ\text{C} + \frac{\left(4.1\times 10^8\ \frac{\text{W}}{\text{m}^3}\right)(0.0050\ \text{m})^2}{(4)\left(1.9\ \frac{\text{W}}{\text{m}\cdot\text{K}}\right)}$$

$$= 306.3^\circ\text{C} + 1348.7\ \text{K} = 1655^\circ\text{C} \quad (1700^\circ\text{C})$$

The answer is (D).

12. *Customary U.S. Solution*

Consider this to be two infinite cylindrical fins with

$$T_b = 450^\circ\text{F}$$
$$T_\infty = 80^\circ\text{F}$$
$$h = 3\ \text{Btu/hr-ft}^2\text{-}^\circ\text{F}$$

From Eq. 17.75 (also *NCEES Handbook:* Planar Geometry—Area and Perimeter), the perimeter length is

$$P = \pi d = \frac{\pi\left(\frac{1}{16}\ \text{in}\right)}{12\ \frac{\text{in}}{\text{ft}}} = 0.01636\ \text{ft}$$

From Eq. 17.74 (also *NCEES Handbook:* Planar Geometry—Area and Perimeter), the cross-sectional area of the fin at its base is

$$A_b = \pi r^2 = \pi\left(\frac{d}{2}\right)^2 = \left(\frac{\pi}{4}\right)d^2$$

$$= \frac{\left(\frac{\pi}{4}\right)\left(\frac{1}{16}\ \text{in}\right)^2}{\left(12\ \frac{\text{in}}{\text{ft}}\right)^2}$$

$$= 2.131\times 10^{-5}\ \text{ft}^2$$

From Eq. 17.73 (also *NCEES Handbook:* Combination of Heat-Transfer Mechanisms), with two fins joined at the middle,

$$Q = 2\sqrt{hPkA_b}\,(T_b - T_\infty)$$

$$= (2)\sqrt{\begin{array}{l}\left(3\ \frac{\text{Btu}}{\text{hr-ft}^2\text{-}^\circ\text{F}}\right)(0.01636\ \text{ft}) \\ \times\left(215\ \frac{\text{Btu-ft}}{\text{hr-ft}^2\text{-}^\circ\text{F}}\right)(2.131\times 10^{-5}\ \text{ft}^2)\end{array}}$$
$$\times\ (450^\circ\text{F} - 80^\circ\text{F})$$

$$= 11.1\ \text{Btu/hr} \quad (11\ \text{Btu/hr})$$

This disregards radiation.

SI Solution

Consider this an infinite cylindrical fin with

$$T_b = 230^\circ\text{C}$$
$$T_\infty = 27^\circ\text{C}$$
$$h = 17\ \text{W/m}^2\cdot\text{K}$$

From Eq. 17.75 (also *NCEES Handbook:* Planar Geometry—Area and Perimeter), the perimeter length is

$$P = \pi d = \frac{\pi(1.6\ \text{mm})}{1000\ \frac{\text{mm}}{\text{m}}} = 5.027\times 10^{-3}\ \text{m}$$

Heat Transfer

From Eq. 17.74 (also *NCEES Handbook:* Planar Geometry—Area and Perimeter), the cross-sectional area of the fin at its base is

$$A_b = \pi r^2 = \pi\left(\frac{d}{2}\right)^2 = \left(\frac{\pi}{4}\right)d^2$$
$$= \frac{\left(\frac{\pi}{4}\right)(1.6\ \text{mm})^2}{\left(1000\ \frac{\text{mm}}{\text{m}}\right)^2}$$
$$= 2.011 \times 10^{-6}\ \text{m}^2$$

From Eq. 17.73 (also *NCEES Handbook:* Combination of Heat-Transfer Mechanisms), with two fins joined at the middle,

$$Q = 2\sqrt{hPkA_b}\,(T_b - T_\infty)$$
$$= (2)\sqrt{\begin{array}{l}\left(17\ \frac{\text{W}}{\text{m}^2\cdot\text{K}}\right)(5.027 \times 10^{-3}\ \text{m}) \\ \quad \times \left(372.1\ \frac{\text{W}}{\text{m}\cdot\text{K}}\right)(2.011 \times 10^{-6}\ \text{m}^2)\end{array}}$$
$$\times (230^\circ\text{C} - 27^\circ\text{C})$$
$$= 3.25\ \text{W} \quad (3.3\ \text{W})$$

This disregards radiation.

The answer is (B).

18 Natural Convection, Evaporation, and Condensation

PRACTICE PROBLEMS

1. The density of 87% wet steam at 50 psia (350 kPa) is most nearly

(A) 0.75 lbm/ft^3 (12 kg/m^3)

(B) 0.89 lbm/ft^3 (14 kg/m^3)

(C) 0.94 lbm/ft^3 (15 kg/m^3)

(D) 1.07 lbm/ft^3 (17 kg/m^3)

2. The viscosity of 100°F (38°C) water in units of lbm/hr-ft (kg/s·m) is most nearly

(A) 1.2 lbm/ft-hr (0.00052 kg/m·s)

(B) 1.4 lbm/ft-hr (0.00060 kg/m·s)

(C) 1.6 lbm/ft-hr (0.00068 kg/m·s)

(D) 1.8 lbm/ft-hr (0.00077 kg/m·s)

3. A fluid in a tank is maintained at 85°F (29°C) by a copper tube carrying hot water. The water decreases in temperature from 190°F (88°C) to 160°F (71°C) as it flows through the tube. The fluid's film coefficient must be calculated at the average temperature differential of this system. The temperature that this film coefficient should be calculated at is most nearly

(A) 130°F (54°C)

(B) 160°F (71°C)

(C) 175°F (79°C)

(D) 190°F (88°C)

4. A bare, horizontal conductor with a circular cross section with an outside diameter of 0.6 in (1.5 cm) dissipates 8.0 W/ft (25 W/m). The conductor is cooled by free convection, and the surrounding air temperature is 60°F (15°C). The film temperature is 100 °F (38°C), the heat capacity ratio (C) is 0.85, and the thermal expansion coefficient of air (β) can be estimated at 0.0018 1/°F (0.00318 1/°C). Using the given relation,

$$Nu = C(GrPr)^{0.25}$$

the conductor's surface temperature is most nearly

(A) 85°F (29°C)

(B) 110°F (43°C)

(C) 130°F (50°C)

(D) 160°F (67°C)

SOLUTIONS

1. *Customary U.S. Solution*

The steam is 87% wet, so the quality is $x = 0.13$.

From App. 10B (also *NCEES Handbook* "Saturated Steam (U.S. Units)—Temperature Table") at 50 psia,

$$\upsilon_f = 0.01727 \text{ ft}^3/\text{lbm}$$
$$\upsilon_g = 8.517 \text{ ft}^3/\text{lbm}$$

From Eq. 10.43 (also *NCEES Handbook:* Properties for Two-Phase (Vapor-Liquid) Systems), the specific volume of steam is

$$\begin{aligned}\upsilon &= \upsilon_f + x\upsilon_{fg}\\ &= 0.01727 \ \frac{\text{ft}^3}{\text{lbm}} + (0.13)\left(8.517 \ \frac{\text{ft}^3}{\text{lbm}} - 0.01727 \ \frac{\text{ft}^3}{\text{lbm}}\right)\\ &= 1.122 \text{ ft}^3/\text{lbm}\end{aligned}$$

The density is

$$\begin{aligned}\rho &= \frac{1}{\upsilon}\\ &= \frac{1}{1.122 \ \frac{\text{ft}^3}{\text{lbm}}}\\ &= 0.891 \text{ lbm/ft}^3 \quad (0.89 \text{ lbm/ft}^3)\end{aligned}$$

SI Solution

The steam is 87% wet, so the quality is $x = 0.13$.

From App. 10J (also *NCEES Handbook* "Saturated Steam (SI Units)—Temperature Table") at 350 kPa (350 bars),

$$\upsilon_f = 1.0786 \text{ cm}^3/\text{g}$$
$$\upsilon_g = 524.2 \text{ cm}^3/\text{g}$$

From Eq. 10.43 (also *NCEES Handbook:* Properties for Two-Phase (Vapor-Liquid) Systems), the specific volume of steam is

$$\begin{aligned}\upsilon &= \upsilon_f + x\upsilon_{fg}\\ &= 1.0786 \ \frac{\text{cm}^3}{\text{g}} + (0.13)\left(524.2 \ \frac{\text{cm}^3}{\text{g}}\right)\\ &= 69.22 \text{ cm}^3/\text{g}\end{aligned}$$

The density is

$$\begin{aligned}\rho &= \frac{1}{\upsilon}\\ &= \left(\frac{1}{69.22 \ \frac{\text{cm}^3}{\text{g}}}\right)\left(\frac{\left(100 \ \frac{\text{cm}}{\text{m}}\right)^3}{1000 \ \frac{\text{g}}{\text{kg}}}\right)\\ &= 14.45 \text{ kg/m}^3 \quad (14 \text{ kg/m}^3)\end{aligned}$$

The answer is (B).

2. *Customary U.S. Solution*

From App. 18A (also *NCEES Handbook* table "Physical Properties of Liquid Water), the viscosity of water at 100°F is

$$\begin{aligned}\mu &= \left(0.458 \times 10^{-3} \ \frac{\text{lbm}}{\text{ft-sec}}\right)\left(3600 \ \frac{\text{sec}}{\text{hr}}\right)\\ &= 1.6488 \text{ lbm/ft-hr} \quad (1.6 \text{ lbm/ft-hr})\end{aligned}$$

SI Solution

From App. 18B (also *NCEES Handbook* table "Physical Properties of Liquid Water (SI Units)"), the viscosity of water at 38°C (use 37.8°C) is

$$\mu = 0.682 \times 10^{-3} \text{ kg/m·s} \quad (0.00068 \text{ kg/m·s})$$

The answer is (C).

3. *Customary U.S. Solution*

The midpoint tube temperature is

$$T_s = \left(\frac{1}{2}\right)(190°\text{F} + 160°\text{F}) = 175°\text{F}$$
$$T_\infty = 85°\text{F} \quad \text{[given]}$$

Per Eq. 18.11, the film coefficient should be evaluated at a temperature of

$$\begin{aligned}T_h &= \tfrac{1}{2}(T_s + T_\infty)\\ &= \left(\frac{1}{2}\right)(175°\text{F} + 85°\text{F})\\ &= 130°\text{F}\end{aligned}$$

The answer is (A).

Heat Transfer

SI Solution

The midpoint tube temperature is

$$T_s = \left(\frac{1}{2}\right)(88°\text{C} + 71°\text{C}) = 79.5°\text{C}$$

$$T_\infty = 29°\text{C} \quad \text{[given]}$$

Per Eq. 18.11, the film coefficient should be evaluated at a temperature of

$$\begin{aligned} T_h &= \tfrac{1}{2}(T_s + T_\infty) \\ &= \left(\frac{1}{2}\right)(79.5°\text{F} + 29°\text{C}) \\ &= 54.3°\text{C} \quad (54°\text{C}) \end{aligned}$$

The answer is (A).

4. *Customary U.S. Solution*

The heat loss per unit length is

$$\frac{Q}{L} = \left(8.0\ \frac{\text{W}}{\text{ft}}\right)\left(3.413\ \frac{\text{Btu}}{\text{hr-W}}\right) = 27.3\ \text{Btu/hr-ft}$$

From App. 18C (also *NCEES Handbook:* Temperature-Dependent Properties of Air (U.S. Customary Units)), the air properties at 100°F are

$$\text{Pr} = 0.72$$

$$\frac{g\beta\rho^2}{\mu^2} = 1.76 \times 10^6\ \frac{1}{\text{ft}^3\text{-°F}}$$

From Eq. 18.4 (also *NCEES Handbook:* Similitude), the Grashof number is

$$\text{Gr} = L^3\left(\frac{g\beta\rho^2}{\mu^2}\right)\Delta T$$

The characteristic length is the wire diameter.

$$L = \frac{0.6\ \text{in}}{12\ \frac{\text{in}}{\text{ft}}} = 0.05\ \text{ft}$$

The temperature gradient is

$$\Delta T = T_s - T_\infty = T_{\text{wire}} - 60°\text{F}$$

T_{wire} is unknown. Start by assuming $T_{\text{wire}} = 150°\text{F}$.

$$\begin{aligned} \text{Gr} &= (0.05\ \text{ft})^3\left(1.76 \times 10^6\ \frac{1}{\text{ft}^3\text{-°F}}\right)(150°\text{F} - 60°\text{F}) \\ &= 19{,}800 \end{aligned}$$

$$\text{PrGr} = (0.72)(19{,}800) = 14{,}256$$

As in *NCEES Handbook* table "Forced Convection—External Flow," the heat capacity ratio is $C = 0.85$ for this GrPr value. As in *NCEES Handbook* table "Temperature-Dependent Properties of Air at 14.7 psia (U.S. Units)," the thermal conductivity for air is $k = 0.015$ Btu/hr-ft-°F. The correlation equation for free convection over a horizontal cylinder can be found in *NCEES Handbook*: Free/Forced Heat-Transfer Coefficients/Correlations. Rearrange the correlation to find the film coefficient. Use the diameter for the characteristic length.

$$\begin{aligned} \text{Nu} &= C(\text{GrPr})^{1/4} = \frac{hd}{k} \\ h &= \left(\frac{k}{d}\right)C(\text{GrPr})^{1/4} \\ &= \left(\frac{0.015\ \frac{\text{Btu}}{\text{hr-ft°F}}}{0.05\ \text{ft}}\right)(0.85)(14{,}256)^{1/4} \\ &= 2.78\ \text{Btu/hr-ft}^2\text{-°F} \end{aligned}$$

The heat transfer from the wire is found from Eq. 18.1 (also *NCEES Handbook:* Convection).

$$\begin{aligned} Q &= \pi dLh(T_{\text{wire}} - T_\infty) \\ T_{\text{wire}} &= \frac{\frac{Q}{L}}{\pi dh} + T_\infty \\ &= \frac{27.3\ \frac{\text{Btu}}{\text{hr-ft}}}{\pi(0.05\ \text{ft})\left(2.78\ \frac{\text{Btu}}{\text{hr-ft}^2\text{-°F}}\right)} + 60°\text{F} \\ &= 122.5°\text{F} \end{aligned}$$

Perform one more iteration with $T_{\text{wire}} = 122.5°\text{F}$.

$$\begin{aligned} \text{Gr} &= (0.05)^3\left(1.76 \times 10^6\ \frac{1}{\text{ft}^3\text{-°F}}\right)(122.5°\text{F} - 60°\text{F}) \\ &= 13\,750 \end{aligned}$$

$$\text{PrGr} = (0.72)(13\,750) = 9900$$

As in *NCEES Handbook:* Free/Forced Heat-Transfer Coefficients/Correlations,

$$\begin{aligned} h &= \left(\frac{k}{d}\right)C(\text{GrPr})^{1/4} \\ &= \left(\frac{0.015\ \frac{\text{Btu}}{\text{hr-ft°F}}}{0.05\ \text{ft}}\right)(0.85)(9900)^{1/4} \\ &= 2.55\ \text{Btu/hr-ft}^2\text{-°F} \end{aligned}$$

Heat Transfer

$$
\begin{aligned}
T_{\text{wire}} &= \frac{\frac{Q}{L}}{\pi d h} + T_\infty \\
&= \frac{27.3\ \frac{\text{Btu}}{\text{hr-ft}}}{\pi(0.05\ \text{ft})\left(2.55\ \frac{\text{Btu}}{\text{hr-ft}^2\text{-}^\circ\text{F}}\right)} + 60^\circ\text{F} \\
&= 128.2^\circ\text{F} \quad (130^\circ\text{F})
\end{aligned}
$$

There is no need to repeat iterations since the assumed temperature and the calculated temperature are about the same.

The answer is (C).

SI Solution

From App. 18D (also *NCEES Handbook:* Temperature-Dependent Properties of Air (SI Units)), the air properties at 38°C are

$$
\begin{aligned}
\text{Pr} &= 0.705 \\
\frac{g\beta\rho^2}{\mu^2} &= 1.12 \times 10^8\ \frac{1}{\text{K}\cdot\text{m}^3}
\end{aligned}
$$

From Eq. 18.4 (also *NCEES Handbook:* Similitude), the Grashof number is

$$
\text{Gr} = L^3\left(\frac{g\beta\rho^2}{\mu^2}\right)\Delta T
$$

The characteristic length is the wire diameter.

$$
L = \frac{1.5\ \text{cm}}{100\ \frac{\text{cm}}{\text{m}}} = 0.015\ \text{m}
$$

The temperature gradient is

$$
\begin{aligned}
\Delta T &= T_s - T_\infty = T_{\text{wire}} - 15^\circ\text{C} \\
\text{Gr} &= (0.015\ \text{m})^3\left(1.12 \times 10^8\ \frac{1}{\text{K}\cdot\text{m}^3}\right)(T_{\text{wire}} - 15^\circ\text{C})
\end{aligned}
$$

T_{wire} is unknown. Start by assuming $T_{\text{wire}} = 50^\circ\text{C}$.

$$
\begin{aligned}
\text{Gr} &= (0.015\ \text{m})^3\left(1.12 \times 10^8\ \frac{1}{\text{K}\cdot\text{m}^3}\right)(50^\circ\text{C} - 15^\circ\text{C}) \\
&= 13\,230 \\
\text{PrGr} &= (0.705)(13\,230) \\
&= 9327
\end{aligned}
$$

As in *NCEES Handbook* table "Forced Convection—External Flow," the heat capacity ratio is C = 0.85 for this GrPr value. As in *NCEES Handbook* table "Temperature-Dependent Properties of Air at 0.1 MPa (SI Units)," the thermal conductivity for air is $k = 0.03$ W/m·k. The correlation for free convection over a horizontal cylinder can be found in *NCEES Handbook:* Free/Forced Heat-Transfer Coefficients/Correlations. Rearrange the correlation equation to find the film coefficient. Use the diameter for the characteristic length.

$$
\begin{aligned}
\text{Nu} &= C(\text{GrPr})^{1/4} = \frac{hd}{k} \\
h &= \left(\frac{k}{d}\right)C(\text{GrPr})^{1/4} \\
&= \left(\frac{0.03\ \frac{\text{W}}{\text{m}\cdot\text{K}}}{0.015\ \text{m}}\right)(0.85)(9327)^{1/4} \\
&= 16.7\ \text{W/m}^2\cdot\text{K}
\end{aligned}
$$

The heat transfer from the wire is found from Eq. 18.1 (also *NCEES Handbook:* Free/Forced Heat-Transfer Coefficients/Correlations).

$$
\begin{aligned}
Q &= \pi d L h (T_{\text{wire}} - T_\infty) \\
T_{\text{wire}} &= \frac{\frac{Q}{L}}{\pi d h} + T_\infty \\
&= \frac{25\ \frac{\text{W}}{\text{m}}}{\pi(0.015\ \text{m})\left(16.7\ \frac{\text{W}}{\text{m}^2\cdot\text{K}}\right)} + 15^\circ\text{C} \\
&= 46.8^\circ\text{C} \quad (50^\circ\text{C})
\end{aligned}
$$

There is no need to perform another iteration since the assumed temperature and the calculated temperature are about the same.

The answer is (C).

19 Forced Convection and Heat Exchangers

PRACTICE PROBLEMS

1. Water is heated from 55°F to 87°F (15°C to 30°C) by stack gases that are cooled from 350°F to 270°F (175°C to 130°C). The logarithmic mean temperature difference is most nearly

(A) 190°F (105°C)

(B) 210°F (117°C)

(C) 235°F (130°C)

(D) 270°F (150°C)

2. A fluid in a tank is maintained at 85°F (30°C) by an immersed hot water coil. The hot water enters at 190°F (90°C) and leaves at 160°F (70°C). The logarithmic mean temperature difference is most nearly

(A) 89°F (49°C)

(B) 96°F (53°C)

(C) 111°F (62°C)

(D) 127°F (71°C)

3. Light no. 10 oil is heated from 95°F to 105°F (35°C to 41°C) in a 0.6 in (1.52 cm) inside diameter tube. The average velocity of the oil is 2.0 ft/sec (0.6 m/s). The viscosity of the oil at 100°F (38°C) is 45 centistokes (cS). The Reynolds number is most nearly

(A) 200

(B) 2500

(C) 4600

(D) 19,000

4. A white, uninsulated, rectangular duct passes through a 50 ft (15 m) wide room. The duct is 18 in (45 cm) wide and 12 in (30 cm) high. The room and its contents are at 70°F (21°C). Air at 100°F (40°C) enters the duct flowing at 800 ft/min (4.0 m/s). The film coefficient for the outside of the duct is 2.0 Btu/hr-ft^2-°F (11 W/m^2·K). All theoretical hydraulic properties are applicable. What is most nearly the temperature of the air after it has traveled in the duct the full 50 ft (15 m)?

(A) 85°F (29°C)

(B) 88°F (31°C)

(C) 91°F (33°C)

(D) 94°F (36°C)

5. A steel pipe carrying 350°F (175°C) air is 100 ft (30 m) long. The outside and inside diameters are 4.00 in and 3.50 in (10 cm and 9.0 cm), respectively. The pipe is covered with 2.0 in (5.0 cm) of insulation with a thermal conductivity of 0.05 Btu-ft/hr-ft^2-°F (0.086 W/m·K). The pipe passes through a 50°F (10°C) basement. Flow is laminar and fully developed. The heat loss is most nearly

(A) 3100 Btu/hr (0.93 kW)

(B) 3500 Btu/hr (1.1 kW)

(C) 4200 Btu/hr (1.3 kW)

(D) 8700 Btu/hr (2.0 kW)

6. An uninsulated horizontal pipe with 4.00 in (10 cm) outside diameter carries saturated 300 psia (2.1 MPa) steam through a 70°F (21°C) room. The steam flow rate is 5000 lbm/hr (0.63 kg/s). The pipe emissivity is 0.80. Any thermal gradients within the pipe are negligible. Approximately what decrease in quality will occur in the first 50 ft (15 m) of length?

(A) 1.5%

(B) 2.2%

(C) 2.8%

(D) 4.3%

7. A crossflow tubular feedwater heater is being designed to heat 2940 lbm/hr (0.368 kg/s) of water from 70°F (21°C) to 190°F (90°C). Saturated steam at 134 psia (923 kPa) is condensing on the outside of the tubes. The tubes are made from constantan, a copper alloy containing 70% Cu and 30% Ni. Each tube has a 1 in (2.54 cm) outside diameter and a 0.9 in (2.29 cm) inside diameter. The water velocity inside the tubes is 3 ft/sec (0.9 m/s). The outside tube surface area required is most nearly

(A) 3.5 ft^2 (0.33 m^2)

(B) 5.5 ft^2 (0.51 m^2)

(C) 10 ft^2 (0.93 m^2)

(D) 18 ft^2 (1.7 m^2)

8. A U-tube surface feedwater heater with one shell pass and two tube passes is being designed to heat 500,000 lbm/hr (60 kg/s) of water from 200°F to 390°F (100°C to 200°C). The water flows at 5 ft/sec (1.5 m/s) through the tubes. Dry, saturated steam at 400°F (205°C) is to be used as the heating medium. The heater is to operate straight condensing (i.e., the condensed steam will not be mixed with the heated water). Saturated water at 400°F (205°C) is removed from the heater. The tubes in the heater are $^7/_8$ in (2.2 cm) outside diameter with $^1/_{16}$ in (1.6 mm) walls. The overall heat transfer coefficient is estimated as 700 Btu/hr-ft^2-°F (4 kW/m^2·°C). Approximately how many tubes are required?

(A) 80

(B) 110

(C) 140

(D) 170

9. A single-pass heat exchanger is tested in a clean condition and is found to heat 100 gal/min (6.3 L/s) of 70°F (21°C) water to 140°F (60°C). The hot side uses 230°F (110°C) steam. The tube's inner surface area is 50 ft^2 (4.7 m^2). After being used in the field for several months, the exchanger heats 100 gal/min (6.3 L/s) of 70°F (21°C) water to 122°F (50°C). The fouling factor is most nearly

(A) 0.0004 hr-ft^2-°F/Btu (0.00008 m^2·K/W)

(B) 0.0008 hr-ft^2-°F/Btu (0.0001 m^2·K/W)

(C) 0.001 hr-ft^2-°F/Btu (0.0002 m^2·K/W)

(D) 0.002 hr-ft^2-°F/Btu (0.0004 m^2·K/W)

10. Cold water is used to cool hot water in a single-pass, tube-in-tube, counterflow heat exchanger. Compared to a plot of heat transfer rate-versus-distance traveled for the hot water, the plot of heat transfer rate-versus-time traveled for the hot water is

(A) identical

(B) reversed end-to-end

(C) shifted (delayed)

(D) inverted top-to-bottom

11. Cold water is used to cool hot water in a single-pass, tube-in-tube, counterflow heat exchanger. Compared to a plot of heat transfer rate-versus-distance traveled by the cold water, the heat transfer rate-versus-distance traveled plot for the hot water is

(A) identical

(B) reversed end-to-end

(C) shifted (delayed)

(D) inverted top-to-bottom

12. Cold water is used to cool hot water in a single-pass, tube-in-tube, parallel-flow (co-current) heat exchanger. Compared to a plot of temperature-versus-location by the hot water, the temperature-versus-location traveled plot of the cold water is

(A) identical

(B) reversed end-to-end

(C) shifted (delayed)

(D) inverted top-to-bottom

13. Cold water is used to cool hot water in a single-pass, tube-in-tube, counterflow heat exchanger. The temperature gradient experienced by the hot water along the length of the heat exchanger will

(A) essentially be constant

(B) decrease nonlinearly

(C) increase nonlinearly

(D) decrease linearly

14. Cold water is used to condense saturated steam in a single-pass, tube-in-tube, counterflow heat exchanger without subcooling. The temperature of the steam along the length of the heat exchanger from entrance to exit

(A) decreases linearly

(B) decreases parabolically

(C) decreases logarithmically

(D) is constant

15. Heat exchanger "duty" is best defined as the

(A) entering enthalpy difference of hot and cold fluids

(B) non-adiabatic heat loss

(C) heat transfer rate per unit mass of cold fluid

(D) heat transfer rate per unit time

16. A 2 ft long cylindrical fin with a diameter of 0.5 in is attached at one end to a heat source. The temperature of the heat source is 350°F, and the temperature of the ambient air is 75°F. The thermal conductivity of the fin is 128 Btu/hr-ft-°F, and the average film coefficient along the length of fin is 1.3 Btu/hr-ft^2-°F. No heat is transferred from the exposed fin tip. Heat transfer from the heat source into the base of the fin is most nearly

(A) 22 Btu/hr

(B) 29 Btu/hr

(C) 36 Btu/hr

(D) 46 Btu/hr

17. A heat-conducting rod has an outside diameter of 0.35 in (8.9 mm). Its uniform temperature is 100°F (38°C). The rod is inserted perpendicularly into a 100 ft/sec (30 m/s) airflow. The air temperature is 150°F (66°C). The film coefficient on the outside of the rod is most nearly

(A) 36 Btu/hr-ft^2-°F (210 W/m^2·K)

(B) 45 Btu/hr-ft^2-°F (260 W/m^2·K)

(C) 66 Btu/hr-ft^2-°F (380 W/m^2·K)

(D) 91 Btu/hr-ft^2-°F (530 W/m^2·K)

SOLUTIONS

1. *Customary U.S. Solution*

The logarithmic mean temperature difference will be different for different types of flow.

Parallel flow:

$$\Delta T_A = 350°\text{F} - 55°\text{F} = 295°\text{F}$$
$$\Delta T_B = 270°\text{F} - 87°\text{F} = 183°\text{F}$$

From Eq. 19.68 (also *NCEES Handbook:* Log-Mean Temperature Difference), the logarithmic mean temperature difference is

$$\Delta T_{\text{lm}} = \frac{\Delta T_A - \Delta T_B}{\ln \dfrac{\Delta T_A}{\Delta T_B}} = \frac{295°\text{F} - 183°\text{F}}{\ln \dfrac{295°\text{F}}{183°\text{F}}} = 234.6°\text{F}$$

Counterflow:

$$\Delta T_A = 270°\text{F} - 55°\text{F} = 215°\text{F}$$
$$\Delta T_B = 350°\text{F} - 87°\text{F} = 263°\text{F}$$

From Eq. 19.68 (also *NCEES Handbook:* Log-Mean Temperature Difference), the logarithmic mean temperature difference is

$$\Delta T_{\text{lm}} = \frac{\Delta T_A - \Delta T_B}{\ln \dfrac{\Delta T_A}{\Delta T_B}} = \frac{215°\text{F} - 263°\text{F}}{\ln \dfrac{215°\text{F}}{263°\text{F}}} = 238.2°\text{F} \quad (235°\text{F})$$

SI Solution

The logarithmic mean temperature difference will be different for different types of flow.

Parallel flow:

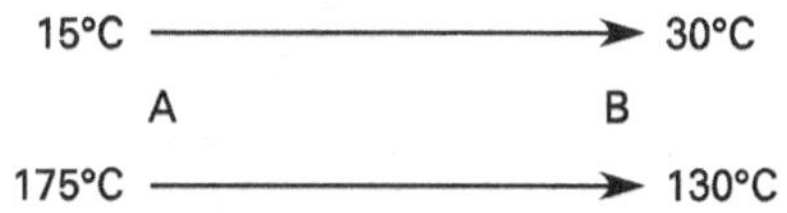

$$\Delta T_A = 175°\text{C} - 15°\text{C} = 160°\text{C}$$
$$\Delta T_B = 130°\text{C} - 30°\text{C} = 100°\text{C}$$

From Eq. 19.68 (also *NCEES Handbook:* Log-Mean Temperature Difference), the logarithmic mean temperature difference is

$$\Delta T_{\text{lm}} = \frac{\Delta T_A - \Delta T_B}{\ln \dfrac{\Delta T_A}{\Delta T_B}} = \frac{160°\text{C} - 100°\text{C}}{\ln \dfrac{160°\text{C}}{100°\text{C}}} = 127.7°\text{C}$$

Counterflow:

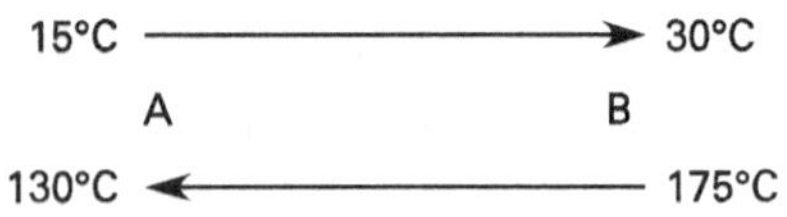

$$\Delta T_A = 130°\text{C} - 15°\text{C} = 115°\text{C}$$
$$\Delta T_B = 175°\text{C} - 30°\text{C} = 145°\text{C}$$

From Eq. 19.68 (also *NCEES Handbook:* Log-Mean Temperature Difference), the logarithmic mean temperature difference is

$$\Delta T_{\text{lm}} = \frac{\Delta T_A - \Delta T_B}{\ln \dfrac{\Delta T_A}{\Delta T_B}} = \frac{115°\text{C} - 145°\text{C}}{\ln \dfrac{115°\text{C}}{145°\text{C}}} = 129.4°\text{C} \quad (130°\text{C})$$

The answer is (C).

2. *Customary U.S. Solution*

$$\Delta T_A = 190°\text{F} - 85°\text{F} = 105°\text{F}$$
$$\Delta T_B = 160°\text{F} - 85°\text{F} = 75°\text{F}$$

From Eq. 19.68 (also *NCEES Handbook:* Log-Mean Temperature Difference), the logarithmic mean temperature difference is

$$\Delta T_{\text{lm}} = \frac{\Delta T_A - \Delta T_B}{\ln \dfrac{\Delta T_A}{\Delta T_B}} = \frac{105°\text{F} - 75°\text{F}}{\ln \dfrac{105°\text{F}}{75°\text{F}}} = 89.2°\text{F} \quad (89°\text{F})$$

SI Solution

$$\Delta T_A = 90°\text{C} - 30°\text{C} = 60°\text{C}$$
$$\Delta T_B = 70°\text{C} - 30°\text{C} = 40°\text{C}$$

From Eq. 19.68 (also *NCEES Handbook:* Log-Mean Temperature Difference), the logarithmic mean temperature difference is

$$\Delta T_{\text{lm}} = \frac{\Delta T_A - \Delta T_B}{\ln \dfrac{\Delta T_A}{\Delta T_B}} = \frac{60°\text{C} - 40°\text{C}}{\ln \dfrac{60°\text{C}}{40°\text{C}}} = 49.3°\text{C} \quad (49°\text{C})$$

The answer is (A).

3. *Customary U.S. Solution*

As in *NCEES Handbook* table "Dimensionless Numbers," the Reynolds number is

$$\text{Re}_d = \frac{\text{v}D}{\nu}$$
$$D = \frac{0.6 \text{ in}}{12 \dfrac{\text{in}}{\text{ft}}} = 0.05 \text{ ft}$$
$$\text{v} = 2 \text{ ft/sec}$$

$$\nu = \frac{45 \text{ cS}}{\left(100 \dfrac{\text{cS}}{\text{S}}\right)\left(929 \dfrac{\text{sec-stoke}}{\text{ft}^2}\right)} = 4.84 \times 10^{-4} \text{ ft}^2/\text{sec}$$

$$\text{Re} = \frac{\left(2 \dfrac{\text{ft}}{\text{sec}}\right)(0.05 \text{ ft})}{4.84 \times 10^{-4} \dfrac{\text{ft}^2}{\text{sec}}} = 206.6 \quad (200)$$

SI Solution

As in *NCEES Handbook* table "Dimensionless Numbers," the Reynolds number is

$$\mathrm{Re}_d = \frac{\mathrm{v}D}{\nu}$$

$$D = \frac{1.52\ \mathrm{cm}}{100\ \frac{\mathrm{cm}}{\mathrm{m}}} = 0.0152\ \mathrm{m}$$

$$\mathrm{v} = 0.6\ \mathrm{m/s}$$

$$\nu = \frac{(45\ \mathrm{cS})\left(1\ \frac{\frac{\mu\mathrm{m}^2}{\mathrm{s}}}{\mathrm{cS{\cdot}s}}\right)}{10^6\ \frac{\mu\mathrm{m}^2}{\mathrm{m}^2}} = 45\times 10^{-6}\ \mathrm{m^2/s}$$

$$\mathrm{Re}_d = \frac{\left(0.6\ \frac{\mathrm{m}}{\mathrm{s}}\right)(0.0152\ \mathrm{m})}{45\times 10^{-6}\ \frac{\mathrm{m}^2}{\mathrm{s}}} = 202.7 \quad (200)$$

The answer is (A).

4. *Customary U.S. Solution*

The exposed duct area is

$$A = (2W + 2H)L = \frac{\big((2)(18\ \mathrm{in}) + (2)(12\ \mathrm{in})\big)(50\ \mathrm{ft})}{12\ \frac{\mathrm{in}}{\mathrm{ft}}} = 250\ \mathrm{ft}^2$$

The duct is noncircular; therefore, the hydraulic diameter of the duct will be used. From Eq. 19.48 (also *NCEES Handbook*: Free/Forced Heat-Transfer Coefficients/Correlations), the theoretical hydraulic diameter is

$$d_H = 4\left(\frac{\text{area in flow}}{\text{wetted perimeter}}\right) = 4\left(\frac{WH}{2(W+H)}\right) = \frac{(4)\left(\frac{(18\ \mathrm{in})(12\ \mathrm{in})}{(2)(18\ \mathrm{in} + 12\ \mathrm{in})}\right)}{12\ \frac{\mathrm{in}}{\mathrm{ft}}} = 1.2\ \mathrm{ft}$$

From App. 18C, for air at 100°F,

$$\nu = 18.0\times 10^{-5}\ \mathrm{ft^2/sec}$$
$$\rho = 0.0710\ \mathrm{lbm/ft^3}$$
$$\mathrm{Pr} = 0.71$$

The Reynolds number is

$$\mathrm{Re} = \frac{\mathrm{v}D}{\nu} = \frac{\left(800\ \frac{\mathrm{ft}}{\mathrm{min}}\right)(1.2\ \mathrm{ft})}{\left(18.0\times 10^{-5}\ \frac{\mathrm{ft}^2}{\mathrm{sec}}\right)\left(60\ \frac{\mathrm{sec}}{\mathrm{min}}\right)} = 8.90\times 10^4$$

This is a turbulent flow. The thermal conductivity for air can be found in *NCEES Handbook:* Temperature-Dependent Properties of Air (U.S. Customary Units). The correlation for forced convection in a pipe can be found in *NCEES Handbook:* Free/Forced Heat-Transfer Coefficients/Correlations. Rearrange the correlation to find the film coefficient for the inside of the duct. (Ignore the viscosity term for such a small temperature change.)

$$\mathrm{Nu} = 0.023\mathrm{Re}^{0.8}\mathrm{Pr}^{1/3}\left(\frac{\mu_\infty}{\mu_s}\right) = \frac{h_i d}{k}$$

$$h_i = \left(\frac{k}{d}\right)0.023\mathrm{Re}^{0.8}\mathrm{Pr}^{1/3}\left(\frac{\mu_\infty}{\mu_s}\right) = \left(\frac{0.015\ \frac{\mathrm{Btu}}{\mathrm{hr\text{-}ft\text{-}°F}}}{1.2\ \mathrm{ft}}\right)(0.23)(8.9\times 10^4)^{0.8}(0.71)^{1/3} = 2.34\ \mathrm{Btu/hr\text{-}ft^2\text{-}°F}$$

Disregarding the duct thermal resistance, the overall heat transfer coefficient from Eq. 19.73 (also *NCEES Handbook:* Overall Heat-Transfer Coefficient) is

$$\frac{1}{U} = \frac{1}{h_i} + \frac{1}{h_o} = \frac{1}{2.34\ \frac{\mathrm{Btu}}{\mathrm{hr\text{-}ft^2\text{-}°F}}} + \frac{1}{2.0\ \frac{\mathrm{Btu}}{\mathrm{hr\text{-}ft^2\text{-}°F}}} = 0.927\ \mathrm{hr\text{-}ft^2\text{-}°F/Btu}$$

$$U = 1.08\ \mathrm{Btu/hr\text{-}ft^2\text{-}°F}$$

The heat transfer to the room is

$$Q = UA(T_{\text{ave}} - T_\infty) = UA\left(\tfrac{1}{2}(T_{\text{in}} + T_{\text{out}}) - T_\infty\right)$$

Since T_{out} is unknown, an iteration procedure may be required. Assume $T_{\text{out}} \approx 95°\text{F}$.

$$\begin{aligned} Q &= \left(1.08\ \frac{\text{Btu}}{\text{hr-ft}^2\text{-°F}}\right)(250\ \text{ft}^2) \\ &\quad \times \left(\left(\frac{1}{2}\right)(100°\text{F} + 95°\text{F}) - 70°\text{F}\right) \\ &= 7425\ \text{Btu/hr} \quad (7400\ \text{Btu/hr}) \end{aligned}$$

(Notice that ΔT (not ΔT_{lm}) is used in accordance with standard conventions in the HVAC industry.)

Temperature T_{out} can be verified by using

$$\begin{aligned} Q &= \dot{m}c_p(T_{\text{in}} - T_{\text{out}}) \\ \dot{m} &= GA_{\text{flow}} = GWH \\ &= \frac{\left(3408.0\ \frac{\text{lbm}}{\text{hr-ft}^2}\right)(18\ \text{in})(12\ \text{in})}{\left(12\ \frac{\text{in}}{\text{ft}}\right)^2} \\ &= 5112\ \text{lbm/hr} \\ c_p &= 0.240\ \text{Btu/lbm-°F} \end{aligned}$$

As in *NCEES Handbook* table "Temperature-Dependent Properties of Air at 14.7 psia (U.S. Units),"

$$\begin{aligned} 7425\ \frac{\text{Btu}}{\text{hr}} &= \left(5112\ \frac{\text{lbm}}{\text{hr}}\right)\left(0.240\ \frac{\text{Btu}}{\text{lbm-°F}}\right) \\ &\quad \times (100°\text{F} - T_{\text{out}}) \\ T_{\text{out}} &= 94°\text{F} \quad [\text{close enough}] \end{aligned}$$

SI Solution

The exposed duct area is

$$\begin{aligned} A &= (2W + 2H)L \\ &= \frac{\left((2)(45\ \text{cm}) + (2)(30\ \text{cm})\right)(15\ \text{m})}{100\ \frac{\text{cm}}{\text{m}}} \\ &= 22.5\ \text{m}^2 \end{aligned}$$

The duct is noncircular; therefore, the hydraulic diameter of the duct will be used. From Eq. 19.48 (also *NCEES Handbook:* Free/Forced Heat-Transfer Coefficients/Correlations), the hydraulic diameter is

$$\begin{aligned} d_H &= 4\left(\frac{\text{area in flow}}{\text{wetted perimeter}}\right) = 4\left(\frac{WH}{2(W+H)}\right) \\ &= \frac{(4)\left(\frac{(45\ \text{cm})(30\ \text{cm})}{(2)(45\ \text{cm} + 30\ \text{cm})}\right)}{100\ \frac{\text{cm}}{\text{m}}} \\ &= 0.36\ \text{m} \end{aligned}$$

From App. 18D (also *NCEES Handbook* table "Temperature-Dependent Properties of Air at 0.1 MPa (SI Units)"), for air at 40°C,

$$\begin{aligned} \mu &= 1.91 \times 10^{-5}\ \text{kg/m·s} \\ \rho &= 1.130\ \text{kg/m}^3 \\ c_p &= 1.0051\ \text{kJ/kg·K} \\ k &= 0.02718\ \text{W/m·K} \\ \text{Pr} &= 0.0709 \end{aligned}$$

As in *NCEES Handbook* table "Dimensionless Numbers," the Reynolds number is

$$\begin{aligned} \text{Re} &= \frac{\rho v D}{\mu} = \frac{\left(1.130\ \frac{\text{kg}}{\text{m}^3}\right)\left(4.0\ \frac{\text{m}}{\text{s}}\right)(0.36\ \text{m})}{1.91 \times 10^{-5}\ \frac{\text{kg}}{\text{m·s}}} \\ &= 8.52 \times 10^4 \end{aligned}$$

This is a turbulent flow. From Eq. 19.34 (also *NCEES Handbook*: Free/Forced Heat-Transfer Coefficients/Correlations), the Nusselt number is

$$\text{Nu} = 0.023\,\text{Re}^{0.8}(\text{Pr})^{1/3}$$

The film coefficient is

$$\begin{aligned} h &= 0.023\,\text{Re}^{0.8}\left(\frac{k}{d}\right)(\text{Pr})^{1/3} \\ &= (0.023)(8.52 \times 10^4)^{0.8}\left(\frac{0.02718\ \frac{\text{W}}{\text{m·K}}}{0.36\ \text{m}}\right)(0.709)^{1/3} \\ &= 13.6\ \text{W/m}^2\text{·K} \end{aligned}$$

Disregarding the duct thermal resistance, the overall heat transfer coefficient from Eq. 19.73 (also *NCEES Handbook:* Overall Heat-Transfer Coefficient) is

$$\begin{aligned}\frac{1}{U} &= \frac{1}{h_i} + \frac{1}{h_o}\\ &= \frac{1}{11\ \frac{\text{W}}{\text{m}^2\text{·K}}} + \frac{1}{13.6\ \frac{\text{W}}{\text{m}^2\text{·K}}}\\ &= 0.1644\ \text{m}^2\text{·K/W}\\ U &= 6.08\ \text{W/m}^2\text{·K}\end{aligned}$$

The heat transfer to the room is

$$Q = UA(T_{\text{ave}} - T_\infty) = UA\left(\tfrac{1}{2}(T_{\text{in}} + T_{\text{out}}) - T_\infty\right)$$

Since T_{out} is unknown, an iterative procedure may be required. Assume $T_{\text{out}} = 36°\text{C}$.

$$\begin{aligned}Q &= \left(6.08\ \frac{\text{W}}{\text{m}^2\text{·K}}\right)(22.5\ \text{m}^2)\\ &\quad\times\left(\left(\frac{1}{2}\right)(40°\text{C} + 36°\text{C}) - 21°\text{C}\right)\\ &= 2325\ \text{W}\quad (2.4\ \text{kW})\end{aligned}$$

(Notice that ΔT (not ΔT_{lm}) is used in accordance with standard conventions in the HVAC industry.)

Temperature T_{out} can be verified by using

$$\begin{aligned}Q &= \dot{m}c_p(T_{\text{in}} - T_{\text{out}})\\ \dot{m} &= \rho A_{\text{flow}}\text{v}\\ &= \frac{\left(1.130\ \frac{\text{kg}}{\text{m}^3}\right)(45\ \text{cm})(30\ \text{cm})\left(4\ \frac{\text{m}}{\text{s}}\right)}{\left(100\ \frac{\text{cm}}{\text{m}}\right)^2}\\ &= 0.6102\ \text{kg/s}\end{aligned}$$

$$\begin{aligned}2325\ \text{W} &= \left(0.6102\ \frac{\text{kg}}{\text{s}}\right)\left(1.0051\ \frac{\text{kJ}}{\text{kg·K}}\right)\left(1000\ \frac{\text{J}}{\text{kJ}}\right)\\ &\quad\times(40°\text{C} - T_{\text{out}})\\ T_{\text{out}} &= 36.2°\text{C}\quad (36°\text{C})\end{aligned}$$

The answer is (D).

5. *Customary U.S. Solution*

Refer to Fig. 17.3. The radii are

$$r_a = \frac{d_i}{2} = \frac{3.5\ \text{in}}{(2)\left(12\ \frac{\text{in}}{\text{ft}}\right)} = 0.1458\ \text{ft}$$

$$r_b = \frac{d_o}{2} = \frac{4\ \text{in}}{(2)\left(12\ \frac{\text{in}}{\text{ft}}\right)} = 0.1667\ \text{ft}$$

$$\begin{aligned}r_c &= r_b + t_{\text{insulation}} = 0.1667\ \text{ft} + \frac{2\ \text{in}}{12\ \frac{\text{in}}{\text{ft}}}\\ &= 0.3334\ \text{ft}\end{aligned}$$

From App. 17B (also *NCEES Handbook* table "Physical Properties of Metals at 68°F (U.S. Units)"), for steel, $k_{\text{pipe}} \approx 24.8$ Btu-ft/hr-ft^2-°F.

Initially assume a typical value of $h_c = 1.5$ Btu/hr-ft^2-°F.

For fully developed laminar flow, from Eq. 19.28 (also *NCEES Handbook:* Free/Forced Heat-Transfer Coefficients/Correlations), $\text{Nu}_d = 3.658$.

From App. 18C (also *NCEES Handbook* table "Temperature-Dependent Properties of Air at 14.7 psia (U.S. Units)"), for air at 350°F,

$$\begin{aligned}k_{\text{air}} &\approx 0.0203\ \text{Btu/hr-ft-°F}\\ \text{Nu}_d &= \frac{h_a d_i}{k_{\text{air}}} = 3.658\\ h_a &= \frac{3.658 k_{\text{air}}}{d_i}\\ &= \frac{(3.658)\left(0.0203\ \frac{\text{Btu}}{\text{hr-ft-°F}}\right)\left(12\ \frac{\text{in}}{\text{ft}}\right)}{3.5\ \text{in}}\\ &= 0.255\ \text{Btu/hr-ft}^2\text{-°F}\end{aligned}$$

Heat Transfer

Neglect thermal resistance between pipe and insulation. From Eq. 17.41 (also *NCEES Handbook:* Combination of Heat-Transfer Mechanisms), the heat transfer is

$$Q = \frac{2\pi L(T_i - T_\infty)}{\frac{1}{r_a h_a} + \frac{\ln\left(\frac{r_b}{r_a}\right)}{k_{\text{pipe}}} + \frac{\ln\left(\frac{r_c}{r_b}\right)}{k_{\text{insulation}}} + \frac{1}{r_c h_c}}$$

$$= \frac{2\pi(100 \text{ ft})(350°\text{F} - 50°\text{F})}{\begin{array}{l}\frac{1}{(0.1458 \text{ ft})\left(0.255 \frac{\text{Btu}}{\text{hr-ft}^2\text{-°F}}\right)} + \frac{\ln\left(\frac{0.1667 \text{ ft}}{0.1458 \text{ ft}}\right)}{24.8 \frac{\text{Btu-ft}}{\text{hr-ft}^2\text{-°F}}} \\ + \frac{\ln\left(\frac{0.3334 \text{ ft}}{0.1667 \text{ ft}}\right)}{0.05 \frac{\text{Btu-ft}}{\text{hr-ft}^2\text{-°F}}} + \frac{1}{(0.3334 \text{ ft})\left(1.5 \frac{\text{Btu}}{\text{hr-ft}^2\text{-°F}}\right)}\end{array}}$$

$$= \frac{188{,}496 \text{ ft-°F}}{\begin{array}{l}26.90 \frac{\text{hr-ft-°F}}{\text{Btu}} + 0.00523 \frac{\text{hr-ft-°F}}{\text{Btu}} \\ + 13.863 \frac{\text{hr-ft-°F}}{\text{Btu}} + 2.00 \frac{\text{hr-ft-°F}}{\text{Btu}}\end{array}}$$

$$= 4408 \text{ Btu/hr}$$

Use Eq. 17.41 (also *NCEES Handbook:* Combination of Heat-Transfer Mechanisms) to find T_2 by using all resistances except the outer $(T_2 - T_\infty)$ resistance.

$$\begin{aligned} T_i - T_2 &= \left(\frac{4408 \frac{\text{Btu}}{\text{hr}}}{2\pi(100 \text{ ft})}\right) \times \left(\begin{array}{l}26.9 \frac{\text{hr-ft-°F}}{\text{Btu}} + 0.00523 \frac{\text{hr-ft-°F}}{\text{Btu}} \\ +13.863 \frac{\text{hr-ft-°F}}{\text{Btu}}\end{array}\right) \\ &= 286.0°\text{F} \\ T_2 &= T_i - 286.0°\text{F} = 350°\text{F} - 286.0°\text{F} \\ &= 64.0°\text{F} \end{aligned}$$

To evaluate h_c, use film temperature.

$$\begin{aligned} T_{\text{film}} &= \tfrac{1}{2}(T_2 + T_\infty) = \left(\frac{1}{2}\right)(64.0°\text{F} + 50°\text{F}) \\ &= 57°\text{F} \end{aligned}$$

From App. 18C (also *NCEES Handbook* table "Temperature-Dependent Properties of Air at 14.7 psia (U.S. Units)"), for air at 57°F,

$$\text{Pr} = 0.72$$

$$\frac{g\beta\rho^2}{\mu^2} = 2.645 \times 10^6 \ \frac{1}{\text{ft}^3\text{-°F}}$$

From Eq. 18.4 (also *NCEES Handbook*: Similitude), the Grashof number is

$$\text{Gr} = \frac{L^3 g\beta\rho^2(T_2 - T_\infty)}{\mu^2}$$

For pipe,

$$\begin{aligned} L &= d_c = 2r_c = (2)(0.3334 \text{ ft}) = 0.6668 \text{ ft} \\ \text{Gr} &= (0.6668 \text{ ft})^3\left(2.645 \times 10^6 \ \frac{1}{\text{ft}^3\text{-°F}}\right) \\ &\quad \times (64.0°\text{F} - 50°\text{F}) \\ &= 1.1 \times 10^7 \\ \text{Gr}\,\text{Pr} &= (1.1 \times 10^7)(0.72) \\ &= 7.9 \times 10^6 \end{aligned}$$

As in *NCEES Handbook* table "Forced Convection—External Flow," $C = 0.48$ for this GrPr value. The correlation for free convection over a horizontal cylinder can be found in *NCEES Handbook:* Free/Forced Heat-Transfer Coefficients/Correlations. Rearrange the correlation to find the film coefficient.

$$\begin{aligned} \text{Nu} &= C(\text{GrPr})^{1/4} = \frac{hd}{k} \\ h &= \left(\frac{k}{d}\right)C(\text{GrPr})^{1/4} \\ &= \left(\frac{0.0203 \frac{\text{Btu}}{\text{hr-ft-°F}}}{0.6668 \text{ ft}}\right)(0.48)(7.9 \times 10^6)^{1/4} \\ &= 0.775 \text{ Btu/hr-ft}^2\text{-°F} \end{aligned}$$

At the second iteration, the heat transfer is

$$Q = \frac{188{,}496\ \text{ft-°F}}{\begin{array}{l} 26.90\ \dfrac{\text{hr-ft-°F}}{\text{Btu}} + 0.00523\ \dfrac{\text{hr-ft-°F}}{\text{Btu}} \\ \quad + 13.863\ \dfrac{\text{hr-ft-°F}}{\text{Btu}} \\ \quad + \dfrac{1}{(0.3334\ \text{ft})\left(0.775\ \dfrac{\text{Btu}}{\text{hr-ft-°F}}\right)} \end{array}}$$

$$= 4222\ \text{Btu/hr} \quad (4200\ \text{Btu/hr})$$

Additional iterations will improve the accuracy further.

SI Solution

Refer to Fig. 17.3. The radii are

$$r_a = \frac{d_i}{2} = \frac{9.0\ \text{cm}}{(2)\left(100\ \dfrac{\text{cm}}{\text{m}}\right)} = 0.045\ \text{m}$$

$$r_b = \frac{d_o}{2} = \frac{10\ \text{cm}}{(2)\left(100\ \dfrac{\text{cm}}{\text{m}}\right)} = 0.050\ \text{m}$$

$$r_c = r_b + t_{\text{insulation}} = 0.050\ \text{m} + \frac{5.0\ \text{cm}}{100\ \dfrac{\text{cm}}{\text{m}}}$$

$$= 0.100\ \text{m}$$

From App. 17B (also *NCEES Handbook:* Physical Properties of Metal at 20°C (SI Units), for steel, the thermal conductivity is 42.9 W/m·K.

Initially assume a typical value of $h_c = 3.5\ \text{W/m}^2\text{·K}$.

For fully developed laminar flow from Eq. 19.28 (also *NCEES Handbook:* Free/Forced Heat-Transfer Coefficients/Correlations), $\text{Nu}_d = 3.658$.

From App. 18D (also *NCEES Handbook* table "Temperature-Dependent Properties of Air at 0.1 MPa (SI Units)"), for air at 175°C,

$$k_{\text{air}} \approx 0.03709\ \text{W/m·K}$$

$$\text{Nu}_d = \frac{h_a d_i}{k_{\text{air}}} = 3.658$$

$$h_a = \frac{3.658 k_{\text{air}}}{d_i} = \frac{(3.658)\left(0.03709\ \dfrac{\text{W}}{\text{m·K}}\right)\left(100\ \dfrac{\text{cm}}{\text{m}}\right)}{9.0\ \text{cm}}$$

$$= 1.510\ \text{W/m}^2\text{·K}$$

Neglect thermal resistance between pipe and insulation. From Eq. 17.41 (also *NCEES Handbook:* Combination of Heat-Transfer Mechanisms), the heat transfer is

$$Q = \frac{2\pi L(T_i - T_\infty)}{\dfrac{1}{r_a h_a} + \dfrac{\ln\left(\dfrac{r_b}{r_a}\right)}{k_{\text{pipe}}} + \dfrac{\ln\left(\dfrac{r_c}{r_b}\right)}{k_{\text{insulation}}} + \dfrac{1}{r_c h_c}}$$

$$= \frac{2\pi(30\ \text{m})(175°\text{C} - 10°\text{C})}{\begin{array}{l} \dfrac{1}{(0.045\ \text{m})\left(1.510\ \dfrac{\text{W}}{\text{m}^2\text{·K}}\right)} + \dfrac{\ln\left(\dfrac{0.050\ \text{m}}{0.045\ \text{m}}\right)}{42.9\ \dfrac{\text{W}}{\text{m·K}}} \\ \quad + \dfrac{\ln\left(\dfrac{0.10\ \text{m}}{0.050\ \text{m}}\right)}{0.086\ \dfrac{\text{W}}{\text{m·K}}} + \dfrac{1}{(0.10\ \text{m})\left(3.5\ \dfrac{\text{W}}{\text{m}^2\text{·K}}\right)} \end{array}}$$

$$= \frac{31\,101.8\ \text{m·°C}}{\begin{array}{l} 14.74\ \dfrac{\text{m·K}}{\text{W}} + 0.00238\ \dfrac{\text{m·K}}{\text{W}} \\ \quad + 8.06\ \dfrac{\text{m·K}}{\text{W}} + 2.86\ \dfrac{\text{m·K}}{\text{W}} \end{array}}$$

$$= 1212\ \text{W}$$

Use Eq. 17.41 (also *NCEES Handbook:* Combination of Heat-Transfer Mechanisms) to find T_2 by using all resistances except the outer $(T_2 - T_\infty)$ resistance.

$$T_i - T_2 = \left(\frac{1212\ \text{W}}{2\pi(30\ \text{m})}\right) \times \left(14.74\ \frac{\text{m·K}}{\text{W}} + 0.00238\ \frac{\text{m·K}}{\text{W}} + 8.06\ \frac{\text{m·K}}{\text{W}}\right)$$

$$= 146.6°\text{C}$$

$$T_2 = T_i - 146.6°\text{C} = 175°\text{C} - 146.6°\text{C}$$

$$= 28.4°\text{C}$$

To evaluate h_c, use film temperature.

$$T_{\text{film}} = \tfrac{1}{2}(T_2 + T_\infty) = \left(\frac{1}{2}\right)(28.4°\text{C} + 10°\text{C})$$

$$= 19.2°\text{C}$$

From App. 18D (also *NCEES Handbook* table "Temperature-Dependent Properties of Air at 0.1 MPa (SI Units)"), for air at 19.3°C,

$$\text{Pr} = 0.710$$

$$\frac{g\beta\rho^2}{\mu^2} = 1.52 \times 10^8\ \frac{1}{\text{K·m}^3}$$

Heat Transfer

From Eq. 18.4 (also *NCEES Handbook:* Similitude), the Grashof number is

$$\mathrm{Gr} = \frac{L^3 g\beta\rho^2(T_2 - T_\infty)}{\mu^2}$$

For pipe,

$$L = d_c = 2r_c = (2)(0.10\ \mathrm{m}) = 0.20\ \mathrm{m}$$

$$\mathrm{Gr} = (0.20\ \mathrm{m})^3\left(1.52\times10^8\ \frac{1}{\mathrm{K{\cdot}m^3}}\right)(28.4^\circ\mathrm{C} - 10^\circ\mathrm{C})$$
$$= 2.24\times10^7$$
$$\mathrm{Gr\,Pr} = (2.24\times10^7)(0.710)$$
$$= 1.6\times10^7$$

As in *NCEES Handbook* table "Forced Convection—External Flow," $C = 0.48$ for this GrPr value. The correlation for free convection over a horizontal cylinder can be found in *NCEES Handbook:* Free/Forced Heat-Transfer Coefficients/Correlations. Rearrange the correlation to find the film coefficient.

$$\mathrm{Nu} = C(\mathrm{GrPr})^{1/4} = \frac{hd}{k}$$
$$h = \left(\frac{k}{d}\right)C(\mathrm{GrPr})^{1/4}$$
$$= \left(\frac{0.03709\ \frac{\mathrm{W}}{\mathrm{m{\cdot}K}}}{0.2\ \mathrm{m}}\right)(0.48)(1.6\times10^7)^{1/4}$$
$$= 5.6\ \mathrm{W/m^2{\cdot}K}$$

Further iteration will improve accuracy.

$$Q = \frac{31\,101.8\ \mathrm{m{\cdot}^\circ C}}{14.74\ \frac{\mathrm{m{\cdot}K}}{\mathrm{W}} + 0.00238\ \frac{\mathrm{m{\cdot}K}}{\mathrm{W}} + 8.06\ \frac{\mathrm{m{\cdot}K}}{\mathrm{W}} + \frac{1}{(0.10\ \mathrm{m})\left(5.6\ \frac{\mathrm{W}}{\mathrm{m^2{\cdot}K}}\right)}}$$
$$= 1264.9\ \mathrm{W}\quad (1.3\ \mathrm{kW})$$

The answer is (C).

6. *Customary U.S. Solution*

Neglecting pipe resistance (no information for pipe is given), $T_{\mathrm{pipe}} = T_{\mathrm{sat}}$.

From App. 10B (also *NCEES Handbook* table "Saturated Steam (U.S. Units)—Temperature Table"), for 300 lbf/in^2 steam, $T_{\mathrm{sat}} = 417.35^\circ\mathrm{F}$. When a vapor condenses, the vapor and condensed liquid are at the same temperature. Therefore, the entire pipe is assumed to be at 417.35°F. The outside film coefficient should be evaluated from Eq. 19.11.

$$T_{\mathrm{film}} = \tfrac{1}{2}(T_s + T_\infty) = \left(\frac{1}{2}\right)(417.35^\circ\mathrm{F} + 70^\circ\mathrm{F})$$
$$= 243.7^\circ\mathrm{F}$$

From App. 18C (also *NCEES Handbook* table "Temperature-Dependent Properties of Air at 14.7 psia (U.S. Units)"), for air at 243.7°F,

$$\mathrm{Pr} = 0.715$$
$$\frac{g\beta\rho^2}{\mu^2} = 0.673\times10^6\ \frac{1}{\mathrm{ft^3\text{-}^\circ F}}$$

From Eq. 18.4 (also *NCEES Handbook:* Similitude), the Grashof number is

$$\mathrm{Gr} = \frac{L^3 g\beta\rho^2(T_s - T_\infty)}{\mu^2}$$

$$L = d_{\mathrm{outside}} = \frac{4\ \mathrm{in}}{12\ \frac{\mathrm{in}}{\mathrm{ft}}} = 0.3333\ \mathrm{ft}$$

$$\mathrm{Gr} = (0.3333\ \mathrm{ft})^3\left(0.673\times10^6\ \frac{1}{\mathrm{ft^3\text{-}^\circ F}}\right)\times(417.35^\circ\mathrm{F} - 70^\circ\mathrm{F})$$
$$= 8.66\times10^6$$
$$\mathrm{Gr\,Pr} = (8.66\times10^6)(0.715)$$
$$= 6.19\times10^6$$

As in *NCEES Handbook* table "Forced Convection—External Flow," $C = 0.48$ for this GrPr value. The correlation for free convection over a horizontal cylinder can be found in *NCEES Handbook:* Free/Forced Heat-Transfer Coefficients/Correlations. Rearrange the correlation to find the film coefficient.

$$\mathrm{Nu} = C(\mathrm{GrPr})^{1/4} = \frac{hd}{k}$$
$$h = \left(\frac{k}{d}\right)C(\mathrm{GrPr})^{1/4}$$
$$= \left(\frac{0.019\ \frac{\mathrm{Btu}}{\mathrm{hr\text{-}ft\text{-}^\circ F}}}{0.3333\ \mathrm{ft}}\right)(0.48)(6.19\times10^6)^{1/4}$$
$$= 1.37\ \mathrm{Btu/hr\text{-}ft^2\text{-}^\circ F}$$

From Eq. 18.1 (also *NCEES Handbook:* Convection), the heat transfer for the first 50 ft due to convection is

$$\begin{aligned} Q &= h_c A_{\text{outside}}(T_s - T_\infty) \\ &= h_c(\pi d_{\text{outside}} L)(T_s - T_\infty) \\ &= \left(1.37\ \frac{\text{Btu}}{\text{hr-ft}^2\text{-°F}}\right)\pi(0.3333\ \text{ft})(50\ \text{ft}) \\ &\quad \times (417.35\text{°F} - 70\text{°F}) \\ Q_{\text{convection}} &= 24{,}914\ \text{Btu/hr} \end{aligned}$$

To determine heat transfer due to radiation, assume oxidized steel pipe, completely enclosed.

$$F_a = 1$$

As in *NCEES Handbook*: Temperature, the absolute temperatures are

$$\begin{aligned} T_1 &= 417.35\text{°F} + 460° = 877.35\text{°R} \\ T_2 &= 70\text{°F} + 460° = 530\text{°R} \\ F_e &= \epsilon_{\text{pipe}} = 0.80 \end{aligned}$$

The radiation heat transfer is from *NCEES Handbook:* Radiation.

$$\begin{aligned} E_{\text{net}} &= \sigma F_a F_e (T_1^4 - T_2^4) \\ &= \left(0.1713 \times 10^{-8}\ \frac{\text{Btu}}{\text{hr-ft}^2\text{-°R}^4}\right) \\ &\quad \times (1)(0.80)\left((877.35\text{°R})^4 - (530\text{°R})^4\right) \\ &= 704\ \text{Btu/hr-ft}^2 \\ Q_{\text{radiation}} &= E_{\text{net}} A = E_{\text{net}}(\pi d_{\text{outside}} L) \\ &= \left(704\ \frac{\text{Btu}}{\text{hr-ft}^2}\right)\pi(0.3333\ \text{ft})(50\ \text{ft}) \\ &= 36{,}858\ \text{Btu/hr} \end{aligned}$$

The total heat loss is

$$\begin{aligned} Q_{\text{total}} &= Q_{\text{convection}} + Q_{\text{radiation}} \\ &= 24{,}914\ \frac{\text{Btu}}{\text{hr}} + 36{,}858\ \frac{\text{Btu}}{\text{hr}} \\ &= 61{,}772\ \text{Btu/hr} \\ Q_{\text{total}} &= \dot{m}\Delta h \end{aligned}$$

The enthalpy decrease per pound is

$$\Delta h = \frac{Q_{\text{total}}}{\dot{m}_{\text{steam}}} = \frac{61{,}772\ \frac{\text{Btu}}{\text{hr}}}{5000\ \frac{\text{lbm}}{\text{hr}}} = 12.35\ \text{Btu/lbm}$$

This is a quality loss of

$$\begin{aligned} \Delta x &= \frac{\Delta h}{h_{fg}} = \frac{12.35\ \frac{\text{Btu}}{\text{lbm}}}{809.4\ \frac{\text{Btu}}{\text{lbm}}} \\ &= 0.0153\quad (1.5\%) \end{aligned}$$

SI Solution

From the customary U.S. solution, T_{pipe} is the same for the entire length.

From App. 10J (also *NCEES Handbook* table "Saturated Steam (SI Units)—Temperature Table") for 2.1 MPa, $T_{\text{sat}} = 214.72\text{°C}$. The outside film coefficient should be evaluated from Eq. 19.11 as

$$\begin{aligned} T_{\text{film}} &= \tfrac{1}{2}(T_s + T_\infty) = \left(\frac{1}{2}\right)(214.72\text{°C} + 21\text{°C}) \\ &= 117.9\text{°C} \end{aligned}$$

From App. 18D (also *NCEES Handbook* table "Temperature-Dependent Properties of Air at 0.1 MPa (SI Units)"), for air at 117.9°C,

$$\begin{aligned} \text{Pr} &= 0.692 \\ \frac{g\beta\rho^2}{\mu^2} &= 0.403 \times 10^8\ \frac{1}{\text{K}\cdot\text{m}^3} \end{aligned}$$

From Eq. 18.4 (also *NCEES Handbook:* Similitude), the Grashof number is

$$\begin{aligned} \text{Gr} &= \frac{L^3 g\beta\rho^2(T_s - T_\infty)}{\mu^2} \\ L &= d_{\text{outside}} = \frac{10\ \text{cm}}{100\ \frac{\text{cm}}{\text{m}}} = 0.10\ \text{m} \\ \text{Gr} &= (0.10\ \text{m})^3\left(0.403 \times 10^8\ \frac{1}{\text{K}\cdot\text{m}^3}\right) \\ &\quad \times (214.72\text{°C} - 21\text{°C}) \\ &= 7.807 \times 10^6 \\ \text{Gr Pr} &= (7.807 \times 10^6)(0.692) \\ &= 5.40 \times 10^6 \end{aligned}$$

As in *NCEES Handbook* table "Forced Convection—External Flow," $C = 0.48$ for this GrPr value. The correlation for free convection over a horizontal cylinder can be found in *NCEES Handbook:* Free/Forced Heat-Transfer Coefficients/Correlations. Rearrange the correlation to find the film coefficient.

$$\begin{aligned} \mathrm{Nu} &= C(\mathrm{GrPr})^{1/4} = \frac{hd}{k} \\ h &= \left(\frac{k}{d}\right)C(\mathrm{GrPr})^{1/4} \\ &= \left(\frac{0.0328\ \frac{\mathrm{W}}{\mathrm{m\cdot K}}}{0.1\ \mathrm{m}}\right)(0.48)(5.4\times 10^6)^{1/4} \\ &= 7.6\ \mathrm{W/m^2{\cdot}K} \end{aligned}$$

From Eq. 18.1, the heat transfer for the first 15 m due to convection is

$$\begin{aligned} Q_{\mathrm{convection}} &= h_c(\pi d_{\mathrm{outside}}L)(T_s - T_\infty) \\ &= \left(7.6\ \frac{\mathrm{W}}{\mathrm{m^2{\cdot}K}}\right)\pi(0.10\ \mathrm{m})(15\ \mathrm{m}) \\ &\quad \times (214.72^\circ\mathrm{C} - 21^\circ\mathrm{C}) \\ &= 6938\ \mathrm{W} \end{aligned}$$

To determine heat transfer due to radiation, assume oxidized steel pipe, completely enclosed.

$$F_a = 1$$

As in *NCEES Handbook* table "Dimensionless Numbers," the absolute temperatures are

$$\begin{aligned} T_1 &= 214.72^\circ\mathrm{C} + 273^\circ = 487.72\mathrm{K} \\ T_2 &= 21^\circ\mathrm{C} + 273^\circ = 294\mathrm{K} \\ F_e &= \epsilon_{\mathrm{pipe}} = 0.80 \end{aligned}$$

As in *NCEES Handbook:* Radiation, the radiation heat transfer is

$$\begin{aligned} E_{\mathrm{net}} &= \sigma F_a F_e (T_1^4 - T_2^4) \\ &= \left(5.67\times 10^{-8}\ \frac{\mathrm{W}}{\mathrm{m^2{\cdot}K^4}}\right) \\ &\quad \times (1)(0.80)\left((487.72\mathrm{K})^4 - (294\mathrm{K})^4\right) \\ &= 2228\ \mathrm{W/m^2} \\ Q_{\mathrm{radiation}} &= E_{\mathrm{net}}(\pi d_{\mathrm{outside}}L) \\ &= \left(2228\ \frac{\mathrm{W}}{\mathrm{m^2}}\right)\pi(0.10\ \mathrm{m})(15\ \mathrm{m}) \\ &= 10\,499\ \mathrm{W} \end{aligned}$$

The total heat loss is

$$\begin{aligned} Q_{\mathrm{total}} &= Q_{\mathrm{convection}} + Q_{\mathrm{radiation}} = 6938\ \mathrm{W} + 10\,499\ \mathrm{W} \\ &= 17\,437\ \mathrm{W} \\ Q_{\mathrm{total}} &= \dot{m}\Delta h \end{aligned}$$

The enthalpy decrease per kilogram is

$$\begin{aligned} \Delta h &= \frac{Q_{\mathrm{total}}}{\dot{m}_{\mathrm{steam}}} = \frac{17\,437\ \mathrm{W}}{\left(0.63\ \frac{\mathrm{kg}}{\mathrm{s}}\right)\left(1000\ \frac{\mathrm{J}}{\mathrm{kJ}}\right)} \\ &= 27.68\ \mathrm{kJ/kg} \end{aligned}$$

This is a quality loss of

$$\begin{aligned} \Delta x &= \frac{\Delta h}{h_{fg}} = \frac{27.68\ \frac{\mathrm{kJ}}{\mathrm{kg}}}{1879.8\ \frac{\mathrm{kJ}}{\mathrm{kg}}} \\ &= 0.0147 \quad (1.5\%) \end{aligned}$$

The answer is (A).

7. *Customary U.S. Solution*

The bulk temperature of the water is

$$\begin{aligned} T_b &= \tfrac{1}{2}(T_{\mathrm{in}} + T_{\mathrm{out}}) = \left(\frac{1}{2}\right)(70^\circ\mathrm{F} + 190^\circ\mathrm{F}) \\ &= 130^\circ\mathrm{F} \end{aligned}$$

From App. 18A (also *NCEES Handbook* table "Physical Properties of Liquid Water (U.S. Units)") the properties of water at 130°F are

$$\begin{aligned} c_p &= 0.999\ \mathrm{Btu/lbm\text{-}^\circ F} \\ \nu &= 0.582\times 10^{-5}\ \mathrm{ft^2/sec} \\ \mathrm{Pr} &= 3.45 \\ k &= 0.376\ \mathrm{Btu/hr\text{-}ft\text{-}^\circ F} \end{aligned}$$

The heat transfer is found from the temperature gain of the water.

$$\begin{aligned} Q &= \dot{m}c_p\Delta T \\ &= \left(2940\ \frac{\mathrm{lbm}}{\mathrm{hr}}\right)\left(0.999\ \frac{\mathrm{Btu}}{\mathrm{lbm\text{-}^\circ F}}\right)(190^\circ\mathrm{F} - 70^\circ\mathrm{F}) \\ &= 352{,}447\ \mathrm{Btu/hr} \end{aligned}$$

As in *NCEES Handbook*: Similitude, the Reynolds number is

$$\text{Re} = \frac{\text{v}D}{\nu} = \frac{\left(3\ \frac{\text{ft}}{\text{sec}}\right)(0.9\ \text{in})}{\left(0.582\times10^{-5}\ \frac{\text{ft}^2}{\text{sec}}\right)\left(12\ \frac{\text{in}}{\text{ft}}\right)} = 3.87\times10^4$$

From Eq. 19.33 (also *NCEES Handbook:* Free/Forced Heat-Transfer Coefficients/Correlations), the film coefficient is

$$\begin{aligned} h &= 0.023\text{Re}^{0.8}\text{Pr}^{\frac{1}{3}}\left(\frac{k}{d}\right) \\ &= (0.023)(3.87\times10^4)^{0.8}(3.45)^{1/3} \\ &\quad\times\left(\frac{\left(0.376\ \frac{\text{Btu}}{\text{hr-ft-°F}}\right)\left(12\ \frac{\text{in}}{\text{ft}}\right)}{0.9\ \text{in}}\right) \\ &= 815\ \text{Btu/hr-ft}^2\text{-°F} \end{aligned}$$

The saturation temperature for 134 psia steam is approximately 350°F. Assume the wall is 20°F lower (≈ 330°F). The film properties are evaluated at the average of the wall and saturation temperatures.

$$\begin{aligned} T_h &= \tfrac{1}{2}(T_{\text{sat},v} + T_s) = \left(\frac{1}{2}\right)(350°\text{F} + 330°\text{F}) \\ &= 340°\text{F} \end{aligned}$$

Film properties are obtained from App. 18A (also *NCEES Handbook* table "Physical Properties of Liquid Water (U.S. Units)") for liquid water and from App. 10A and App. 10B (also *NCEES Handbook* table "Saturated Steam (U.S. Units)—Temperature Table") for vapor.

$$\begin{aligned} k_{340°\text{F}} &= 0.392\ \text{Btu/hr-ft-°F} \\ \mu_{340°\text{F}} &= \left(0.109\times10^{-3}\ \frac{\text{lbm}}{\text{sec-ft}}\right)\left(3600\ \frac{\text{sec}}{\text{hr}}\right) \\ &= 0.392\ \text{lbm/ft-hr} \\ \rho_{l,340°\text{F}} &= \frac{1}{v_{f,340°\text{F}}} = \frac{1}{0.01787\ \frac{\text{ft}^3}{\text{lbm}}} \\ &= 55.96\ \text{lbm/ft}^3 \\ \rho_{v,340°\text{F}} &= \frac{1}{v_{g,340°\text{F}}} = \frac{1}{3.788\ \frac{\text{ft}^3}{\text{lbm}}} \\ &= 0.2640\ \text{lbm/ft}^3 \end{aligned}$$

$$\begin{aligned} h_{fg,134\text{psia}} &= 871.3\ \text{Btu/lbm} \\ d &= \frac{1.0\ \text{in}}{12\ \frac{\text{in}}{\text{ft}}} = 0.083\ \text{ft} \\ g &= \left(32.2\ \frac{\text{ft}}{\text{sec}^2}\right)\left(3600\ \frac{\text{sec}}{\text{hr}}\right)^2 \\ &= 4.17\times10^8\ \text{ft/hr}^2 \end{aligned}$$

From Eq. 18.27 (also *NCEES Handbook:* Condensation), the film coefficient is

$$\begin{aligned} h_o &= 0.725\left(\frac{\rho_l(\rho_l-\rho_v)gh_{fg}k_l^3}{d\mu_l(T_{\text{sat},v}-T_s)}\right)^{1/4} \\ &= (0.725) \\ &\quad\times\left(\frac{\begin{array}{c}\left(55.96\ \frac{\text{lbm}}{\text{ft}^3}\right)\left(55.96\ \frac{\text{lbm}}{\text{ft}^3} - 0.2640\ \frac{\text{lbm}}{\text{ft}^3}\right) \\ \times\left(4.17\times10^8\ \frac{\text{ft}}{\text{hr}^2}\right)\left(871.3\ \frac{\text{Btu}}{\text{lbm}}\right) \\ \times\left(0.392\ \frac{\text{Btu}}{\text{hr-ft-°F}}\right)^3\end{array}}{(0.083\ \text{ft})\left(0.392\ \frac{\text{lbm}}{\text{ft-hr}}\right)(350°\text{F} - 330°\text{F})}\right)^{1/4} \\ &= 2320\ \text{Btu/hr-ft}^2\text{-°F} \end{aligned}$$

From Table 19.7 (also *NCEES Handbook* table "Physical Properties of Metals at 68°F (U.S. Units)"), for constantan, $k = 13.1$ Btu/hr-ft-°F.

From Eq. 19.70 (also *NCEES Handbook:* Combination of Heat-Transfer Mechanisms), assuming the fouling factors are both equal to zero, the overall heat transfer coefficient based on the outside area is

$$\frac{1}{U_{ov}} = \frac{1}{h_i}\left(\frac{D_o}{D_i}\right) + \frac{D_o}{2k}\ln\left(\frac{D_o}{D_i}\right) + \frac{1}{h_o}$$

$$\begin{aligned} \frac{1}{U_{ov}} &= \frac{1}{815\frac{\text{Btu}}{\text{hr·ft}^2\text{·°F}}}\left(\frac{1\,\text{in}}{0.9\,\text{in}}\right) \\ &\quad+\frac{1\,\text{in}\left(\frac{1\,\text{ft}}{12\,\text{in}}\right)}{(2)\left(13.1\frac{\text{Btu}}{\text{hr·ft·°F}}\right)}\ln\left(\frac{1\,\text{in}}{0.9\,\text{in}}\right) + \frac{1}{2320\frac{\text{Btu}}{\text{hr·ft}^2\text{·°F}}} \end{aligned}$$

$$\frac{1}{U_{ov}} = \frac{1}{733.5\,\frac{\text{Btu}}{\text{hr-ft}^2\text{-}^\circ\text{F}}} + \frac{1}{2984\,\frac{\text{Btu}}{\text{hr-ft}^2\text{-}^\circ\text{F}}} + \frac{1}{2320\,\frac{\text{Btu}}{\text{hr-ft}^2\text{-}^\circ\text{F}}}$$

$$U_{ov} = 469\,\frac{\text{Btu}}{\text{hr-ft}^2\text{-}^\circ\text{F}}$$

For crossflow operation,

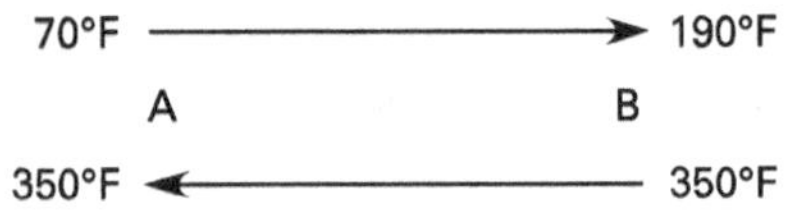

$$\Delta T_A = 350^\circ\text{F} - 70^\circ\text{F} = 280^\circ\text{F}$$
$$\Delta T_B = 350^\circ\text{F} - 190^\circ\text{F} = 160^\circ\text{F}$$

From Eq. 19.68 (also *NCEES Handbook:* Log-Mean Temperature Difference), the logarithmic mean temperature difference is

$$\Delta T_{\text{lm}} = \frac{\Delta T_A - \Delta T_B}{\ln \frac{\Delta T_A}{\Delta T_B}} = \frac{280^\circ\text{F} - 160^\circ\text{F}}{\ln \frac{280^\circ\text{F}}{160^\circ\text{F}}} = 214.4^\circ\text{F}$$

The heat transfer is known; therefore, the outside area can be calculated from Eq. 19.69 (also *NCEES Handbook:* Combination of Heat-Transfer Mechanisms).

$$Q = U_o A_o F_c \Delta T_{\text{lm}}$$

For steam condensation, the temperature of steam remains constant. Therefore, $F_c = 1$.

$$A_o = \frac{352{,}447\,\frac{\text{Btu}}{\text{hr}}}{\left(469\,\frac{\text{Btu}}{\text{hr-ft}^2\text{-}^\circ\text{F}}\right)(1)(214.4^\circ\text{F})} = 3.5\ \text{ft}^2$$

At this point, the assumption that $T_{\text{sat},v} - T_s = 20^\circ\text{F}$ could be checked using $Q = U_{\text{partial}} A_o \Delta T_{\text{partial}}$, working from the inside (at 130°F) to the outside (at T_s).

SI Solution

The bulk temperature of the water is

$$T_b = \tfrac{1}{2}(T_{\text{in}} + T_{\text{out}}) = \left(\frac{1}{2}\right)(21^\circ\text{C} + 90^\circ\text{C}) = 55.5^\circ\text{C}$$

From App. 18B (also *NCEES Handbook* table "Physical Properties of Liquid Water (SI Units)"), the properties of water at 55.5°C are

$$c_p = 4.186\ \text{kJ/kg·K}$$
$$\rho = 986.6\ \text{kg/m}^3$$
$$\mu = 0.523 \times 10^{-3}\ \text{kg/m·s}$$
$$k = 0.6503\ \text{W/m·K}$$
$$\text{Pr} = 3.37$$

The heat transfer is found from the temperature gain of the water.

$$\begin{aligned} Q &= \dot{m} c_p \Delta T \\ &= \left(0.368\,\frac{\text{kg}}{\text{s}}\right)\left(4.186\,\frac{\text{kJ}}{\text{kg·K}}\right)\left(1000\,\frac{\text{J}}{\text{kJ}}\right) \\ &\quad \times (90^\circ\text{C} - 21^\circ\text{C}) \\ &= 106\,291\ \text{W} \end{aligned}$$

As in *NCEES Handbook* table "Dimensionless Numbers," the Reynolds number is

$$\begin{aligned} \text{Re} &= \frac{\rho v D}{\mu} \\ &= \frac{\left(986.6\,\frac{\text{kg}}{\text{m}^3}\right)\left(0.9\,\frac{\text{m}}{\text{s}}\right)(2.29\ \text{cm})}{\left(0.523 \times 10^{-3}\,\frac{\text{kg}}{\text{m·s}}\right)\left(100\,\frac{\text{cm}}{\text{m}}\right)} \\ &= 3.89 \times 10^4 \end{aligned}$$

From Eq. 19.33 (also *NCEES Handbook:* Free/Forced Heat-Transfer Coefficients/Correlations), the film coefficient is

$$\begin{aligned} h &= 0.023\text{Re}^{0.8}\text{Pr}^{\frac{1}{3}}\left(\frac{k}{d}\right) \\ &= (0.023)(3.89 \times 10^4)^{0.8}(3.37)^{1/3} \\ &\quad \times \left(\frac{\left(0.6503\,\frac{\text{W}}{\text{m·K}}\right)\left(100\,\frac{\text{cm}}{\text{m}}\right)}{2.29\ \text{cm}}\right) \\ &= 4601\ \text{W/m}^2\text{·K} \end{aligned}$$

The saturation temperature for 923 kPa steam is 176.4°C. Assume the wall is 10°C lower (or 166.4°C). The film properties are evaluated at the average of the wall and saturation temperatures.

$$T_h = \tfrac{1}{2}(T_{\text{sat},v} + T_s) = \left(\frac{1}{2}\right)(176.4°\text{C} + 166.4°\text{C})$$
$$= 171.4°\text{C}$$

Film properties are obtained from App. 18B (also *NCEES Handbook* table "Physical Properties of Liquid Water (SI Units)"), for liquid water and App. 10I (also *NCEES Handbook* table "Saturated Steam (SI Units)—Temperature Table") for vapor.

$$k_{171.4°\text{C}} = 0.6745 \text{ W/m·K}$$
$$\mu_{171.4°\text{C}} = 0.1712 \times 10^{-3} \text{ kg/m·s}$$
$$\rho_{l,171.4°\text{C}} = \frac{1}{v_{f,171.4°\text{C}}} = \frac{(1)\left(100 \ \frac{\text{cm}}{\text{m}}\right)^3}{\left(1.1161 \ \frac{\text{cm}^3}{\text{g}}\right)\left(1000 \ \frac{\text{g}}{\text{kg}}\right)}$$
$$= 896.0 \text{ kg/m}^3$$
$$\rho_{v,171.4°\text{C}} = \frac{1}{v_{g,171.4°\text{C}}} = \frac{(1)\left(100 \ \frac{\text{cm}}{\text{m}}\right)^3}{\left(235.3 \ \frac{\text{cm}^3}{\text{g}}\right)\left(1000 \ \frac{\text{g}}{\text{kg}}\right)}$$
$$= 4.25 \text{ kg/m}^3$$

$$h_{fg,923 \text{ kPa}} = \left(2026.8 \ \frac{\text{kJ}}{\text{kg}}\right)\left(1000 \ \frac{\text{J}}{\text{kg}}\right)$$
$$= 2.0268 \times 10^6 \text{ J/kg}$$
$$d = \frac{2.54 \text{ cm}}{100 \ \frac{\text{cm}}{\text{m}}} = 0.0254 \text{ m}$$

From Eq. 18.27 (also *NCEES Handbook:* Condensation), the film coefficient is

$$h_o = 0.725\left(\frac{\rho_l(\rho_l - \rho_v)gh_{fg}k_l^3}{d\mu_l(T_{\text{sat},v} - T_s)}\right)^{1/4}$$
$$= (0.725) \times \left(\frac{\left(896 \ \frac{\text{kg}}{\text{m}^3}\right)\left(896 \ \frac{\text{kg}}{\text{m}^3} - 4.25 \ \frac{\text{kg}}{\text{m}^3}\right)\left(9.81 \ \frac{\text{m}}{\text{s}^2}\right) \times \left(2.0268 \times 10^6 \ \frac{\text{J}}{\text{kg}}\right)\left(0.6745 \ \frac{\text{W}}{\text{m·K}}\right)^3}{(0.0254 \text{ m})\left(0.1712 \times 10^{-3} \ \frac{\text{kg}}{\text{m·s}}\right) \times (176.4°\text{C} - 166.4°\text{C})}\right)^{1/4}$$
$$= 13\,266 \text{ W/m}^2\text{·K}$$

From Table 19.7 and the table footnote (also *NCEES Handbook* table "Physical Properties of Metals at 20°C (SI Units)"), for constantan, the thermal conductivity is 22.7 W/m·K.

From Eq. 19.70 (also *NCEES Handbook:* Combination of Heat-Transfer Mechanisms), assuming the fouling factors are both equal to zero, the overall heat transfer coefficient based on outside area is

$$\frac{1}{U_{ov}} = \frac{1}{h_i}\left(\frac{D_o}{D_i}\right) + \frac{D_o}{2k}\ln\left(\frac{D_o}{D_i}\right) + \frac{1}{h_o}$$

$$\frac{1}{U_{ov}} = \frac{1}{4601 \frac{\text{W}}{\text{m}^2\text{·K}}}\left(\frac{2.54 \text{ cm}}{2.29 \text{ cm}}\right) + \frac{0.0254 \text{ m}}{(2)\left(22.7 \frac{\text{W}}{\text{m·K}}\right)}\ln\left(\frac{2.54 \text{ cm}}{2.29 \text{ cm}}\right) + \frac{1}{13\,266 \frac{\text{W}}{\text{m}^2\text{·K}}}$$

$$\frac{1}{U_{ov}} = \frac{1}{4148 \frac{W}{\text{m}^2\text{·K}}} + \frac{1}{17\,251 \frac{W}{\text{m}^2\text{·K}}} + \frac{1}{13\,266 \frac{W}{\text{m}^2\text{·K}}}$$

$$U_{ov} = 2671 \frac{W}{\text{m}^2\text{·K}}$$

For crossflow operation,

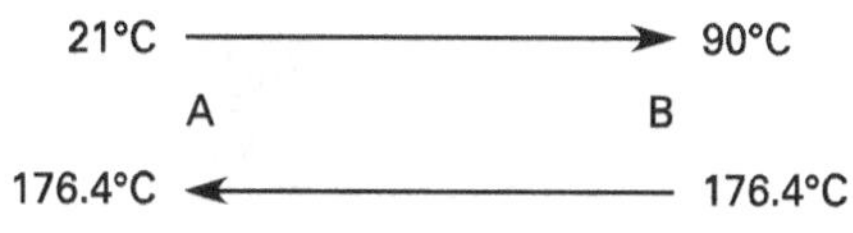

$$\Delta T_A = 176.4°\text{C} - 21°\text{C} = 155.4°\text{C}$$
$$\Delta T_B = 176.4°\text{C} - 90°\text{C} = 86.4°\text{C}$$

From Eq. 19.68 (also *NCEES Handbook:* Log-Mean Temperature Difference), the logarithmic mean temperature difference is

$$\Delta T_{\text{lm}} = \frac{\Delta T_A - \Delta T_B}{\ln \frac{\Delta T_A}{\Delta T_B}} = \frac{155.4°\text{C} - 86.4°\text{C}}{\ln \frac{155.4°\text{C}}{86.4°\text{C}}}$$
$$= 117.5°\text{C}$$

The heat transfer is known; therefore, the outside area can be calculated from Eq. 19.69 (also *NCEES Handbook:* Combination of Heat-Transfer Mechanisms).

$$Q = U_o A_o F_c \Delta T_{\text{lm}}$$

For steam condensation, the temperature of steam remains constant. Therefore, $F_c = 1$.

$$A_o = \frac{106\,291 \text{ W}}{\left(2671 \ \frac{\text{W}}{\text{m}^2\text{·K}}\right)(1)(117.5°\text{C})}$$
$$= 0.339 \text{ m}^2 \quad (0.33 \text{ m}^2)$$

At this point, the assumption that $T_{\text{sat},v} - T_s = 10°\text{C}$ could be checked using $Q = U_{\text{partial}} A_o \Delta T_{\text{partial}}$, working from the inside (at 55°C) to the outside (at T_s).

The answer is (A).

8. Use the illustration shown for both the customary U.S. and SI solutions.

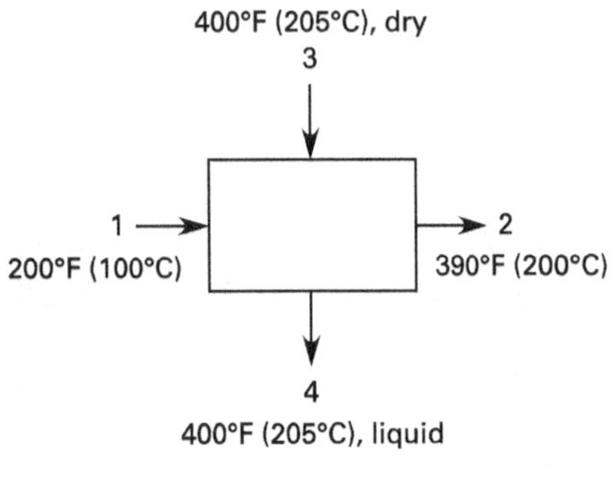

Customary U.S. Solution

From App. 10A (also *NCEES Handbook* table "Saturated Steam (U.S. Units)—Temperature Table") the enthalpy of each point is

$$h_1 = 168.13 \text{ Btu/lbm}$$
$$h_2 = 364.3 \text{ Btu/lbm}$$
$$h_3 = 1201.4 \text{ Btu/lbm}$$
$$h_4 = 375.1 \text{ Btu/lbm}$$

The heat transfer is due to the temperature gain of water.

$$Q = \dot{m}(h_2 - h_1)$$
$$= \left(500{,}000 \ \frac{\text{lbm}}{\text{hr}}\right)\left(364.3 \ \frac{\text{Btu}}{\text{lbm}} - 168.13 \ \frac{\text{Btu}}{\text{lbm}}\right)$$
$$= 9.809 \times 10^7 \text{ Btu/hr}$$

The mass flow rate per tube is

$$\dot{m}_{\text{tube}} = \rho A_{\text{tube}} \text{v}$$

Select ρ where the specific volume is greatest (at 390°F). From App. 10A (also *NCEES Handbook* table "Saturated Steam (U.S. Units)—Temperature Table"),

$$v_f = 0.01850 \text{ ft}^3\text{/lbm}$$
$$\rho = \frac{1}{v_f} = \frac{1}{0.01850 \ \frac{\text{ft}^3}{\text{lbm}}}$$
$$= 54.05 \text{ lbm/ft}^3$$

The inside diameter of the tube is

$$d_i = d_o - 2(\text{wall})$$
$$= \frac{7}{8} \text{ in} - (2)\left(\frac{1}{16} \text{ in}\right)$$
$$= 0.750 \text{ in}$$

The area per tube is

$$A_{\text{tube}} = \left(\frac{\pi}{4}\right)d_i^2 = \frac{\left(\frac{\pi}{4}\right)(0.750\ \text{in})^2}{\left(12\ \frac{\text{in}}{\text{ft}}\right)^2}$$
$$= 0.003068\ \text{ft}^2$$
$$\dot{m}_{\text{tube}} = \left(54.05\ \frac{\text{lbm}}{\text{ft}^3}\right)(0.003068\ \text{ft}^2) \times \left(5\ \frac{\text{ft}}{\text{sec}}\right)\left(3600\ \frac{\text{sec}}{\text{hr}}\right)$$
$$= 2985\ \text{lbm/hr}$$

The required number of tubes is

$$N = \frac{500{,}000\ \frac{\text{lbm}}{\text{hr}}}{2985\ \frac{\text{lbm}}{\text{hr}}} = 167.5 \quad (170)$$

SI Solution

From App. 10I (also *NCEES Handbook* table "Saturated Steam (SI Units)—Temperature Table"), the enthalpy of each point is

$$h_1 = 419.17\ \text{kJ/kg}$$
$$h_2 = 852.27\ \text{kJ/kg}$$
$$h_3 = 2794.8\ \text{kJ/kg}$$
$$h_4 = 874.88\ \text{kJ/kg}$$

The heat transfer is due to the temperature gain of water.

$$Q = \dot{m}(h_2 - h_1) = \left(60\ \frac{\text{kg}}{\text{s}}\right)\left(852.27\ \frac{\text{kJ}}{\text{kg}} - 419.17\ \frac{\text{kJ}}{\text{kg}}\right) = 25\,986\ \text{kW}$$

The mass flow rate per tube is

$$\dot{m}_{\text{tube}} = \rho A_{\text{tube}} \text{v}$$

Select ρ where the specific volume is greatest (at 200°C). From App. 10I (also *NCEES Handbook* table "Saturated Steam (SI Units)—Temperature Table"),

$$v_f = \frac{\left(1.1565\ \frac{\text{cm}^3}{\text{g}}\right)\left(1000\ \frac{\text{g}}{\text{kg}}\right)}{\left(100\ \frac{\text{cm}}{\text{m}}\right)^3} = 1.1565 \times 10^{-3}\ \text{m}^3/\text{kg}$$

$$\rho = \frac{1}{v_f} = \frac{1}{1.1565 \times 10^{-3}\ \frac{\text{m}^3}{\text{kg}}} = 864.7\ \text{kg/m}^3$$

The inside diameter of the tube is

$$d_i = d_o - 2(\text{wall}) = \frac{2.2\ \text{cm} - \frac{(2)(1.6\ \text{mm})}{10\ \frac{\text{mm}}{\text{cm}}}}{100\ \frac{\text{cm}}{\text{m}}} = 0.0188\ \text{m}$$

The area per tube is

$$A_{\text{tube}} = \left(\frac{\pi}{4}\right)d_i^2 = \left(\frac{\pi}{4}\right)(0.0188\ \text{m})^2 = 2.776 \times 10^{-4}\ \text{m}^2$$

$$\dot{m}_{\text{tube}} = \left(864.7\ \frac{\text{kg}}{\text{m}^3}\right)(2.776 \times 10^{-4}\ \text{m}^2)\left(1.5\ \frac{\text{m}}{\text{s}}\right) = 0.360\ \text{kg/s}$$

The required number of tubes is

$$N = \frac{60\ \frac{\text{kg}}{\text{s}}}{0.360\ \frac{\text{kg}}{\text{s}}} = 166.7 \quad (170)$$

The answer is (D).

9. *Customary U.S. Solution*

The water's bulk temperature is

$$T_{b,\text{water}} = \left(\frac{1}{2}\right)(70°\text{F} + 140°\text{F}) = 105°\text{F}$$

The fluid properties at 105°F are obtained from App. 18A (also *NCEES Handbook* table "Physical Properties of Liquid Water (U.S. Units)").

$$\rho_{105°F} = 61.92\ \text{lbm/ft}^3$$
$$c_{p,105°F} = 0.998\ \text{Btu/lbm-°F}$$

The mass flow rate of water is

$$\dot{m}_{\text{water}} = \dot{V}\rho = \frac{\left(100\ \frac{\text{gal}}{\text{min}}\right)\left(60\ \frac{\text{min}}{\text{hr}}\right)\left(61.92\ \frac{\text{lbm}}{\text{ft}^3}\right)}{7.48\ \frac{\text{gal}}{\text{ft}^3}} = 49{,}668\ \text{lbm/hr}$$

The heat transfer is found from the temperature gain of the water.

$$Q_{\text{clean}} = \dot{m}c_p\Delta T = \left(49{,}668\ \frac{\text{lbm}}{\text{hr}}\right)\left(0.998\ \frac{\text{Btu}}{\text{lbm-°F}}\right) \times (140°\text{F} - 70°\text{F}) = 3.470 \times 10^6\ \text{Btu/hr}$$

$$\Delta T_A = 230°\text{F} - 70°\text{F} = 160°\text{F}$$
$$\Delta T_B = 230°\text{F} - 140°\text{F} = 90°\text{F}$$

From Eq. 19.68 (also *NCEES Handbook:* Log-Mean Temperature Difference), the logarithmic mean temperature difference is

$$\Delta T_{\text{lm}} = \frac{\Delta T_A - \Delta T_B}{\ln\frac{\Delta T_A}{\Delta T_B}} = \frac{160°\text{F} - 90°\text{F}}{\ln\frac{160°\text{F}}{90°\text{F}}} = 121.66°\text{F}$$

The heat transfer is known. Therefore, the overall heat transfer coefficient can be calculated from Eq. 19.69 (also *NCEES Handbook:* Combination of Heat-Transfer Mechanisms).

$$Q_{\text{clean}} = U_{\text{clean}}AF_c\Delta T_{\text{lm}}$$

For steam condensation, the temperature of steam remains constant. Therefore, $F_c = 1$.

$$U_{\text{clean}} = \frac{Q_{\text{clean}}}{AF_c\Delta T_{\text{lm}}} = \frac{3.470 \times 10^6\ \frac{\text{Btu}}{\text{hr}}}{(50\ \text{ft}^2)(1)(121.66°\text{F})} = 570.4\ \text{Btu/hr-ft}^2\text{-°F}$$

After fouling,

$$Q_{\text{fouled}} = \left(49{,}668\ \frac{\text{lbm}}{\text{hr}}\right)\left(0.998\ \frac{\text{Btu}}{\text{lbm-°F}}\right) \times (122°\text{F} - 70°\text{F}) = 2.578 \times 10^6\ \text{Btu/hr}$$

$$\Delta T_B = 230°\text{F} - 122°\text{F} = 108°\text{F}$$
$$\Delta T_A = 230°\text{F} - 70°\text{F} = 160°\text{F}$$

From Eq. 19.68 (also *NCEES Handbook:* Combination of Heat-Transfer Mechanisms), the logarithmic mean temperature difference is

$$\Delta T_{\text{lm}} = \frac{\Delta T_A - \Delta T_B}{\ln\frac{\Delta T_A}{\Delta T_B}} = \frac{160°\text{F} - 108°\text{F}}{\ln\frac{160°\text{F}}{108°\text{F}}} = 132.3°\text{F}$$

$$U_{\text{fouled}} = \frac{2.578 \times 10^6\ \frac{\text{Btu}}{\text{hr}}}{(50\ \text{ft}^2)(1)(132.3°\text{F})} = 389.7\ \text{Btu/hr-ft}^2\text{-°F}$$

From Eq. 19.75 (also *NCEES Handbook:* Fouling Factors), the fouling factor is

$$R_f = \frac{1}{U_{\text{fouled}}} - \frac{1}{U_{\text{clean}}} = \frac{1}{389.7\ \frac{\text{Btu}}{\text{hr-ft}^2\text{-°F}}} - \frac{1}{570.4\ \frac{\text{Btu}}{\text{hr-ft}^2\text{-°F}}} = 0.000813\ \text{hr-ft}^2\text{-°F/Btu} \quad (0.0008\ \text{hr-ft}^2\text{-°F/Btu})$$

SI Solution

The water's bulk temperature is

$$T_{b,\text{water}} = \left(\frac{1}{2}\right)(21°\text{C} + 60°\text{C}) = 40.5°\text{C}$$

The fluid properties at 40.5°C are obtained from App. 18B (also *NCEES Handbook* table "Physical Properties of Liquid Water (SI Units)").

$$\rho_{40.5°C} = 993.5\ \text{kg/m}^3$$
$$c_{p,40.5°C} = 4.183\ \text{kJ/kg·K}$$

The mass flow rate of water is

$$\dot{m}_{\text{water}} = \dot{V}\rho = \frac{\left(6.3\ \frac{\text{L}}{\text{s}}\right)\left(993.5\ \frac{\text{kg}}{\text{m}^3}\right)}{1000\ \frac{\text{L}}{\text{m}^3}} = 6.26\ \text{kg/s}$$

The heat transfer is found from the temperature gain of the water.

$$\begin{aligned}Q_{\text{clean}} &= \dot{m}c_p\Delta T\\ &= \left(6.26\ \frac{\text{kg}}{\text{s}}\right)\left(4.183\ \frac{\text{kJ}}{\text{kg}\cdot\text{K}}\right)\left(1000\ \frac{\text{J}}{\text{kJ}}\right)\\ &\quad\times(60^\circ\text{C}-21^\circ\text{C})\\ &= 1.021\times10^6\ \text{W}\\ \Delta T_A &= 110^\circ\text{C}-21^\circ\text{C} = 89^\circ\text{C}\\ \Delta T_B &= 110^\circ\text{C}-60^\circ\text{C} = 50^\circ\text{C}\end{aligned}$$

From Eq. 19.68 (also *NCEES Handbook:* Log-Mean Temperature Difference), the logarithmic mean temperature difference is

$$\Delta T_{\text{lm}} = \frac{\Delta T_A - \Delta T_B}{\ln\frac{\Delta T_A}{\Delta T_B}} = \frac{89^\circ\text{C}-50^\circ\text{C}}{\ln\frac{89^\circ\text{C}}{50^\circ\text{C}}} = 67.64^\circ\text{C}$$

The heat transfer is known. Therefore, the overall heat transfer coefficient can be calculated from Eq. 19.69 (also *NCEES Handbook:* Combination of Heat-Transfer Mechanisms).

$$Q_{\text{clean}} = U_{\text{clean}}AF_c\Delta T_{\text{lm}}$$

For steam condensation, the temperature of steam remains constant. Therefore, $F_c = 1$.

$$\begin{aligned}U_{\text{clean}} &= \frac{Q_{\text{clean}}}{AF_c\Delta T_{\text{lm}}}\\ &= \frac{1.021\times10^6\ \text{W}}{(4.7\ \text{m}^2)(1)(67.64^\circ\text{C})}\\ &= 3211.6\ \text{W/m}^2\cdot\text{K}\end{aligned}$$

After fouling,

$$\begin{aligned}Q_{\text{fouled}} &= \left(6.26\ \frac{\text{kg}}{\text{s}}\right)\left(4.183\ \frac{\text{kJ}}{\text{kg}\cdot\text{K}}\right)\left(1000\ \frac{\text{J}}{\text{kJ}}\right)\\ &\quad\times(50^\circ\text{C}-21^\circ\text{C})\\ &= 7.594\times10^5\ \text{W}\\ \Delta T_A &= 110^\circ\text{C}-21^\circ\text{C} = 89^\circ\text{C}\\ \Delta T_B &= 110^\circ\text{C}-50^\circ\text{C} = 60^\circ\text{C}\end{aligned}$$

From Eq. 19.68 (also *NCEES Handbook:* Log-Mean Temperature Difference), the logarithmic mean temperature difference is

$$\Delta T_{\text{lm}} = \frac{\Delta T_A - \Delta T_B}{\ln\frac{\Delta T_A}{\Delta T_B}} = \frac{89^\circ\text{C}-60^\circ\text{C}}{\ln\frac{89^\circ\text{C}}{60^\circ\text{C}}} = 73.55^\circ\text{C}$$

$$U_{\text{fouled}} = \frac{7.594\times10^5\ \text{W}}{(4.7\ \text{m}^2)(1)(73.55^\circ\text{C})} = 2196.8\ \text{W/m}^2\cdot\text{K}$$

From Eq. 19.75 (also *NCEES Handbook:* Fouling Factors), the fouling factor is

$$\begin{aligned}R_f &= \frac{1}{U_{\text{fouled}}} - \frac{1}{U_{\text{clean}}}\\ &= \frac{1}{2196.8\ \frac{\text{W}}{\text{m}^2\cdot\text{K}}} - \frac{1}{3211.6\ \frac{\text{W}}{\text{m}^2\cdot\text{K}}}\\ &= 0.000144\ \text{m}^2\cdot\text{K/W}\quad(0.0001\ \text{m}^2\cdot\text{K/W})\end{aligned}$$

The answer is (B).

10. Heat exchanger tubing has a constant cross-sectional area, so the flow velocity within a heat exchanger is constant. Since distance traveled at constant velocity is proportional to time, and since both fluids have the same specific heat, the heat transfer rate-distance and heat transfer rate-time profiles are identical.

The answer is (A).

11. At any point in the heat exchanger, the instantaneous heat gain by the cold water is equal to the instantaneous heat loss by the hot water. The heat transfer rates are the same. At that point, the two fluids will have traveled different distances. For example, at one entrance, the hot water will have traveled very little distance, while the cold water will have traveled almost the entire length of the heat exchanger (e.g., reversed end-to-end). However, the heat transfer rates will be the same. (It could be argued that from the standpoint of the thermodynamic sign convention, hot water has a negative heat transfer rate while cold water has a positive heat transfer rate. In that case, the temperature profiles would be reversed end-to-end *and* inverted top-to-bottom. However, that is not one of the options.)

The answer is (B).

12. At the entrance, as a result of the large temperature difference, the hot water loses heat rapidly, and the cold water gains heat rapidly. At the exit, the hot water loses heat more slowly while the cold water gains heat more slowly. Thus, the plot is inverted top-to-bottom.

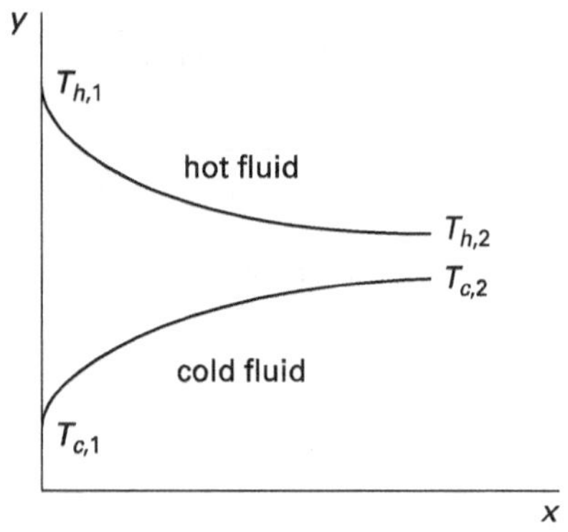

The answer is (D).

13. The temperature difference between the two water flows will essentially be constant.

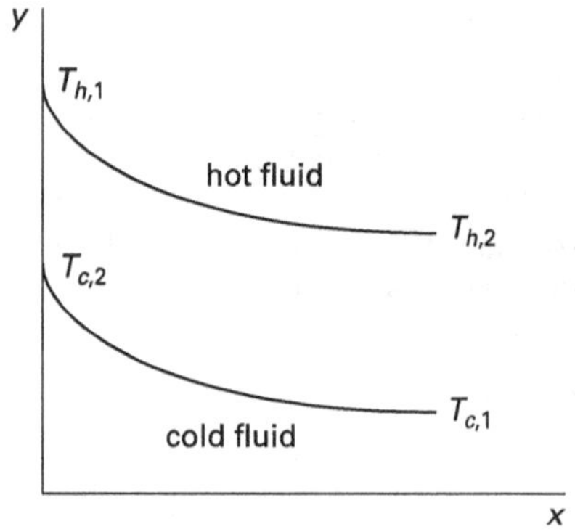

The answer is (A).

14. Since the steam enters the heat exchanger saturated and is not subcooled, the steam temperature remains constant at the saturation temperature throughout.

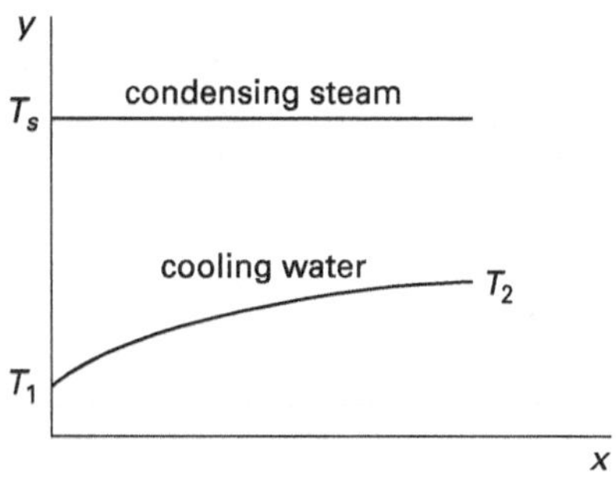

The answer is (D).

15. "Duty" is the amount of energy that the heat exchanger transfers per unit time.

The answer is (D).

16. The perimeter of the cylindrical fin is

$$P = \pi d = \frac{\pi(0.5 \text{ in})}{12 \frac{\text{in}}{\text{ft}}} = 0.13 \text{ ft}$$

The cross-sectional area at the base of the cylindrical fin is

$$A_b = \frac{\pi d^2}{4} = \frac{\pi(0.5 \text{ in})^2}{(4)\left(12 \frac{\text{in}}{\text{ft}}\right)^2} = 0.0014 \text{ ft}^2$$

From Eq. 17.80 (also *NCEES Handbook:* Combination of Heat-Transfer Mechanisms), the heat transfer is

$$\begin{aligned} Q &= \sqrt{hPkA}\,(T_b - T_\infty)\tanh\left(\left(L + \frac{A}{P}\right)\sqrt{\frac{hP}{kA}}\right) \\ &= \sqrt{\left(1.3 \frac{\text{Btu}}{\text{hr-ft}^2\text{-°F}}\right)(0.13 \text{ ft})\left(128 \frac{\text{Btu}}{\text{hr-ft-°F}}\right)(0.0014 \text{ ft}^2)} \\ &\quad \times (350°\text{F} - 75°\text{F}) \\ &\quad \times \left(\tanh\left(\left(2 \text{ ft} + \frac{(0.0014 \text{ ft}^2)}{(0.13 \text{ ft})}\right)\sqrt{\frac{\left(1.3 \frac{\text{Btu}}{\text{hr-ft}^2\text{-°F}}\right) \times (0.13 \text{ ft})}{\left(128 \frac{\text{Btu}}{\text{hr-ft-°F}}\right) \times (0.0014 \text{ ft}^2)}}\right)\right) \\ &= 45.9 \text{ Btu/hr} \quad (46 \text{ Btu/hr}) \end{aligned}$$

The answer is (D).

17. *Customary U.S. Solution*

From Eq. 18.11, the film temperature of the air is

$$T_h = \tfrac{1}{2}(T_s + T_\infty) = \left(\frac{1}{2}\right)(100°\text{F} + 150°\text{F}) = 125°\text{F}$$

From App. 18C (also *NCEES Handbook* table "Temperature-Dependent Properties of Air at 14.7 psia (U.S. Units)"), the air properties at 125°F are

$$\begin{aligned} \nu &= 0.195 \times 10^{-3} \text{ ft}^2\text{/sec} \\ k &= 0.0159 \text{ Btu/hr-ft-°F} \\ \text{Pr} &= 0.72 \end{aligned}$$

As in *NCEES Handbook* table "Dimensionless Numbers," the Reynolds number based on the diameter, d, is

$$\mathrm{Re}_d = \frac{\mathrm{v}d}{\nu} = \frac{\left(100\ \frac{\mathrm{ft}}{\mathrm{sec}}\right)(0.35\ \mathrm{in})}{\left(0.195\times 10^{-3}\ \frac{\mathrm{ft}^2}{\mathrm{sec}}\right)\left(12\ \frac{\mathrm{in}}{\mathrm{ft}}\right)} = 1.50\times 10^4$$

From Eq. 19.51 (also *NCEES Handbook:* Free/Forced Heat-Transfer Coefficients/Correlations), the film coefficient is

$$h = C_1(\mathrm{Re}_d)^n \mathrm{Pr}^{1/3}\left(\frac{k}{d}\right)$$

From Table 19.3 (also *NCEES Handbook:* Free/Forced Heat-Transfer Coefficients/Correlations), $C_1 = 0.193$ and $n = 0.618$.

$$\begin{aligned} h &= (0.193)(1.50\times 10^4)^{0.618}(0.72)^{1/3} \\ &\quad\times\left(\frac{\left(0.0159\ \frac{\mathrm{Btu}}{\mathrm{hr\text{-}ft\text{-}^\circ F}}\right)\left(12\ \frac{\mathrm{in}}{\mathrm{ft}}\right)}{0.35\ \mathrm{in}}\right) \\ &= 35.9\ \mathrm{Btu/hr\text{-}ft^2\text{-}^\circ F}\quad (36\ \mathrm{Btu/hr\text{-}ft^2\text{-}^\circ F}) \end{aligned}$$

SI Solution

From Eq. 18.11, the film temperature of the air is

$$T_h = \tfrac{1}{2}(T_s + T_\infty) = \left(\frac{1}{2}\right)(38^\circ\mathrm{C} + 66^\circ\mathrm{C}) = 52^\circ\mathrm{C}$$

From App. 18D (also *NCEES Handbook* table "Temperature-Dependent Properties of Air at 0.1 MPa (SI Units)"), the air properties at 52°C are

$$\begin{aligned} \rho &= 1.09\ \mathrm{kg/m^3} \\ \mu &= 1.966\times 10^{-5}\ \mathrm{kg/m\cdot s} \\ k &= 0.02815\ \mathrm{W/m\cdot K} \\ \mathrm{Pr} &= 0.703 \end{aligned}$$

As in *NCEES Handbook* table "Dimensionless Numbers," the Reynolds number based on the diameter, d, is

$$\begin{aligned} \mathrm{Re}_d &= \frac{\rho \mathrm{v} d}{\mu} \\ &= \frac{\left(1.09\ \frac{\mathrm{kg}}{\mathrm{m^3}}\right)\left(30\ \frac{\mathrm{m}}{\mathrm{s}}\right)(8.9\ \mathrm{mm})}{\left(1.966\times 10^{-5}\ \frac{\mathrm{kg}}{\mathrm{m\cdot s}}\right)\left(1000\ \frac{\mathrm{mm}}{\mathrm{m}}\right)} \\ &= 1.48\times 10^4 \end{aligned}$$

From Eq. 19.51 (also *NCEES Handbook:* Free/Forced Heat-Transfer Coefficients/Correlations), the film coefficient is

$$h = C_1(\mathrm{Re}_d)^h \mathrm{Pr}^{1/3}\left(\frac{k}{d}\right)$$

From Table 19.3 (also *NCEES Handbook:* Free/Forced Heat-Transfer Coefficients/Correlations), $C_1 = 0.193$ and $n = 0.618$.

$$\begin{aligned} h &= (0.193)(1.48\times 10^4)^{0.618}(0.703)^{1/3} \\ &\quad\times\left(\frac{\left(0.02815\ \frac{\mathrm{W}}{\mathrm{m\cdot K}}\right)\left(1000\ \frac{\mathrm{mm}}{\mathrm{m}}\right)}{8.9\ \mathrm{mm}}\right) \\ &= 205.0\ \mathrm{W/m^2\cdot K}\quad (210\ \mathrm{W/m^2\cdot K}) \end{aligned}$$

The answer is (A).

20 Radiation and Combined Heat Transfer

PRACTICE PROBLEMS

1. A 6 in (15 cm) thick furnace wall has a 3 in (8 cm) square peephole. The interior of the furnace is at 2200°F (1200°C). The shape factor is 0.38. The surrounding air temperature is 70°F (20°C). The heat loss due to radiation when the peephole is open is most nearly

(A) 450 Btu/hr (150 W)

(B) 1300 Btu/hr (440 W)

(C) 2000 Btu/hr (680 W)

(D) 7900 Btu/hr (2.7 kW)

2. Dry air at 1 atmospheric pressure flows at 500 ft^3/min (0.25 m^3/s) through 50 ft (15 m) of 12 in (30 cm) diameter uninsulated duct. The emissivity of the duct surface is 0.28. Air enters the duct at 45°F (7°C). The walls, air, and contents of the room through which the duct passes are at 80°F (27°C). An engineer states that the air leaving the duct will be at 50°F (10°C). Considering both convection and radiation, is the engineer's statement that the air leaving the duct will be at 50°F (10°C) correct?

(A) The temperature is 45°F (7.2°C); this does not agree with the engineer's estimate.

(B) The temperature is 49°F (9.7°C); this does agree with the engineer's estimate.

(C) The temperature is 51°F (11°C); this does agree with the engineer's estimate.

(D) The temperature is 53°F (12°C); this does not agree with the engineer's estimate.

3. A steel pipe is painted on the outside with dull gray (oil-based) paint. The pipe is 35 ft (10 m) long. The pipe has a 4.00 in (10.2 cm) inside diameter and 4.25 in (10.8 cm) outside diameter. The pipe carries 200 ft^3/min (0.1 m^3/s) of 500°F (260°C), 25 psig (170 kPa) air through a 70°F (20°C) room. The conditions of the air at the end of the pipe are 350°F (180°C) and 15 psig (100 kPa). Using theoretical methods or empirical correlations, the calculated overall coefficient of heat transfer is most nearly

(A) 1.0 Btu/hr-ft^2-°F (6.1 W/m^2·K)

(B) 1.8 Btu/hr-ft^2-°F (11 W/m^2·K)

(C) 3.8 Btu/hr-ft^2-°F (22 W/m^2·K)

(D) 54 Btu/hr-ft^2-°F (330 W/m^2·K)

4. The temperature of a gas in a duct with 600°F (315°C) walls is evaluated with a 0.5 in (13 mm) diameter thermocouple probe. The emissivity of the probe is 0.8. The gas flow rate is 3480 lbm/hr-ft^2 (4.7 kg/s·m^2). The gas velocity is 400 ft/min (2 m/s). The film coefficient on the probe is given empirically as

$$h = \frac{0.024G^{0.8}}{D^{0.4}}$$

h in Btu/hr-ft^2-°F

G in lbm/hr-ft^2

D in ft

$$h = \frac{17G^{0.8}}{D^{0.4}}$$

h in W/m^2·K

G in kg/s·m^2

D in m

If the probe reading indicates that the gas temperature is 300°F (150°C), the actual gas temperature is most nearly

(A) 650°R (360K)

(B) 740°R (410K)

(C) 770°R (430K)

(D) 810°R (450K)

5. A 9 in (23 cm) diameter duct is painted with white enamel. The surface of the duct is at 200°F (95°C). The duct carries hot air through a room whose walls are 70°F (20°C). The air in the room is at 80°F (27°C). The unit heat transfer is most nearly

(A) 685 Btu/hr-ft length (685 W/m length)

(B) 900 Btu/hr-ft length (900 W/m length)

(C) 1100 Btu/hr-ft length (1100 W/m length)

(D) 1500 Btu/hr-ft length (1500 W/m length)

SOLUTIONS

1. *Customary U.S. Solution*

As in *NCEES Handbook:* Temperature, the absolute temperatures are

$$T_{\text{furnace}} = 2200°\text{F} + 460° = 2660°\text{R}$$
$$T_{\infty} = 70°\text{F} + 460° = 530°\text{R}$$

The radiation heat loss is

$$Q = AE_{\text{net}} = A\sigma F_{12}(T_{\text{furnace}}^4 - T_{\infty}^4)$$
$$= \frac{(3\text{ in})^2\left(0.1713 \times 10^{-8}\ \frac{\text{Btu}}{\text{hr-ft}^2\text{-°R}^4}\right) \times (0.38)\left((2660°\text{R})^4 - (530°\text{R})^4\right)}{\left(12\ \frac{\text{in}}{\text{ft}}\right)^2}$$
$$= 2033.6\text{ Btu/hr} \quad (2000\text{ Btu/hr})$$

SI Solution

As in *NCEES Handbook:* Temperature, the absolute temperatures are

$$T_{\text{furnace}} = 1200°\text{C} + 273° = 1473\text{K}$$
$$T_{\infty} = 20°\text{C} + 273° = 293\text{K}$$

The radiation heat loss is

$$Q = AE_{\text{net}} = A\sigma F_{12}(T_{\text{furnace}}^4 - T_{\infty}^4)$$
$$= \frac{(8\text{ cm})^2\left(5.67 \times 10^{-8}\ \frac{\text{W}}{\text{m}^2\cdot\text{K}^4}\right) \times (0.38)\left((1473\text{K})^4 - (293\text{K})^4\right)}{\left(100\ \frac{\text{cm}}{\text{m}}\right)^2}$$
$$= 648.2\text{ W} \quad (680\text{ W})$$

The answer is (C).

2. *Customary U.S. Solution*

As in *NCEES Handbook:* Temperature, the absolute temperature of air entering the duct is

$$45°\text{F} + 460° = 505°\text{R}$$

From the ideal gas law, the density of air entering the duct is

$$\rho = \frac{pM}{RT} = \frac{\left(14.7\ \frac{\text{lbf}}{\text{in}^2}\right)\left(12\ \frac{\text{in}}{\text{ft}}\right)^2\left(29\ \frac{\text{lbm}}{\text{lbmol}}\right)}{\left(1545.35\ \frac{\text{ft-lbf}}{\text{lbmol-°R}}\right)(505°\text{R})}$$
$$= 0.07857\text{ lbm/ft}^3$$

The mass flow rate of entering air is

$$\dot{m} = \rho\dot{V} = \left(0.07857\ \frac{\text{lbm}}{\text{ft}^3}\right)\left(500\ \frac{\text{ft}^3}{\text{min}}\right)\left(60\ \frac{\text{min}}{\text{hr}}\right)$$
$$= 2357.1\text{ lbm/hr}$$

The mass velocity entering the duct is

$$G = \frac{\dot{m}}{A_{\text{flow}}} = \frac{\left(2357.1\ \frac{\text{lbm}}{\text{hr}}\right)\left(12\ \frac{\text{in}}{\text{ft}}\right)^2}{\left(\frac{\pi}{4}\right)(12\text{ in})^2}$$
$$= 3001.1\text{ lbm/hr-ft}^2$$

To calculate the initial film coefficients, estimate the temperature based on the claim. The film coefficients are not highly sensitive to small temperature differences.

$$T_{\text{bulk,air}} = \tfrac{1}{2}(T_{\text{air,in}} + T_{\text{air,out}}) = \left(\frac{1}{2}\right)(45°\text{F} + 50°\text{F})$$
$$= 47.5°\text{F}$$
$$T_{\text{surface}} = 70°\text{F} \quad [\text{estimate}]$$

As in *NCEES Handbook:* Temperature-Dependent Properties of Air (U.S. Customary Units), the viscosity and Prandtl number are

$$\mu = 1.28 \times 10^{-5}\ \frac{\text{lbm}}{\text{ft-sec}}$$

$$\text{Pr} = 0.708$$

Using the volumetric flow rate and cross-sectional area of the duct, the velocity is

$$\text{v} = \frac{\dot{V}}{A} = \frac{\dot{V}}{\frac{\pi D^2}{4}} = \frac{500\ \frac{\text{ft}^3}{\text{min}}}{\left(\frac{\pi(1\text{ ft})^2}{4}\right)\left(60\ \frac{\text{s}}{\text{min}}\right)} = 10.6\text{ ft/sec}$$

From *NCEES Handbook* table "Dimensionless Numbers," the Reynolds number is

$$\mathrm{Re} = \frac{\rho u D}{\mu} = \frac{\left(0.07857\ \frac{\text{lbm}}{\text{ft}^3}\right)\left(10.6\ \frac{\text{ft}}{\text{sec}}\right)(1\ \text{ft})}{\left(1.28\times 10^{-5}\ \frac{\text{lbm}}{\text{ft-sec}}\right)} = 65{,}000$$

As in *NCEES Handbook* table "Temperature-Dependent Properties of Air at 14.7 psia (U.S. Units)," the thermal conductivity for air is 0.0156. Ignore the viscosity term for such a small temperature change. The correlation for turbulent flow in circular tubes can be found in *NCEES Handbook* table "Forced Convection—Internal Flow." Rearrange the correlation to find the film coefficient for air flowing inside the duct.

$$\mathrm{Nu_D} = 0.023\mathrm{Re_D}^{0.8}\mathrm{Pr}^{1/3}\left(\frac{\mu_\infty}{\mu_s}\right)^{0.14} = \frac{h_i D}{k}$$

$$h_i = \left(\frac{k}{D}\right)0.023\mathrm{Re_D}^{0.8}\mathrm{Pr}^{1/3}\left(\frac{\mu_\infty}{\mu_s}\right)^{0.14}$$

$$= \left(\frac{0.0156\ \frac{\text{Btu}}{\text{hr-ft-°F}}}{1\ \text{ft}}\right)(0.023)(65{,}000)^{0.8}(0.708)^{1/3}$$

$$= 2.27\ \text{Btu/hr-ft-°F}$$

For natural convection on the outside of the duct, estimate the film temperature.

$$T_{\text{film}} = \tfrac{1}{2}(T_{\text{surface}} + T_\infty) = \left(\frac{1}{2}\right)(70°\text{F} + 80°\text{F}) = 75°\text{F}$$

From App. 18C (also *NCEES Handbook:* Temperature-Dependent Properties of Air (U.S. Customary Units)), the properties of air at 75°F are

$$\mathrm{Pr} \approx 0.72$$

$$\frac{g\beta\rho^2}{\mu^2} = 2.27\times 10^6\ \frac{1}{\text{ft}^3\text{-°F}}$$

The characteristic length is the diameter of the duct.

$$L = \frac{12\ \text{in}}{12\ \frac{\text{in}}{\text{ft}}} = 1\ \text{ft}$$

As in *NCEES Handbook* table "Dimensionless Numbers," the Grashof number is

$$\mathrm{Gr} = L^3\left(\frac{\rho^2\beta g}{\mu^2}\right)(T_\infty - T_s)$$

$$= (1\ \text{ft})^3\left(2.27\times 10^6\ \frac{1}{\text{ft}^3\text{-°F}}\right)(80°\text{F} - 70°\text{F})$$

$$= 2.27\times 10^7$$

$$\mathrm{PrGr} = (0.72)(2.27\times 10^7) = 1.63\times 10^7$$

As in *NCEES Handbook* table "Forced Convection—External Flow," $C = 0.125$ and $n = 0.333$ for this GrPr value. The correlation equation for free convection over a horizontal cylinder can be found in *NCEES Handbook:* Natural (Free) Convection. Rearrange the correlation to find the film coefficient for a horizontal cylinder.

$$\mathrm{Nu_D} = C(\mathrm{GrPr})^n = \frac{h_o D}{k}$$

$$h_o = \left(\frac{k}{D}\right)C(\mathrm{GrPr})^{1/3}$$

$$= \left(\frac{0.015\ \frac{\text{Btu}}{\text{hr-ft-°F}}}{1\ \text{ft}}\right)(0.125)(1.63\times 10^7)^{1/3}$$

$$= 0.48\ \text{Btu/hr-ft-°F}$$

Neglecting the wall resistance, the overall film coefficient from Eq. 19.72 (also *NCEES Handbook:* Overall Heat-Transfer Coefficient) is

$$\frac{1}{U} = \frac{1}{h_o} + \frac{1}{h_i} = \frac{1}{2.27\ \frac{\text{Btu}}{\text{hr-ft}^2\text{-°F}}} + \frac{1}{0.48\ \frac{\text{Btu}}{\text{hr-ft}^2\text{-°F}}}$$

$$= 2.524\ \text{hr-ft}^2\text{-°F/Btu}$$

$$U = \frac{1}{2.524\ \frac{\text{hr-ft}^2\text{-°F}}{\text{Btu}}} = 0.393\ \text{Btu/hr-ft}^2\text{-°F}$$

The heat transfer due to convection is

$$Q_{\text{convection}} = UA_{\text{surface}}(T_\infty - T_{\text{bulk,air}})$$

$$= U(\pi dL)(T_\infty - T_{\text{bulk,air}})$$

$$= \left(0.393\ \frac{\text{Btu}}{\text{hr-ft}^2\text{-°F}}\right) \times \pi(1\ \text{ft})(50\ \text{ft})(80°\text{F} - 47.5°\text{F})$$

$$= 2006.3\ \text{Btu/hr}$$

The heat transfer due to radiation is as shown in *NCEES Handbook:* Radiation.

$$Q_{\text{radiation}} = \sigma F_e F_a A_{\text{surface}}(T_\infty^4 - T_{\text{surface}}^4)$$

Assume the room and duct have an unobstructed view of each other. Then $F_a = 1.0$ and $F_e = \epsilon = 0.28$. As in *NCEES Handbook:* Temperature, the absolute temperatures are

$$\begin{aligned} T_\infty &= 80°\text{F} + 460° = 540°\text{R} \\ T_{\text{surface}} &= 70°\text{F} + 460° = 530°\text{R} \\ Q_{\text{radiation}} &= \left(0.1713 \times 10^{-8}\ \frac{\text{Btu}}{\text{hr-ft}^2\text{-°R}^4}\right)(0.28)(1.0) \\ &\quad \times \pi(1\ \text{ft})(50\ \text{ft})\left((540°\text{R})^4 - (530°\text{R})^4\right) \\ &= 461.5\ \text{Btu/hr} \end{aligned}$$

The total heat transfer to the air is

$$\begin{aligned} Q_{\text{total}} &= Q_{\text{convection}} + Q_{\text{radiation}} \\ &= 2006.3\ \frac{\text{Btu}}{\text{hr}} + 461.5\ \frac{\text{Btu}}{\text{hr}} \\ &= 2467.8\ \text{Btu/hr} \end{aligned}$$

At 47.5°F, the specific heat of air is approximately 0.240 Btu/lbm-°F. Since the heat transfer is known, the temperature of air leaving the duct can be calculated from

$$\begin{aligned} Q_{\text{total}} &= \dot{m}c_p(T_{\text{air,out}} - T_{\text{air,in}}) \\ T_{\text{air,out}} &= T_{\text{air,in}} + \frac{Q_{\text{total}}}{\dot{m}c_p} \\ &= 45°\text{F} + \frac{2467.8\ \frac{\text{Btu}}{\text{hr}}}{\left(2357.1\ \frac{\text{lbm}}{\text{hr}}\right)\left(0.24\ \frac{\text{Btu}}{\text{lbm-°F}}\right)} \\ &= 49.4°\text{F} \quad (49°\text{F}) \end{aligned}$$

This agrees with the engineer's estimate.

SI Solution

As in *NCEES Handbook:* Temperature, the absolute temperature of air entering the duct is from

$$7°\text{C} + 273° = 280\text{K}$$

From the ideal gas law, the density of air entering the duct is

$$\begin{aligned} \rho &= \frac{pM}{RT} = \frac{(101.3\ \text{kPa})\left(29\ \frac{\text{kg}}{\text{kmol}}\right)}{\left(8.314\ \frac{\text{kJ}}{\text{kmol·K}}\right)(280\text{K})} \\ &= 1.2604\ \text{kg/m}^3 \end{aligned}$$

The mass flow rate of air entering the duct is

$$\dot{m} = \rho\dot{V} = \left(1.2604\ \frac{\text{kg}}{\text{m}^3}\right)\left(0.25\ \frac{\text{m}^3}{\text{s}}\right) = 0.3151\ \text{kg/s}$$

The diameter of the duct is

$$d = \frac{30\ \text{cm}}{100\ \frac{\text{cm}}{\text{m}}} = 0.30\ \text{m}$$

The velocity of air entering the duct is

$$\begin{aligned} \text{v} &= \frac{\dot{V}}{A_{\text{flow}}} = \frac{0.25\ \frac{\text{m}^3}{\text{s}}}{\left(\frac{\pi}{4}\right)(0.30\ \text{m})^2} \\ &= 3.537\ \text{m/s} \end{aligned}$$

To calculate the initial film coefficients, estimate the temperatures based on the claim. The film coefficients are not highly sensitive to small temperature differences.

$$\begin{aligned} T_{\text{bulk,air}} &= \tfrac{1}{2}(T_{\text{air,in}} + T_{\text{air,out}}) = \left(\frac{1}{2}\right)(7°\text{C} + 10°\text{C}) \\ &= 8.5°\text{C} \\ T_{\text{surface}} &= 20°\text{C} \quad [\text{estimate}] \end{aligned}$$

As in *NCEES Handbook:* Temperature-Dependent Properties of Air (SI Units), the viscosity and Prandtl number are

$$\mu = 20.9\ \mu\text{Pa·s}$$

$$\text{Pr} = 0.707$$

Using the volumetric flow rate and cross-sectional area of the duct, the velocity is

$$\mathrm{v} = \frac{\dot{V}}{A} = \frac{\dot{V}}{\frac{\pi D^2}{4}} = \frac{0.25\ \frac{\mathrm{m}^3}{\mathrm{s}}}{\frac{\pi(0.3\ \mathrm{m})^2}{4}} = 3.54\ \mathrm{m/s}$$

As in *NCEES Handbook:* Dimensional Analysis, the Reynolds number is

$$\mathrm{Re} = \frac{\rho u D}{\mu} = \frac{\left(1.2604\ \frac{\mathrm{kg}}{\mathrm{m}^3}\right)\left(3.54\ \frac{\mathrm{m}}{\mathrm{s}}\right)(0.3\ \mathrm{m})}{(20.9\ \mu\mathrm{Pa{\cdot}s})\left(10^{-6}\ \frac{\mathrm{Pa}}{\mu\mathrm{Pa}}\right)} = 64{,}045$$

Refer to *NCEES Handbook:* Temperature-Dependent Properties of Air (SI Units) for the thermal conductivity of air. The correlation for forced convection in a pipe can be found in *NCEES Handbook:* Free/Forced Heat-Transfer Coefficients/Correlations. Rearrange the correlation to find the film coefficient for air flowing inside the duct. (Ignore the viscosity term for such a small temperature change.)

$$\mathrm{Nu_D} = 0.023\mathrm{Re_D}^{0.8}\mathrm{Pr}^{1/3}\left(\frac{\mu_\infty}{\mu_s}\right)^{0.14} = \frac{h_i D}{k}$$

$$\begin{aligned} h_i &= \left(\frac{k}{D}\right)0.023\mathrm{Re_D}^{0.8}\mathrm{Pr}^{1/3}\left(\frac{\mu_\infty}{\mu_s}\right)^{0.14} \\ &= \left(\frac{0.03\ \frac{\mathrm{W}}{\mathrm{m{\cdot}K}}}{0.3\ \mathrm{m}}\right)(0.023)(64045)^{0.8}(0.707)^{1/3} \\ &= 14.34\ \mathrm{W/m^2{\cdot}K} \end{aligned}$$

For natural convection on the outside of the duct, estimate the film coefficient.

$$\begin{aligned} T_{\mathrm{film}} &= \tfrac{1}{2}(T_{\mathrm{surface}} + T_\infty) = \left(\frac{1}{2}\right)(20^\circ\mathrm{C} + 27^\circ\mathrm{C}) \\ &= 23.5^\circ\mathrm{C} \end{aligned}$$

From App. 18D (also *NCEES Handbook:* Temperature-Dependent Properties of Air (SI Units)), the properties of air at 23.5°C are

$$\mathrm{Pr} = 0.709$$

$$\frac{g\beta\rho^2}{\mu^2} = 1.43 \times 10^8\ \frac{1}{\mathrm{K{\cdot}m^3}}$$

The characteristic length, L, is the diameter of the duct, which is 0.30 m. The Grashof number is

$$\begin{aligned} \mathrm{Gr} &= L^3\left(\frac{\rho^2\beta g}{\mu^2}\right)(T_\infty - T_{\mathrm{surface}}) \\ &= (0.30\ \mathrm{m})^3\left(1.43 \times 10^8\ \frac{1}{\mathrm{K{\cdot}m^3}}\right)(27^\circ\mathrm{C} - 20^\circ\mathrm{C}) \\ &= 2.70 \times 10^7 \\ \mathrm{PrGr} &= (0.709)(2.70 \times 10^7) = 1.9 \times 10^7 \end{aligned}$$

As in *NCEES Handbook:* Free/Forced Heat-Transfer Coefficients/Correlations, $C = 0.125$ and $n = 0.333$ for this GrPr value. The correlation for free convection over a horizontal cylinder can be found in *NCEES Handbook:* Natural (Free) Convection. Rearrange the correlation to find the film coefficient for a horizontal cylinder.

$$\mathrm{Nu_D} = C(\mathrm{Gr_D PR})^n = \frac{h_o D}{k}$$

$$\begin{aligned} h_o &= \left(\frac{k}{D}\right)C(\mathrm{Gr_D PR})^{1/3} \\ &= \left(\frac{0.03\ \frac{\mathrm{W}}{\mathrm{m{\cdot}K}}}{0.3\ \mathrm{m}}\right)(0.125)(1.9 \times 10^7)^{1/3} \\ &= 3.33\ \mathrm{W/m^2{\cdot}K} \end{aligned}$$

Neglecting the wall resistance, the overall film coefficient from Eq. 19.72 (also *NCEES Handbook:* Overall Heat-Transfer Coefficient) is

$$\begin{aligned} \frac{1}{U} &= \frac{1}{h_o} + \frac{1}{h_i} = \frac{1}{3.3\ \frac{\mathrm{W}}{\mathrm{m^2{\cdot}K}}} + \frac{1}{14.34\ \frac{\mathrm{W}}{\mathrm{m^2{\cdot}K}}} \\ &= 0.372\ \mathrm{m^2{\cdot}K/W} \\ U &= \frac{1}{0.372\ \frac{\mathrm{m^2{\cdot}K}}{\mathrm{W}}} = 2.70\ \mathrm{W/m^2{\cdot}K} \end{aligned}$$

The heat transfer due to convection is

$$\begin{aligned} Q_{\mathrm{convection}} &= UA_{\mathrm{surface}}(T_\infty - T_{\mathrm{bulk,air}}) \\ &= U\pi dL(T_\infty - T_{\mathrm{bulk,air}}) \\ &= \left(2.70\ \frac{\mathrm{W}}{\mathrm{m^2{\cdot}K}}\right)\pi(0.30\ \mathrm{m})(15\ \mathrm{m}) \\ &\quad \times (27^\circ\mathrm{C} - 8.5^\circ\mathrm{C}) \\ &= 706\ \mathrm{W} \end{aligned}$$

The heat transfer due to radiation is

$$Q_{\text{radiation}} = \sigma F_e F_a A_{\text{surface}}(T_\infty^4 - T_{\text{surface}}^4)$$

Assume the room and duct have an unobstructed view of each other. Then $F_a = 1.0$ and $F_e = \epsilon = 0.28$. The absolute temperatures are as in *NCEES Handbook:* Temperature.

$$\begin{aligned} T_\infty &= 27°\text{C} + 273° = 300\text{K} \\ T_{\text{surface}} &= 20°\text{C} + 273° = 293\text{K} \\ Q_{\text{radiation}} &= \left(5.67 \times 10^{-8}\ \frac{\text{W}}{\text{m}^2\text{·K}^4}\right)(0.28)(1)\pi \\ &\quad \times (0.30\text{ m})(15\text{ m})\left((300\text{K})^4 - (293\text{K})^4\right) \\ &= 163.8\text{ W} \end{aligned}$$

The total heat transfer to the air is

$$\begin{aligned} Q_{\text{total}} &= Q_{\text{convection}} + Q_{\text{radiation}} = 706\text{ W} + 163.8\text{ W} \\ &= 869.8\text{ W} \end{aligned}$$

At 8.5°C, the specific heat of air is

$$\left(1.0048\ \frac{\text{kJ}}{\text{kg·K}}\right)\left(1000\ \frac{\text{J}}{\text{kJ}}\right) = 1004.8\text{ J/kg·K}$$

Since the heat transfer is known, the temperature of air leaving the duct can be calculated from

$$\begin{aligned} Q_{\text{total}} &= \dot{m}c_p(T_{\text{air,out}} - T_{\text{air,in}}) \\ T_{\text{air,out}} &= T_{\text{air,in}} + \frac{Q_{\text{total}}}{\dot{m}c_p} \\ &= 7°\text{C} + \frac{869.8\text{ W}}{\left(0.3151\ \frac{\text{kg}}{\text{s}}\right)\left(1004.8\ \frac{\text{J}}{\text{kg·K}}\right)} \\ &= 9.7°\text{C} \end{aligned}$$

This agrees with the engineer's estimate.

The answer is (B).

3. *Customary U.S. Solution*

As in *NCEES Handbook:* Temperature, the absolute temperature of entering air is

$$500°\text{F} + 460° = 960°\text{R}$$

The absolute pressure of entering air is

$$25\text{ psig} + 14.7\ \frac{\text{lbf}}{\text{in}^2} = 39.7\text{ psia}$$

The density of air entering, from the ideal gas law, is

$$\begin{aligned} \rho &= \frac{pM}{RT} = \frac{\left(39.7\ \frac{\text{lbf}}{\text{in}^2}\right)\left(12\ \frac{\text{in}}{\text{ft}}\right)^2\left(29\ \frac{\text{lbm}}{\text{lbmol}}\right)}{\left(1545.35\ \frac{\text{ft-lbf}}{\text{lbmol-°R}}\right)(960°\text{R})} \\ &= 0.1116\text{ lbm/ft}^3 \end{aligned}$$

The mass flow rate of entering air is

$$\begin{aligned} \dot{m} &= \rho\dot{V} = \left(0.1116\ \frac{\text{lbm}}{\text{ft}^3}\right)\left(200\ \frac{\text{ft}^3}{\text{min}}\right)\left(60\ \frac{\text{min}}{\text{hr}}\right) \\ &= 1339.2\text{ lbm/hr} \end{aligned}$$

The absolute temperature of leaving air is

$$350°\text{F} + 460° = 810°\text{R}$$

The heat loss is

$$\begin{aligned} \dot{Q} &= \dot{m}(h_1 - h_2) = \dot{m}c_p(T_1 - T_2) \\ &= \left(1339.2\ \frac{\text{lbm}}{\text{hr}}\right)\left(0.242\ \frac{\text{Btu}}{\text{lbm-°F}}\right)(500°\text{F} - 350°\text{F}) \\ &= 48{,}613\text{ Btu/hr} \end{aligned}$$

Assuming midpoint pipe surface temperature,

$$\tfrac{1}{2}(T_{\text{in}} + T_{\text{out}}) = \left(\frac{1}{2}\right)(500°\text{F} + 350°\text{F}) = 425°\text{F}$$

Since the heat loss is known, the overall heat transfer coefficient can be determined from

$$\begin{aligned} Q &= UA\Delta T = U(\pi dL)\Delta T \\ U &= \frac{Q}{(\pi dL)\Delta T} \\ &= \frac{\left(48{,}613\ \frac{\text{Btu}}{\text{hr}}\right)\left(12\ \frac{\text{in}}{\text{ft}}\right)}{\pi(4.25\text{ in})(35\text{ ft})(425°\text{F} - 70°\text{F})} \\ &= 3.52\text{ Btu/hr-ft}^2\text{-°F} \end{aligned}$$

To calculate the overall heat transfer coefficient, disregard the pipe thermal resistance and the inside film coefficient (small compared with outside film and radiation). Work with the midpoint pipe temperature of 425°F.

The absolute temperatures are

$$\begin{aligned} T_1 &= 425°\text{F} + 460° = 885°\text{R} \\ T_2 &= 70°\text{F} + 460° = 530°\text{R} \end{aligned}$$

For radiation heat loss, assume $F_a = 1$ and use *NCEES Handbook:* Emissivity (ε). For 500°F enamel paint of any color,

$$F_e = \epsilon \approx 0.9$$

$$\begin{aligned}\frac{Q_{\text{net}}}{A} &= E_{\text{net}} = \sigma F_e F_a(T_1^4 - T_2^4) \\ &= \left(0.1713 \times 10^{-8}\ \frac{\text{Btu}}{\text{hr-ft}^2\text{-}^\circ\text{R}^4}\right) \\ &\quad \times (0.9)(1.0)\left((885^\circ\text{R})^4 - (530^\circ\text{R})^4\right) \\ &= 824.1\ \text{Btu/hr-ft}^2\end{aligned}$$

From Eq. 20.23, the radiant heat transfer coefficient is

$$\begin{aligned}h_{\text{radiation}} &= \frac{E_{\text{net}}}{T_1 - T_2} = \frac{824.1\ \frac{\text{Btu}}{\text{hr-ft}^2}}{885^\circ\text{R} - 530^\circ\text{R}} \\ &= 2.32\ \text{Btu/hr-ft}^2\text{-}^\circ\text{F}\end{aligned}$$

For the outside film coefficient, evaluate the film at the pipe midpoint. The film temperature is

$$T_f = \left(\frac{1}{2}\right)(425^\circ\text{F} + 70^\circ\text{F}) = 247.5^\circ\text{F}$$

From App. 18C (also *NCEES Handbook:* Temperature-Dependent Properties of Air (U.S. Customary Units)),

$$\begin{aligned}\text{Pr} &= 0.72 \\ \frac{g\beta\rho^2}{\mu^2} &= 0.657 \times 10^6\ \frac{1}{\text{ft}^3\text{-}^\circ\text{F}}\end{aligned}$$

The characteristic length, L, is $d_o = 4.25$ in. The Grashof number is

$$\begin{aligned}\text{Gr} &= L^3\left(\frac{\rho^2\beta g}{\mu^2}\right)\Delta T \\ &= \frac{(4.25\ \text{in})^3\left(0.657 \times 10^6\ \frac{1}{\text{ft}^3\text{-}^\circ\text{F}}\right)(425^\circ\text{F} - 70^\circ\text{F})}{\left(12\ \frac{\text{in}}{\text{ft}}\right)^3} \\ &= 1.04 \times 10^7\end{aligned}$$

$$\text{PrGr} = (0.72)(1.04 \times 10^7) = 7.5 \times 10^6$$

As in *NCEES Handbook* table "Natural (Free) Convection," $n = 1/4$, $C = 0.480$ for this GrPr value. The correlation for free convection over a horizontal cylinder can be found in *NCEES Handbook:* Natural (Free) Convection. Rearrange the correlation to find the film coefficient for a horizontal cylinder.

$$\begin{aligned}\text{Nu}_\text{D} &= C(\text{GrPr})^n = \frac{h_o D}{k} \\ h_o &= \left(\frac{k}{D}\right)C(\text{GrPr})^n \\ &= \left(\frac{0.02\ \frac{\text{Btu}}{\text{hr-ft-}^\circ\text{F}}}{0.354\ \text{ft}}\right)(0.480)(7.5 \times 10^6)^{1/4} \\ &= 1.42\ \text{Btu/hr-ft}^2\text{-}^\circ\text{F}\end{aligned}$$

The overall film coefficient is

$$\begin{aligned}U &= h_{\text{total}} = h_{\text{radiation}} + h_o \\ &= 2.32\ \frac{\text{Btu}}{\text{hr-ft}^2\text{-}^\circ\text{F}} + 1.42\ \frac{\text{Btu}}{\text{hr-ft}^2\text{-}^\circ\text{F}} \\ &= 3.74\ \text{Btu/hr-ft}^2\text{-}^\circ\text{F} \quad (3.8\ \text{Btu/hr-ft}^2\text{-}^\circ\text{F})\end{aligned}$$

SI Solution

The absolute temperature of entering air is

$$260^\circ\text{C} + 273^\circ = 533\text{K}$$

The absolute pressure of entering air is

$$170\ \text{kPa} + 101.3\ \text{kPa} = 271.3\ \text{kPa}$$

The density of air entering, from the ideal gas law, is

$$\begin{aligned}\rho &= \frac{pM}{RT} = \frac{(271.3\ \text{kPa})\left(29\ \frac{\text{kg}}{\text{kmol}}\right)}{\left(8.314\ \frac{\text{kJ}}{\text{kmol}\cdot\text{K}}\right)(533\text{K})} \\ &= 1.773\ \text{kg/m}^3\end{aligned}$$

The mass flow rate of entering air is

$$\begin{aligned}\dot{m} &= \rho\dot{V} = \left(1.773\ \frac{\text{kg}}{\text{m}^3}\right)\left(0.1\ \frac{\text{m}^3}{\text{s}}\right) \\ &= 0.1773\ \text{kg/s}\end{aligned}$$

The absolute temperature of leaving air is

$$180^\circ\text{C} + 273^\circ = 453\text{K}$$

The heat loss is

$$\begin{aligned} Q &= \dot{m}(h_1 - h_2) = \dot{m}c_p(T_1 - T_2) \\ &= \left(0.1773\ \frac{\text{kg}}{\text{s}}\right)\left(1\ \frac{\text{kJ}}{\text{kg·K}}\right)(533\text{K} - 453\text{K}) \\ &= 14.2\ \text{kW} \end{aligned}$$

Assuming midpoint pipe surface temperature,

$$\tfrac{1}{2}(T_{\text{in}} - T_{\text{out}}) = \left(\frac{1}{2}\right)(260^\circ\text{C} + 180^\circ\text{C}) = 220^\circ\text{C}$$

Since the heat loss is known, the overall heat transfer coefficient can be determined from

$$Q = UA\Delta T = U(\pi dL)\Delta T$$

$$\begin{aligned} U &= \frac{Q}{(\pi dL)\Delta T} \\ &= \frac{(14\,200\ \text{W})\left(100\ \frac{\text{cm}}{\text{m}}\right)}{\pi(10.8\ \text{cm})(10\ \text{m})(220^\circ\text{C} - 20^\circ\text{C})} \\ &= 20.92\ \text{W/m}^2\text{·K} \end{aligned}$$

From the customary U.S. solution, work with the midpoint pipe temperature of 220°C.

As in *NCEES Handbook:* Temperature, the absolute temperatures are

$$\begin{aligned} T_1 &= 220^\circ\text{C} + 273^\circ = 493\text{K} \\ T_2 &= 20^\circ\text{C} + 273^\circ = 293\text{K} \end{aligned}$$

For radiation heat loss, assume $F_a = 1$ and use *NCEES Handbook:* Emissivity (ε). For 260°C enamel paint of any color,

$$F_e = \epsilon \approx 0.9$$

From Eq. 20.23, the radiant heat transfer coefficient is

$$\begin{aligned} h_{\text{radiation}} &= \frac{\sigma F_a F_e (T_1^4 - T_2^4)}{T_1 - T_2} \\ &= \frac{\left(5.67 \times 10^{-8}\ \frac{\text{W}}{\text{m}^2\text{·K}^4}\right)(1)(0.9) \times \left((493\text{K})^4 - (293\text{K})^4\right)}{493\text{K} - 293\text{K}} \\ &= 13.2\ \text{W/m}^2\text{·K} \end{aligned}$$

For the outside film coefficient, evaluate the film at the pipe midpoint. The film temperature is

$$T_f = \left(\frac{1}{2}\right)(220^\circ\text{C} + 20^\circ\text{C}) = 120^\circ\text{C}$$

From App. 18D (also *NCEES Handbook* table "Temperature-Dependent Properties of Air (SI Units)"), at 120°C,

$$\begin{aligned} \text{Pr} &\approx 0.692 \\ \frac{g\beta\rho^2}{\mu^2} &= 0.528 \times 10^8\ \frac{1}{\text{K·m}^3} \end{aligned}$$

The characteristic length is

$$L = \text{outside diameter} = \frac{10.8\ \text{cm}}{100\ \frac{\text{cm}}{\text{m}}} = 0.108\ \text{m}$$

As in *NCEES Handbook:* Similitude, the Grashof number is

$$\begin{aligned} \text{Gr} &= L^3\left(\frac{\rho^2 g\beta}{\mu^2}\right)\Delta T \\ &= (0.108\ \text{m})^3\left(0.528 \times 10^8\ \frac{1}{\text{K·m}^3}\right)(220^\circ\text{C} - 20^\circ\text{C}) \\ &= 1.33 \times 10^7 \end{aligned}$$

$$\text{PrGr} = (0.692)(1.33 \times 10^7) = 9.20 \times 10^6$$

As in *NCEES Handbook* table "Natural (Free) Convection," $C = 0.480$ and $n = 0.25$ for this GrPr value. The correlation for free convection over a horizontal cylinder can be found in *NCEES Handbook:* Natural (Free) Convection. Rearrange the correlation to find the film coefficient for a horizontal cylinder.

$$\begin{aligned} \text{Nu}_\text{D} &= C(\text{Gr}_\text{D}\text{PR})^n = \frac{h_o D}{k} \\ h_o &= \left(\frac{k}{D}\right)C(\text{Gr}_\text{D}\text{PR})^n \\ &= \left(\frac{0.034\ \frac{\text{W}}{\text{m·K}}}{0.108\ \text{m}}\right)(0.480)(9.2 \times 10^6)^{1/4} \\ &= 8.32\ \text{W/m}^2\text{·K} \end{aligned}$$

The overall film coefficient is

$$\begin{aligned} U = h_{\text{total}} &= h_{\text{radiation}} + h_o \\ &= 13.2\ \frac{\text{W}}{\text{m}^2\text{·K}} + 8.32\ \frac{\text{W}}{\text{m}^2\text{·K}} \\ &= 21.52\ \text{W/m}^2\text{·K} \quad (22\ \text{W/m}^2\text{·K}) \end{aligned}$$

The answer is (C).

4. *Customary U.S. Solution*

The velocity is relatively low, so incompressible flow can be assumed.

The film coefficient on the probe is

$$h = \frac{0.024G^{0.8}}{D^{0.4}} = \frac{(0.024)\left(3480\ \dfrac{\text{lbm}}{\text{hr-ft}^2}\right)^{0.8}}{\left(\dfrac{0.5\ \text{in}}{12\ \dfrac{\text{in}}{\text{ft}}}\right)^{0.4}}$$

$$= 58.3\ \text{Btu/hr-ft}^2\text{-°F}$$

As in *NCEES Handbook:* Temperature, the absolute temperature of the walls is

$$T_{\text{walls}} = 600°\text{F} + 460° = 1060°\text{R}$$

Neglect conduction and the insignificant kinetic energy loss. The thermocouple gains heat through radiation from the walls and loses heat through convection to the gas.

$$\begin{aligned} Q_{\text{convection}} &= AE_{\text{radiation}} \\ hA(T_{\text{probe}} - T_{\text{gas}}) &= A\sigma\epsilon(T_{\text{walls}}^4 - T_{\text{probe}}^4) \\ h(T_{\text{probe}} - T_{\text{gas}}) &= \sigma\epsilon(T_{\text{walls}}^4 - T_{\text{probe}}^4) \end{aligned}$$

If $T_{\text{probe}} = 300°\text{F} + 460° = 760°\text{R}$,

$$\begin{aligned} &\left(58.3\ \frac{\text{Btu}}{\text{hr-ft}^2\text{-°F}}\right) \\ &\quad \times(760°\text{R} - T_{\text{gas}}) = \left(0.1713\times10^{-8}\ \frac{\text{Btu}}{\text{hr-ft}^2\text{-°R}^4}\right) \\ &\qquad\qquad \times(0.8)\big((1060°\text{R})^4 - (760°\text{R})^4\big) \\ &\quad T_{\text{gas}} = 738.2°\text{R} \quad (740°\text{R}) \end{aligned}$$

SI Solution

The velocity is relatively low, so incompressible flow can be assumed.

The film coefficient on the probe is

$$h = \frac{17G^{0.8}}{D^{0.4}} = \frac{(17)\left(4.7\ \dfrac{\text{kg}}{\text{s·m}^2}\right)^{0.8}}{\left(\dfrac{13\ \text{mm}}{1000\ \dfrac{\text{mm}}{\text{m}}}\right)^{0.4}}$$

$$= 333.1\ \text{W/m}^2\text{·K}$$

As in *NCEES Handbook:* Temperature, the absolute temperature of the walls is

$$T_{\text{walls}} = 315°\text{C} + 273° = 588\text{K}$$

Neglect conduction and the insignificant kinetic energy loss. The thermocouple gains heat through radiation from the walls and loses heat through convection to the gas.

$$\begin{aligned} \frac{Q_{\text{convection}}}{A} &= E_{\text{radiation}} \\ h(T_{\text{probe}} - T_{\text{gas}}) &= \sigma\epsilon(T_{\text{walls}}^4 - T_{\text{probe}}^4) \end{aligned}$$

If $T_{\text{probe}} = 150°\text{C} + 273° = 423\text{K}$,

$$\begin{aligned} &\left(333.1\ \frac{\text{W}}{\text{m}^2\text{·K}}\right) \\ &\quad \times(423\text{K} - T_{\text{gas}}) = \left(5.67\times10^{-8}\ \frac{\text{W}}{\text{m}^2\text{·K}^4}\right)(0.8) \\ &\qquad\qquad \times\big((588\text{K})^4 - (423\text{K})^4\big) \\ &\quad T_{\text{gas}} = 411.1\text{K} \quad (410\text{K}) \end{aligned}$$

The answer is (B).

5. *Customary U.S. Solution*

As in *NCEES Handbook:* Temperature, the absolute temperatures are

$$\begin{aligned} T_\infty &= 80°\text{F} + 460° = 540°\text{R} \\ T_{\text{duct}} &= 200°\text{F} + 460° = 660°\text{R} \\ T_{\text{wall}} &= 70°\text{F} + 460° = 530°\text{R} \end{aligned}$$

As in *NCEES Handbook:* Temperature-Dependent Properties of Air (U.S. Customary Units),

$$\rho = 0.071\ \frac{\text{lbm}}{\text{ft}^3}$$

$$\mu = 1.28 \times 10^{-5}\ \frac{\text{lbm}}{\text{ft-sec}}$$

$$\text{Pr} = 0.708$$

$$k = 0.0155\ \frac{\text{Btu}}{\text{hr-ft-°F}}$$

The average temperature is

$$\begin{aligned} T_{\text{avg}} &= \frac{1}{\beta} = \frac{T_\infty + T_{\text{duct}}}{2} = \frac{540°\text{R} + 660°\text{R}}{2} \\ &= 600°\text{R} \end{aligned}$$

The characteristic length of the cylinder is its diameter, which is 0.75 ft. Substituting the definition of the Grashof number (see *NCEES Handbook* table "Dimensionless Numbers"),

$$\text{GrPr} = \frac{g\beta\rho^2}{\mu^2}(T_s - T_\infty)L^3\text{Pr} = 5.9 \times 10^7$$

As in *NCEES Handbook* table "Natural (Free) Convection," $C = 0.125$ and $n = 0.333$ for this GrPr value. The correlation for free convection over a horizontal cylinder can be found in *NCEES Handbook:* Natural (Free) Convection. Rearrange the correlation to find the convective film coefficient for the outside of the duct.

$$\text{Nu}_\text{D} = C(\text{Gr}_\text{D}\text{PR})^n = \frac{hD}{k}$$

$$\begin{aligned} h &= \left(\frac{k}{D}\right)C(\text{Gr}_\text{D}\text{PR})^{1/3} \\ &= \left(\frac{0.0155\ \frac{\text{Btu}}{\text{hr-ft-°F}}}{0.75\ \text{ft}}\right)(0.125)(5.9 \times 10^7)^{1/3} \\ &= 1.00\ \text{Btu/hr-ft}^2\text{-°F} \end{aligned}$$

The duct area per unit length is

$$\frac{A}{L} = \pi D = \frac{\pi(9\ \text{in})}{12\ \frac{\text{in}}{\text{ft}}} = 2.356\ \text{ft}^2/\text{ft}$$

The convection losses (per unit length) are

$$\begin{aligned} \frac{Q_{\text{convection}}}{L} &= h\left(\frac{A}{L}\right)\Delta T = h\left(\frac{A}{L}\right)(T_{\text{duct}} - T_\infty) \\ &= \left(1.00\ \frac{\text{Btu}}{\text{hr-ft}^2\text{-°F}}\right)\left(2.356\ \frac{\text{ft}^2}{\text{ft}}\right) \\ &\quad \times (660°\text{R} - 540°\text{R}) \\ &= 282.7\ \text{Btu/hr-ft} \end{aligned}$$

Assume $\epsilon_{\text{duct}} \approx 0.90$. Then $F_e = \epsilon_{\text{duct}} = 0.90$. $F_a = 1$ since the duct is enclosed. The radiation losses (per unit length) are

$$\begin{aligned} \frac{E_{\text{net}}}{L} &= \left(\frac{A}{L}\right)\sigma F_a F_e(T_{\text{duct}}^4 - T_{\text{wall}}^4) \\ &= (2.356\ \text{ft}^2)\left(0.1713 \times 10^{-8}\ \frac{\text{Btu}}{\text{hr-ft}^2\text{-°R}^4}\right) \\ &\quad \times (1)(0.90)\left((660°\text{R})^4 - (530°\text{R})^4\right) \\ &= 402.6\ \text{Btu/hr-ft} \end{aligned}$$

The total heat transfer per unit length is

$$\begin{aligned} \frac{Q_{\text{total}}}{L} &= \frac{Q_{\text{convection}}}{L} + \frac{E_{\text{net}}}{L} = 282.7\ \frac{\text{Btu}}{\text{hr-ft}} + 402.6\ \frac{\text{Btu}}{\text{hr-ft}} \\ &= 685\ \text{Btu/hr-ft length} \quad (685\ \text{Btu/hr-ft length}) \end{aligned}$$

SI Solution

As in *NCEES Handbook:* Temperature, the absolute temperatures are

$$\begin{aligned} T_{\text{duct}} &= 95°\text{C} + 273° = 368\text{K} \\ T_{\text{wall}} &= 20°\text{C} + 273° = 293\text{K} \\ T_\infty &= 27°\text{C} + 273° = 300\text{K} \end{aligned}$$

The characteristic length of a cylinder is its diameter. From *NCEES Handbook:* Temperature-Dependent Properties of Air (SI Units),

$$\rho = 1.067\ \frac{\text{kg}}{\text{m}^3}$$

$$\mu = 20.2\ \mu\text{Pa·s}$$

$$\text{Pr} = 0.714$$

$$k = 0.0285\ \frac{\text{W}}{\text{m·K}}$$

The average temperature is

$$\begin{aligned} T_{\text{avg}} &= \frac{1}{\beta} = \frac{T_\infty + T_{\text{duct}}}{2} = \frac{300\text{K} + 368\text{K}}{2} \\ &= 334\text{K} \end{aligned}$$

Substituting the definition of the Grashof number (see *NCEES Handbook* table "Dimensionless Numbers"),

$$\text{GrPr} = \frac{g\beta\rho^2}{\mu^2}(T_s - T_\infty)L^3\text{Pr} = 5 \times 10^7$$

The characteristic length of the cylinder is its diameter, which is 0.23 m. As in *NCEES Handbook* table "Natural (Free) Convection," $C = 0.125$ and $n = 0.333$ for this GrPr value. The correlation for free convection over a horizontal cylinder can be found in the same table. Rearrange the correlation to find the convective film coefficient for the outside of the duct.

$$\begin{aligned}
\mathrm{Nu_D} &= C(\mathrm{Gr_D PR})^n = \frac{hD}{k} \\
h &= \left(\frac{k}{D}\right) C(\mathrm{Gr_D PR})^{1/3} \\
&= \left(\frac{0.0285\ \frac{\mathrm{W}}{\mathrm{m\cdot K}}}{0.23\ \mathrm{m}}\right)(0.125)(5\times 10^7)^{1/3} \\
&= 5.70\ \mathrm{W/m^2\cdot K}
\end{aligned}$$

The duct area per unit length is

$$\frac{A}{L} = \pi D = \frac{\pi(23\ \mathrm{cm})}{100\ \frac{\mathrm{cm}}{\mathrm{m}}} = 0.723\ \mathrm{m^2/m}$$

The convective losses per unit length are

$$\begin{aligned}
\frac{Q_{\text{convective}}}{L} &= h\left(\frac{A}{L}\right)(T_{\text{duct}} - T_\infty) \\
&= \left(5.70\ \frac{\mathrm{W}}{\mathrm{m^2\cdot K}}\right)\left(0.723\ \frac{\mathrm{m^2}}{\mathrm{m}}\right)(368\ \mathrm{K} - 300\mathrm{K}) \\
&= 280.2\ \mathrm{W/m}
\end{aligned}$$

Assuming $\epsilon_{\text{duct}} \approx 0.90$, $F_e = \epsilon_{\text{duct}} = 0.90$. $F_a = 1$ since the duct is enclosed. The radiation losses per unit length are

$$\begin{aligned}
\frac{E_{\text{net}}}{L} &= \left(\frac{A}{L}\right)\sigma F_a F_e (T_{\text{duct}}^4 - T_{\text{wall}}^4) \\
&= \left(0.723\ \frac{\mathrm{m^2}}{\mathrm{m}}\right)\left(5.67\times 10^{-8}\ \frac{\mathrm{W}}{\mathrm{m^2\cdot K^4}}\right) \\
&\quad \times (1)(0.90)\left((368\mathrm{K})^4 - (293\mathrm{K})^4\right) \\
&= 404.7\ \mathrm{W/m}
\end{aligned}$$

The total heat transfer per unit length is

$$\begin{aligned}
\frac{Q_{\text{total}}}{L} &= \frac{Q_{\text{convection}}}{L} + \frac{E_{\text{net}}}{L} = 280.2\ \frac{\mathrm{W}}{\mathrm{m}} + 404.7\ \frac{\mathrm{W}}{\mathrm{m}} \\
&= 684.9\ \mathrm{W/m\ length} \quad (685\ \mathrm{W/m\ length})
\end{aligned}$$

The answer is (A).

21 Water Supply Treatment and Distribution

PRACTICE PROBLEMS

There are no problems in this book corresponding to Chap. 21 of the *PE Chemical Review*.

22 Environmental Engineering

PRACTICE PROBLEMS

There are no problems in this book corresponding to Chap. 22 of the *PE Chemical Review*.

23 Basic Principles of Mass Transfer

PRACTICE PROBLEMS

1. An ethanol-water mixture at 1 atm total pressure and 95.5°C contains 10% ethanol (mole basis).

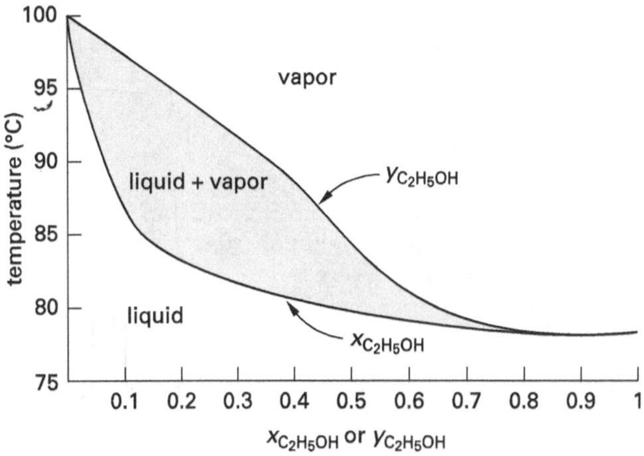

Using the *Txy* diagram provided, the amount of vapor in 1000 mol of the mixture at equilibrium is

(A) 83.3 mol

(B) 100 mol

(C) 467 mol

(D) 533 mol

2. A *Txy* diagram is provided for a methanol-carbon tetrachloride mixture consisting of 845 mol of mixture with 10% methanol (mole basis) at 60°C and 1 atm total pressure.

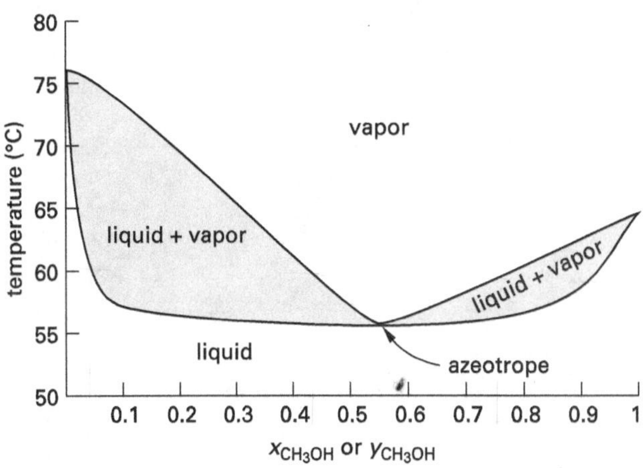

At equilibrium, the amount of vapor is

(A) 55.1 mol

(B) 84.5 mol

(C) 130 mol

(D) 715 mol

3. The equilibrium data in the *xy* diagram provided describe a methanol-carbon tetrachloride mixture.

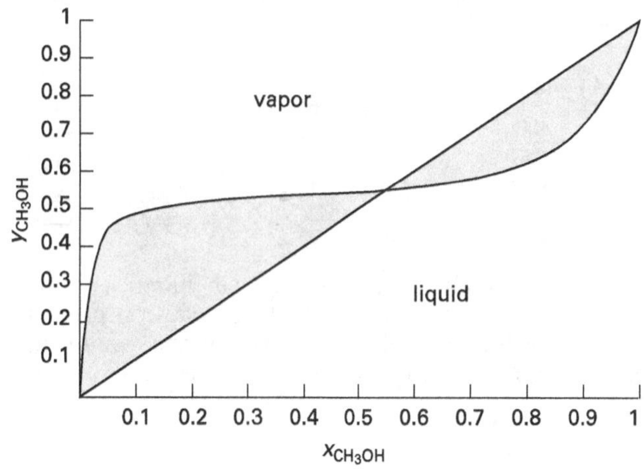

What is the relative volatility of methanol to carbon tetrachloride at the azeotropic composition?

(A) 0.55

(B) 0.67

(C) 1

(D) 1.49

4. An equilibrium methanol-carbon tetrachloride mixture with a liquid mole fraction, $x_{CH_3OH} = 0.40$ is described by the plotted data shown.

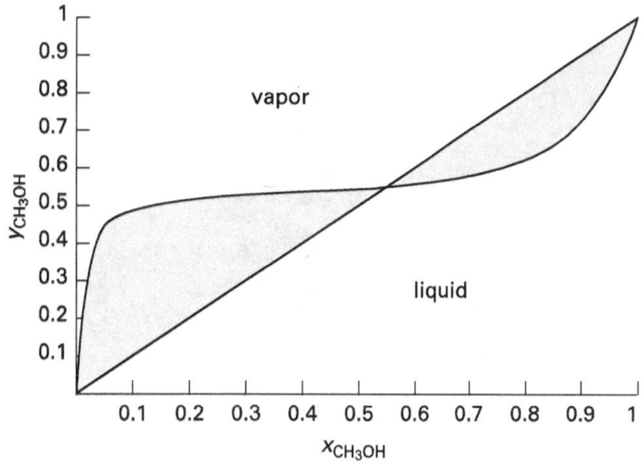

What is the relative volatility of methanol to carbon tetrachloride?

(A) 0.57

(B) 0.78

(C) 1.28

(D) 1.75

5. 800 mol of a liquid mixture contain 40 mol% acetone ($(CH_3)_2CO$) in water at 62°C and 1 atm total pressure. The following *Txy* diagram presents the equilibrium data for the acetone-water system at 1 atm total pressure.

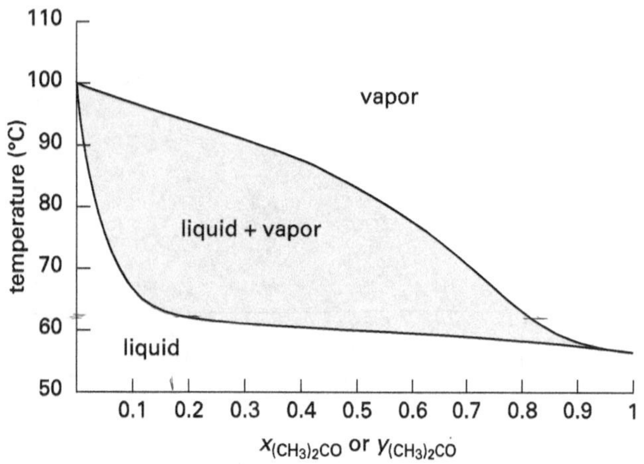

At equilibrium, how many moles of vapor are there?

(A) $V = 270$ mol

(B) $V = 250$ mol

(C) $V = 230$ mol

(D) $V = 210$ mol

6. An aqueous solution of methanol in water contains 24.4 wt% water. The solution follows Raoult's law. The flash point of methanol is 54°F, and the vapor pressure of methanol at this temperature is 62 mm Hg. The saturation vapor pressure of methanol at different temperatures is shown.

temperature (°F)	saturation pressure (mm Hg)
54.00	62.00
59.00	73.91
66.16	92.10
67.96	97.58
69.75	103.1
73.33	114.0
80.49	140.2
87.66	171.5
94.82	208.4

The molecular weight of water is 18 lbm/lbmol, and that of methanol is 32 lbm/lbmol. Most nearly, the flash point of the aqueous mixture is

(A) 54°F

(B) 59°F

(C) 68°F

(D) 80°F

SOLUTIONS

1.

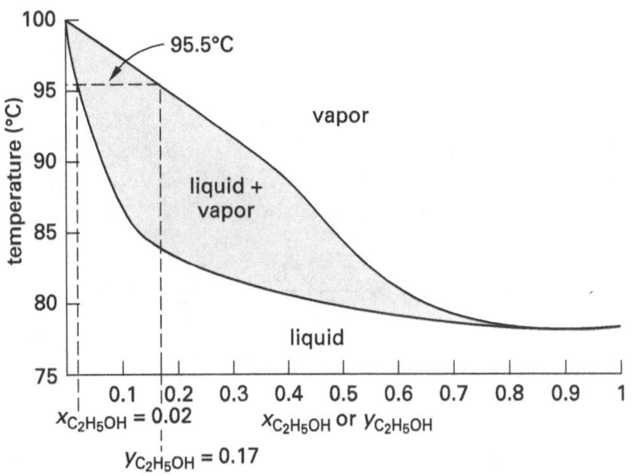

For an equilibrium liquid and vapor mixture at 95.5°C, the isotherm's intersection with the saturated liquid line indicates that the equilibrium composition of the liquid phase, $x_{C_2H_5OH}$, is 0.02. The isotherm's intersection with the saturated vapor line gives the equilibrium composition of the vapor phase, $y_{C_2H_5OH}$, as 0.17. The mole fraction of water in each phase is calculated using Eq. 23.4 and Eq. 23.5 (also *NCEES Handbook:* Fractions).

$$x_{C_2H_5OH} + x_{H_2O} = 1$$

$$x_{H_2O} = 1 - x_{C_2H_5OH} = 1 - 0.02 = 0.98$$

$$y_{C_2H_5OH} + y_{H_2O} = 1$$

$$y_{H_2O} = 1 - y_{C_2H_5OH} = 1 - 0.17 = 0.83$$

From the problem statement,

$$F = 1000 \text{ mol}$$

$$z_{C_2H_5OH} = 0.10$$

As in *NCEES Handbook:* Single-Stage Flash,

$$\frac{V}{F} = A_v = \frac{x_F - x_A}{y_A - x_A} = \frac{0.1 - 0.02}{0.17 - 0.02} = 0.533$$

$$V = A_v F = (0.533)(1000 \text{ mol}) = 533 \text{ mol}$$

The answer is (D).

2.

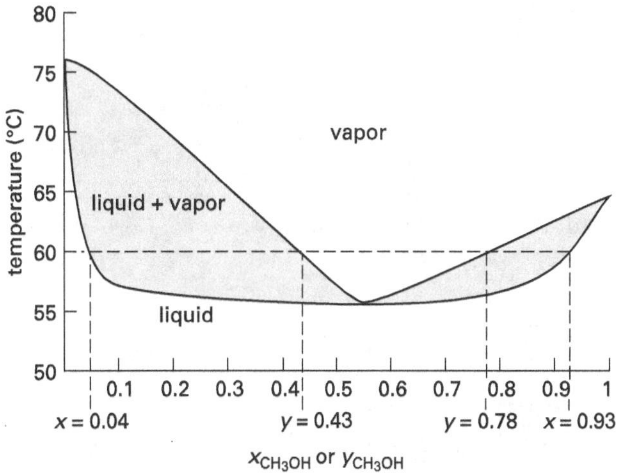

Draw the isotherm at 60°C and read the equilibrium composition of the liquid and the vapor phases. There are two possible equilibrium compositions. These are

$$x_{CH_3OH} = 0.04$$

$$y_{CH_3OH} = 0.43$$

and

$$x_{CH_3OH} = 0.93$$

$$y_{CH_3OH} = 0.78$$

The feed composition is given as $z_{CH_3OH} = 0.10$. Plotting the feed composition on the equilibrium diagram reveals that the equilibrium composition of the liquid and the vapor phases lies on the left side of the two-phase envelope. So,

$$x_{CH_3OH} = 0.04$$

$$y_{CH_3OH} = 0.43$$

From the problem statement,

$$F = 845 \text{ mol}$$

As in *NCEES Handbook:* Single-Stage Flash,

$$\frac{V}{F} = A_v = \frac{x_F - x_A}{y_A - x_A} = \frac{0.1 - 0.04}{0.43 - 0.04} = 0.154$$

$$V = A_v F = (0.154)(845 \text{ mol}) = 130 \text{ mol}$$

The answer is (C).

3. From the xy diagram, at the azeotrope, $x_{CH_3OH} = 0.55$, $y_{CH_3OH} = 0.55$. From Eq. 23.7, the equilibrium ratio for methanol is

$$K_{CH_3OH} = \frac{y_{CH_3OH}}{x_{CH_3OH}} = \frac{0.55}{0.55} = 1.0$$

The mole fractions for carbon tetrachloride are calculated by using Eq. 23.4 and Eq. 23.5 (also *NCEES Handbook:* Fractions).

$$x_{CCl_4} + x_{CH_3OH} = 1$$
$$x_{CCl_4} = 1 - x_{CH_3OH} = 1 - 0.55 = 0.45$$
$$y_{CCl_4} + y_{CH_3OH} = 1$$
$$y_{CCl_4} = 1 - y_{CH_3OH} = 1 - 0.55 = 0.45$$

The equilibrium ratio from *NCEES Handbook:* Distribution of Components Between Phases in a Vapor/Liquid Equilibrium for carbon tetrachloride is

$$K_{CCl_4} = \frac{y_{CCl_4}}{x_{CCl_4}} = \frac{0.45}{0.45} = 1.0$$

The relative volatility from *NCEES Handbook:* Distribution of Components Between Phases in a Vapor/Liquid Equilibrium of methanol to carbon tetrachloride is

$$\alpha_{CH_3OH-CCl_4} = \frac{K_{CH_3OH}}{K_{CCl_4}} = \frac{1.0}{1.0} = 1$$

The relative volatility of carbon tetrachloride to methanol is also equal to 1.

The answer is (C).

4. From the xy diagram, at $x_{CH_3OH} = 0.40$, $y_{CH_3OH} = 0.54$. The equilibrium ratio is

$$\begin{aligned} K_{CH_3OH} &= \frac{y_{CH_3OH}}{x_{CH_3OH}} \\ &= \frac{0.54}{0.40} \\ &= 1.35 \end{aligned}$$

Using Eq. 23.4 and Eq. 23.5 (also *NCEES Handbook:* Fractions), the mole fractions of carbon tetrachloride are

$$x_{CCl_4} + x_{CH_3OH} = 1$$
$$x_{CCl_4} = 1 - x_{CH_3OH} = 1 - 0.40 = 0.60$$
$$y_{CCl_4} + y_{CH_3OH} = 1$$
$$y_{CCl_4} = 1 - y_{CH_3OH} = 1 - 0.54 = 0.46$$

The equilibrium ratio from *NCEES Handbook:* Distribution of Components Between Phases in a Vapor/Liquid Equilibrium for carbon tetrachloride is

$$K_{CCl_4} = \frac{y_{CCl_4}}{x_{CCl_4}} = \frac{0.46}{0.60} = 0.77$$

The relative volatility from *NCEES Handbook:* Distribution of Components Between Phases in a Vapor/Liquid Equilibrium of methanol to carbon tetrachloride is

$$\alpha_{CH_3OH-CCl_4} = \frac{K_{CH_3OH}}{K_{CCl_4}} = \frac{1.35}{0.77} = 1.75$$

The answer is (D).

5. The vapor equilibrium composition is found from the points where the 62°C isotherm intersects the liquid and the vapor equilibrium lines.

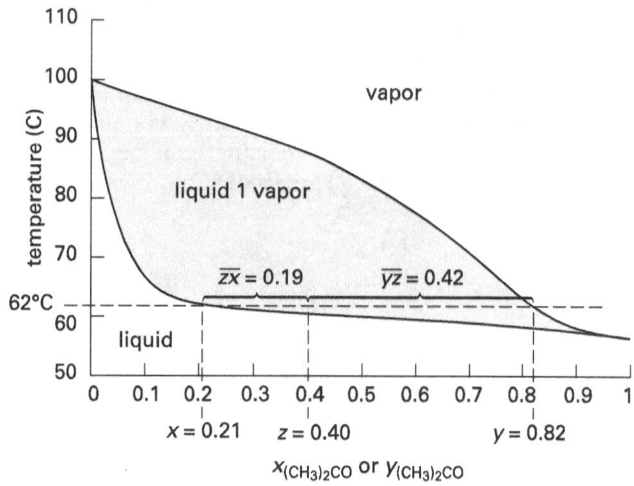

For a 40 mol% acetone solution, $z_{(CH_3)_2CO} = 0.40$. The equilibrium compositions of acetone are $x_{(CH_3)_2CO} = 0.21$ and $y_{(CH_3)_2CO} = 0.82$.

As in *NCEES Handbook:* Single-Stage Flash,

$$\frac{V}{F} = A_v = \frac{x_F - x_A}{y_A - x_A} = \frac{0.4 - 0.21}{0.82 - 0.21} = 0.311$$

$$V = A_v F = (0.311)(800 \text{ mol}) = 250 \text{ mol}$$

The answer is (B).

6. Let the basis of calculation, w, be 100 lbm of aqueous solution. The weight of water in the solution, w_{water}, is given as 24.4 lbm. The weight of methanol in the aqueous solution is

$$\begin{aligned} w_{methanol} &= w - w_{water} = 100 \text{ lbm} - 24.4 \text{ lbm} \\ &= 75.6 \text{ lbm} \end{aligned}$$

The number of moles of water in the solution is

$$n_{\text{water}} = \frac{w_{\text{water}}}{\text{MW}_{\text{water}}} = \frac{24.4 \text{ lbm}}{18 \frac{\text{lbm}}{\text{lbmol}}} = 1.36 \text{ lbmol}$$

The number of moles of methanol in the solution is

$$n_{\text{methanol}} = \frac{w_{\text{methanol}}}{\text{MW}_{\text{methanol}}} = \frac{75.6 \text{ lbm}}{32 \frac{\text{lbm}}{\text{lbmol}}} = 2.36 \text{ lbmol}$$

The total number of moles in the aqueous solution is

$$\begin{aligned} n_{\text{total}} &= n_{\text{water}} + n_{\text{methanol}} = 1.36 \text{ lbmol} + 2.36 \text{ lbmol} \\ &= 3.72 \text{ lbmol} \end{aligned}$$

The fraction mole of methanol in the aqueous solution is

$$x_{\text{methanol}} = \frac{n_{\text{methanol}}}{n_{\text{total}}} = \frac{2.36 \text{ lbmol}}{3.72 \text{ lbmol}} = 0.63$$

The pressure of methanol at the flash point, $p°$, is given as 62 mm Hg. Raoult's law gives the saturation pressure of methanol in the aqueous solution.

$$p = \frac{p°}{x_{\text{methanol}}} = \frac{62 \text{ mm Hg}}{0.63} = 98 \text{ mm Hg}$$

From the data given about saturation pressure versus temperature, the flash point temperature of the aqueous solution is approximately 68°F.

The answer is (C).

24 Vapor-Liquid Processes

PRACTICE PROBLEMS

1. A distillation column with a total condenser and a partial reboiler is used to separate an acetone-acetic acid mixture. The feed, F, is saturated liquid containing 50 mol% acetone ($z_F = 0.5$). The desired distillate composition is 95 mol% acetone ($x_D = 0.95$), and the desired bottoms composition is 2 mol% acetone ($x_B = 0.02$). The external reflux ratio, L_0/D, is 2, and the reflux is a saturated liquid. The column pressure is 1 atm. The equilibrium data are shown.

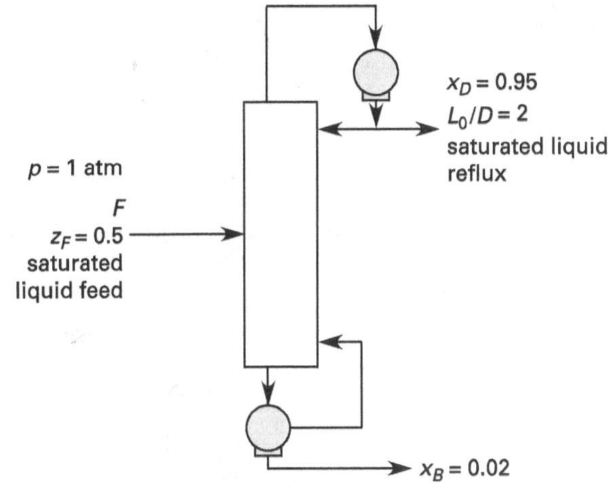

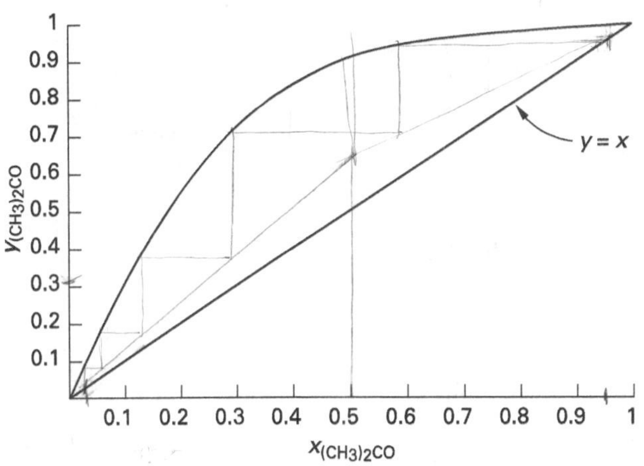

Assuming constant molar overflow, what is the optimum number of stages required to achieve the separation?

(A) 4 equilibrium stages and 1 partial reboiler

(B) 5 equilibrium stages and 1 partial reboiler

(C) 5 equilibrium stages

(D) 6 equilibrium stages and 1 partial reboiler

2. A distillation column with a total condenser and a partial reboiler is used to separate an ethanol-water mixture at 40°C. The feed, F, is saturated liquid containing 30 mol% ethanol ($z_F = 0.3$). The design distillate composition is 85 mol% ethanol ($x_D = 0.85$), and the design bottoms composition is 3 mol% ethanol ($x_B = 0.03$). The external reflux ratio, L_0/D, is 3, and the reflux is a saturated liquid. The column pressure is 1 atm. Refer to the illustration for additional data, including enthalpies, h, of the streams.

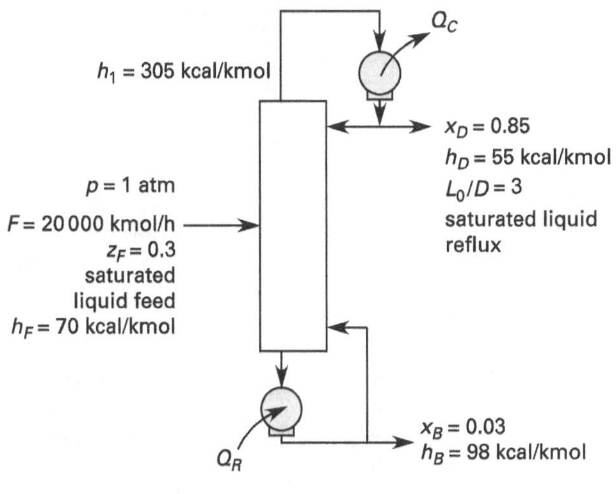

What is the heat duty of the reboiler?

(A) 0.28×10^6 kcal/h

(B) 5.47×10^6 kcal/h

(C) 6.87×10^6 kcal/h

(D) 8.27×10^6 kcal/h

3. A byproduct stream contains 20 mol% methanol in water. An equilibrium-stage distillation column is needed to remove the methanol from this stream and to recycle the water stream to another process. The column operates at 1 atm, using 50% more than the minimum internal reflux ratio, a total condenser, and a partial reboiler. The other process can tolerate a maximum of 2 mol% methanol in the water stream. The methanol, containing as much as 10 mol% water, can also be sold to a commercial recycler. The byproduct stream flows at 80 000 mol/h and is approximately 30% vapor at 1 atm. The equilibrium data are shown.

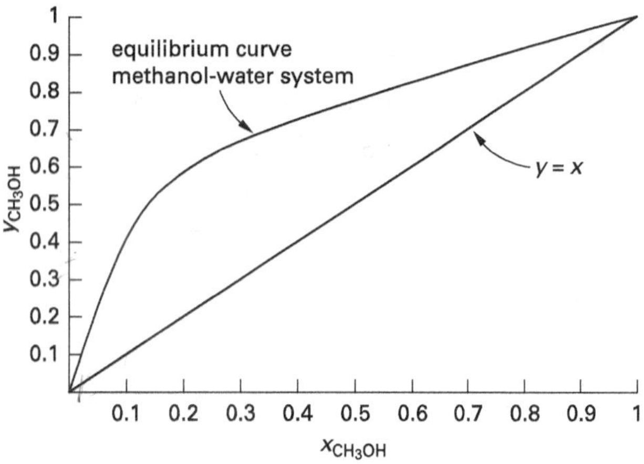

The optimum feed location is at stage

(A) 1

(B) 2

(C) 3

(D) 4

4. A new material, $ARBC^3$, leaves the synthesis process mixed with 55 mol% water. The temperature of the $ARBC^3$-water mixture is 80°C, and the molar flow rate is 5000 kmol/h. Distillation is being evaluated as a means of removing the water from this process stream. Empirical equilibrium data and several enthalpies of the feed composition are available. The enthalpy of the feed, H_F, at 80°C and 1 atm is 45 kcal/kmol. The enthalpy of the saturated vapor, H_V, at 1 atm pressure is 900 kcal/kmol. At 1 atm, the saturated liquid enthalpy, H_L, is 175 kcal/kmol. The target distillate composition, x_D, is 90 mol% $ARBC^3$, and the desired bottoms composition, x_B, is 3 mol% $ARBC^3$. As a starting point, a column with a total condenser and a partial reboiler is being evaluated. The external reflux ratio, L_0/D, is $4/3$, and the liquid reflux is saturated. Constant molar overflow is a valid assumption, as is operation at atmospheric pressure.

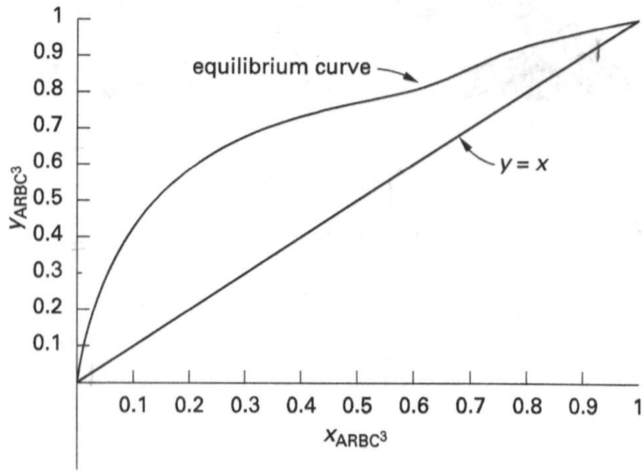

Counting from the top of the column, where is the optimum feed location?

(A) stage 3

(B) stage 4

(C) stage 6

(D) stage 8

5. A liquid stream is fed into a flash chamber. The output stream is composed of a vapor component and a liquid component.

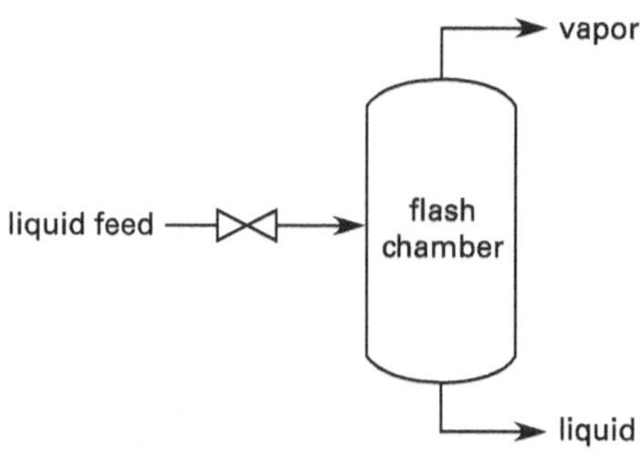

This process is

(A) distillation

(B) leaching

(C) filtration

(D) sublimination

6. A mixture of hydrocarbons is flash distilled at 400°C and 40 atm. The feed rate is 100 mol/min, and the vapor flow rate is 55 mol/min. The feed composition for component A in the hydrocarbon mixture is 0.30, and the vapor-liquid equilibrium constant is 1.25. Most nearly, the vapor composition of component A is

(A) 0.26

(B) 0.33

(C) 0.41

(D) 0.53

7. An absorption tower uses water to remove ammonia from air. The tower operates at 1 atm and 27°C. Its inlet water is pure ($x_0 = 0$), and its inlet gas has 0.045 mole fraction of ammonia ($y_{n+1} = 0.045$). The tower removes 95.75% of the ammonia from the air. The feed rate is 15 000 mol/h, and the liquid-to-gas flow-rate ratio, L/G, is 2.25.

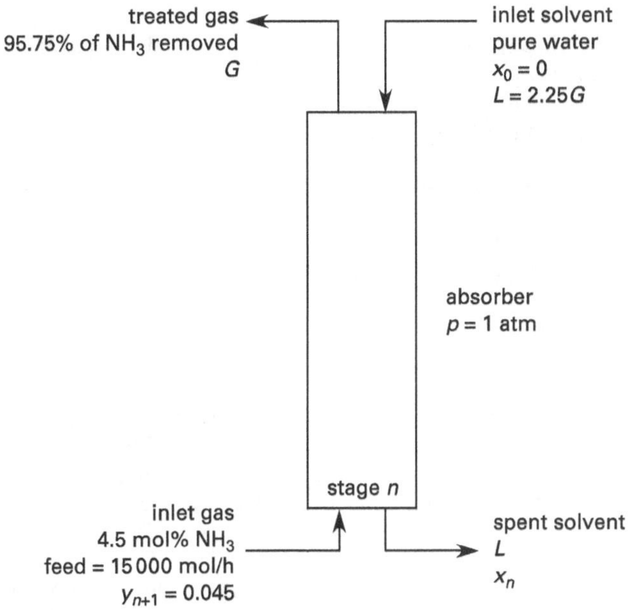

The equilibrium data for the ammonia-water system at 1 atm, in terms of mole ratios of ammonia, X and Y, are given by the following diagram.

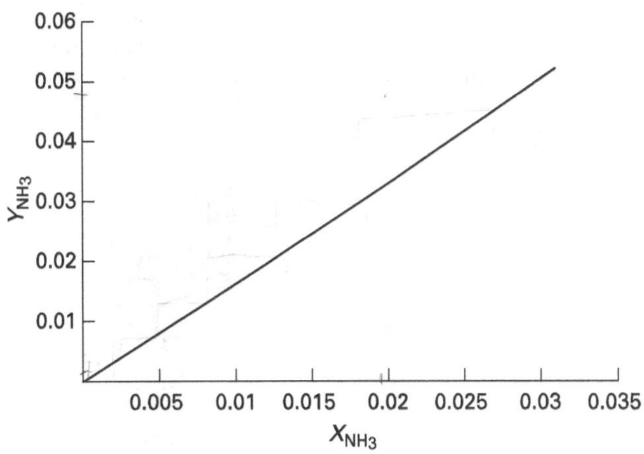

How many equilibrium states does this separation require?

(A) 6 equilibrium stages

(B) 7 equilibrium stages

(C) 8 equilibrium stages

(D) 9 equilibrium stages

8. An absorption tower uses water to remove ammonia from air. The tower operates at 1 atm and 27°C. The inlet water is pure ($x_0 = 0$). The inlet gas is 0.03 mole fraction ammonia ($y_{n+1} = 0.03$). The tower is required to remove 95% of the ammonia from the air. The inlet gas feed rate is 38 000 mol/h, and the liquid-to-gas flow rate ratio, L/G, is 3.25.

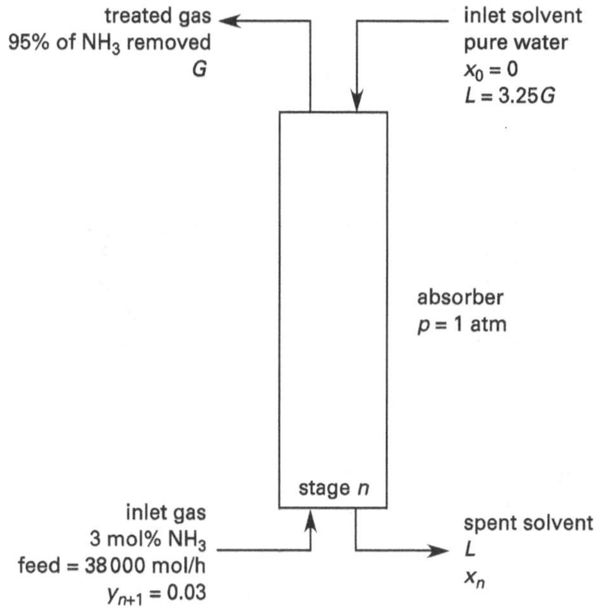

The equilibrium data for the ammonia-water system at 1 atm, in terms of the mole ratios of ammonia, X and Y, are given in the following diagram.

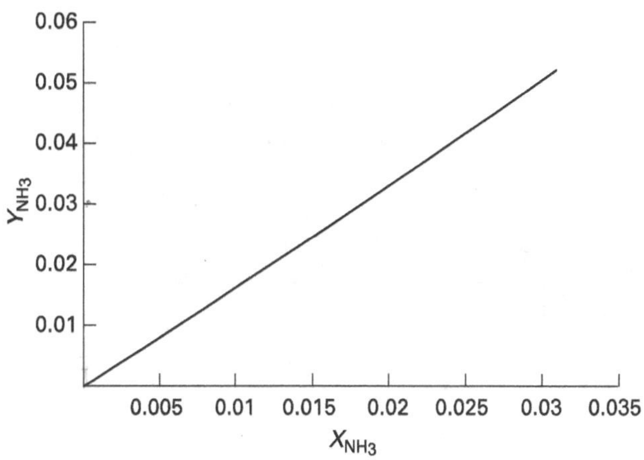

Approximately how many equilibrium stages does this separation require?

(A) 2 1/4

(B) 3 3/5

(C) 4

(D) 4 1/2

9. A mixture of water (component A) and ethanol (component B) containing 20 mol% of ethanol is maintained at a fixed temperature. At saturation, the vapor pressure of water is 0.552 psi, and that of ethanol is 2.848 psi. Assume the system is ideal and that it follows Raoult's law. Most nearly, the total pressure of the system is

(A) 0.47 psi

(B) 1.0 psi

(C) 2.3 psi

(D) 3.4 psi

10. A gas stripper uses air to remove component A from a water stream. The tower operates at 1 atm and 30°C. The water feed rate, L_F, is 1000 mol/h. The inlet water has 9.1 mol% of component A. The stripper must reduce the mole fraction of component A in the water to 0.008. The mole fraction of component A in the air may be increased to 0.074. The constant molar overflow assumption is valid.

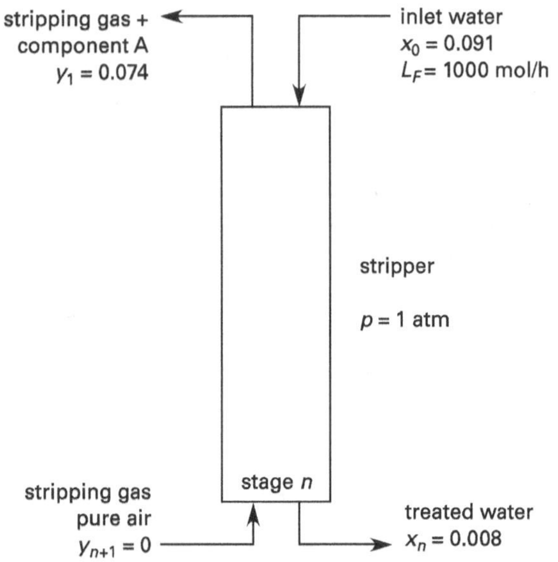

The equilibrium data in terms of mole ratios of component A, X and Y, are given in the following diagram.

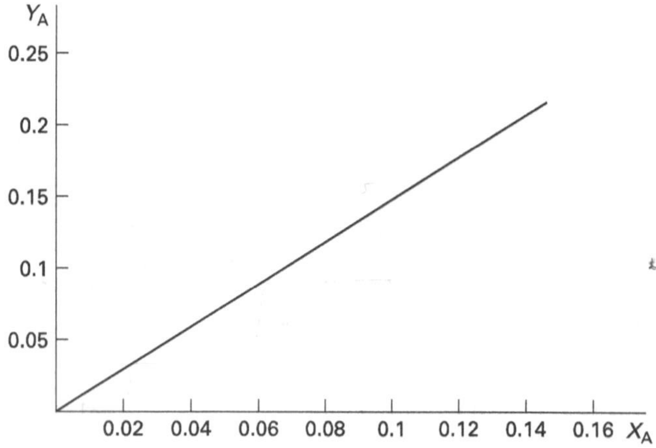

How many equilibrium stages does this separation require?

(A) 4 equilibrium stages

(B) 5 equilibrium stages

(C) 6 equilibrium stages

(D) 7 equilibrium stages

11. A stripping column is used to extract carbon dioxide from an aqueous solution. The mole fraction of carbon dioxide in the solution is 8.2×10^{-6}. The solution enters the column at 8962 lbmol/hr. The stripping is done with pure nitrogen that enters the bottom of the column at 6.2 lbmol/hr. The lean carbon dioxide stream leaves the stripping column from the bottom; the mole fraction of carbon dioxide in this stream is 1.9×10^{-6}. The stripping column is kept at 760 mm Hg and 250°F. At these conditions, the straight line for the equilibrium carbon dioxide–water system follows Henry's law with a slope of 3622 and an intercept of 0. Assume steady state. With x_1 the mole fraction of carbon dioxide in the stream entering the column, x_2 the mole fraction of carbon dioxide in the stream leaving the column, and x_2^* the mole fraction of carbon dioxide (in the aqueous phase) in equilibrium with the mole fraction of carbon dioxide in the gas phase, the number of theoretical stages, N, is given by the Kremser equation.

$$N = \frac{\ln\left(\left(1 - \frac{\dot{L}}{m\dot{G}}\right)\left(\frac{x_1 - x_2^*}{x_2 - x_2^*}\right) + \frac{\dot{L}}{m\dot{G}}\right)}{\ln\left(\frac{m\dot{G}}{\dot{L}}\right)}$$

Using the Kremser equation, the required theoretical number of stages is

(A) 0.084

(B) 1.0

(C) 1.2

(D) 2.0

SOLUTIONS

1. **Feed Line**

Since the feed is a saturated liquid, the feed quality, q, is 1. Therefore, the feed line is vertical at the feed composition, $z_F = x_F = 0.5$.

Top (Rectifying Section) Operating Line

From Eq. 24.35 (also *NCEES Handbook:* Column Material Balance), the operating equation for the rectifying section is

$$y = \left(\frac{L}{V}\right)x_n + \frac{x_D D}{V} = \left(\frac{L}{V}\right)x_n + \left(1 - \frac{L}{V}\right)x_D$$

$$\frac{L}{V} \quad \text{[slope]}$$

$$\left(1 - \frac{L}{V}\right)x_D \quad [y\text{-intercept}]$$

Using Eq. 24.36 (also *NCEES Handbook:* Column Material Balance) to calculate the internal reflux ratio, L/V, from the external reflux ratio, R,

$$\frac{L}{V} = \frac{R}{1+R} = \frac{2}{1+2} = 2/3 \quad \text{[slope]}$$

The y-intercept is

$$\left(1 - \frac{L}{V}\right)x_D = \left(1 - \frac{2}{3}\right)(0.95) = 0.317$$

The top operating line can be plotted on the equilibrium diagram using the slope and y-intercept.

Bottom (Stripping Section) Operating Line

The operating equation for the stripping section is found from Eq. 24.37 (also *NCEES Handbook:* Column Material Balance).

$$y = \left(\frac{L'}{V'}\right)x_n + \frac{x_B B}{V'} = \left(\frac{L'}{V'}\right)x_n + \left(1 - \frac{L'}{V'}\right)x_B$$

One point on the line is the intersection of the bottom operating line with the top operating line at the feed line. One additional point is needed to define the bottom operating line. Rather than determine the slope of the bottom operating line, $\overline{L}/\overline{V}$, the second point can be found by determining where the bottom operating line intersects the $y = x$ line. Substitute $y = x$ into the operating line equation.

$$y = \left(\frac{L'}{V'}\right)x + \left(1 - \frac{L'}{V'}\right)x_B$$

$$x = \left(\frac{L'}{V'}\right)x + \left(1 - \frac{L'}{V'}\right)x_B$$

$$x - \left(\frac{L'}{V'}\right)x = \left(1 - \frac{L'}{V'}\right)x_B$$

$$\left(1 - \frac{L'}{V'}\right)x = \left(1 - \frac{L'}{V'}\right)x_B$$

$$x = x_B = 0.02$$

Refer to the accompanying illustration. The separation requires 4 equilibrium stages and 1 partial reboiler.

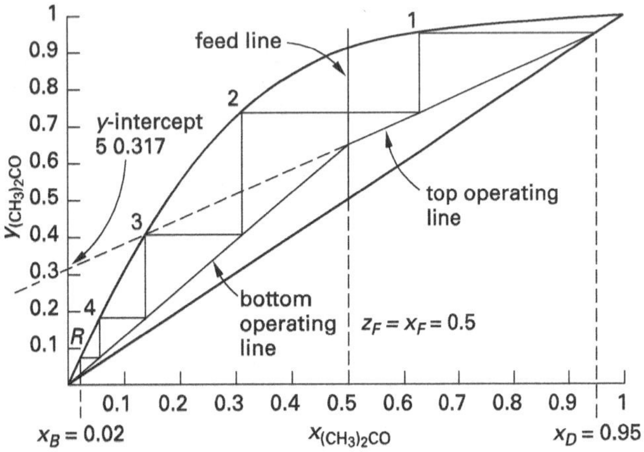

The answer is (A).

2. The distillate, D, and the bottoms, B, flow rates can be determined from the overall and ethanol mass balances.

$$F = B + D$$

$$B = F - D$$

$$Fz_F = Bx_B + Dx_D$$

$$Fz_F = (F - D)x_B + Dx_D$$

Rearrange the ethanol mass-balance equation.

$$D = F\left(\frac{z_F - x_B}{x_D - x_B}\right) = \left(20\,000\ \frac{\text{kmol}}{\text{h}}\right)\left(\frac{0.3 - 0.03}{0.85 - 0.03}\right)$$

$$= 6585\ \text{kmol/h}$$

From the overall mass balance, the bottoms flow rate is

$$B = F - D = 20\,000\ \frac{\text{kmol}}{\text{h}} - 6585\ \frac{\text{kmol}}{\text{h}}$$

$$= 13\,415\ \text{kmol/h}$$

The energy balance is used to determine the condenser heat duty. For the condenser, with Eq. 24.17,

$$Q_C = \left(1 + \frac{L_0}{D}\right)\left(\frac{z_F - x_B}{x_D - x_B}\right)F(h_1 - h_D)$$

$$= (1 + 3)\left(\frac{0.3 - 0.03}{0.85 - 0.03}\right)\left(20\,000\ \frac{\text{kmol}}{\text{h}}\right) \times \left(305\ \frac{\text{kcal}}{\text{kmol}} - 55\ \frac{\text{kcal}}{\text{kmol}}\right)$$

$$= 6.59 \times 10^6\ \text{kcal/h}$$

The condenser heat duty is positive, indicating that heat is removed from the overhead stream.

For the reboiler, from Eq. 24.13, the heat duty is found from the energy balance.

$$Q_R = Dh_D + Bh_B - Fh_F + Q_C$$

$$= \left(6585\ \frac{\text{kmol}}{\text{h}}\right)\left(55\ \frac{\text{kcal}}{\text{kmol}}\right) + \left(13\,415\ \frac{\text{kmol}}{\text{h}}\right)\left(98\ \frac{\text{kcal}}{\text{kmol}}\right) - \left(20\,000\ \frac{\text{kmol}}{\text{h}}\right)\left(70\ \frac{\text{kcal}}{\text{kmol}}\right) + 6.59 \times 10^6\ \frac{\text{kcal}}{\text{h}}$$

$$= 6.87 \times 10^6\ \text{kcal/h}$$

The reboiler duty is positive, indicating that heat is added by the reboiler.

The answer is (C).

3. The bottoms constitute the "water stream." With a maximum of 2 mol% methanol, $x_B = 0.02$. The overhead stream (the distillate) will contain the more volatile, concentrated methanol. With a maximum of 10 mol%

water allowed in this stream, the target mole fraction of methanol is $x_D = 0.90$.

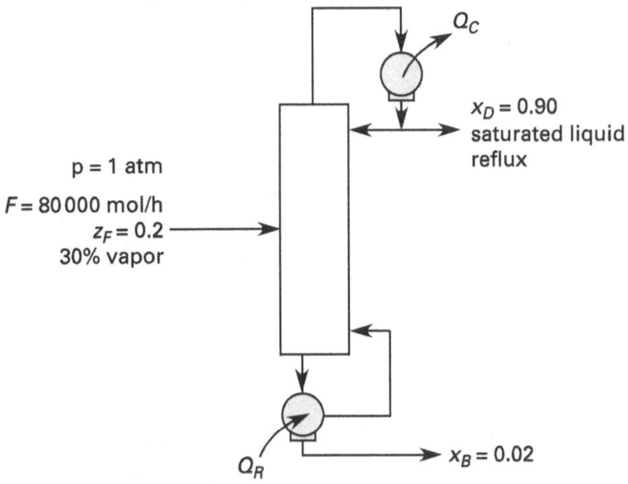

Feed Line

One point on the feed line is the intersection of the liquid feed composition with the $y = x$ line, or $y = x = z_F = 0.2$. Since the feed is 30% vapor, the feed is 70% liquid. The feed quality, q, is 0.7. From Eq. 24.44 (also *NCEES Handbook:* Graphical Solution for Binary Distillation (McCabe-Thiele Diagram)), the feed equation is

$$y = \left(\frac{q}{q-1}\right)x + \left(\frac{1}{1-q}\right)z_F$$

$$\frac{q}{q-1} \quad \text{[slope]}$$

$$\left(\frac{1}{1-q}\right)z_F = \left(\frac{1}{1-0.7}\right)(0.2) = 0.67 \quad [y\text{-intercept}]$$

Using the points (0.2, 0.2) and (0, 0.67), the feed line can be plotted on the equilibrium diagram.

Top (Rectifying Section) Operating Line

The intersection of the top operating line with the feed line at the equilibrium data creates a pinch point, which corresponds to the minimum internal reflux. The top operating line for the rectifying section at the minimum reflux ratio can be plotted on the equilibrium diagram by connecting the distillate composition ($y = x = x_D = 0.9$) to the intersection of the feed line at the equilibrium data curve (at (0.1, 0.42)). The slope of this line is 0.6.

$$\left(\frac{L}{V}\right)_{\text{min}} = \frac{\Delta y}{\Delta x} = \frac{0.9 - 0.42}{0.9 - 0.1} = 0.6$$

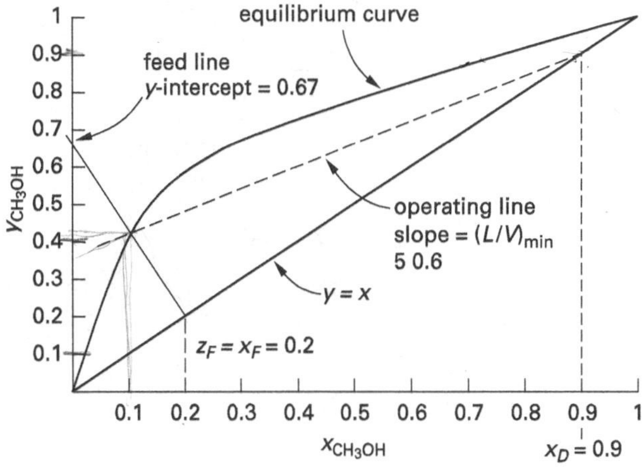

The problem statement specifies an internal reflux ratio 50% greater than the minimum internal reflux ratio.

$$\frac{L}{V} = 1.5\left(\frac{L}{V}\right)_{\text{min}} = (1.5)(0.6) = 0.9$$

With one point ($y = x = x_D = 0.9$) and the operating line slope ($L/V = 0.9$), the top operating line can be plotted on the equilibrium diagram. To check the operating line, calculate the y-intercept. As in *NCEES Handbook:* Column Material Balance, the top operating equation is

$$y = \left(\frac{L}{V}\right)x_n + \frac{x_D D}{V} = \left(\frac{L}{V}\right)x_n + \left(1 - \frac{L}{V}\right)x_D$$

$$\frac{L}{V} = 0.9 \quad \text{[slope]}$$

$$\left(1 - \frac{L}{V}\right)x_D = (1 - 0.9)(0.9) = 0.09 \quad [y\text{-intercept}]$$

Bottom (Stripping Section) Operating Line

The bottom operating line intersects the top operating line at the feed line. The bottom operating line also intersects the $y = x$ line at $x_B = 0.02$. Using these two points, the bottom operating line can be plotted on the equilibrium diagram.

Starting at the distillate and stepping off stages shows that this separation requires 4 equilibrium stages and a partial reboiler to achieve a bottoms composition of just under 2 mol%.

From the illustration, the optimum feed location is at stage 3, since the stages were stepped off to minimize the total number of stages.

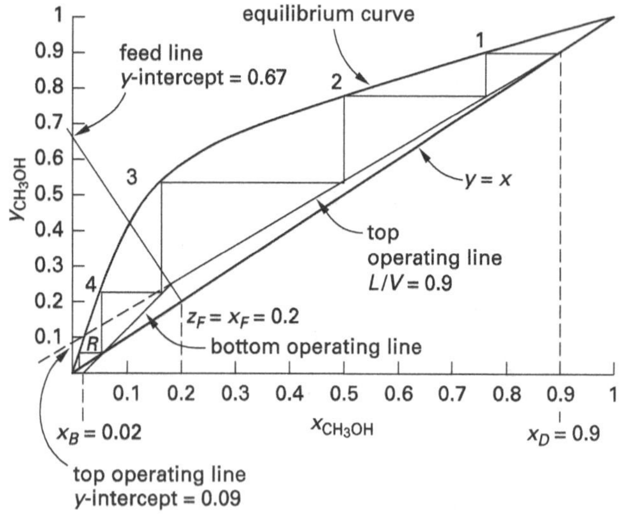

The answer is (C).

4. Calculate the feed quality, q, from *NCEES Handbook:* Feed Conditions.

$$
\begin{aligned}
q &= 1+\frac{c_{pL}(T_b-T_F)}{\Delta h_{\text{vap}}} = \frac{h_V-h_F}{h_V-h_L} \\
&= \frac{900\ \frac{\text{kcal}}{\text{kmol}} - 45\ \frac{\text{kcal}}{\text{kmol}}}{900\ \frac{\text{kcal}}{\text{kmol}} - 175\ \frac{\text{kcal}}{\text{kmol}}} \\
&= 1.18
\end{aligned}
$$

The feed is subcooled, which is evident from $H_F < H_L$.

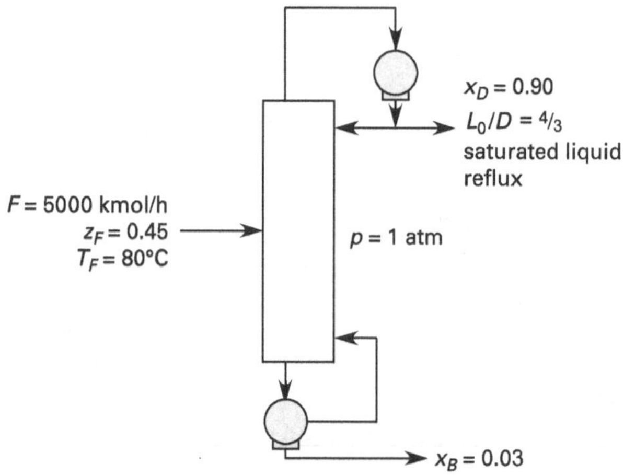

Feed Line

The slope of feed line can be calculated using Eq. 24.44.

$$\frac{q}{q-1} = \frac{1.18}{1.18-1} = 6.56$$

Since the process stream is 55 mol% water, the mole fraction of water in the feed, z_{H_2O}, is 0.55. The mole fraction of ARBC[3] in the feed, z_F, is

$$
\begin{aligned}
1 &= z_{H_2O} + z_F \\
z_F &= 1 - z_{H_2O} = 1 - 0.55 = 0.45
\end{aligned}
$$

The feed line intersects the $y = x$ line at the feed composition, z_F, providing (0.45, 0.45) as one point on the feed line. With this point and the slope, the feed line can be plotted on the equilibrium diagram. Alternatively, a second feed line point can be obtained by calculating the y-intercept of the feed line from the feed equation, Eq. 24.44.

$$\left(\frac{1}{1-q}\right)z_F = \left(\frac{1}{1-1.18}\right)(0.45) = -2.5$$

These two points, $x = y = z_F$ (0.45, 0.45) and the y-intercept (0, −2.5), can also be used to plot the feed line.

Top (Rectifying Section) Operating Line

The slope of the top operating line is calculated from Eq. 24.36, using the given external reflux ratio.

$$\frac{L}{V} = \frac{\frac{L_0}{D}}{1+\frac{L_0}{D}} = \frac{\frac{4}{3}}{1+\frac{4}{3}} = 0.57$$

The y-intercept can be obtained from Eq. 24.35.

$$\frac{x_D}{R+1} = \left(1-\frac{L}{V}\right)x_D = (1-0.57)(0.9) = 0.39$$

Alternatively, the top operating line can be plotted using two points, the intersection of the operating line with the $y = x$ line at x_D (0.9, 0.9) and the y-intercept (0, 0.39).

Bottom (Stripping Section) Operating Line

Knowing that the bottom operating line intersects the top operating line at the feed line provides one point on the line. A second point needed to plot the line can be identified as the intersection of the bottom operating line with the $y = x$ line, at x_B (0.03, 0.03). The feed line and the operating lines are plotted on the equilibrium diagram as shown.

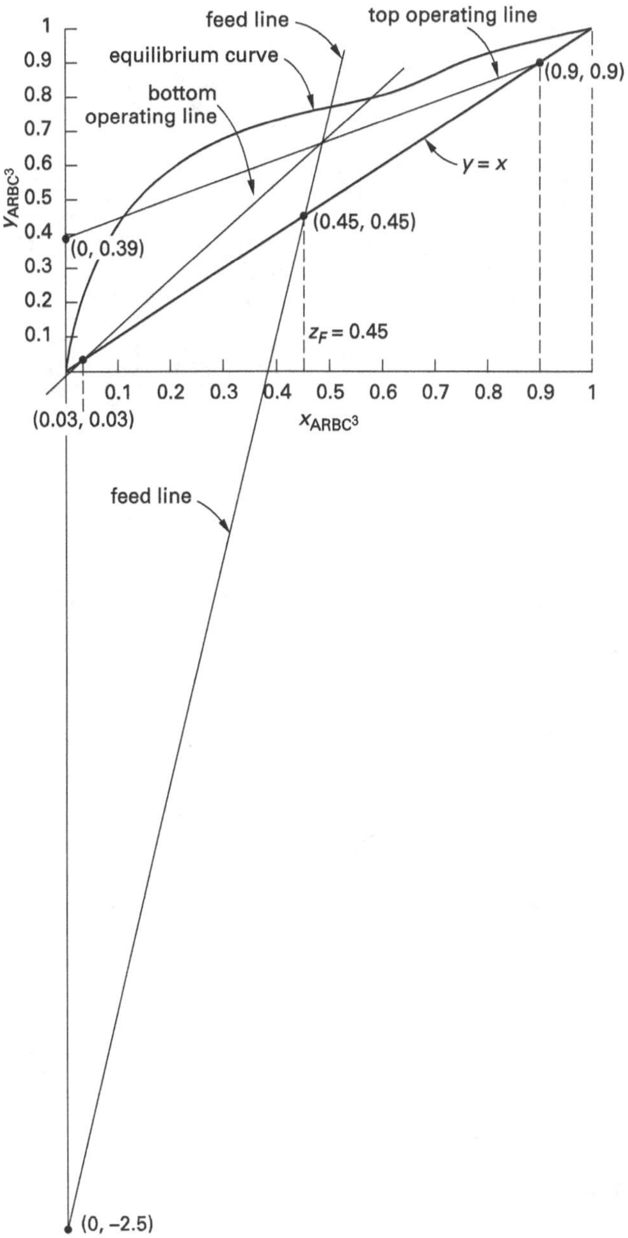

Stepping off stages shows that the total number of equilibrium stages required is approximately 4 plus a reboiler.

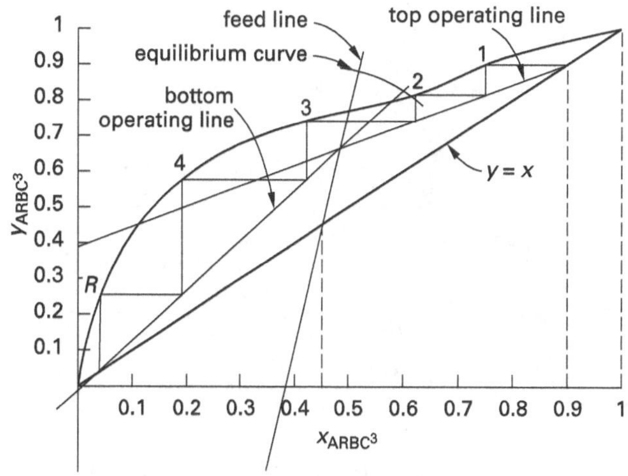

The optimum feed location is where the operating lines intersect the feed line, at stage 3.

The answer is (A).

5. Distillation occurs when a liquid converts to a vapor by vaporization. The feed stream undergoes distillation when it converts to a vapor within the flash chamber. Option A is correct.

Leaching occurs when a liquid washes away part of a solid. Filtration occurs when a fluid component is partially or completely blocked by a filter. Sublimination occurs when a solid converts to a vapor.

The answer is (A).

6. Take an overall material balance, and solve for the liquid flow rate.

$$\begin{aligned} F &= L + V \\ L &= F - V \\ &= 100\ \frac{\text{mol}}{\text{min}} - 55\ \frac{\text{mol}}{\text{min}} \\ &= 45\ \text{mol/min} \end{aligned}$$

The vapor-liquid equilibrium constant is

$$k = \frac{y}{x} = 1.25$$

Therefore, $x = y/1.25$. Take a material balance for component A, and solve for the vapor composition of component A.

$$\begin{aligned}0.30F &= xL + yV \\ &= \left(\frac{y}{1.25}\right)L + yV\end{aligned}$$

$$y = \frac{0.30F}{\dfrac{L}{1.25} + V} = \frac{(0.30)\left(100\ \dfrac{\text{mol}}{\text{min}}\right)}{\dfrac{45\ \dfrac{\text{mol}}{\text{min}}}{1.25} + 55\ \dfrac{\text{mol}}{\text{min}}}$$
$$= 0.3297 \quad (0.33)$$

The answer is (B).

7. The inlet gas and exit (treated) gas compositions in terms of mole ratios, Y_i, and the air flow rate, G, are needed to plot the operating line. The inlet gas stream has 4.5 mol% ammonia. From Eq. 24.53 (also *NCEES Handbook:* Absorption and Stripping), the mole ratio of the inlet gas, Y_{n+1}, is

$$Y_{n+1} = \frac{y_{NH_3}}{1 - y_{NH_3}} = \frac{0.045}{1 - 0.045} = 0.047$$

The amount of ammonia in the feed is

$$\begin{aligned}G_{NH_3,\text{in}} &= y_{n+1}G_F = (0.045)\left(15\,000\ \frac{\text{mol}}{\text{h}}\right) \\ &= 675\ \text{mol/h}\end{aligned}$$

The absorber removes 95.75% of the ammonia from the inlet air stream, leaving 4.25% of the inlet ammonia remaining in the gas.

$$\begin{aligned}G_{NH_3,\text{out}} &= (4.25\%)G_{NH_3,\text{in}} = (0.0425)\left(675\ \frac{\text{mol}}{\text{h}}\right) \\ &= 28.7\ \text{mol/h}\end{aligned}$$

The airflow rate is

$$\begin{aligned}G_{\text{air}} &= G_F - G_{NH_3,\text{in}} = 15\,000\ \frac{\text{mol}}{\text{h}} - 675\ \frac{\text{mol}}{\text{h}} \\ &= 14\,325\ \text{mol/h}\end{aligned}$$

From Eq. 24.53 (also *NCEES Handbook:* Absorption and Stripping), the mole ratio of ammonia in the treated gas stream is

$$\begin{aligned}Y_1 &= \frac{G_{NH_3,\text{out}}}{G} = \frac{28.7\ \dfrac{\text{mol NH}_3}{\text{h}}}{14\,325\ \dfrac{\text{mol air}}{\text{h}}} \\ &= 0.002\ \text{mol NH}_3/\text{mol air}\end{aligned}$$

The operating equation, from Eq. 24.57 (also *NCEES Handbook:* Absorption and Stripping), is

$$Y_{j+1} = \left(\frac{L}{G}\right)X_j + \left(Y_1 - \left(\frac{L}{G}\right)X_0\right)$$

Since $L/G = 2.25$, the y-intercept is

$$Y_1 - \left(\frac{L}{G}\right)X_0 = 0.002 - (2.25)(0) = 0.002$$

Plot the operating line on the equilibrium diagram and step off stages. This separation requires approximately 7 equilibrium stages.

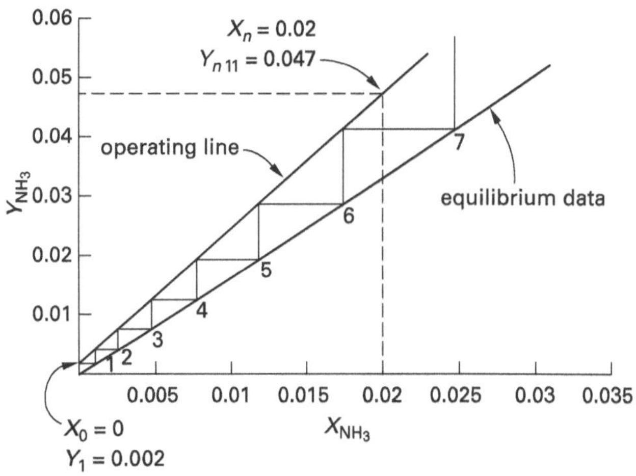

The answer is (B).

8. The inlet gas stream has a mole fraction of ammonia of 0.03. The molar flow rate of ammonia in the inlet (feed) gas is

$$\begin{aligned}G_{NH_3,\text{in}} &= y_{n+1}G_F = (0.03)\left(38\,000\ \frac{\text{mol}}{\text{h}}\right) \\ &= 1140\ \text{mol/h}\end{aligned}$$

The absorber removes 95% of the ammonia from the inlet air stream. The molar flow rate of ammonia in the exit water is

$$\begin{aligned} L_{NH_3,\text{out}} &= (95\%)G_{NH_3,\text{in}} = (0.95)\left(1140\ \frac{\text{mol}}{\text{h}}\right) \\ &= 1083\ \text{mol/h} \end{aligned}$$

The air (carrier gas) flow rate, G, is

$$\begin{aligned} G_{\text{air}} &= G_F - G_{NH_3,\text{in}} \\ &= 38\,000\ \frac{\text{mol gas}}{\text{h}} - 1140\ \frac{\text{mol NH}_3}{\text{h}} \\ &= 36\,860\ \text{mol/h} \end{aligned}$$

Given the ratio $L/G = 3.25$, the solvent flow rate, L, is

$$\begin{aligned} L_{\text{water}} &= 3.25G = (3.25)\left(36\,860\ \frac{\text{mol air}}{\text{h}}\right) \\ &= 119\,800\ \text{mol/h} \end{aligned}$$

The mole ratio of ammonia in the exiting water stream, using Eq. 24.52, is

$$\begin{aligned} X_n &= \frac{L_{NH_3,\text{out}}}{L} = \frac{1083\ \dfrac{\text{mol NH}_3}{\text{h}}}{119\,800\ \dfrac{\text{mol water}}{\text{h}}} \\ &= 0.009\ \text{mol NH}_3/\text{mol air} \end{aligned}$$

The operating line can be plotted by identifying the two points (X_0, Y_1) and (X_n, Y_{n+1}). From the problem statement, the pure water feed means that $X_0 = 0$. Also from the problem statement, the inlet gas stream is 0.03 mole fraction ammonia or

$$y_{n+1} = 0.03$$

The corresponding mole ratio of ammonia in the inlet gas stream, Y_{n+1}, using Eq. 24.53 (also *NCEES Handbook:* Absorption and Stripping), is

$$Y_{n+1} = \frac{y_{NH_3}}{1 - y_{NH_3}} = \frac{0.03}{1 - 0.03} = 0.031$$

One point on the operating line, (X_n, Y_{n+1}), is $(0, 0.031)$.

As previously calculated, the molar flow rate of ammonia in the inlet gas is 1140 mol/h. The absorber removes 95% of the ammonia from the inlet air stream, leaving $100\% - 95\% = 5\%$ of the ammonia in the treated gas stream.

$$\begin{aligned} G_{NH_3,\text{out}} &= (5\%)G_{NH_3,\text{in}} = (0.05)\left(1140\ \frac{\text{mol}}{\text{h}}\right) \\ &= 57\ \text{mol/h} \end{aligned}$$

As previously calculated, the air (carrier gas) flow rate, G, is 36 660 mol/h. The mole ratio of ammonia in the treated gas stream, using Eq. 24.53 (also *NCEES Handbook:* Absorption and Stripping), is

$$\begin{aligned} Y_1 &= \frac{G_{NH_3,\text{out}}}{G} \\ &= \frac{57\ \dfrac{\text{mol NH}_3}{\text{h}}}{36\,860\ \dfrac{\text{mol air}}{\text{h}}} \\ &= 0.0015 \end{aligned}$$

As in *NCEES Handbook:* Absorption and Stripping, the operating equation is

$$\begin{aligned} Y_{\text{out}} &= Y_{\text{in}} - \left(\frac{L_S}{G_S}\right)(X_{\text{out}} - X_{\text{in}}) \\ &= \left(\frac{L_S}{G_S}\right)X_{\text{out}} + \left(Y_{\text{in}} - \left(\frac{L_S}{G_S}\right)X_{\text{in}}\right) \end{aligned}$$

The problem statement specifies that the liquid-to-gas ratio, L/G, is 3.25, which is the slope of the operating line. The y-intercept of this line is

$$\begin{aligned} Y_{\text{in}} - \left(\frac{L_S}{G_S}\right)X_{\text{in}} &= 0.0015 - (3.25)(0) \\ &= 0.0015 \end{aligned}$$

This separation requires approximately 3⅗ equilibrium stages.

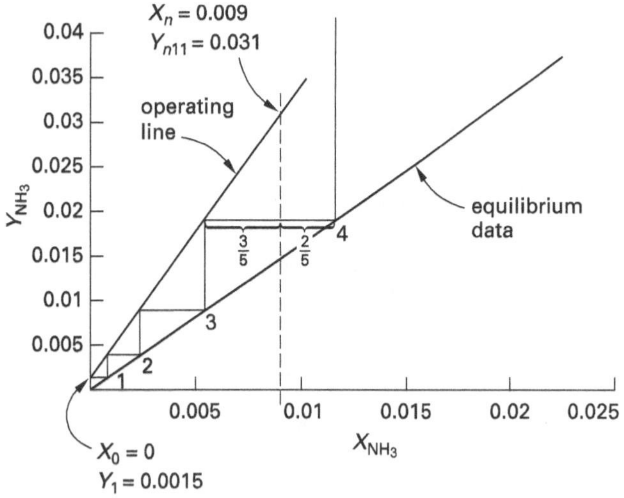

The answer is (B).

9. Use Raoult's law to calculate the pressure of each component.

$$\begin{aligned} P_A &= x_A P_A^\circ \\ &= (0.8)\left(0.552\ \frac{\text{lbf}}{\text{in}^2}\right) \\ &= 0.441\ \text{lbf/in}^2 \\ P_B &= x_B P_B^\circ \\ &= (0.2)\left(2.848\ \frac{\text{lbf}}{\text{in}^2}\right) \\ &= 0.570\ \text{lbf/in}^2 \end{aligned}$$

Calculate the total pressure of the system using Dalton's law of partial pressure.

$$\begin{aligned} P &= P_A + P_B \\ &= 0.441\ \frac{\text{lbf}}{\text{in}^2} + 0.570\ \frac{\text{lbf}}{\text{in}^2} \\ &= 1.011\ \text{lbf/in}^2 \quad (1.0\ \text{psi}) \end{aligned}$$

The answer is (B).

10. The inlet water stream has 9.1 mol% A. The mole ratio of component A in the inlet solvent stream is from *NCEES Handbook:* Absorption and Stripping.

$$X_0 = \frac{x_0}{1 - x_0} = \frac{0.091}{1 - 0.091} = 0.10$$

Given the desired mole fraction of component A of 0.008 in the treated water, the mole ratio of component A needed in this stream is from *NCEES Handbook:* Absorption and Stripping.

$$X_n = \frac{x_n}{1 - x_n} = \frac{0.008}{1 - 0.008} = 0.008$$

The inlet gas stream is pure air. The mole ratio of solute in the inlet air is

$$Y_{n+1} = 0$$

With a target mole fraction of A in the exit gas stream of 0.074, the corresponding mole ratio of component A in this stream is from *NCEES Handbook:* Absorption and Stripping.

$$Y_1 = \frac{y_1}{1 - y_1} = \frac{0.074}{1 - 0.074} = 0.08$$

In contrast to absorption processes, the operating line lies below the equilibrium data. The operating line can be plotted using the conditions at the entrance and the exit of the stripping tower; one point is (X_n, Y_{n+1}), and another point is (X_0, Y_1). Plotting the operating line with the equilibrium data and stepping off stages shows that this separation requires approximately 4 (3.5) stages.

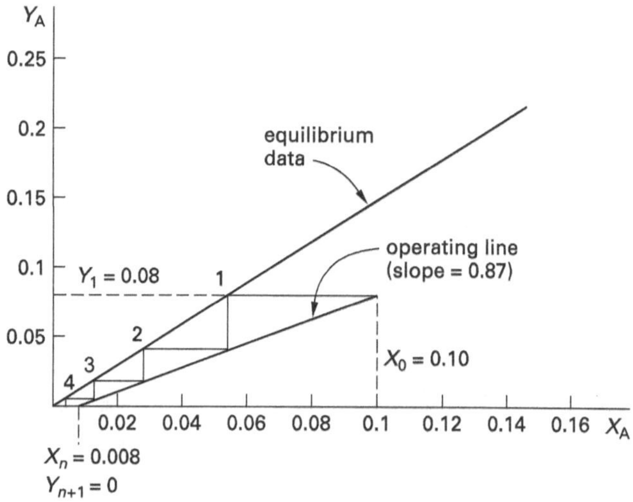

The answer is (A).

11. Draw the stripping column.

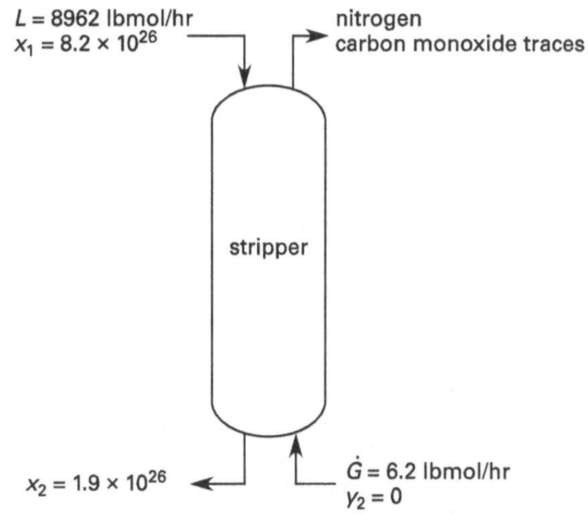

The intercept, b, of the straight-line equation for the Henry's law is given as 0. The equation of this straight line is

$$y = mx + b$$

Applying the preceding equation at equilibrium gives

$$y_1 = mx_2^* + b$$

Solving for the equilibrium concentration of carbon dioxide in the aqueous phase gives

$$x_2^* = \frac{y_1 - b}{m} = \frac{0 - 0}{m} = 0$$

Using the Kremser equation, the required theoretical number of stages is

$$N = \frac{\ln\left(\left(1 - \frac{\dot{L}}{m\dot{G}}\right)\left(\frac{x_1 - x_2^*}{x_2 - x_2^*}\right) + \frac{\dot{L}}{m\dot{G}}\right)}{\ln\left(\frac{m\dot{G}}{\dot{L}}\right)}$$

$$= \frac{\ln\left(\begin{array}{l}\left(1 - \frac{8962\ \frac{\text{lbmol}}{\text{hr}}}{(3622)\left(6.2\ \frac{\text{lbmol}}{\text{hr}}\right)}\right) \\ \times \left(\frac{8.2 \times 10^{-6} - 0}{1.9 \times 10^{-6} - 0}\right) \\ + \frac{8962\ \frac{\text{lbmol}}{\text{hr}}}{(3622)\left(6.2\ \frac{\text{lbmol}}{\text{hr}}\right)}\end{array}\right)}{\ln \frac{(3622)\left(6.2\ \frac{\text{lbmol}}{\text{hr}}\right)}{8962\ \frac{\text{lbmol}}{\text{hr}}}}$$

$$= 1.19 \quad (1.2)$$

The answer is (C).

25 Liquid-Liquid Extraction

PRACTICE PROBLEMS

1. 100 g of 45 wt% acetic acid in water and 180 g of isopropyl ether are combined in a beaker.

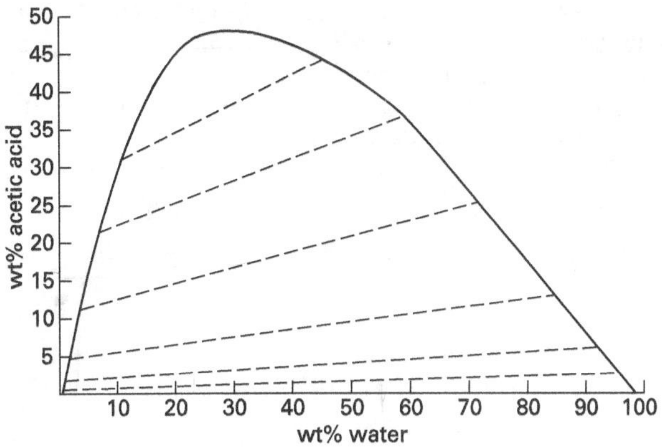

Given the phase diagram shown, the equilibrium weight percentage of isopropyl ether in each phase is most nearly

(A) 0% and 80%

(B) 0% and 100%

(C) 4% and 83%

(D) 64% and 64%

2. A byproduct stream flowing at the rate of 2323 kg/h is water containing 1 wt% nicotine. The nicotine content needs to be reduced to 0.1 wt% using pure kerosene. The equilibrium data is provided.

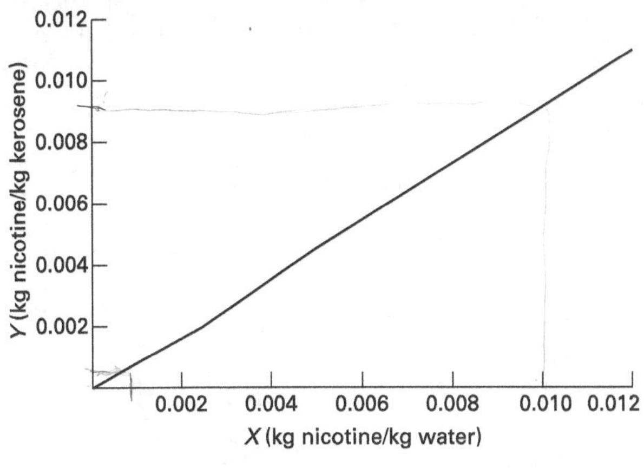

What is the minimum solvent flow rate needed to achieve the separation?

(A) 23 kg/h

(B) 30 kg/h

(C) 2300 kg/h

(D) 3000 kg/h

3. Equal amounts of a solution of 50 wt% 1,1,2-trichloroethane in water and pure acetone are mixed.

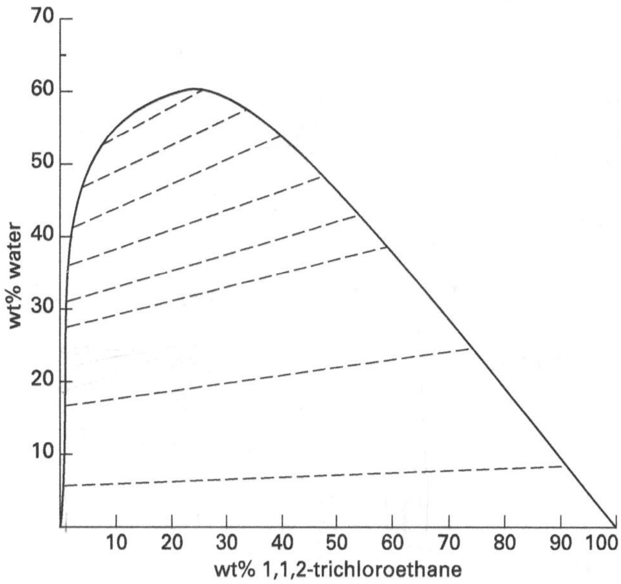

Given the phase diagram shown, the equilibrium weight percentage of acetone in each phase is most nearly

(A) 0% and 100%

(B) 1% and 77%

(C) 3% and 74%

(D) 50% and 50%

SOLUTIONS

1. This is a mixing problem.

The composition of the diluent, D (water), and of the solute, A (acetic acid), for the mixing point can be calculated using Eq. 25.8 and Eq. 25.9. See also *NCEES Handbook:* Measures of Composition.

$$x_{D,M} = \frac{F_1 x_{D,F_1} + F_2 x_{D,F_2}}{F_1 + F_2} = \frac{(180\text{ g})(0) + (100\text{ g})(0.55)}{180\text{ g} + 100\text{ g}} = 0.196$$

$$x_{A,M} = \frac{F_1 x_{A,F_1} + F_2 x_{A,F_2}}{F_1 + F_2} = \frac{(180\text{ g})(0) + (100\text{ g})(0.45)}{180\text{ g} + 100\text{ g}} = 0.161$$

In terms of wt%, the mixing point is at (19.6, 16.1).

By drawing the tie line that passes through the mixing point, the equilibrium compositions can be read. For each phase,

Phase I at (69.1,27.1): 69.1 wt% water,
27.1 wt% acetic acid,
3.8 wt% (4 wt%) isopropyl ether

Phase II at (4.3,12.7): 4.3 wt% water,
12.7 wt% acetic acid,
83 wt% isopropyl ether

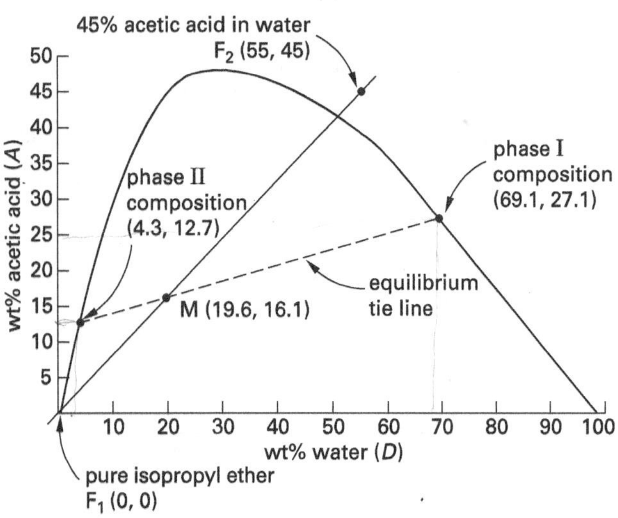

The answer is (C).

2. Since both streams are dilute, x is approximately the same as X, and y is approximately the same as Y.

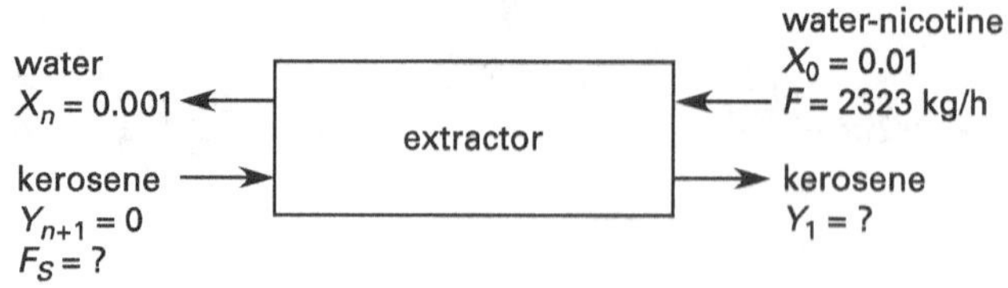

The minimum solvent flow rate can be determined graphically or from the operating equation, Eq. 25.4. The slope of the operating line is the ratio of the diluent feed to the solvent feed, F_D/F_S. The minimum solvent flow rate will be at the maximum slope of this line. One point on the operating line is at (X_n, Y_{n+1}) or (0.001, 0), which is plotted on the equilibrium diagram. The operating line is drawn from this point to intersect the equilibrium curve at $X_0 = 0.01$. From the illustration, $X_0 = 0.01$ at $Y_{1,\text{max}} = 0.0092$. The slope of this line is

$$\frac{F_D}{F_{S,\text{min}}} = \frac{\Delta Y}{\Delta X} = \frac{Y_1 - Y_{n+1}}{X_0 - X_n} = \frac{0.0092 - 0}{0.01 - 0.001} = 1$$

$$F_D = F_{S,\text{min}}$$

Since the by-product stream feed is 99% water (diluent),

$$F_D = F(0.99) = \left(2323\ \frac{\text{kg}}{\text{h}}\right)(0.99) = 2300\text{ kg/h}$$

$$F_D = F_{S,\text{min}} = 2300\text{ kg/h}$$

This separation requires a minimum solvent flow rate, F_S, of 2300 kg/h kerosene.

The answer is (C).

3. This is a mixing problem.

The composition of the diluent, D (1,1,2-trichloroethane) and of the solute, A (water), at the mixing point can be calculated using Eq. 25.8 and Eq. 25.9. See also *NCEES Handbook:* Measures of Composition.

$$x_{D,M} = \frac{F_1 x_{D,F_1} + F_2 x_{D,F_2}}{F_1 + F_2} = \frac{(100\text{ g})(0) + (100\text{ g})(0.50)}{100\text{ g} + 100\text{ g}} = 0.25$$

$$x_{A,M} = \frac{F_1 x_{A,F_1} + F_2 x_{A,F_2}}{F_1 + F_2} = \frac{(100\text{ g})(0) + (100\text{ g})(0.50)}{100\text{ g} + 100\text{ g}} = 0.25$$

In terms of wt%, the mixing point is at (25, 25).

The equilibrium compositions can be read by drawing the tie line that passes through the mixing point. For each phase,

Phase I at (67,32): 67 wt% 1,1,2-trichloroethane,
32 wt% water,
1 wt% acetone

Phase II at (1,22): 1 wt% 1,1,2-trichloroethane,
22 wt% water,
77 wt% acetone

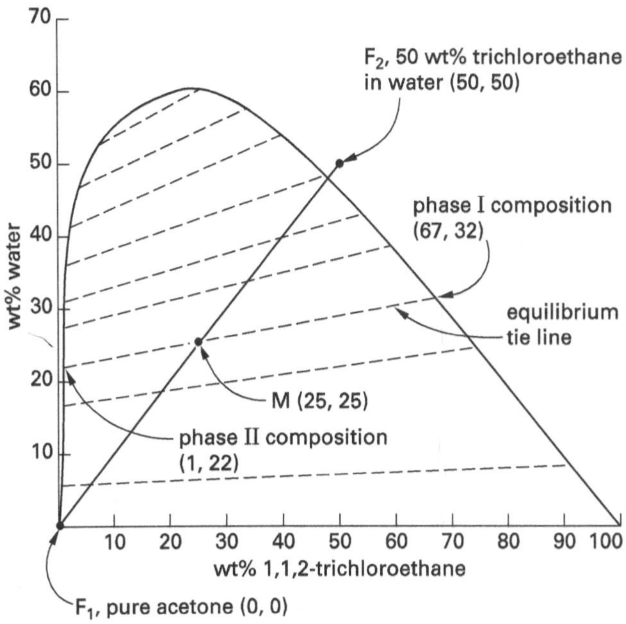

The answer is (B).

26 Solid-Liquid Processes

PRACTICE PROBLEMS

1. One method of recovering phosphorus from wastewater streams is to precipitate magnesium ammonium phosphate ($NH_4MgPO_4 \cdot 6H_2O$), commonly known as struvite, which is a potentially troublesome precipitate in pipelines. Struvite can be recovered in a controlled manner through various processes and can then be collected and applied as a phosphorus-based fertilizer in farmlands. One process for recovering struvite is shown.

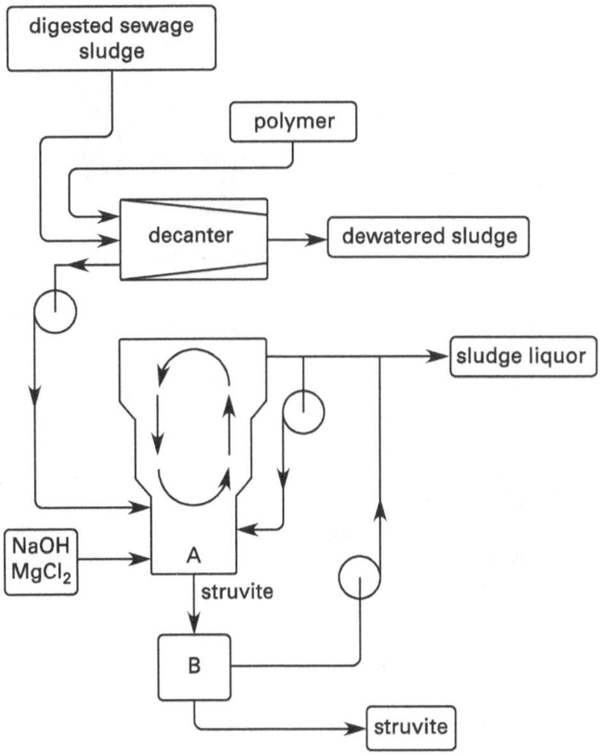

In this process, struvite is precipitated by dosing $MgCl_2$ and increasing the pH by adding caustic soda (NaOH). Process equipment A and B are, respectively,

(A) a dryer and a crystallizer

(B) a crystallizer and a dryer

(C) a flotation column and a microfiltration/reverse osmosis package

(D) a microfiltration/reverse osmosis package and a flotation column

2. A batch of material with 20% moisture (wet basis) must be dried to a 2.5% moisture content (dry basis). The drying curve is shown. The initial damp solids weight is 700 lbm, and the drying surface is 0.2 ft^2/lbm dry weight. Estimate the total drying time given that the drying rate in falling rate period II is e^X.

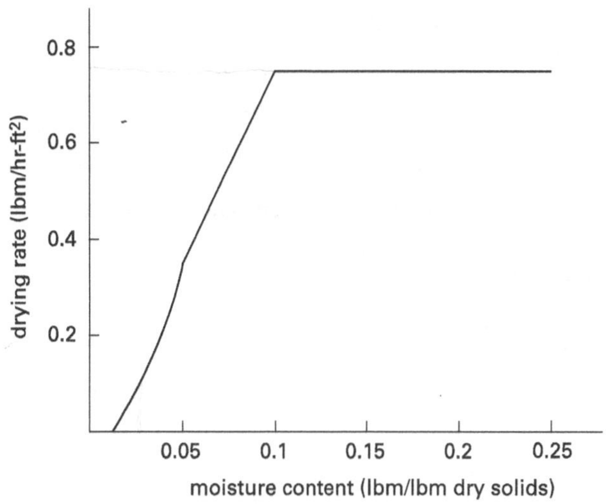

(A) 1 hr

(B) 1.6 hr

(C) 2 hr

(D) 2.4 hr

3. A batch of material that has a 22.5% moisture content must be dried to a 2.5% moisture content. The drying curve is shown. The initial, total moist solids mass is 1000 lbm, and the drying surface is 0.4 ft^2/lbm (dry mass basis). Estimate the total drying time given that the drying rate in falling rate period II is e^{5X}.

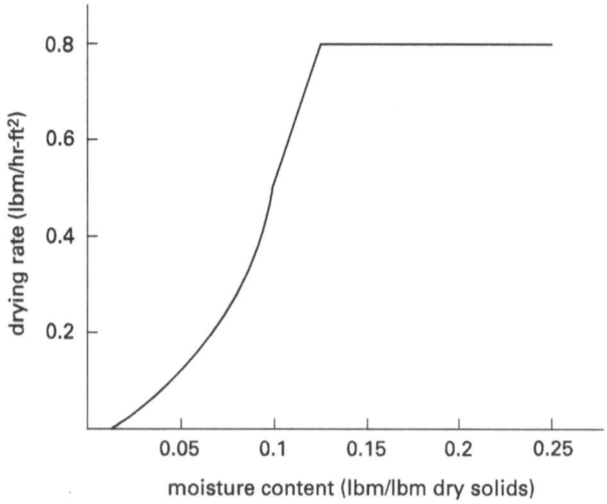

(A) 0.36 hr

(B) 0.52 hr

(C) 0.76 hr

(D) 0.91 hr

4. Stormwater washes over a farm field and into an adjacent watershed. Which of the following phenomena is most likely to occur?

(A) leaching

(B) distillation

(C) sublimation

(D) chemosynthesis

5. A proprietary solvent leaches caffeine from coffee beans. Analyses indicate that nothing else dissolves in this solvent except caffeine. The equilibrium relationship is known to be $Y = 0.8X$ for caffeine in the solvent, where

$$Y = \frac{\text{g caffeine}}{\text{kg solvent}}$$

$$X = \frac{\text{g caffeine}}{\text{kg insoluble solids}}$$

Starting with pure solvent and coffee beans that have 20 g caffeine/kg beans, the goal is to reduce the caffeine content in the beans to 5% of the original content (i.e., 1 g caffeine/kg beans).

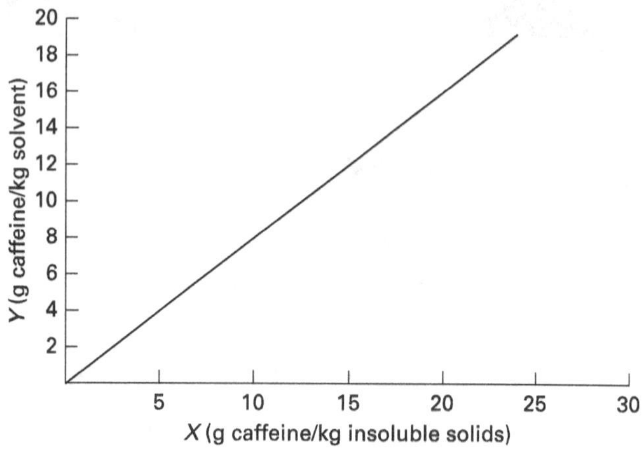

Determine the minimum amount of solvent required to achieve this separation using a countercurrent, equilibrium-staged system.

(A) 0.8 kg solvent/kg insoluble solids

(B) 0.84 kg solvent/kg insoluble solids

(C) 1.2 kg solvent/kg insoluble solids

(D) 1.25 kg solvent/kg insoluble solids

6. Which of the following statements about the leaching process is NOT correct?

(A) Solution mining is a leaching process used to recover minerals through boreholes drilled into a deposit that occurs *in situ.*

(B) If the density and viscosity of liquid is constant from stage to stage, the overflow and underflow rates are both constant and the operating line is straight.

(C) In leaching, equilibrium occurs when the concentration of the underflow at a given stage equals the concentration of the overflow at the same stage.

(D) A Pachuca tank is an air-lift reactor used in oxidative leaching processes where the solids settling velocity is small.

7. A saturated solution contains 3300 lbm of potassium chloride (KCl) at 212°F. The solution is slowly cooled in a crystallizer to 176°F. The solubility of potassium chloride is 56.37 g per 100 g of solution at 212°F and 48.5 g per 100 g of solution at 176°F. The loss of water by evaporation is negligible. The mass of KCl that crystallizes is most nearly

(A) 0.14 lbm

(B) 460 lbm

(C) 3200 lbm

(D) 3300 lbm

8. Which pair of membrane filtration processes can remove both dissolved solids and salts (ions) from a water or wastewater stream?

(A) microfiltration and ultrafiltration

(B) ultrafiltration and nanofiltration

(C) nanofiltration and reverse osmosis

(D) reverse osmosis and electrodialysis reversal

SOLUTIONS

1. Equipment A is a crystallizer and equipment B is a dryer. Option A puts the equipment in the wrong order. Options C and D are not appropriate for this purpose.

The answer is (B).

2. *Constant Rate Period:*

NCEES Handbook: Specific Drying Applications contains equations for each step in this problem. The moisture content of the material is reduced from the initial moisture content to the critical moisture content, X_c, in the constant rate period. Since the initial material has 20% moisture on a wet basis, we can determine the initial moisture content, X_1, on a dry basis as

$$X_1 = \frac{\text{mass of water}}{\text{mass of dry solids}} = \frac{m_W}{m_S}$$

The mass of water (m_W) can be calculated from the initial moisture content on a wet basis as

$$\begin{aligned} m_W &= \left(0.2 \ \frac{\text{lbm water}}{\text{lbm damp solids}}\right)(700 \text{ lbm damp solids}) \\ &= 140 \text{ lbm water} \end{aligned}$$

The mass of dry solids (m_S) can be calculated from the difference between the total mass of damp solids and the mass of water; therefore

$$\begin{aligned} X_1 &= \frac{m_W}{m_S} = \frac{140 \text{ lbm}}{700 \text{ lbm} - 140 \text{ lbm}} \\ &= 0.25 \text{ lbm moisture/lbm dry solids} \end{aligned}$$

From the given drying curve, the critical moisture content, X_c, is

$$X_c = 0.1 \text{ lbm moisture/lbm dry solids}$$

Reading from the drying curve, in the constant rate period, the drying rate, N_c, is

$$N_c = 0.75 \text{ lbm moisture/hr-ft}^2$$

In the constant rate drying period, the drying time, t_c, is

$$t_c = \left(\frac{L_s}{A}\right)\left(\frac{X_1 - X_c}{N_c}\right) = \left(\frac{\text{lbm dry solids}}{0.2 \text{ ft}^2}\right) \times \left(\frac{0.25 \frac{\text{lbm moisture}}{\text{lbm dry solids}} - 0.1 \frac{\text{lbm moisture}}{\text{lbm dry solids}}}{0.75 \frac{\text{lbm moisture}}{\text{hr-ft}^2}}\right) = 1 \text{ hr}$$

Falling Rate Period I:

In falling rate period I, the initial drying rate, N_I, is N_c.

$$N_1 = N_c = 0.75 \text{ lbm moisture/hr-ft}^2$$

From the drying curve, the final drying rate, N_2, is

$$N_2 = 0.35 \text{ lbm moisture/hr-ft}^2$$

The initial moisture content, X_1, is the critical moisture content, X_c.

$$X_1 = X_c = 0.1 \text{ lbm moisture/lbm dry solids}$$

From the drying curve, the final moisture content, X_2, is

$$X_2 = 0.05 \text{ lbm moisture/lbm dry solids}$$

Since the drying curve in falling rate period I is linear, the total drying time, t_I, is

$$t_\text{I} = \left(\frac{L_s}{A}\right)\left(\frac{X_1 - X_2}{N_1 - N_2}\right)\ln\frac{N_1}{N_2} = \left(\frac{\text{lbm dry solids}}{0.2 \text{ ft}^2}\right) \times \left(\frac{0.1 \frac{\text{lbm moisture}}{\text{lbm dry solids}} - 0.05 \frac{\text{lbm moisture}}{\text{lbm dry solids}}}{0.75 \frac{\text{lbm moisture}}{\text{hr-ft}^2} - 0.35 \frac{\text{lbm moisture}}{\text{hr-ft}^2}}\right) \times \left(\ln\frac{0.75 \frac{\text{lbm moisture}}{\text{hr-ft}^2}}{0.35 \frac{\text{lbm moisture}}{\text{hr-ft}^2}}\right) = 0.48 \text{ hr}$$

Falling Rate Period II:

In falling rate period II, the drying rate, N_II, is given as

$$N_\text{II} = e^X$$

From the drying curve, the initial moisture content, X_1, is

$$X_1 = 0.05 \text{ lbm moisture/lbm dry solids}$$

From the problem statement, the final material has 2.5% moisture. The final moisture content, X_2, is

$$X_2 = \frac{X_T}{L_s} = \frac{0.025}{1 - 0.025} = \frac{0.025 \text{ lbm moisture}}{0.975 \text{ lbm dry solids}} = 0.026 \text{ lbm moisture/lbm dry solids}$$

In falling rate period II, the drying time, t_II, is

$$t_\text{II} = \left(\frac{L_s}{A}\right)\int_{X_2}^{X_1}\frac{dX}{N} = \left(\frac{1}{0.2}\right)\int_{0.026}^{0.05} e^{-X}dX = (5)\left(-e^{-X}\Big|_{0.026}^{0.05}\right) = (5)\left(-e^{-0.05} - (-e^{-0.026})\right) = 0.12 \text{ hr}$$

The total drying time, t_D, is the sum of the drying times for each period.

$$t_D = t_c + t_\text{I} + t_\text{II} = 1 \text{ hr} + 0.48 \text{ hr} + 0.12 \text{ hr} = 1.6 \text{ hr}$$

The answer is (B).

3. *Constant Rate Period:*

NCEES Handbook: Specific Drying Applications contains equations for each step in this problem. The moisture content of the material is reduced from the initial moisture content of 22.5% to the critical moisture content, X_c, in the constant rate period. The initial material has 22.5% moisture, so by definition the initial moisture content, X_1, is

$$X_1 = \%\,\text{moisture in solids} = \frac{m_w}{m_s} = \frac{\left(0.225 \frac{\text{lbm moisture}}{\text{lbm total}}\right)(1000 \text{ lbm total})}{\left((1 - 0.225) \frac{\text{lbm dry solids}}{\text{lbm total}}\right)(1000 \text{ lbm total})} = 0.29 \text{ lbm moisture/lbm dry solids}$$

From the given drying curve, the critical moisture content, X_c, is

$$X_c = 0.125 \text{ lbm moisture/lbm dry solids}$$

Reading from the drying curve, in the constant rate period, the drying rate, N_c, is

$$N_c = 0.8 \text{ lbm moisture/hr-ft}^2$$

In the constant rate drying period, the drying time, t_c, is

$$t_c = \left(\frac{m_s}{A}\right)\left(\frac{X_1 - X_2}{N_c}\right) = \left(\frac{m_s}{A}\right)\left(\frac{X_1 - X_c}{N_c}\right)$$

The ratio, m_s/A, which is the drying surface per 1 lb_m (dry weight basis), is equal to (1 lb_m dry solids)/0.4 ft^2.

$$\begin{aligned} t_c &= \left(\frac{1 \text{ lb}_\text{m} \text{ dry solids}}{0.4 \text{ ft}^2}\right)\left(\frac{0.29 \frac{\text{lb}_\text{m} \text{ moisture}}{\text{lb}_\text{m} \text{ dry solids}} - 0.125 \frac{\text{lb}_\text{m} \text{ moisture}}{\text{lb}_\text{m} \text{ dry solids}}}{0.8 \frac{\text{lb}_\text{m} \text{ moisture}}{\text{hr-ft}^2}}\right) \\ &= 0.52 \text{ hr} \end{aligned}$$

Linear Falling Rate Period I:

In falling rate period I, the initial drying rate, N_1, is N_c.

$$N_1 = N_c = 0.8 \text{ lbm moisture/hr-ft}^2$$

From the drying curve, the final drying rate, N_2, is

$$N_2 = 0.5 \text{ lbm moisture/hr-ft}^2$$

The initial moisture content, X_1, is the critical moisture content, X_c.

$$X_1 = X_c = 0.125 \text{ lbm moisture/lbm dry solids}$$

From the drying curve, the final moisture content, X_2, is

$$X_2 = 0.10 \text{ lbm moisture/lbm dry solids}$$

Since the drying curve in falling rate period I is linear, the total drying time, t_I, is

$$\begin{aligned} t_I &= \left(\frac{m_s}{A}\right)\left(\frac{X_1 - X_2}{N_1 - N_2}\right)\ln\frac{N_1}{N_2} \\ &= \left(\frac{1 \text{ lb}_\text{m} \text{ dry solids}}{0.4 \text{ ft}^2}\right)\left(\frac{0.125 \frac{\text{lb}_\text{m} \text{ moisture}}{\text{lb}_\text{m} \text{ dry solids}} - 0.10 \frac{\text{lb}_\text{m} \text{ moisture}}{\text{lb}_\text{m} \text{ dry solids}}}{0.8 \frac{\text{lb}_\text{m} \text{ moisture}}{\text{hr-ft}^2} - 0.5 \frac{\text{lb}_\text{m} \text{ moisture}}{\text{hr-ft}^2}}\right)\left(\ln\frac{0.8 \frac{\text{lb}_\text{m} \text{ moisture}}{\text{hr-ft}^2}}{0.5 \frac{\text{lb}_\text{m} \text{ moisture}}{\text{hr-ft}^2}}\right) \\ &= 0.098 \text{ hr} \quad (0.10 \text{ hr}) \end{aligned}$$

Falling Rate Period II:

In falling rate period II, the drying rate N_II, is given as

$$N_\text{II} = e^{5X}$$

From the drying curve, the initial moisture content, X_1, is

$$X_1 = 0.10 \text{ lbm moisture/lbm dry solids}$$

From the problem statement, the final material has 2.5% moisture. The final moisture content, X_2, is

$$\begin{aligned} X_1 &= \%\text{moisture in solids} = \frac{m_w}{m_s} = \frac{0.025 \text{ lb}_\text{m} \text{ moisture}}{(1 - 0.025) \text{ lb}_\text{m} \text{ dry solids}} \\ &= 0.026 \text{ lb}_\text{m}\text{moisture/lb}_\text{m}\text{dry solids} \end{aligned}$$

In falling rate period II, the drying time, t_II, is

$$\begin{aligned} t_{II} &= \left(\frac{m_s}{A}\right)\int_{X_2}^{X_1} \frac{dX}{N} = \left(\frac{m_s}{A}\right)\int_{X_2}^{X_1} N^{-1}\, dX \\ &= \left(\frac{1 \text{ lb}_\text{m} \text{ dry solids}}{0.4 \text{ ft}^2}\right)\left(\int_{0.26}^{0.10} e^{-5X}\, dX \;\frac{\text{hr-ft}^2}{\text{lb}_\text{m} \text{ moisture}}\right) \\ &= \left(\frac{1}{0.4}\right)\left(\left(-\frac{1}{5}\right)e^{-5X}\Big|_{0.026}^{0.10}\right) \\ &= \left(\frac{1}{0.4}\right)\left(\left(-\frac{1}{5}\right)e^{-(5)(0.10)} - \left(\left(-\frac{1}{5}\right)e^{-(5)(0.026)}\right)\right) \\ &= \left(\frac{1}{0.4}\right)(-0.1213 - (-0.1756)) \\ &= 0.1358 \text{ hr} \quad (0.14 \text{ hr}) \end{aligned}$$

The total drying time, t_D, is the sum of the drying times for each period.

$$\begin{aligned} t_D &= t_c + t_\text{I} + t_\text{II} = 0.52 \text{ hr} + 0.10 \text{ hr} + 0.14 \text{ hr} \\ &= 0.76 \text{ hr} \end{aligned}$$

The answer is (C).

4. The stormwater is a solvent. The nutrients in the farmland's soil are soluble solutes. The water washes away some or all of the solutes into the watershed. This process is called leaching. Nutrients like nitrogen-rich fertilizers are leached into the watershed. Option A is correct.

Distillation occurs when a liquid is converted to a vapor by vaporization. Sublimation occurs when a solid directly converts to a vapor. Chemosynthesis is the biological conversion of one or more carbon-containing molecules (usually carbon dioxide or methane) and nutrients into organic matter using the oxidation of inorganic compounds.

The answer is (A).

5. To find the minimum solvent required, use the McCabe-Thiele method to draw the operating curve.

The data given in the problem are shown in the following illustration.

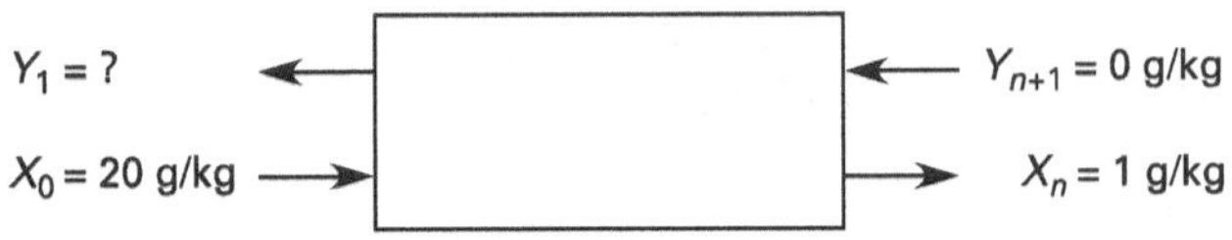

The operating equation for leaching is

$$Y_{n+1} = \left(\frac{L_n}{V_{n+1}}\right)X_n + \frac{V_a Y_a - L_a X_a}{V_{n+1}}$$

The slope of the operating line represents the ratio of insoluble solids to solvent flow rates. For the operating line, only enough information is available to plot one point at $(X_1, Y_{n+1}) = (1, 0)$. However, the minimum solvent flow rate is related to the slope of the operating line that intersects the equilibrium curve (to create a pinch point) at the point where $X_0 = 20$ g caffeine/kg insoluble solids. Using these points and drawing this operating line on the equilibrium diagram reveals the corresponding value of $Y_1 = 16$ g caffeine/kg solvent.

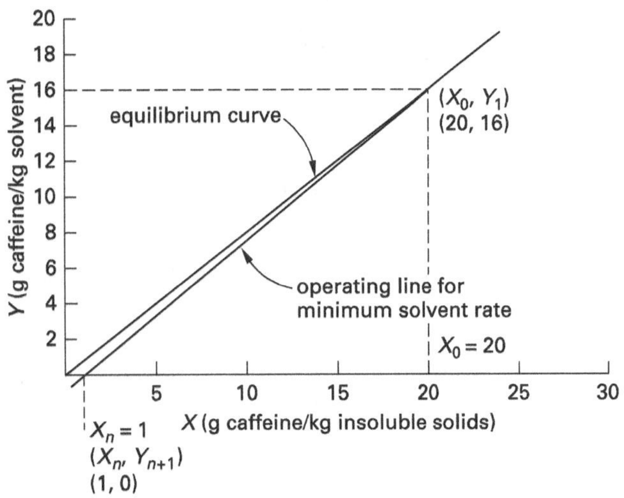

The slope is

$$\text{slope} = \frac{L_n}{V_{n+1}} = \frac{\Delta Y}{\Delta X} = \frac{Y_1 - Y_{n+1}}{X_0 - X_n}$$

$$= \frac{16\ \frac{\text{g caffeine}}{\text{kg solvent}} - 0\ \frac{\text{g caffeine}}{\text{kg solvent}}}{\left(20\ \frac{\text{g caffeine}}{\text{kg insoluble solids}} - 1\ \frac{\text{g caffeine}}{\text{kg insoluble solids}}\right)}$$

$$= 0.84 \text{ kg insoluble solids/kg solvent}$$

The minimum amount of solvent needed is the reciprocal of the minimum slope.

$$\frac{V_{n+1}}{L_n} = \frac{1}{\text{slope}} = \frac{1}{0.84\ \frac{\text{kg insoluble solids}}{\text{kg solvent}}}$$

$$= 1.2 \text{ kg solvent/kg insoluble solids}$$

Therefore, the separation requires a minimum of 1.2 kg of solvent per kg of coffee beans.

The answer is (C).

6. Equilibrium in a leaching process is irrelevant to the underflow or overflow concentration at any stage; equilibrium is reached when all the solute is solved within the solvent in the liquid phase.

The answer is (C).

7. The mass of potassium chloride in the saturated solution at the original temperature of 212°F, $m_{\text{KCl},212}$, is known to be 3300 lbm. The solubility of potassium chloride at 212°F, $K_{\text{KCl},212}$, is 56.37 g per 100 g of solution. The mass of water needed to dissolve the potassium chloride in the original solution is

$$m_{\text{water}} = \frac{m_{\text{KCl},212}}{K_{\text{KCl},212}} = \frac{3300 \text{ lbm}}{\frac{56.37 \text{ g}}{100 \text{ g}}} = 5854 \text{ lbm}$$

The mass of potassium chloride dissolved in the same amount of water at 176°F is

$$m_{\text{KCl},176} = m_{\text{water}} K_{\text{KCl},176} = (5854 \text{ lbm})\left(\frac{48.5 \text{ g}}{100 \text{ g}}\right)$$

$$= 2839 \text{ lbm}$$

The loss in the dissolved mass from 212°F to 176°F is the mass of potassium chloride that crystallizes.

$$\begin{aligned} m_{\text{KCl,crystal}} &= m_{\text{KCl},212} - m_{\text{KCl},176} \\ &= 3300 \text{ lbm} - 2839 \text{ lbm} \\ &= 461 \text{ lbm} \quad (460 \text{ lbm}) \end{aligned}$$

The answer is (B).

8. Dissolved solids and salts contain very fine particles. Of the filtration processes listed, only reverse osmosis and electrodialysis reversal can remove particles as small as these from a liquid.

The answer is (D).

27 Kinetics

PRACTICE PROBLEMS

1. The rate constant for a reaction doubles when the temperature is raised from 37°C to 57°C. The temperature at which the rate constant again doubles is most nearly

(A) 77°C

(B) 80°C

(C) 97°C

(D) 124°C

2. Compound A reacts to form compound P in a constant-volume, isothermal batch reactor.

time (h)	concentration of reactant A (mol/L)
0	5.00
0.5	0.69
1.0	0.37
1.5	0.25
2.0	0.19
2.5	0.16

The rate that best fits the data is

(A) k

(B) kC_A

(C) kC_A^2

(D) $kC_A/(1+C_A)$

3. Consider the reversible reaction

$$A + B \leftrightarrow C + D$$

The feed stream contains 3 mol of A, 2 mol of B, and one mol of both C and D. The equilibrium constant, K_{eq}, is 1.45. At equilibrium, the amount of A is most nearly

(A) 1.81 mol

(B) 2.02 mol

(C) 2.13 mol

(D) 2.36 mol

4. An endothermic first-order reaction, A→P, is carried out in a batch reactor.

heat of reaction	97 000 kJ/kmol
reactor volume	1 m^3
heat capacity of mixture	2.5 kJ/kg·K
total mass	975 kg
initial number of moles	10.5 kmol
rate constant	$4750e^{-8500/T}\ s^{-1}$

After the reaction mixture is heated to 300°C, it is allowed to proceed adiabatically. 15 mol% of A is converted during the heat-up. What is the temperature in the reactor at 75% conversion?

(A) 155K

(B) 260K

(C) 322K

(D) 510K

5. The reaction $A \xrightarrow{k_1} P \xrightarrow{k_2} S$ is carried out in a tubular plug-flow reactor. Both reactions are first order.

reactor diameter	10 cm
feed rate	$0.15\ m^3/h$ pure A
feed concentration	$0.10\ kmol\ A/m^3$
rate constants	$k_1 = 175\ h^{-1}$
	$k_2 = 12.5\ h^{-1}$

Determine the reactor length that will maximize the yield of P.

(A) 0.08 m

(B) 0.31 m

(C) 0.97 m

(D) 1.5 m

6. In a liquid-phase batch reactor, the elementary reaction $2A \rightarrow B$ occurs. The initial concentration of A is 3 mol/L, and 90% conversion is achieved in 2 hr. Assume density and temperature are constant during the reaction. Most nearly, the reaction rate constant is

(A) 0.67 L/mol·h

(B) 1.0 L/mol·h

(C) 1.5 L/mol·h

(D) 2.3 L/mol·h

7. A CSTR is used to produce B at a rate of 55 kmol/h by the first-order reaction, $A \rightarrow B$, with a rate constant of $7.0\ h^{-1}$. The feed concentration of reactant A is $1\ kmol/m^3$. What total reactor volume is required to achieve 70% conversion in a series of two CSTRs?

(A) $9.27\ m^3$

(B) $13.10\ m^3$

(C) $18.55\ m^3$

(D) $26.20\ m^3$

8. The activation energy of a first-order reaction is 60.3 kJ/mol at 35°C. The temperature at which the rate constant will triple is most nearly

(A) 35°C

(B) 38°C

(C) 41°C

(D) 50°C

9. The hydrolysis of propene (C_3H_6) produces a mixture of the two isomers of propanol (C_3H_7OH). The by-product of the reaction is propyl ether ($(C_3H_7)_2O$). The feed into a reactor contains 58 mol% propene, 35 mol% water, and 7 mol% inert substances. The product from the reactor contains 4.89 mol% isomers of propanol, 48.26 mol% propene, 0.21 mol% propyl ether, 35.87 mol% water, and 10.77 mol% inert substances. Most nearly, the yield of propanol based on propene is

(A) 0.12%

(B) 3.8%

(C) 12%

(D) 300%

10. The parallel competing reactions $M \rightarrow N$ and $M \rightarrow P$ are to be carried out in an ideal continuous stirred tank reactor (CSTR) operating at steady state. The conversion of species M is 92%. The initial concentration of species M is 4.8 mol/L. The reaction to produce species N is a zero-order reaction, having a reaction constant of 2.8 mol/L·min. The reaction to produce species P is a first-order reaction, having a rate constant of 1.2 1/min. The overall yield of formation of species N is most nearly

(A) 0.31

(B) 0.63

(C) 0.73

(D) 0.86

11. The dehydration of ethanol (C_2H_5OH) produces mainly diethyl ether ($C_4H_{10}O$). The by-product of the reaction is ethene (C_2H_4). The main reaction, having a reaction rate constant of 20 1/min, is first order in the concentration of C_2H_5OH. The reaction rate of the independent parallel reaction, having a reaction rate constant of 10 1/min is first order in the concentration of C_2H_5OH. The selectivity is defined as the ratio of the moles of diethyl ether produced to the moles of ethene produced. Most nearly, the selectivity is

(A) 0.5

(B) 1.0

(C) 2.0

(D) 5.0

12. The hydrolysis of 1-butene (C_4H_8) produces mainly an undistinguishable mixture of the two isomers of butanol with the molecular formula C_4H_9OH. The by-product of the reaction is butyl ether ($(C_4H_9)_2O$). The reactor is fed with a stream containing 62 mol% 1-butene, 32 mol% water, and 6 mol% inert. The product from the reactor contains 5.26 mol% of the isomers mixture, 43.25 mol% 1-butene, 0.32 mol% butyl ether, 36.89 mol% water, and 14.28 mol% inert.

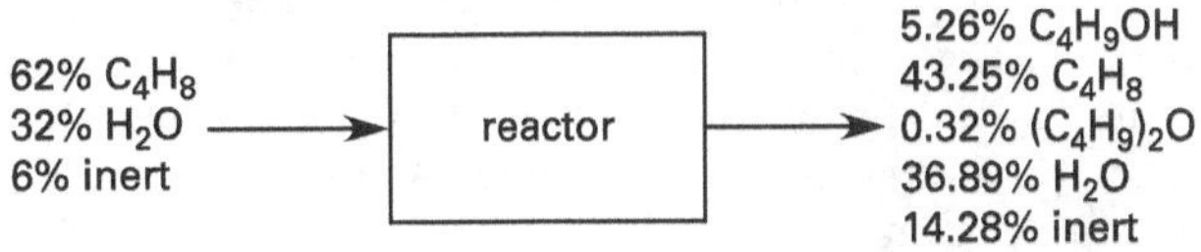

The selectivity of butanol is most nearly

(A) 0.06

(B) 2.2

(C) 7.4

(D) 16

13. A reactor is fed with a stream containing toluene ($C_6H_5CH_3$), hydrogen (H_2), methane (CH_4), and benzene (C_6H_6). In addition to the desirable product of this reaction, C_6H_6, the subproduct, diphenyl ($C_{12}H_{10}$) is also obtained according to the following balanced parallel reactions.

$$C_6H_5CH_3 + H_2 \rightarrow C_6H_6 + CH_4$$
$$2C_6H_6 \rightarrow C_{12}H_{10} + H_2$$

The following table shows the molar flow rate of the reactor feed and the effluent streams.

species	feed flow rate, n_f (kmol/h)	effluent flow rate, n_o (kmol/h)
H_2	1875	1548
CH_4	798	1012
C_6H_6	16	375
$C_6H_5CH_3$	365	112
$C_{12}H_{10}$	0	5

The C_6H_6 yield from $C_6H_5CH_3$ is most nearly

(A) 0.75

(B) 0.98

(C) 1.4

(D) 1.5

14. Two simultaneous reactions take place in a CSTR at a temperature of 200°C. B is the desired product of the first reaction, and C is the undesired product of the second reaction. The two gas phase reactions, and the rate expressions for the formation of B and C are

$$\mathrm{A} \rightarrow \mathrm{B} \quad r_\mathrm{B} = 0.015e^{(8000)(1/400-1/T)}C_\mathrm{A}$$
$$\mathrm{B} \rightarrow \mathrm{C} \quad r_\mathrm{C} = 0.019e^{(7000)(1/400-1/T)}C_\mathrm{A}^{1.5}$$

Temperature is in kelvins. The initial concentration of A is 3.4 mol/L. The desired fractional conversion of A is 0.90. Most nearly, the selectivity for the formation of B and C is

(A) 0.65

(B) 1.3

(C) 0.1

(D) 2.0

15. The following reversible reaction takes place in air at 1 atm and 1500°C.

$$\tfrac{1}{2}N_2 + \tfrac{1}{2}O_2 \rightleftharpoons NO$$

The correct expression for the equilibrium constant, K_eq, for this reaction is

(A) $K_\mathrm{eq} = \dfrac{[NO]}{[N_2][O_2]}$

(B) $K_\mathrm{eq} = \dfrac{[N_2]^{1/2}[O_2]^{1/2}}{[NO]}$

(C) $K_\mathrm{eq} = \dfrac{[NO]}{[N_2]^{1/2}[O_2]^{1/2}}$

(D) $K_\mathrm{eq} = \dfrac{[N_2][O_2]}{[NO]}$

16. The following system is at equilibrium.

$$CO(g) + 2H_2(g) \rightleftharpoons CH_3OH(g)$$

Pressure is increased by compressing the system. As a consequence, the concentration of methanol (CH_3OH)

(A) remains the same

(B) decreases

(C) increases

(D) decreases for a while and then increases

17. The gas-phase reaction shown takes place at 500°C with a Gibbs free energy of −13 000 cal/mol.

$$CO + 3H_2 \rightleftharpoons CH_4 + H_2O$$

Most nearly, the equilibrium constant for the reaction is

(A) 4700

(B) 5900

(C) 7300

(D) 9100

SOLUTIONS

1. From the problem statement, $k_2 = 2k_1$. The activation energy, E, can be found by taking a ratio of rate constants as defined by the Arrhenius law equation. First, all temperatures must be absolute. As in *NCEES Handbook* table "Conversion Table for Temperature Units,"

$$T_1 = 37°C + 273° = 310K$$
$$T_2 = 57°C + 273° = 330K$$

From *NCEES Handbook:* Temperature Dependence,

$$\ln\frac{k_1}{k_2} = -\left(\frac{E}{R^*}\right)\left(\frac{1}{T_1} - \frac{1}{T_2}\right)$$

$$\ln\frac{1}{2} = -\left(\frac{E}{8.314\ \frac{J}{mol{\cdot}K}}\right)\left(\frac{1}{310K} - \frac{1}{330K}\right)$$

$$E = 29\,477\ J/mol \quad (29\,500\ J/mol)$$

Doubling the rate constant again implies that

$$k_2' = 2k_2 = (2)(2k_1) = 4k_1$$

$$\ln\frac{k_1}{k_2'} = -\left(\frac{E}{R^*}\right)\left(\frac{1}{T_1} - \frac{1}{T_2'}\right)$$

$$\ln\frac{1}{4} = -\left(\frac{29\,500\ \frac{J}{mol}}{8.314\ \frac{J}{mol{\cdot}K}}\right)\left(\frac{1}{310K} - \frac{1}{T_2'}\right)$$

$$T_2' = 353K \quad (80°C)$$

The answer is (B).

2. Find k by substituting data points into the integrated form of the material balance for each rate. The rate that best fits the data is the rate that gives the least divergent values of k. For this problem, it suffices to find the value of k for the first two data points—(0, 5) and (0.5, 0.69)—and compare to the value of k for the first and last data points—(0, 5) and (2.5, 0.16).

Use *NCEES Handbook:* Zero-Order Reactions to find the integrated form for each rate. Solving for both values of k in option A,

$$k = \frac{C_{Ao} - C_A}{t} = \frac{5 - 0.69}{0.5} = 8.62$$

$$k = \frac{C_{Ao} - C_A}{t} = \frac{5 - 0.16}{2.5} = 1.94$$

These two values of k are not close. For option B, the integrated form can be found in *NCEES Handbook:* First-Order Reactions.

$$k = \frac{\ln \frac{C_{\text{Ao}}}{C_{\text{A}}}}{t} = \frac{\ln \frac{5}{0.69}}{0.5} = 3.96$$

$$k = \frac{\ln \frac{C_{\text{Ao}}}{C_{\text{A}}}}{t} = \frac{\ln \frac{5}{0.16}}{2.5} = 1.38$$

These two values are closer, but not close enough. For option C, the integrated form can be found in *NCEES Handbook:* Second-Order Reactions.

$$\begin{aligned} k &= \left(\frac{1}{t}\right)\left(\frac{1}{C_{\text{A}}} - \frac{1}{C_{\text{Ao}}}\right) \\ &= \left(\frac{1}{0.5}\right)\left(\frac{1}{0.69} - \frac{1}{5}\right) \\ &= 2.5 \end{aligned}$$

$$\begin{aligned} k &= \left(\frac{1}{t}\right)\left(\frac{1}{C_{\text{A}}} - \frac{1}{C_{\text{Ao}}}\right) \\ &= \left(\frac{1}{2.5}\right)\left(\frac{1}{0.16} - \frac{1}{5}\right) \\ &= 2.42 \end{aligned}$$

Option C values of k diverge less than option A or option B values. For option D, the integrated form looks like a combination of the integrated forms of option A and option B.

$$k = \frac{C_{\text{Ao}} - C_{\text{A}}}{t} + \frac{\ln \frac{C_{\text{Ao}}}{C_{\text{A}}}}{t} = \frac{5 - 0.69}{0.5} + \frac{\ln \frac{5}{0.69}}{0.5} = 12.58$$

$$k = \frac{C_{\text{Ao}} - C_{\text{A}}}{t} + \frac{\ln \frac{C_{\text{Ao}}}{C_{\text{A}}}}{t} = \frac{5 - 0.16}{2.5} + \frac{\ln \frac{5}{0.16}}{2.5} = 3.32$$

Option D values of k diverge the most. The rate that best fits the data is option C.

The answer is (C).

3. As in *NCEES Handbook:* Chemical Equilibrium Constants from Rate Constants for Reversible Reactions, the equilibrium constant is

$$K_{\text{eq}} = \frac{C_{\text{C}}^1 C_{\text{D}}^1}{C_{\text{A}}^1 C_{\text{B}}^1} = \frac{(1+\epsilon)^2}{(2-\epsilon)(3-\epsilon)} = 1.45$$

Rearranging into a quadratic equation,

$$-0.45\epsilon^2 + 9.25\epsilon - 7.7 = 0$$

Using the properties of quadratic equations found in *NCEES Handbook:* Polynomials, the extent of reaction is

$$\begin{aligned} \epsilon &= \frac{-b \pm \sqrt{b^2 - 4ac}}{2a} \\ &= \frac{-9.25 \pm \sqrt{(9.25)^2 - (4)(-0.45)(-7.7)}}{2(-0.45)} \\ &= 0.87 \end{aligned}$$

Assume a base volume of 1 L for ease of calculation. The amount of A at equilibrium is

$$\begin{aligned} C_{\text{A}} &= C_{\text{Ao}} - \epsilon = 3 \text{ mol} - \epsilon \\ &= 3 \text{ mol} - 0.87 \text{ mol} \\ &= 2.13 \text{ mol} \end{aligned}$$

The answer is (C).

4. The mass balance and the energy balance around the batch reactor are solved simultaneously to determine the final temperature in the reactor. See *NCEES Handbook* section titled "Heat Effects in Reactors." For a batch reactor,

$$\frac{dX_{\text{A}}}{d\theta} = \left(\frac{V}{N_{\text{A}_0}}\right)(-r_{\text{A}})$$

$$m_t c_p \frac{dT}{d\theta} = V(-\Delta H_r)(-r_{\text{A}}) + qA_k$$

In an adiabatic reactor, the heat flux, q, is 0. Since the reaction is endothermic, ΔH_r is positive. The negative term, $-\Delta H_r$, indicates that the temperature will decrease during the reaction. Combining the two equations results in the following expression.

$$\frac{dT}{dX} = \frac{(-\Delta H_r)N_{\text{A}_0}}{m_t c_p}$$

Upon integration,

$$\begin{aligned}T &= T_0 + \left(\frac{-\Delta H_r N_{A_0}}{m_t c_p}\right)(X_A - X_{A_0}) \\ &= 300°C + 273° \\ &\quad + \left(\frac{\left(-97\,000\ \frac{kJ}{mol}\right)(10.5\ kmol)}{(975\ kg)\left(2.5\ \frac{kJ}{kg\cdot K}\right)}\right)(0.75 - 0.15) \\ &= 322K\end{aligned}$$

The answer is (C).

5. From *NCEES Handbook:* Two First-Order Irreversible Reactions in Series, the residence time for maximum yield of an intermediate is

$$\tau_{max} = \frac{\ln \frac{k_2}{k_1}}{k_2 - k_1} = \frac{\ln \frac{12.5\ h^{-1}}{175\ h^{-1}}}{12.5\ h^{-1} - 175\ h^{-1}} = 0.0162\ h$$

From *NCEES Handbook:* Plug-Flow Reactor, the residence time is

$$\tau = \frac{C_{Ao} V_{PFR}}{F_{Ao}}$$

Substituting $V = \pi(D^2/4)L$ into the previous equation and rearranging to solve for reactor length,

$$\begin{aligned}L &= \tau_{max}\left(\frac{F_{A_0}}{C_{A_0}}\right)\left(\frac{4}{\pi D^2}\right) = \tau_{max}\dot{V}_{A_0}\left(\frac{4}{\pi D^2}\right) \\ &= (0.0162\ h)\left(0.15\ \frac{m^3}{h}\right) \\ &\quad \times \left(\frac{4}{\pi\left((10\ cm)\left(\frac{1\ m}{100\ cm}\right)\right)^2}\right) \\ &= 0.31\ m\end{aligned}$$

The answer is (B).

6. The final reactant A concentration is found from the conversion in *NCEES Handbook* Reaction Parameters Nomenclature.

$$\begin{aligned}C_A &= C_{A0}(1 - X_A) = \left(3\ \frac{mol}{L}\right)(1 - 0.90) \\ &= 0.3\ mol/L\end{aligned}$$

The rate constant is found in *NCEES Handbook* Second-Order Reactions: Constant Volume—Batch Reactor.

$$\begin{aligned}k &= \frac{\frac{1}{C_A} - \frac{1}{C_{A0}}}{t} = \frac{\frac{1}{0.3\ \frac{mol}{L}} - \frac{1}{3\ \frac{mol}{L}}}{2\ h} \\ &= 1.5\ L/mol\cdot h\end{aligned}$$

The answer is (C).

7. From the problem statement, B is produced at 55 kmol/h with a 70% conversion ($X_A = 0.7$) of A to B. Therefore, the feed concentration of A is

$$\begin{aligned}F_{A_0} X_A &= F_B \\ F_{A_0} &= \frac{F_B}{X_A} \\ &= \frac{55\ \frac{kmol}{h}}{0.7} \\ &= 78.6\ kmol/h\end{aligned}$$

As in *NCEES Handbook:* Continuous Stirred Tank Reactors in Series, the total residence time of N is

$$\begin{aligned}\tau &= \left(\frac{N}{k}\right)\left(\left(\frac{C_{Ao}}{C_A}\right)^{1/N} - 1\right) \\ &= \left(\frac{2}{7\ h^{-1}}\right)\left(\left(\frac{1\ \frac{kmol}{m^3}}{0.3\ \frac{kmol}{m^3}}\right)^{1/2} - 1\right) \\ &= 0.236\ h\end{aligned}$$

As in *NCEES Handbook:* Continuous Stirred Tank Reactor (CSTR), the total reactor volume required to achieve 70% conversion in a series of two CSTRs is

$$\begin{aligned}\tau &= \frac{C_{Ao} V}{F_{Ao}} \\ V &= \frac{\tau F_{Ao}}{C_{Ao}} \\ &= \frac{(0.236\ h)\left(78.6\ \frac{kmol}{h}\right)}{1\ \frac{kmol}{m^3}} \\ &= 18.55\ m^3\end{aligned}$$

The answer is (C).

8. The equation for a first-order reaction at two temperatures is

$$\ln\frac{k_2}{k_1} = \frac{E}{R}\left(\frac{1}{T_1} - \frac{1}{T_2}\right)$$

To find the temperature at which the rate constant will triple, set k_2 to three times k_1 and solve for T_2.

$$\begin{aligned}
\ln\frac{3k_1}{k_1} &= \frac{E}{R}\left(\frac{1}{T_1} - \frac{1}{T_2}\right) \\
T_2 &= \frac{1}{\frac{1}{T_1} - \frac{R\ln 3}{E}} \\
&= \frac{1}{\frac{1}{35^\circ\text{C} + 273^\circ} - \frac{\left(8.314\ \frac{\text{J}}{\text{mol·K}}\right)(1.0986)}{60\,300\ \frac{\text{J}}{\text{mol}}}} \\
&= 323\text{K}
\end{aligned}$$

Converting to degrees Celsius gives

$$323\text{K} - 273^\circ = 50^\circ\text{C}$$

The answer is (D).

9. The balanced reactions are

$$C_3H_6 + H_2O \rightarrow C_3H_7OH$$

$$2C_3H_7OH \rightarrow (C_3H_7)_2O + H_2O$$

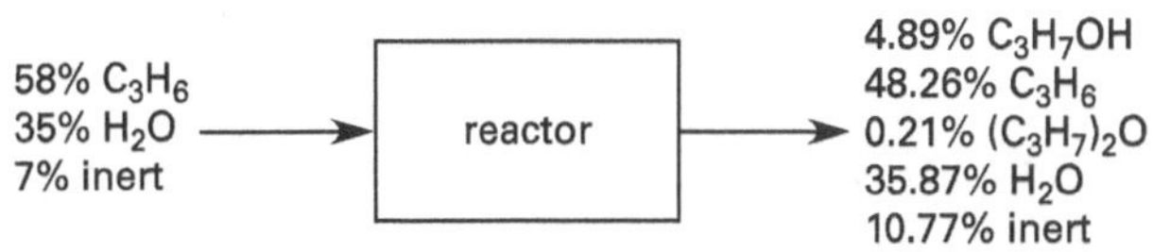

Each 100 mol in the feed stream contains 58 mol of propene and 7 mol of inert substances. From the hydrolysis reaction, the stoichiometric coefficient is 1. As the inert substances remain constant, the ratio of moles of inert substances in the feed stream to moles of inert substances in the product remains constant. This ratio is applicable to the other components in the product stream. The number of moles of propanol in the product stream, assuming 100 mol of feed, is

$$\begin{aligned}
(n_{C_3H_7OH})_p &= (4.89)\left(\frac{n_{\text{inert,feed}}}{10.77}\right) \\
&= (4.89)\left(\frac{7\ \text{mol}}{10.77}\right) \\
&= 3.18\ \text{mol}
\end{aligned}$$

The number of moles of propene in the product stream, assuming 100 mol of feed, is

$$\begin{aligned}
(n_{C_3H_6})_p &= (48.26)\left(\frac{n_{\text{inert,feed}}}{10.77}\right) \\
&= (48.26)\left(\frac{7\ \text{mol}}{10.77}\right) \\
&= 31.37\ \text{mol}
\end{aligned}$$

A mole balance around the reactor gives the number of moles of C_3H_6 that react.

$$\begin{aligned}
n_{C_3H_6,\text{reacted}} &= n_{C_3H_6,\text{feed}} - n_{C_3H_6,\text{product}} \\
&= 58\ \text{mol} - 31.37\ \text{mol} \\
&= 26.63\ \text{mol}
\end{aligned}$$

The yield of the hydrolysis reaction is the ratio of the number of moles of desired product (propanol) to the number of moles of propene that reacted. The percent yield of propanol based on propene is

$$\begin{aligned}
y &= \frac{n_{C_3H_7OH,\text{product}}{}^{s}}{n_{C_3H_6,\text{reacted}}} \times 100\% \\
&= \frac{(3.18\ \text{mol})(1)}{26.63\ \text{mol}} \times 100\% \\
&= 11.93\% \quad (12\%)
\end{aligned}$$

The answer is (C).

10. The final concentration of species M is

$$\begin{aligned}
C_M &= (1 - x_M)C_{M,o} \\
&= (1 - 0.92)\left(4.8\ \frac{\text{mol}}{\text{L}}\right) \\
&= 0.384\ \text{mol/L}
\end{aligned}$$

Because the formation of species N is given as a zero-order reaction, the rate of formation of species N is $r_N = k_N = 2.8$ mol/L·min. A mole balance of species M gives the rate of reaction of disappearance of species M.

$$\begin{aligned}
-r_M &= k_N + k_P C_M \\
&= 2.8\ \frac{\text{mol}}{\text{L·min}} + \left(1.2\ \frac{1}{\text{min}}\right)\left(0.384\ \frac{\text{mol}}{\text{L}}\right) \\
&= 3.261\ \text{mol/L·min}
\end{aligned}$$

For a CSTR, the residence time is

$$\begin{aligned}\theta &= \frac{C_{M,o} - C_M}{-r_M} \\ &= \frac{4.8\ \frac{\text{mol}}{\text{L}} - 0.384\ \frac{\text{mol}}{\text{L}}}{3.261\ \frac{\text{mol}}{\text{L·min}}} \\ &= 1.354\ \text{min}\end{aligned}$$

The overall yield is

$$\begin{aligned}y_N &= \frac{r_N\theta}{C_{M,o} - C_M} \\ &= \frac{\left(2.8\ \frac{\text{mol}}{\text{L·min}}\right)(1.354\ \text{min})}{4.8\ \frac{\text{mol}}{\text{L}} - 0.384\ \frac{\text{mol}}{\text{L}}} \\ &= 0.8585 \quad (0.86)\end{aligned}$$

The answer is (D).

11. The balanced reactions are

$$2C_2H_5OH \rightarrow C_4H_{10}O + H_2O$$

$$C_2H_5OH \rightarrow C_2H_4 + H_2O$$

In the first reaction, the reaction rate constant for the disappearance of ethanol, k_1, is given as 20 1/min. In the second reaction, k_2 is given as 10 1/min. Let C_A be the instantaneous concentration of ethanol at a given time. Because the first reaction is first order, by definition the rate of disappearance of ethanol in the first reaction is

$$r_1 = -k_1C_A$$

Because diethyl ether is produced in the first reaction, the rate of formation of diethyl ether equals the rate of disappearance of ethanol in the first reaction. The rate of formation of the desired product, diethyl ether, is

$$r_D = -r_1 = k_1C_A$$

Ethanol is also consumed in the second reaction. Because the second reaction is a first-order reaction, by definition the rate of disappearance of ethanol in the second reaction is

$$r_2 = -k_2C_A$$

The rate of formation of ethene in the second reaction equals the rate of disappearance of ethanol in the second reaction. The rate of formation of ethene in the second reaction is

$$r_U = -r_2 = k_2C_A$$

By definition the selectivity of ethanol toward the formation of diethyl ether is

$$\begin{aligned}S_{D/U} &= \frac{r_D}{r_U} = \frac{k_1C_A}{k_2C_A} = \frac{k_1}{k_2} \\ &= \frac{20\ \frac{1}{\text{min}}}{10\ \frac{1}{\text{min}}} \\ &= 2.0\end{aligned}$$

The answer is (C).

12. The balanced reactions are

$$C_4H_8 + H_2O \rightarrow C_4H_9OH$$

$$2C_4H_9OH \rightarrow (C_4H_9)_2O + H_2O$$

Let the basis of calculation be 100 mol of feed. The number of moles of 1-butene in the feed stream, $n_{C_4H_8,f}$, is 62 mol. From the hydrolysis reaction, the stoichiometric coefficient, s, is 1. The number of moles of inert in the feed stream, $n_{\text{inert},f}$, is 6 mol. As the inert material remains constant, the ratio of moles inert in the feed stream to moles inert in the product remains constant. This ratio is also applicable to the other components in the product stream. The number of moles of butanol in the product stream is

$$\begin{aligned}n_{C_4H_8,p} &= (5.26)\left(\frac{n_{\text{inert},f}}{14.28}\right) \\ &= (5.26)\left(\frac{6\ \text{mol}}{14.28}\right) \\ &= 2.21\ \text{mol}\end{aligned}$$

The number of moles of butyl ether in the product stream is

$$\begin{aligned}n_{(C_4H_8)_2O,p} &= (0.32)\left(\frac{n_{\text{inert},f}}{14.28}\right) \\ &= (0.32)\left(\frac{6\ \text{mol}}{14.28}\right) \\ &= 0.1345\ \text{mol}\end{aligned}$$

The selectivity of a reaction is the ratio of the desired product (butanol) to the undesired product (butyl ether). The selectivity of butanol is

$$\begin{aligned} y &= \frac{n_{C_4H_8,p}s}{n_{(C_4H_8)_2O,p}} \\ &= \frac{(2.21\ \text{mol})(1)}{0.1345\ \text{mol}} \\ &= 16.438 \quad (16) \end{aligned}$$

The answer is (D).

13. Perform a molar flow rate balance to find the molar flow rate of $C_6H_5CH_3$ consumed in the reactor.

$$\begin{aligned} n_{C_6H_5CH_3,c} &= n_{C_6H_5CH_3,f} - n_{C_6H_5CH_3,o} \\ &= 365\ \frac{\text{kmol}}{\text{h}} - 112\ \frac{\text{kmol}}{\text{h}} \\ &= 253\ \text{kmol/hr} \end{aligned}$$

Perform a molar flow rate balance to find the molar flow rate of C_6H_6 produced in the reactor.

$$\begin{aligned} n_{C_6H_6,p} &= n_{C_6H_6,o} - n_{C_6H_6,f} \\ &= 375\ \frac{\text{kmol}}{\text{h}} - 16\ \frac{\text{kmol}}{\text{h}} \\ &= 359\ \text{kmol/hr} \end{aligned}$$

The yield of C_6H_6 from $C_6H_5CH_3$ is

$$\begin{aligned} Y &= \frac{n_{C_6H_6,p}}{n_{C_6H_5CH_3,c}} \\ &= \frac{359\ \frac{\text{kmol}}{\text{h}}}{253\ \frac{\text{kmol}}{\text{h}}} \\ &= 1.4189 \quad (1.4) \end{aligned}$$

The answer is (C).

14. Find the amount of A converted in the reaction by multiplying the initial concentration of A by the desired fractional conversion of A.

$$\left(3.4\ \frac{\text{mol}}{\text{L}}\right)(0.90) = 3.06\ \frac{\text{mol}}{\text{L}}$$

Find the final concentration of A by subtracting the converted amount of A from the initial amount of A.

$$3.4\ \frac{\text{mol}}{\text{L}} - 3.06\ \frac{\text{mol}}{\text{L}} = 0.34\ \frac{\text{mol}}{\text{L}}$$

The selectivity for the formation of B and C is the molar ratio of desired product B to undesired product C formed by the parallel reactions.

$$S = \frac{r_B}{r_C}$$

Use the final concentration of A to find the selectivity for the formation of B and C.

$$\begin{aligned} S = \frac{r_B}{r_C} &= \frac{0.015e^{(8000)(1/400-1/T)}C_A}{0.019e^{(7000)(1/400-1/T)}C_A^{1.5}} \\ &= \frac{0.789e^{(1000)(1/400-1/T)}}{C_A^{0.5}} \\ &= \frac{0.789e^{(1000)(1/400-1/(200^\circ\text{C}+273^\circ))}}{(0.34)^{0.5}} \\ &= 1.99 \quad (2.0) \end{aligned}$$

The answer is (D).

15. For the reaction $a\text{A} + b\text{B} \rightleftharpoons c\text{C} + d\text{D}$, the equilibrium constant is

$$K_{eq} = \frac{[\text{C}]^c[\text{D}]^d}{[\text{A}]^a[\text{B}]^b}$$

For the given reaction, then, the expression for the equilibrium constant is

$$K_{eq} = \frac{[\text{NO}]}{[\text{N}_2]^{1/2}[\text{O}_2]^{1/2}}$$

The answer is (C).

16. According to Le Châtelier's principle, when a stress (such as change in pressure, temperature, or composition) is applied to a system already in equilibrium, the equilibrium shifts in a way that tends to relieve the stress. When pressure is increased from compressing a system in equilibrium, the system moves in the direction of relieving or decreasing the pressure. For a gaseous system, the system pressure is proportional to the number of moles. In the forward reaction of the system specified, 3 mol of reactants result in the formation of 1 mol of product. Therefore, the system pressure decreases as the forward reaction takes place. Since the forward reaction produces methanol, the methanol concentration increases when the system is compressed.

The answer is (C).

17. Convert the temperature to kelvins.

$$T = 500^\circ\text{C} + 273.15^\circ = 773.15\text{K}$$

Calculate the equilibrium constant for the reaction from the Gibbs free energy.

$$\Delta \hat{g}_r = -RT \ln K_{\text{eq}}$$

$$\ln K_{\text{eq}} = -\frac{\Delta \hat{g}_r}{RT}$$

$$K_{\text{eq}} = e^{-\Delta \hat{g}_r / RT}$$

$$= e^{-(-13\,000\ \text{cal/mol})/(1.987\ \text{cal/mol})(773.15\text{K})}$$

$$= 4732.31 \quad (4700)$$

The answer is (A).

28 Basic Chemical Plant Design

PRACTICE PROBLEMS

1. Use the laboratory pilot-study data provided to design a reactor with a 35 000 L capacity. The fluid has a viscosity of 0.01 kg/m · s and specific gravity of 1.1.

laboratory data	
vessel diameter	25 cm
impeller diameter	7.5 cm
impeller type	4-bladed, 45° pitched-blade turbine
liquid level	25 cm
baffle width	2 cm
number of baffles	4
impeller speed	690 rpm
vessel volume	13.5 L
power	7.0 W

What is the power requirement of the new impeller?

(A) 96 W

(B) 1314 W

(C) 18 150 W

(D) 3.38×10^6 W

2. Heat transfer in clean, round pipes follows the empirical correlation,

$$\text{Nu} = 0.0225\text{Re}^{0.8}\text{Pr}^{0.4}$$

In a pilot-scale unit, a heat-transfer coefficient of 300 W/m²·K is measured for fluid flowing at 5 m/s through a 3 cm diameter pipe. Calculate the heat-transfer coefficient when the process is scaled up to a geometrically similar 15 cm diameter pipe, using the same fluid.

(A) 22 $W/m^2{\cdot}K$

(B) 60 $W/m^2{\cdot}K$

(C) 83 $W/m^2{\cdot}K$

(D) 300 $W/m^2{\cdot}K$

3. 10 000 kg/h of fresh feed is provided to process 1. As the gross product exits process 1, x is the fraction of the gross product that is recycled product. The gross product then enters process 2. y is the fraction of the gross product that exits process 2 as waste, and z is the fraction of the gross product that exits process 2 as finished product. The remainder of the gross product is recycled product that exits process 2 and is transferred to process 1.

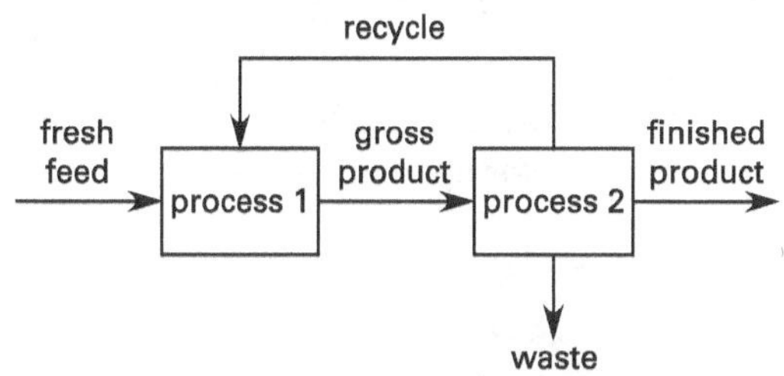

The ratio of recycled to finished product is

(A) 1/2

(B) (gross product/x)/$(1 - x - y)$

(C) $x/(1 - x)$

(D) $x/(1 - x - y)$

SOLUTIONS

1. For geometric similarity, the number of baffles must be the same as the laboratory unit (4) and the impeller type must be the same (4-bladed, 45° pitched-blade turbine). The other geometric quantities are calculated with a scale factor, R, that is estimated from the reactor volumes.

$$R = \frac{D_f}{D_p} \approx \left(\frac{V_f}{V_p}\right)^{1/3} = \left(\frac{35\,000\ \text{L}}{13.5\ \text{L}}\right)^{1/3} = 13.7$$

The required impeller speed for the full-scale reactor is calculated by maintaining kinematic similarity.

$$n_f = n_p\left(\frac{D_p}{D_f}\right) = \frac{n_p}{R} = \frac{690\ \frac{\text{rev}}{\text{min}}}{13.7} = 50.4\ \text{rpm}$$

As in *NCEES Handbook:* Tank Mixing, the power number is held constant for dynamic similarity.

$$\begin{aligned} N_{p,f} &= N_{p,p} \\ \frac{P_f}{\rho n_f^3 D_f^5} &= \frac{P_p}{\rho n_p^3 D_p^5} \\ P_f &= P_p\left(\frac{n_f}{n_p}\right)^3\left(\frac{D_f}{D_p}\right)^5 = P_p\left(\frac{1}{R}\right)^3 R^5 \\ &= P_p R^2 \\ &= (7.0\ \text{W})(13.7)^2 \\ &= 1314\ \text{W} \end{aligned}$$

The answer is (B).

2. Maintaining a constant Reynolds number, as shown in *NCEES Handbook* table "Various Forms of Reynolds Numbers and Their Units in Circular Conduits (Pipes)," provides kinematic similarity.

$$\begin{aligned} \text{Re}_p &= \text{Re}_f \\ \left(\frac{D\text{v}\rho}{\mu}\right)_p &= \left(\frac{D\text{v}\rho}{\mu}\right)_f \\ \text{v}_f = \text{v}_p\left(\frac{D_p}{D_f}\right) &= \left(5\ \frac{\text{m}}{\text{s}}\right)\left(\frac{3\ \text{cm}}{15\ \text{cm}}\right) = 1\ \text{m/s} \end{aligned}$$

The heat-transfer coefficient in the full-scale unit can be obtained by scaling the given design equation. The Nusselt equation is in *NCEES Handbook:* Convective Mass Transfer. The Prandtl equation is in *NCEES Handbook* table "Dimensionless Numbers." The constant terms and fluid properties will cancel.

$$\begin{aligned} \frac{\text{Nu}_p}{\text{Nu}_f} &= \frac{0.0225\text{Re}_p^{0.8}\text{Pr}_p^{0.4}}{0.0225\text{Re}_f^{0.8}\text{Pr}_f^{0.4}} = \left(\frac{\text{Re}_p}{\text{Re}_f}\right)^{0.8}\left(\frac{\text{Pr}_p}{\text{Pr}_f}\right)^{0.4} \\ \frac{h_p D_p}{h_f D_f} &= \left(\frac{\text{v}_p D_p}{\text{v}_f D_f}\right)^{0.8} \\ h_f &= h_p\left(\frac{D_p}{D_f}\right)^{0.2}\left(\frac{\text{v}_f}{\text{v}_p}\right)^{0.8} \\ &= \left(300\ \frac{\text{W}}{\text{m}^2\text{·K}}\right)\left(\frac{3\ \text{cm}}{15\ \text{cm}}\right)^{0.2}\left(\frac{1\ \frac{\text{m}}{\text{s}}}{5\ \frac{\text{m}}{\text{s}}}\right)^{0.8} \\ &= 60\ \text{W/m}^2\text{·K} \end{aligned}$$

The answer is (B).

3. The flow of gross product is

$$\begin{aligned} \text{gross product} &= \text{fresh feed} + \text{recycle} \\ &= 10\,000\ \frac{\text{kg}}{\text{h}} + \text{recycle} \end{aligned}$$

The flow of the recycle is

$$\text{recycle} = x(\text{gross product})$$

The flow of the finished product is

$$\begin{aligned} \text{finished product} &= \text{gross product} - \text{recycle} - \text{waste} \\ &= \text{gross product} - x(\text{gross product}) \\ &\quad - y(\text{gross product}) \end{aligned}$$

The ratio of recycle to finished product is

$$\begin{aligned} \frac{\text{recycle}}{\text{finished product}} &= \frac{x(\text{gross product})}{\text{gross product} - x(\text{gross product}) - y(\text{gross product})} \\ &= \frac{x}{1 - x - y} \end{aligned}$$

The answer is (D).

29 Psychrometrics

PRACTICE PROBLEMS

1. A room contains air at 80°F (27°C) dry-bulb and 67°F (19°C) wet-bulb. The total pressure is 1 atm. Assuming a specific heat capacity for steam of 0.4 Btu/lbm/F, the specific heat of the room's atmosphere is

(A) 0.234 Btu/lbm-°F (0.979 kJ/kg·K)

(B) 0.237 Btu/lbm-°F (0.991 kJ/kg·K)

(C) 0.239 Btu/lbm-°F (0.999 kJ/kg·K)

(D) 0.242 Btu/lbm-°F (1.012 kJ/kg·K)

2. 1000 ft^3/min (0.5 m^3/s) of air at 50°F (10°C) dry-bulb and 95% relative humidity are mixed with 1500 ft^3/min (0.75 m^3/s) of air at 76°F (24°C) dry-bulb and 45% relative humidity. The specific humidity of the mixture is most nearly

(A) 0.008 lbm/lbm (0.008 kg/kg)

(B) 0.009 lbm/lbm (0.009 kg/kg)

(C) 0.010 lbm/lbm (0.010 kg/kg)

(D) 0.011 lbm/lbm (0.011 kg/kg)

3. 95°F (35°C) dry-bulb, 75°F (24°C) wet-bulb air passes through a cooling tower and leaves at 85°F (29°C) dry-bulb and 90% relative humidity. The change in moisture content per cubic foot (meter) of air is most nearly

(A) 1.8×10^{-4} lbm/ft^3 (2.7×10^{-3} kg/m^3)

(B) 3.3×10^{-4} lbm/ft^3 (4.5×10^{-3} kg/m^3)

(C) 6.7×10^{-4} lbm/ft^3 (9.9×10^{-3} kg/m^3)

(D) 9.2×10^{-4} lbm/ft^3 (14×10^{-3} kg/m^3)

4. An air washer receives 1800 ft^3/min (0.85 m^3/s) of air at 70°F (21°C) and 40% relative humidity and discharges the air at 75% relative humidity. A recirculating water spray with a constant temperature of 50°F (10°C) is used. What mass of makeup water is required per minute?

(A) 0.127 lbm/min (0.00141 kg/s)

(B) 0.267 lbm/min (0.00296 kg/s)

(C) 0.374 lbm/min (0.00415 kg/s)

(D) 0.961 lbm/min (0.01067 kg/s)

5. An air washer receives 1800 ft^3/min (0.85 m^3/s) of air. Air is received at 70°F (21°C) and 40% relative humidity and discharges at 75% relative humidity. The air washer recirculates a spray of saturated steam at atmospheric pressure. The mass of makeup water required is most nearly

(A) 0.801 lbm/min (0.0056 kg/s)

(B) 1.437 lbm/min (0.01005 kg/s)

(C) 1.907 lbm/min (0.01333 kg/s)

(D) 3.541 lbm/min (0.02476 kg/s)

6. During performances, a theater experiences a sensible heat load of 500,000 Btu/hr (150 kW) and a moisture load of 175 lbm/hr (80 kg/h). Air enters the theater at 65°F (18°C) and 55% relative humidity and is removed when it reaches 75°F (24°C) or 60% relative humidity, whichever comes first. What is the relative humidity of the air leaving the theater?

(A) 38%

(B) 44%

(C) 58%

(D) 62%

7. 500 ft^3/min (0.25 m^3/s) of air at 80°F (27°C) dry-bulb and 70% relative humidity are removed from a room. 150 ft^3/min (0.075 m^3/s) pass through an air conditioner and leave saturated at 50°F (10°C). The remaining 350 ft^3/min (0.175 m^3/s) bypass the air conditioner and mix with the conditioned air at 1 atm. The heat load (in tons) of the air conditioner is most nearly

(A) 0.90 ton

(B) 1.3 ton

(C) 2.4 tons

(D) 2.9 tons

8. A dehumidifier takes 5000 ft^3/min (2.36 m^3/s) of air at 95°F (35°C) dry-bulb and 70% relative humidity and discharges it at 60°F (16°C) dry-bulb and 95% relative humidity. The dehumidifier uses a wet R-12 refrigeration cycle operating between 100°F (saturated) (38°C) and 50°F (10°C). Find the quantity of heat removed from the air.

(A) 8345 Btu/min (143.0 kW)

(B) 8551 Btu/min (146.5 kW)

(C) 10,508 Btu/min (180.0 kW)

(D) 10,714 Btu/min (183.6 kW)

9. 1500 ft^3/min (0.71 m^3/s) of saturated 25 psia (170 kPa) air is heated from 200°F to 400°F (93°C to 204°C) in a constant pressure, constant moisture drying process. The heat required per unit mass of dry air is most nearly

(A) 31 Btu/lbm (71 kJ/kg)

(B) 57 Btu/lbm (130 kJ/kg)

(C) 99 Btu/lbm (230 kJ/kg)

(D) 120 Btu/lbm (280 kJ/kg)

10. 410 lbm/hr (0.052 kg/s) of dry 800°F (427°C) air pass through a scrubber to reduce particulate emissions. To protect the elastomeric seals in the scrubber, the air temperature is reduced to 350°F (177°C) by passing the air through a spray of 80°F (27°C) water. The pressure in the spray chamber is 20 psia (140 kPa). Approximately how much water is evaporated per hour?

(A) 18 lbm/hr (0.0023 kg/s)

(B) 27 lbm/hr (0.0035 kg/s)

(C) 31 lbm/hr (0.0040 kg/s)

(D) 39 lbm/hr (0.0050 kg/s)

11. Combustion products leaving a gas turbine combustor are released at a temperature of 180°F into the atmosphere. The combustion products have a relative humidity of 20%. What is most nearly the dew point temperature of the combustion products?

(A) 110°F

(B) 120°F

(C) 130°F

(D) 140°F

SOLUTIONS

1. *Customary U.S. Solution*

Locate the intersection of 80°F dry-bulb and 67°F wet-bulb on the psychrometric chart (see App. 29A (also *NCEES Handbook* figure "Psychrometric Chart (U.S. Customary Units)"). Read the value of humidity and enthalpy.

$$\omega = 0.0112 \text{ lbm moisture/lbm dry air} \quad (0.011 \text{ lbm/lbm})$$
$$h = 31.5 \text{ Btu/lbm dry air}$$

c_p is gravimetrically weighted. c_p for air is 0.240 Btu/lbm-°F, and c_p for steam is approximately 0.40 Btu/lbm-°F.

$$G_{\text{air}} = \frac{1}{1+0.0112} = 0.989$$
$$G_{\text{steam}} = \frac{0.0112}{1+0.0112} = 0.011$$
$$\begin{aligned} c_{p,\text{mixture}} &= G_{\text{air}}c_{p,\text{air}} + G_{\text{steam}}c_{p,\text{steam}} \\ &= (0.989)\left(0.240 \ \frac{\text{Btu}}{\text{lbm-°F}}\right) + (0.011)\left(0.40 \ \frac{\text{Btu}}{\text{lbm-°F}}\right) \\ &= 0.242 \text{ Btu/lbm-°F} \end{aligned}$$

SI Solution

Locate the intersection of 27°C dry-bulb and 19°C wet-bulb on the psychrometric chart (see App. 29B (also *NCEES Handbook* figure "Psychrometric Chart (SI Units)"). Read the value of humidity and enthalpy.

$$\omega = \frac{10.5 \ \frac{\text{g}}{\text{kg dry air}}}{1000 \ \frac{\text{g}}{\text{kg}}} = 0.0105 \text{ kg/kg dry air} \quad (0.011 \text{ kg/kg})$$
$$h = 53.9 \text{ kJ/kg dry air}$$

c_p is gravimetrically weighted. c_p for air is 1.0048 kJ/kg·K, and c_p for steam is approximately 1.675 kJ/kg·K.

$$G_{\text{air}} = \frac{1}{1+0.0105} = 0.990$$
$$G_{\text{steam}} = \frac{0.0105}{1+0.0105} = 0.010$$
$$\begin{aligned} c_{p,\text{mixture}} &= G_{\text{air}}c_{p,\text{air}} + G_{\text{air}}c_{p,\text{steam}} \\ &= (0.990)\left(1.0048 \ \frac{\text{kJ}}{\text{kg·K}}\right) + (0.010)\left(1.675 \ \frac{\text{kJ}}{\text{kg·K}}\right) \\ &= 1.0115 \text{ kJ/kg·K} \quad (1.012 \text{ kJ/kg·K}) \end{aligned}$$

The answer is (D).

2. Use the illustration shown for both the customary U.S. and SI solutions.

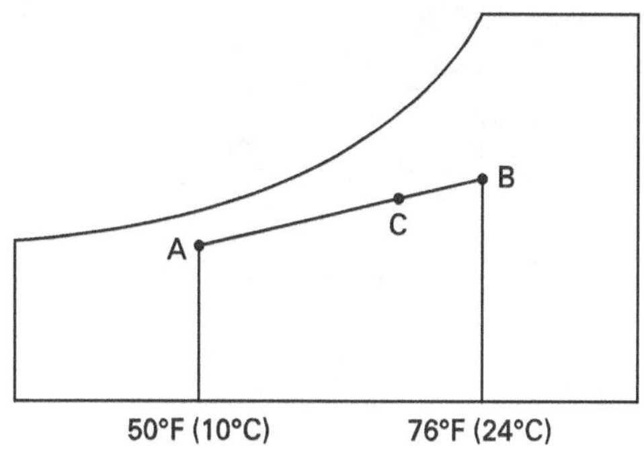

Customary U.S. Solution

Locate the two points on the psychrometric chart in App. 29A (also *NCEES Handbook* figure "Psychrometric Chart (U.S. Customary Units)") and draw a line between them.

Reading from the chart (specific volumes),

$$\upsilon_{\text{A}} = 13.0 \text{ ft}^3\text{/lbm}$$
$$\upsilon_{\text{B}} = 13.7 \text{ ft}^3\text{/lbm}$$

The density at each point is

$$\rho_{\text{A}} = \frac{1}{\upsilon_{\text{A}}} = \frac{1}{13.0 \ \frac{\text{ft}^3}{\text{lbm}}} = 0.0769 \text{ lbm/ft}^3$$
$$\rho_{\text{B}} = \frac{1}{\upsilon_{\text{B}}} = \frac{1}{13.7 \ \frac{\text{ft}^3}{\text{lbm}}} = 0.0730 \text{ lbm/ft}^3$$

As in *NCEES Handbook:* Conservation of Mass, the mass flow at each point is

$$\dot{m}_A = \rho_A \dot{V}_A = \left(0.0769\ \frac{\text{lbm}}{\text{ft}^3}\right)\left(1000\ \frac{\text{ft}^3}{\text{min}}\right) = 76.9\ \text{lbm/min}$$

$$\dot{m}_B = \rho_B \dot{V}_B = \left(0.0730\ \frac{\text{lbm}}{\text{ft}^3}\right)\left(1500\ \frac{\text{ft}^3}{\text{min}}\right) = 109.5\ \text{lbm/min}$$

The gravimetric fraction of flow A is

$$\frac{76.9\ \frac{\text{lbm}}{\text{min}}}{76.9\ \frac{\text{lbm}}{\text{min}} + 109.5\ \frac{\text{lbm}}{\text{min}}} = 0.413$$

Since the scales are all linear,

$$0.413 = \frac{T_B - T_C}{T_B - T_A}$$

$$\begin{aligned} T_C &= T_B - (0.413)(T_B - T_A) \\ &= 76^\circ\text{F} - (0.413)(76^\circ\text{F} - 50^\circ\text{F}) \\ &= 65.3^\circ\text{F} \end{aligned}$$

$$\omega = \begin{matrix} 0.0082\ \text{lbm moisture/lbm dry air} \\ (0.008\ \text{lbm/lbm}) \end{matrix}$$

SI Solution

Locate the two points on the psychrometric chart in App. 29B (also *NCEES Handbook* figure "Psychrometric Chart (SI Units)"), and draw a line between them.

Reading from the chart (specific volumes),

$$\upsilon_A = 0.813\ \text{m}^3\text{/kg dry air}$$

$$\upsilon_B = 0.856\ \text{m}^3\text{/kg dry air}$$

The density at each point is

$$\rho_A = \frac{1}{\upsilon_A} = \frac{1}{0.813\ \frac{\text{m}^3}{\text{kg}}} = 1.23\ \text{kg/m}^3$$

$$\rho_B = \frac{1}{\upsilon_B} = \frac{1}{0.856\ \frac{\text{m}^3}{\text{kg}}} = 1.17\ \text{kg/m}^3$$

As in *NCEES Handbook:* Conservation of Mass, the mass flow at each point is

$$\dot{m}_A = \rho_A \dot{V}_A = \left(1.23\ \frac{\text{kg}}{\text{m}^3}\right)\left(0.5\ \frac{\text{m}^3}{\text{s}}\right) = 0.615\ \text{kg/s}$$

$$\dot{m}_B = \rho_B \dot{V}_B = \left(1.17\ \frac{\text{kg}}{\text{m}^3}\right)\left(0.75\ \frac{\text{m}^3}{\text{s}}\right) = 0.878\ \text{kg/s}$$

The gravimetric fraction of flow A is

$$\frac{0.615\ \frac{\text{kg}}{\text{s}}}{0.615\ \frac{\text{kg}}{\text{s}} + 0.878\ \frac{\text{kg}}{\text{s}}} = 0.412$$

Since the scales are linear,

$$0.412 = \frac{T_B - T_C}{T_B - T_A}$$

$$\begin{aligned} T_C &= T_B - (0.412)(T_B - T_A) \\ &= 24^\circ\text{C} - (0.412)(24^\circ\text{C} - 10^\circ\text{C}) \\ &= 18.2^\circ\text{C} \end{aligned}$$

$$\begin{aligned} \omega &= \frac{8.0\ \frac{\text{g}}{\text{kg dry air}}}{1000\ \frac{\text{g}}{\text{kg}}} \\ &= \begin{matrix} 0.008\ \text{kg moisture/kg dry air} \\ (0.008\ \text{kg/kg}) \end{matrix} \end{aligned}$$

The answer is (A).

3. *Customary U.S. Solution*

Refer to the psychrometric chart (see App. 29A and *NCEES Handbook* figure "Psychrometric Chart (U.S. Customary Units)").

At point 1, properties of air at $T_{db} = 95^\circ\text{F}$ and $T_{wb} = 75^\circ\text{F}$ are

$$\omega_1 = 0.0141\ \text{lbm moisture/lbm air}$$

$$h_1 = 38.4\ \text{Btu/lbm air}$$

$$\upsilon_1 = 14.3\ \text{ft}^3\text{/lbm air}$$

At point 2, properties of air at $T_{db} = 85^\circ\text{F}$ and 90% relative humidity are

$$\omega_2 = 0.0237\ \text{lbm moisture/lbm air}$$

$$h_2 = 46.6\ \text{Btu/lbm air}$$

The enthalpy change is

$$\frac{h_2 - h_1}{\upsilon_1} = \frac{46.6 \ \frac{\text{Btu}}{\text{lbm air}} - 38.4 \ \frac{\text{Btu}}{\text{lbm air}}}{14.3 \ \frac{\text{ft}^3}{\text{lbm air}}}$$
$$= 0.573 \ \text{Btu/ft}^3 \ \text{air}$$

The moisture added is

$$\frac{\omega_2 - \omega_1}{\upsilon_1} = \frac{0.0237 \ \frac{\text{lbm moisture}}{\text{lbm air}} - 0.0141 \ \frac{\text{lbm moisture}}{\text{lbm air}}}{14.3 \ \frac{\text{ft}^3}{\text{lbm air}}}$$
$$= \begin{array}{c} 6.71 \times 10^{-4} \ \text{lbm/ft}^3 \ \text{air} \\ (6.7 \times 10^{-4} \ \text{lbm/ft}^3) \end{array}$$

SI Solution

Refer to the psychrometric chart (see App. 29B and *NCEES Handbook* figure "Psychrometric Chart (SI Units)").

At point 1, properties of air at $T_{\text{db}} = 35°\text{C}$ and $T_{\text{wb}} = 24°\text{C}$ are

$$\omega_1 = 14.3 \ \text{g/kg air}$$
$$h_1 = 71.8 \ \text{kJ/kg air}$$
$$\upsilon_1 = 0.8893 \ \text{m}^3\text{/kg air}$$

At point 2, properties of air at $T_{\text{db}} = 29°\text{C}$ and 90% relative humidity are

$$\omega_2 = 23.1 \ \text{g/kg air}$$
$$h_2 = 88 \ \text{kJ/kg air}$$

The enthalpy change is

$$\frac{h_2 - h_1}{\upsilon_1} = \frac{88 \ \frac{\text{kJ}}{\text{kg air}} - 71.8 \ \frac{\text{kJ}}{\text{kg air}}}{0.8893 \ \frac{\text{m}^3}{\text{kg air}}}$$
$$= 18.2 \ \text{kJ/m}^3 \ \text{air}$$

The moisture added is

$$\frac{\omega_2 - \omega_1}{\upsilon_1} = \frac{23.1 \ \frac{\text{g}}{\text{kg air}} - 14.3 \ \frac{\text{g}}{\text{kg air}}}{\left(0.8893 \ \frac{\text{m}^3}{\text{kg air}}\right)\left(1000 \ \frac{\text{g}}{\text{kg}}\right)}$$
$$= 9.9 \times 10^{-3} \ \text{kg/m}^3$$

The answer is (C).

4. Use the illustration shown for both the customary U.S. and SI solutions.

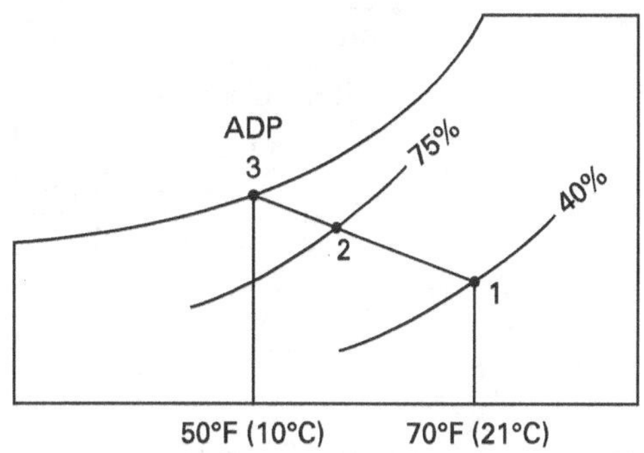

Customary U.S. Solution

Refer to the psychrometric chart (see App. 29A and *NCEES Handbook* figure "Psychrometric Chart (U.S. Customary Units)").

At point 1, properties of air at $T_{\text{db}} = 70°\text{F}$ and $\phi = 40\%$ are

$$h_1 = 23.6 \ \text{Btu/lbm air}$$
$$\omega_1 = 0.00625 \ \text{lbm moisture/lbm air}$$
$$\upsilon_1 = 13.48 \ \text{ft}^3\text{/lbm air}$$

The mass flow rate of incoming air is

$$\dot{m}_{a,1} = \frac{\dot{V}_1}{\upsilon_1} = \frac{1800 \ \frac{\text{ft}^3}{\text{min}}}{13.48 \ \frac{\text{ft}^3}{\text{lbm air}}} = 133.53 \ \text{lbm air/min}$$

Locate point 1 on the psychrometric chart.

Notice that the temperature of the recirculating water is constant but not equal to the air's entering wet-bulb temperature. Therefore, this is not an adiabatic process.

Locate point 3 as 50°F saturated condition (water being sprayed) on the psychrometric chart.

Draw a line from point 1 to point 3. The intersection of this line with 75% relative humidity defines point 2 as

$$h_2 = 21.4 \text{ Btu/lbm air}$$
$$\omega_2 = 0.0072 \text{ lbm moisture/lbm air}$$
$$T_{\text{db},2} = 56^\circ\text{F}$$
$$T_{\text{wb},2} = 51.8^\circ\text{F}$$

The moisture (water) added is

$$\begin{aligned}\dot{m}_w &= \dot{m}_{a,1}(\omega_2 - \omega_1) \\ &= \left(133.53 \ \frac{\text{lbm air}}{\text{min}}\right) \\ &\quad \times \left(\begin{array}{l} 0.0072 \ \dfrac{\text{lbm moisture}}{\text{lbm air}} \\ \quad -0.00625 \ \dfrac{\text{lbm moisture}}{\text{lbm air}} \end{array}\right) \\ &= 0.127 \text{ lbm/min}\end{aligned}$$

SI Solution

Refer to the psychrometric chart (see App. 29B and *NCEES Handbook* figure "Psychrometric Chart (SI Units)").

At point 1, properties of air at $T_{\text{db}} = 21^\circ\text{C}$ and $\phi = 40\%$ are

$$h_1 = 36.75 \text{ kJ/kg air}$$
$$\omega_1 = 6.2 \text{ g moisture/kg air}$$
$$\upsilon_1 = 0.842 \text{ m}^3\text{/kg air}$$

The mass flow rate of incoming air is

$$\begin{aligned}\dot{m}_{a,1} = \frac{\dot{V}_1}{\upsilon_1} &= \frac{0.85 \ \dfrac{\text{m}^3}{\text{s}}}{0.842 \ \dfrac{\text{m}^3}{\text{kg air}}} \\ &= 1.010 \text{ kg air/s}\end{aligned}$$

Locate point 1 on the psychrometric chart.

Notice that the temperature of the recirculating water is constant but not equal to the air's entering wet-bulb temperature. Therefore, this is not an adiabatic process.

Locate point 3 as 10°C saturated condition (water being sprayed) on the psychrometric chart.

Draw a line from point 1 to point 3. The intersection of this line with 75% relative humidity defines point 2 as

$$h_2 = 34.1 \text{ kJ/kg air}$$
$$\omega_2 = 7.6 \text{ g moisture/kg air}$$
$$T_{\text{db},2} = 14.7^\circ\text{C}$$
$$T_{\text{wb},2} = 12.0^\circ\text{C}$$

The water added is

$$\begin{aligned}\dot{m}_w &= \dot{m}_{a,1}(\omega_2 - \omega_1) \\ &= \frac{\left(1.010 \ \dfrac{\text{kg air}}{\text{s}}\right)\left(\begin{array}{l} 7.6 \ \dfrac{\text{g moisture}}{\text{kg air}} \\ \quad -6.2 \ \dfrac{\text{g moisture}}{\text{kg air}} \end{array}\right)}{1000 \ \dfrac{\text{g}}{\text{kg}}} \\ &= 0.00141 \text{ kg/s}\end{aligned}$$

The answer is (A).

5. *Customary U.S. Solution*

(See the SI solution for a trial-and-error solution procedure.)

From App. 29A (also *NCEES Handbook* figure "Psychrometric Chart (U.S. Customary Units)"),

$$\omega_1 = 0.00623 \text{ lbm moisture/lbm air}$$
$$h_1 = 23.6 \text{ Btu/lbm air}$$
$$\dot{m}_{a,1} = 133.53 \text{ lbm air/min}$$

From the steam table in App. 18E (also *NCEES Handbook* table "Saturated Steam (U.S. Units)—Temperature Table"), for 1 atm steam, $h_{\text{steam}} = 1150.3$ Btu/lbm.

From the conservation of energy equation,

$$\dot{m}_{a,1}h_1 + \dot{m}_{\text{steam}}h_{\text{steam}} = \dot{m}_{a,1}h_2$$

$$\begin{aligned}&\left(133.53 \ \frac{\text{lbm air}}{\text{min}}\right)\left(23.6 \ \frac{\text{Btu}}{\text{lbm air}}\right) \\ &\quad + \dot{m}_{\text{steam}}\left(1150.3 \ \frac{\text{Btu}}{\text{lbm}}\right) = \left(133.53 \ \frac{\text{lbm air}}{\text{min}}\right)h_2 \quad \text{[Eq. 1]}\end{aligned}$$

Solve for $\dot{m}_{\text{steam}}$.

$$\dot{m}_{\text{steam}} = 0.1161h_2 - 2.740 \quad \text{[Eq. 3]}$$

From a conservation of mass for the water,

$$\dot{m}_{a,1}\omega_1 + \dot{m}_{\text{steam}} = \dot{m}_{a,2}\omega_2$$

$$\left(133.53\ \frac{\text{lbm air}}{\text{min}}\right)\left(0.00623\ \frac{\text{lbm moisture}}{\text{lbm air}}\right) + \dot{m}_{\text{steam}} = \left(133.53\ \frac{\text{lbm air}}{\text{min}}\right)\omega_2 \quad \text{[Eq. 2]}$$

Solve for $\dot{m}_{\text{steam}}$.

$$\dot{m}_{\text{steam}} = 133.53\omega_2 - 0.8319 \quad \text{[Eq. 4]}$$

Equate Eq. 3 and Eq. 4.

$$0.1161h_2 - 2.740 = 133.53\omega_2 - 0.8319$$
$$h_2 = 1150.1\omega_2 + 16.43$$

Plot this line on the psychrometric chart, and determine properties where it crosses the 75% relative humidity line.

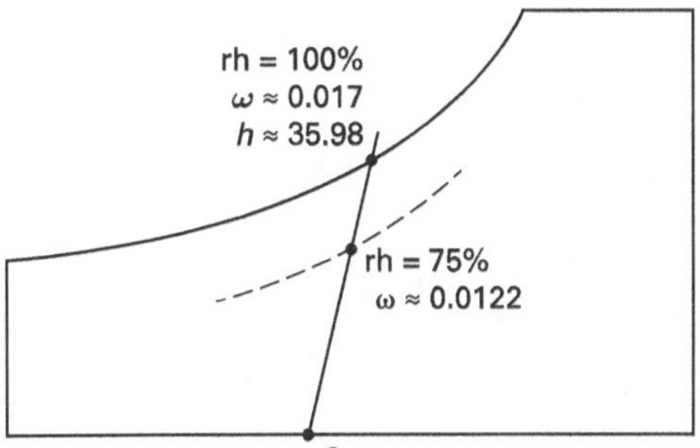

At 75% relative humidity, from the psychrometric chart,

$$\omega_2 = 0.0122\ \text{lbm moisture/lbm air}$$
$$h_2 = 30.5\ \text{Btu/lbm}$$
$$T_{\text{db}} = 72°\text{F}$$
$$T_{\text{wb}} = 66°\text{F}$$

The mass flow rate is

$$\dot{m}_{\text{steam}} = (0.1161)\left(30.5\ \frac{\text{Btu}}{\text{lbm}}\right) - 2.740 = 0.801\ \text{lbm/min}$$

SI Solution

(See the customary U.S. solution for a graphical method.)

As in *NCEES Handbook* figure "Psychrometric Chart (SI Units),"

$$\omega_1 = 6.2\ \text{g moisture/kg air}$$
$$h_1 = 36.75\ \text{kJ/kg air}$$
$$\dot{m}_{a,1} = 1.010\ \text{kg air/s}$$

From the steam table in App. 18F (also *NCEES Handbook* table "Saturated Steam (SI Units)—Temperature Table"), for 1 atm steam, $h_{\text{steam}} = 2675.4$ kJ/kg.

From Eq. 29.33, the conservation of energy equation,

$$\dot{m}_{a,1}h_1 + \dot{m}_{\text{steam}}h_{\text{steam}} = \dot{m}_{a,1}h_2$$

$$\left(1.010\ \frac{\text{kg air}}{\text{s}}\right)\left(36.75\ \frac{\text{kJ}}{\text{kg air}}\right) + \dot{m}_{\text{steam}}\left(2675.4\ \frac{\text{kJ}}{\text{kg}}\right) = \left(1.010\ \frac{\text{kg air}}{\text{s}}\right)h_2 \quad \text{[Eq. 1]}$$

From Eq. 29.34, a conservation of mass for the water,

$$\dot{m}_{a,1}\omega_1 + \dot{m}_{\text{steam}} = \dot{m}_{a,2}\omega_2$$

$$\frac{\left(1.010\ \frac{\text{kg air}}{\text{s}}\right)\left(6.2\ \frac{\text{g moisture}}{\text{kg air}}\right)}{\left(1000\ \frac{\text{kg}}{\text{g}}\right)} + \dot{m}_{\text{steam}} = \left(1.010\ \frac{\text{kg air}}{\text{s}}\right)\omega_2 \quad \text{[Eq. 2]}$$

Since no single relationship exists between ω_2, $\dot{m}_{\text{steam}}$, and h_2, a trial-and-error solution can be used. Once $\dot{m}_{\text{steam}}$ is selected, ω_2 and h_2 can be found from Eq. 1 and Eq. 2 as

$$h_2 = 36.75 + 2648.9\dot{m}_{\text{steam}}$$
$$\omega_2 = 0.0062 + 0.99\dot{m}_{\text{steam}}$$

Once h_2 and ω_2 are known, the relative humidity can be determined from the psychrometric chart. Continue the process until a relative humidity of 75% is achieved.

$\dot{m}_{\text{steam}}$ $\left(\frac{\text{kg}}{\text{s}}\right)$	ω_2 $\left(\frac{\text{kg moisture}}{\text{kg air}}\right)$	h_2 $\left(\frac{\text{kJ}}{\text{kg air}}\right)$	ϕ_2 (%)
0.005	0.0112	49.99	69.5
0.0055	0.0116	51.32	74.5
0.0056	0.0117	51.58	75.0

$$\dot{m}_{\text{steam}} = 0.0056\ \text{kg/s}$$

$$\omega_2 = \left(0.0117\ \frac{\text{kg moisture}}{\text{kg air}}\right)\left(1000\ \frac{\text{g}}{\text{kg}}\right) = 11.7\ \text{g moisture/kg air}$$

$$T_{\text{db}} = 21.3°\text{C}$$
$$T_{\text{wb}} = 18.2°\text{C}$$

The answer is (A).

6. *Customary U.S. Solution*

From the psychrometric chart (see App. 29A and *NCEES Handbook* figure "Psychrometric Chart (U.S. Customary Units)"), for incoming air at 65°F and 55% relative humidity, $\omega_1 = 0.0072$ lbm moisture/lbm air.

With sensible heating as a limiting factor, calculate the mass flow rate of air entering the theater from Eq. 29.25 (ventilation rate).

$$\dot{q} = \dot{m}_a(c_{p,\text{air}} + \omega c_{p,\text{moisture}})(T_2 - T_1)$$

$$500{,}000\ \frac{\text{Btu}}{\text{hr}} = \dot{m}_a \times \left(0.240\ \frac{\text{Btu}}{\text{lbm-°F}} + 0.0072\ \frac{\text{lbm moisture}}{\text{lbm air}} \times \left(0.444\ \frac{\text{Btu}}{\text{lbm-°F}}\right) \right) \times (75\text{°F} - 65\text{°F})$$

$$\dot{m}_a = 2.056 \times 10^5\ \text{lbm air/hr}$$

Assume that this air absorbs all the moisture. Then, the final humidity ratio is given by

$$\dot{m}_w = \dot{m}_a(\omega_2 - \omega_1)$$

$$\omega_2 = \left(\frac{\dot{m}_w}{\dot{m}_a}\right) + \omega_1 = \frac{175\ \frac{\text{lbm moisture}}{\text{hr}}}{2.056 \times 10^5\ \frac{\text{lbm air}}{\text{hr}}} + 0.0072\ \frac{\text{lbm moisture}}{\text{lbm air}}$$

$$= 0.00805\ \text{lbm moisture/lbm air}$$

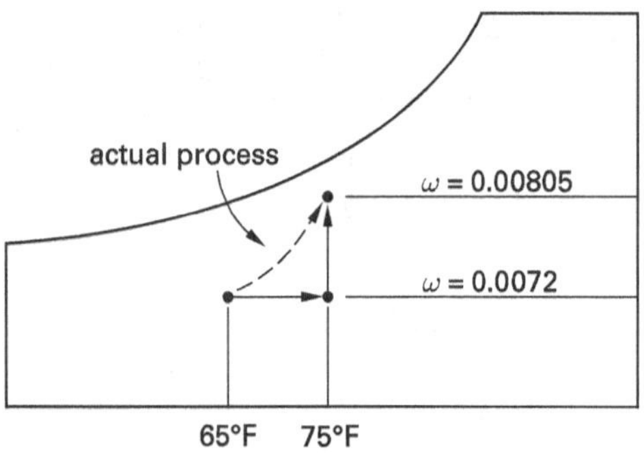

The final conditions are

$$T_{\text{db}} = 75\text{°F} \quad [\text{given}]$$

$$\omega_2 = 0.00805\ \text{lbm moisture/lbm air}$$

From the psychrometric chart (see App. 29A and *NCEES Handbook* figure "Psychrometric Chart (U.S. Customary Units)"), the relative humidity is 44%. This is below 60%.

SI Solution

From the psychrometric chart (see App. 29B and *NCEES Handbook* figure "Psychrometric Chart (SI Units)"), for incoming air at 18°C and 55% relative humidity, $\omega_1 = 7.1$ g moisture/kg air.

With sensible heating as a limiting factor, calculate the mass flow rate of air entering the theater from Eq. 29.25 (ventilation rate).

$$\dot{q} = \dot{m}_a(c_{p,\text{air}} + \omega c_{p,\text{moisture}})(T_2 - T_1)$$

$$(150\ \text{kW}) \times \left(1000\ \frac{\text{W}}{\text{kW}}\right) = \dot{m}_a \left(\left(1.005\ \frac{\text{kJ}}{\text{kg·°C}}\right)\left(1000\ \frac{\text{J}}{\text{kJ}}\right) + \left(1.805\ \frac{\text{kJ}}{\text{kg·°C}}\right)\left(1000\ \frac{\text{J}}{\text{kJ}}\right) \times \left(\frac{7.1\ \frac{\text{g moisture}}{\text{kg air}}}{1000\ \frac{\text{g}}{\text{kg}}}\right) \right) \times (24\text{°C} - 18\text{°C})$$

$$\dot{m}_a = 24.56\ \text{kg/s}$$

Assume that this air absorbs all the moisture. Then, the final humidity ratio is given by

$$\dot{m}_w = \dot{m}_a(\omega_2 - \omega_1)$$

$$\omega_2 = \frac{\dot{m}_w}{\dot{m}_a} + \omega_1 = \frac{80\ \frac{\text{kg}}{\text{h}}}{\left(24.56\ \frac{\text{kg}}{\text{s}}\right)\left(3600\ \frac{\text{s}}{\text{h}}\right)} + \frac{7.0\ \frac{\text{g moisture}}{\text{kg air}}}{1000\ \frac{\text{g}}{\text{kg}}}$$

$$= 0.00790\ \text{kg moisture/kg air}$$

The final conditions are

$$T_{\text{db}} = 24\text{°C}$$

$$\omega_2 = \left(0.00790\ \frac{\text{kg moisture}}{\text{kg air}}\right)\left(1000\ \frac{\text{g}}{\text{kg}}\right) = 7.9\ \text{g moisture/kg air}$$

From the psychrometric chart (see *NCEES Handbook* figure "Psychrometric Chart (SI Units)"), the relative humidity is 44%. This is below 60%.

The answer is (B).

7. Use the illustration shown for both customary U.S. and SI solutions.

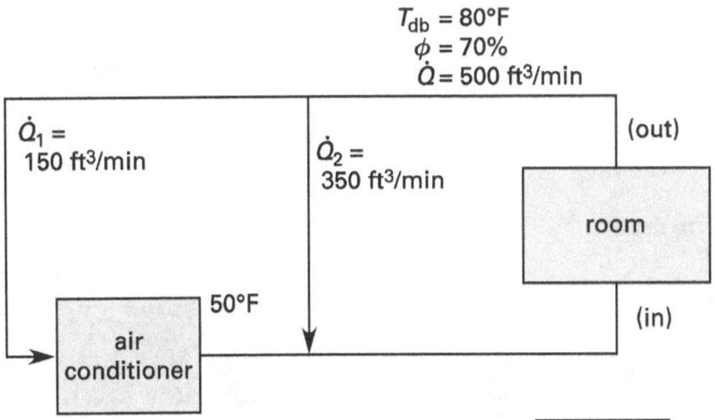

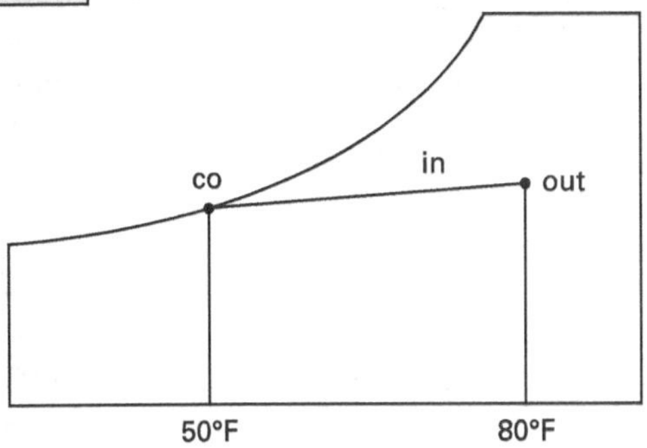

Customary U.S. Solution

Locate point "out" ($T_{db} = 80°F$ and $\phi = 70\%$) and point "co" (saturated at 50°F) on the psychrometric chart. At point "out" from App. 29A (also *NCEES Handbook* figure "Psychrometric Chart (U.S. Customary Units)"),

$$\upsilon_{out} = 13.95\ \text{ft}^3/\text{lbm air}$$
$$h_{out} = 36.2\ \text{Btu/lbm air}$$

At point "co," $h_{co} = 20.3$ Btu/lbm air.

The air conditioner capacity is given by

$$\begin{aligned}\dot{Q} &= \dot{m}_{air}(h_{t,2} - h_{t,1}) = \dot{m}_1(h_{out} - h_{co}) \\ &= \left(10.75\ \frac{\text{lbm air}}{\text{min}}\right)\left(36.2\ \frac{\text{Btu}}{\text{lbm air}} - 20.3\ \frac{\text{Btu}}{\text{lbm air}}\right) \\ &\quad \times \left(\frac{1\ \text{ton}}{200\ \frac{\text{Btu}}{\text{min}}}\right) \\ &= 0.85\ \text{ton} \quad (0.90\ \text{ton})\end{aligned}$$

The answer is (A).

SI Solution

Locate point "out" ($T_{db} = 27°C$, $\phi = 70\%$) and point "co" (saturated at 10°C) on the psychrometric chart. At point "out" from App. 29B (also *NCEES Handbook* figure "Psychrometric Chart (SI Units)"),

$$\upsilon_{out} = 0.872\ \text{m}^3/\text{kg air}$$
$$h_{out} = 67.3\ \text{kJ/kg air}$$

At point "co" from App. 29B (also *NCEES Handbook* figure "Psychrometric Chart (SI Units)"), $h_{co} = 29.26$ kJ/kg air.

As in *NCEES Handbook:* Conservation of Mass, at mass flow rate through the air conditioner,

$$\dot{m}_1 = \frac{\dot{V}_1}{\upsilon} = \frac{0.075\ \frac{\text{m}^3}{\text{s}}}{0.872\ \frac{\text{m}^3}{\text{kg air}}} = 0.0860\ \text{kg air/s}$$

The air conditioner capacity is given by

$$\begin{aligned}\dot{Q} &= \dot{m}_{air}(h_{t,2} - h_{t,1}) = \dot{m}_1(h_{out} - h_{co}) \\ &= \left(0.0860\ \frac{\text{kg air}}{\text{s}}\right)\left(67.3\ \frac{\text{kJ}}{\text{kg air}} - 29.26\ \frac{\text{kJ}}{\text{kg air}}\right) \\ &\quad \times \left(0.2843\ \frac{\text{ton}}{\text{kW}}\right) \\ &= 0.93\ \text{ton} \quad (0.90\ \text{ton})\end{aligned}$$

The answer is (A).

8. Use the illustration shown for both customary U.S. and SI solutions.

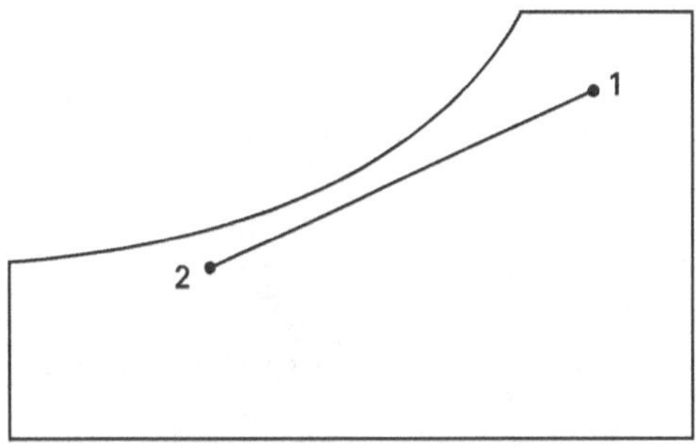

Customary U.S. Solution

At point 1, from the psychrometric chart (see App. 29A and *NCEES Handbook* figure "Psychrometric Chart (U.S. Customary Units)") at $T_{db} = 95°F$ and $\phi = 70\%$,

$$h_1 = 50.7\ \text{Btu/lbm air}$$
$$\upsilon_1 = 14.56\ \text{ft}^3/\text{lbm air}$$
$$\omega_1 = 0.0253\ \text{lbm water/lbm air}$$

At point 2, from the psychrometric chart (see App. 29A and *NCEES Handbook* figure "Psychrometric Chart (U.S. Customary Units)") at $T_{db} = 60°F$ and $\phi = 95\%$,

$$h_2 = 25.8 \text{ Btu/lbm air}$$
$$\omega_2 = 0.0105 \text{ lbm water/lbm air}$$

As in *NCEES Handbook:* Conservation of Mass, the air mass flow rate is

$$\dot{m}_a = \frac{\dot{V}}{v_1} = \frac{5000 \ \frac{\text{ft}^3}{\text{min}}}{14.56 \ \frac{\text{ft}^3}{\text{lbm air}}} = 343.4 \text{ lbm air/min}$$

From Eq. 29.27, the quantity of heat removed is

$$\begin{aligned}\dot{q} &= \dot{m}_a(h_1 - h_2)\\ &= \left(343.4 \ \frac{\text{lbm air}}{\text{min}}\right)\left(50.7 \ \frac{\text{Btu}}{\text{lbm air}} - 25.8 \ \frac{\text{Btu}}{\text{lbm air}}\right)\\ &= 8551 \text{ Btu/min}\end{aligned}$$

SI Solution

At point 1, from the psychrometric chart (see App. 29B and *NCEES Handbook* figure "Psychrometric Chart (SI Units)") at $T_{db} = 35°C$ and $\phi = 70\%$,

$$h_1 = 99.9 \text{ kJ/kg air}$$
$$v_1 = 0.91 \text{ m}^3\text{/kg air}$$
$$\omega_1 = \frac{25.3 \ \frac{\text{g moisture}}{\text{kg air}}}{1000 \ \frac{\text{g}}{\text{kg}}} = 0.0253 \text{ kg moisture/kg air}$$

At point 2, from the psychrometric chart (see App. 29B and *NCEES Handbook* figure "Psychrometric Chart (SI Units)") at $T_{db} = 16°C$ and $\phi = 95\%$,

$$h_2 = 43.4 \text{ kJ/kg air}$$
$$\omega_2 = \frac{10.8 \ \frac{\text{g moisture}}{\text{kg air}}}{1000 \ \frac{\text{g}}{\text{kg}}} = 0.0108 \text{ kg moisture/kg air}$$

As in *NCEES Handbook:* Conservation of Mass, the air mass flow rate is

$$\dot{m}_a = \frac{\dot{V}_1}{v_1} = \frac{\dot{V}_1}{v_{\text{out}}} = \frac{2.36 \ \frac{\text{m}^3}{\text{s}}}{0.91 \ \frac{\text{m}^3}{\text{kg air}}} = 2.593 \text{ kg air/s}$$

From Eq. 29.27, the quantity of heat removed is

$$\begin{aligned}\dot{q} &= \dot{m}_a(h_1 - h_2)\\ &= \left(2.593 \ \frac{\text{kg air}}{\text{s}}\right)\left(99.9 \ \frac{\text{kJ}}{\text{kg air}} - 43.4 \ \frac{\text{kJ}}{\text{kg air}}\right)\\ &= 146.5 \text{ kW}\end{aligned}$$

The answer is (B).

9. *Customary U.S. Solution*

The saturation pressure at 200°F from App. 10A (also *NCEES Handbook* table "Saturated Steam (U.S. Units)—Temperature Table"), is $p_{\text{sat},1} = 11.538$ psia.

Since air is saturated (100% relative humidity), the water vapor pressure is equal to the saturation pressure.

$$p_{w,1} = p_{\text{sat},1} = 11.538 \text{ psia}$$

The partial pressure of the air is

$$\begin{aligned}p_{a,1} &= p_1 - p_{w,1} = 25 \text{ psia} - 11.538 \text{ psia}\\ &= 13.462 \text{ psia}\end{aligned}$$

Use Eq. 29.7 (also *NCEES Handbook:* Air-Water Systems) to determine the humidity ratio.

$$\omega = 0.622\left(\frac{p_{w,1}}{p_{a,1}}\right) = (0.622)\left(\frac{11.538 \text{ psia}}{13.462 \text{ psia}}\right) = 0.533$$

Since it is a constant pressure, constant moisture drying process, mole fractions and partial pressures do not change.

$$p_{w,2} = p_{w,1} = 11.538 \text{ psia}$$

The saturation pressure at 400°F from App. 10A (also *NCEES Handbook* table "Saturated Steam (U.S. Units)—Temperature Table") is $p_{\text{sat},2} = 247.3$ psia.

The relative humidity at state 2 is

$$\phi_2 = \frac{p_{w,2}}{p_{\text{sat},2}} = \frac{11.538 \text{ psia}}{247.3 \text{ psia}} = 0.0467$$

Although the volume may change, the mass does not. The specific humidity remains constant.

$$\omega_2 = \omega_1 = 0.533 \text{ lbm water/lbm air}$$

The heat required consists of two parts.

As in *NCEES Handbook* table "Conversion Table for Temperature Units," the absolute temperatures are

$$T_1 = 200°\text{F} + 460° = 660°\text{R}$$
$$T_2 = 400°\text{F} + 460° = 860°\text{R}$$

As in *NCEES Handbook:* Temperature-Dependent Properties of Air (U.S. Customary Units), the heat capacity is 0.2424 Btu/lbm-°F. The heat absorbed by air is

$$\begin{aligned} q_1 &= h_2 - h_1 = c_p(T_2 - T_1) \\ &= \left(0.2424\ \frac{\text{Btu}}{\text{lbm-°F}}\right)(860°\text{R} - 660°\text{R}) \\ &= 48.48\ \text{Btu/lbm dry air} \end{aligned}$$

From the Mollier diagram (also *NCEES Handbook* figure "Temperature-Entropy (T-S) Diagram (U.S. Customary Units)"), for water, h_1 at 200°F and 11.529 psia is 1146 Btu/lbm (almost saturated).

Follow a constant 11.529 psia pressure curve up to 400°F.

$$h_2 = 1240\ \text{Btu/lbm}$$

The heat absorbed by the steam is

$$\begin{aligned} q_2 &= \omega(h_2 - h_1) \\ &= \left(0.532\ \frac{\text{lbm water}}{\text{lbm air}}\right)\left(1240\ \frac{\text{Btu}}{\text{lbm}} - 1146\ \frac{\text{Btu}}{\text{lbm}}\right) \\ &= 50.01\ \text{Btu/lbm air} \end{aligned}$$

The total heat absorbed is

$$\begin{aligned} q_{\text{total}} &= q_1 + q_2 = 48.48\ \frac{\text{Btu}}{\text{lbm air}} + 50.01\ \frac{\text{Btu}}{\text{lbm air}} \\ &= 98.49\ \text{Btu/lbm air} \quad (99\ \text{Btu/lbm}) \end{aligned}$$

SI Solution

From App. 10I (also *NCEES Handbook* table "Saturated Steam (SI Units)—Temperature Table"), the saturation pressure at 93°C is

$$p_{\text{sat},1} = (0.7884\ \text{bar})\left(100\ \frac{\text{kPa}}{\text{bar}}\right) = 78.84\ \text{kPa}$$

Since air is saturated (100% relative humidity), water vapor pressure is equal to saturation pressure.

$$p_{w,1} = p_{\text{sat},1} = 78.84\ \text{kPa}$$

The partial pressure of air is

$$\begin{aligned} p_{a,1} &= p_1 - p_{w,1} = 170\ \text{kPa} - 78.84\ \text{kPa} \\ &= 91.16\ \text{kPa} \end{aligned}$$

Use Eq. 29.7 (also *NCEES Handbook:* Air-Water Systems) to determine the humidity ratio.

$$\begin{aligned} \omega &= 0.622\left(\frac{p_{w,1}}{p_{a,1}}\right) = (0.622)\left(\frac{78.84\ \text{kPa}}{91.16\ \text{kPa}}\right) \\ &= 0.538\ \text{kg water/kg air} \end{aligned}$$

Since it is a constant pressure, constant moisture drying process, mole fractions and partial pressure do not change.

$$p_{w,2} = p_{w,1} = 78.84\ \text{kPa}$$

The saturation pressure at 204°C from App. 10I (also *NCEES Handbook* table "Saturated Steam (SI Units)") is

$$p_{\text{sat},2} = (16.90\ \text{bar})\left(100\ \frac{\text{kPa}}{\text{bar}}\right) = 1690\ \text{kPa}$$

The relative humidity at state 2 is

$$\phi_2 = \frac{p_{w,2}}{p_{\text{sat},2}} = \frac{78.84\ \text{kPa}}{1690\ \text{kPa}} = 0.0467$$

Although the volume may change, the mass does not. The specific humidity remains constant.

$$\omega_2 = \omega_1 = 0.538\ \text{kg water/kg air}$$

The heat required consists of two parts.

The absolute temperatures are

$$T_1 = 93°\text{C} + 273° = 366\text{K}$$
$$T_2 = 204°\text{C} + 273° = 477\text{K}$$

As in *NCEES Handbook:* Temperature-Dependent Properties of Air (SI Units), the heat capacity is 1.013 kJ/kg·K. The heat absorbed by air is

$$\begin{aligned} q_1 &= h_2 - h_1 = c_p(T_2 - T_1) \\ &= \left(1.013\ \frac{\text{kJ}}{\text{kg·K}}\right)(477\text{K} - 366\text{K}) \\ &= 112.44\ \text{kJ/kg} \end{aligned}$$

For water, h_1 at 93°C and 78.79 kPa is 2670 kJ/kg (almost saturated).

Follow a constant 78.79 kPa pressure curve up to 204°C.

$$h_2 = 2890 \text{ kJ/kg}$$

The heat absorbed by steam is

$$\begin{aligned} q_2 &= \omega(h_2 - h_1) \\ &= \left(0.537 \ \frac{\text{kg water}}{\text{kg air}}\right)\left(2890 \ \frac{\text{kJ}}{\text{kg}} - 2670 \ \frac{\text{kJ}}{\text{kg}}\right) \\ &= 118.14 \text{ kJ/kg air} \end{aligned}$$

The total heat absorbed is

$$\begin{aligned} q_{\text{total}} &= q_1 + q_2 = 112.44 \ \frac{\text{kJ}}{\text{kg air}} + 118.14 \ \frac{\text{kJ}}{\text{kg air}} \\ &= 230.58 \text{ kJ/kg air} \quad (230 \text{ kJ/kg}) \end{aligned}$$

The answer is (C).

10. *Customary U.S. Solution*

As in *NCEES Handbook:* Temperature, the absolute air temperatures are

$$\begin{aligned} T_1 &= 800°\text{F} + 460° = 1260°\text{R} \\ T_2 &= 350°\text{F} + 460° = 810°\text{R} \end{aligned}$$

From App. 10A (also *NCEES Handbook* table "Physical Properties of Liquid Water (U.S. Units)"), the enthalpy of water at 80°F is $h_{w,1} = 48.07$ Btu/lbm (48 Btu/lbm).

From Eq. 29.19(b) (also *NCEES Handbook:* Superheated Steam (U.S. Customary Units)), the enthalpy of steam at 350°F is 1216.4 Btu/lbm.

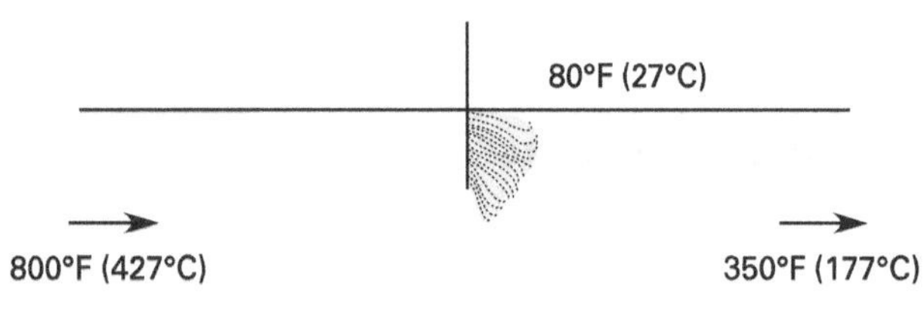

Air temperature is reduced from 800°F to 350°F, and this energy is used to change water at 80°F to steam at 350°F. From the energy balance equation,

$$\dot{m}_w(h_{w,2} - h_{w,1}) = \dot{m}_a(h_1 - h_2)$$

$$\begin{aligned} \dot{m} &= \frac{\dot{m}_a(h_1 - h_2)}{h_{w,2} - h_{w,1}} = \frac{\dot{m}_a c_p(T_1 - T_2)}{h_{w,2} - h_{w,1}} \\ &= \frac{\left(410 \ \frac{\text{lbm}}{\text{hr}}\right)\left(0.25 \ \frac{\text{Btu}}{\text{lbm-°F}}\right) \times (1260°\text{R} - 810°\text{R})}{1216.4 \ \frac{\text{Btu}}{\text{lbm}} - 48.07 \ \frac{\text{Btu}}{\text{lbm}}} \\ &= 39.4 \text{ lbm/hr water} \quad (39 \text{ lbm/hr}) \end{aligned}$$

SI Solution

As in *NCEES Handbook:* Temperature, the absolute temperatures are

$$\begin{aligned} T_1 &= 427°\text{C} + 273° = 700\text{K} \\ T_2 &= 177°\text{C} + 273° = 450\text{K} \end{aligned}$$

From App. 10I (also *NCEES Handbook* table "Physical Properties of Liquid Water (SI Units)"), the enthalpy of water at 27°C is $h_{w,1} = 113.19$ kJ/kg (113 kJ/kg).

From Eq. 29.19(a) (also *NCEES Handbook* table "Superheated Steam (SI Units)"), the enthalpy of steam at 177°C is

$$h_{w,2} = 2820.5 \text{ kJ/kg}$$

Air temperature is reduced from 427°C to 177°C, and this energy is used to change water at 27°C to steam at 177°C. From the energy balance equation,

$$\begin{aligned} \dot{m} &= \frac{\dot{m}_a(h_1 - h_2)}{h_{w,2} - h_{w,1}} = \frac{\dot{m}_a c_p(T_1 - T_2)}{h_{w,2} - h_{w,1}} \\ &= \frac{\left(0.052 \ \frac{\text{kg}}{\text{s}}\right)\left(1.04 \ \frac{\text{kJ}}{\text{kg·K}}\right)(700\text{K} - 450\text{K})}{2820.5 \ \frac{\text{kJ}}{\text{kg}} - 113.19 \ \frac{\text{kJ}}{\text{kg}}} \\ &= 0.00502 \text{ kg/s} \quad (0.0050 \text{ kg/s}) \end{aligned}$$

The answer is (D).

11. Interpolating from App. 10B (also *NCEES Handbook* table "Saturated Steam (U.S. Units)—Temperature Table"), the saturation pressure, p_{sat}, of 180°F water vapor is 7.515 psia. Use Eq. 29.9 (also *NCEES Handbook:* Air-Water Systems) to find the partial pressure of the water vapor, p_w. The relative humidity, ϕ, is given as 20%.

$$\phi = \frac{p_w}{p_{\text{sat}}}$$

$$p_w = \phi p_{\text{sat}} = (0.20)\left(7.515\ \frac{\text{lbf}}{\text{in}^2}\right)$$
$$= 1.5\ \text{lbf/in}^2$$

From App. 10B (also *NCEES Handbook:* Temperature-Dependent Properties of Air (U.S. Customary Units)), the dew point temperature corresponding to 1.5 lbf/in^2 is 115.64°F (120°F).

The answer is (B).

30 Ventilation

PRACTICE PROBLEMS

1. An office room has floor dimensions of 60 ft by 95 ft (18 m by 29 m) and a ceiling height of 10 ft (3 m). Cool air is supplied from ceiling diffusers. 45 people occupy the office. What is most nearly the ventilation rate based on six air changes per hour?

(A) 1300 ft^3/min (35 m^3/min)

(B) 5700 ft^3/min (160 m^3/min)

(C) 13,000 ft^3/min (350 m^3/min)

(D) 34,000 ft^3/min (900 m^3/min)

2. An auditorium is designed to seat 4500 people. The ventilation rate is 1.62×10^7 ft^3/hr (4.54×10^5 m^3/h) of outside air. The outside temperature is 0°F (−18°C) dry-bulb, and the outside pressure is 14.6 psia (100.6 kPa). Air leaves the auditorium at 70°F (21°C) dry-bulb. There is no recirculation. The furnace has a capacity of 1,250,000 Btu/hr (370 kW). The sensible heat generated by each person seated in the theater is 225 Btu/hr. Has the furnace been sized properly?

(A) The furnace is less than half the required capacity.

(B) The furnace is more than half the required capacity.

(C) The furnace is less than twice the required capacity.

(D) The furnace is more than twice the required capacity.

3. The volumetric fraction of carbon dioxide in a submarine is measured to be 0.06%. What is most nearly the concentration measured in parts per million (ppm)?

(A) 0.6 ppm

(B) 60 ppm

(C) 600 ppm

(D) 6000 ppm

4. A room is maintained at design conditions of 75°F (23.9°C) dry-bulb and 50% relative humidity. The air outside is at 95°F (35°C) dry-bulb and 75°F (23.9°C) wet-bulb. The outside air is conditioned and mixed with some room exhaust air. The mixed, conditioned air enters the room and increases 20°F (11.1°C) in temperature before being removed from the room. The sensible and latent loads are 200,000 Btu/hr (60 kW) and 50,000 Btu/hr (15 kW), respectively. Air leaves the coil at 50.8°F (10°C). The volume of air flowing through the coil is most nearly

(A) 4200 ft^3/min (120 m^3/min)

(B) 5800 ft^3/min (160 m^3/min)

(C) 7600 ft^3/min (220 m^3/min)

(D) 9300 ft^3/min (260 m^3/min)

5. How should the pressurization (relative to the surroundings) in clean rooms and laboratories be maintained?

(A) Clean rooms and laboratories should both be at positive pressures.

(B) Clean rooms and laboratories should both be at negative pressures.

(C) Clean rooms should be at positive pressures, while laboratories should be at negative pressures.

(D) Clean rooms should be at negative pressures, while laboratories should be at positive pressures.

SOLUTIONS

1. *Customary U.S. Solution*

The office volume is

$$V = (60\text{ ft})(95\text{ ft})(10\text{ ft}) = 57{,}000\text{ ft}^3$$

Based on six air changes per hour, the flow rate is

$$\dot{V} = \frac{\left(57{,}000\ \frac{\text{ft}^3}{\text{air change}}\right)\left(6\ \frac{\text{air changes}}{\text{hr}}\right)}{60\ \frac{\text{min}}{\text{hr}}} = 5700\text{ ft}^3/\text{min}$$

SI Solution

The office volume is

$$V = (18\text{ m})(29\text{ m})(3\text{ m}) = 1566\text{ m}^3$$

Based on six air changes per hour, the flow rate is

$$\dot{V} = \frac{\left(1566\ \frac{\text{m}^3}{\text{air change}}\right)\left(6\ \frac{\text{air changes}}{\text{h}}\right)}{60\ \frac{\text{min}}{\text{h}}} = 156.6\text{ m}^3/\text{min}\quad(160\text{ m}^3/\text{min})$$

The answer is (B).

2. *Customary U.S. Solution*

As in *NCEES Handbook* table "Conversion Table for Temperature Units," the absolute temperature of outside air is

$$T = 0°\text{F} + 460° = 460°\text{R}$$

From the ideal gas law, the density of outside air is

$$\rho = \frac{pM}{RT} = \frac{\left(14.6\ \frac{\text{lbf}}{\text{in}^2}\right)\left(12\ \frac{\text{in}}{\text{ft}}\right)^2\left(29\ \frac{\text{lbm}}{\text{lbmol}}\right)}{\left(1545.35\ \frac{\text{ft-lbf}}{\text{lbmol-°R}}\right)(460°\text{R})} = 0.08567\text{ lbm/ft}^3$$

As in *NCEES Handbook:* Conservation of Mass, the mass flow rate is

$$\dot{m} = \dot{V}\rho = \left(1.62\times10^7\ \frac{\text{ft}^3}{\text{hr}}\right)\left(0.08567\ \frac{\text{lbm}}{\text{ft}^3}\right) = 1.388\times10^6\text{ lbm/hr}$$

Assume there can be no latent heat (no moisture) at 0°F.

From the problem statement, the sensible heat generated by each person seated in the theater is 225 Btu/hr. Therefore,

$$\dot{q}_{\text{in from people}} = \left(225\ \frac{\frac{\text{Btu}}{\text{hr}}}{\text{person}}\right)(4500\text{ persons}) = 1.01\times10^6\text{ Btu/hr}$$

The air leaves the auditorium at 70°F. From App. 18C (also *NCEES Handbook:* Temperature-Dependent Properties of Air (U.S. Customary Units)), the specific heat of dry air is 0.240 Btu/lbm-°F. (This remains fairly constant over normal temperature ranges.) Since $\dot{q}$ is known, the air temperature entering the auditorium is

$$\dot{q} = \dot{m}c_p(T_{\text{out,air}} - T_{\text{in,air}})$$

$$T_{\text{in,air}} = T_{\text{out,air}} - \frac{\dot{q}}{\dot{m}c_p} = 70°\text{F} - \frac{1.01\times10^6\ \frac{\text{Btu}}{\text{hr}}}{\left(1.388\times10^6\ \frac{\text{lbm}}{\text{hr}}\right)\left(0.240\ \frac{\text{Btu}}{\text{lbm-°F}}\right)} = 67°\text{F}$$

The heat needed to heat dry ventilation air from 0°F to 67°F is

$$\dot{q} = \dot{m}c_p\Delta T = \left(1.388\times10^6\ \frac{\text{lbm}}{\text{hr}}\right)\left(0.240\ \frac{\text{Btu}}{\text{lbm-°F}}\right)(67°\text{F} - 0°\text{F}) = 2.23\times10^7\text{ Btu/hr}$$

Since 1.25×10^6 Btu/hr $< 2.23\times10^7$ Btu/hr, the furnace is less than half the required capacity, so it is too small.

SI Solution

As in *NCEES Handbook* table "Conversion Table for Temperature Units," the absolute temperature of outside air is

$$T = -18°\text{C} + 273° = 255\text{K}$$

From the ideal gas law, the density of outside air is

$$\rho = \frac{pM}{RT} = \frac{(100.6\ \text{kPa})\left(29\ \frac{\text{kg}}{\text{kmol}}\right)}{\left(8.314\ \frac{\text{kJ}}{\text{kmol·K}}\right)(255\text{K})} = 1.374\ \text{kg/m}^3$$

As in *NCEES Handbook:* Conservation of Mass, the mass flow rate is

$$\dot{m} = \rho\dot{V} = \left(1.374\ \frac{\text{kg}}{\text{m}^3}\right)\left(4.54\times 10^5\ \frac{\text{m}^3}{\text{h}}\right)$$
$$= 6.238\times 10^5\ \text{kg/h}$$

Assume there can be no latent heat (no moisture) at −18°C.

From the problem statement, the sensible heat generated by people seated in the theater is

$$\left(225\ \frac{\frac{\text{Btu}}{\text{h}}}{\text{person}}\right)\left(0.293\ \frac{\text{W}}{\frac{\text{Btu}}{\text{h}}}\right) = 65.93\ \text{W/person}$$

Therefore,

$$\dot{q}_{\text{in from people}} = \frac{\left(65.93\ \frac{\text{W}}{\text{person}}\right)(4500\ \text{persons})}{1000\ \frac{\text{W}}{\text{kW}}}$$
$$= 296.7\ \text{kW}$$

The air leaves the auditorium at 21°C. From App. 18D (also *NCEES Handbook:* Temperature-Dependent Properties of Air (SI Units)), the specific heat of dry air is 1.0048 kJ/kg·K. (This remains fairly constant over normal temperature ranges.) Since $\dot{q}$ is known, the air temperature entering the auditorium is

$$\dot{q} = \dot{m}c_p(T_{\text{out,air}} - T_{\text{in,air}})$$
$$T_{\text{in,air}} = T_{\text{out,air}} - \frac{\dot{q}}{\dot{m}c_p}$$
$$= 21°\text{C} - \frac{(296.7\ \text{kW})\left(3600\ \frac{\text{s}}{\text{h}}\right)}{\left(6.238\times 10^5\ \frac{\text{kg}}{\text{h}}\right)\left(1.0048\ \frac{\text{kJ}}{\text{kg·K}}\right)}$$
$$= 19.3°\text{C}\quad (19°\text{C})$$

The heat needed to heat dry ventilation air from −18°C to 19.3°C is

$$\dot{q} = \dot{m}c_p\Delta T$$
$$= \frac{\left(6.238\times 10^5\ \frac{\text{kg}}{\text{h}}\right)\left(1.0048\ \frac{\text{kJ}}{\text{kg·°C}}\right)\times\left(19.3°\text{C} - (-18°\text{C})\right)}{3600\ \frac{\text{s}}{\text{h}}}$$
$$= 6494.28\ \text{kW}\quad (6.5\ \text{MW})$$

Since 370 kW < 6494 kW, the furnace is less than half the required capacity, so it is too small.

The answer is (A).

3. Air contaminants are measured by volume, hence "ppm" are parts per million by volume (ppmv). In this problem, there are 0.06 volumes of carbon dioxide for every 100 volumes of air (i.e., 0.06 parts per 100 parts). In one million volumes of air, there would be $10^6/10^2 = 10^4$ more volumes of carbon dioxide.

$$(0.06\ \text{pphv})\left(10^4\ \frac{\text{ppmv}}{\text{pphv}}\right) = 600\ \text{ppmv}\quad (600\ \text{ppm})$$

The answer is (C).

4. *Customary U.S. Solution*

The mixed conditioned air enters the room at

$$T_{\text{db,in}} = 75°\text{F} - 20°\text{F} = 55°\text{F}$$

From Eq. 30.13 (also *NCEES Handbook:* Latent and Sensible Heat and *NCEES Handbook:* Temperature-Dependent Properties of Air (U.S. Customary Units)), the volumetric flow rate of air entering the room is

$$\dot{V}_{\text{in,ft}^3/\text{min}} = \frac{\dot{q}_{s,\text{Btu/hr}}}{\left(1.08\ \frac{\text{Btu-min}}{\text{ft}^3\text{-hr-°F}}\right)(T_{\text{id,°F}} - T_{\text{in,°F}})}$$
$$= \frac{200{,}000\ \frac{\text{Btu}}{\text{hr}}}{\left(1.08\ \frac{\text{Btu-min}}{\text{ft}^3\text{-hr-°F}}\right)(75°\text{F} - 55°\text{F})}$$
$$= 9259\ \text{ft}^3/\text{min}$$

This is a mixing problem.

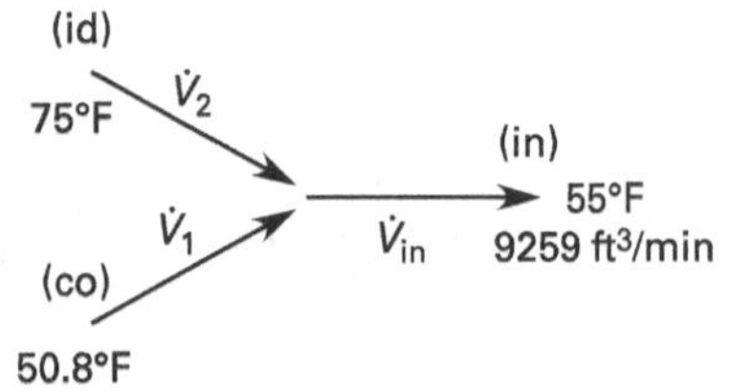

Using the lever rule, and since the temperature scales are all linear, the fraction of air passing through the coil is

$$\frac{T_{id} - T_{in}}{T_{id} - T_{co}} = \frac{75°F - 55°F}{75°F - 50.8°F} = 0.826$$

$$\dot{V}_1 = (0.826)\left(9259\ \frac{ft^3}{min}\right)$$

$$= 7648\ ft^3/min \quad (7600\ ft^3/min)$$

SI Solution

The mixed conditioned air enters the room at

$$T_{db,in} = 23.9°C - 11.1°C = 12.8°C$$

From Eq. 30.12 (also *NCEES Handbook:* Latent and Sensible Heat and *NCEES Handbook:* Temperature-Dependent Properties of Air (SI Units)),

$$\dot{V}_{in,m^3/min} = \frac{\dot{q}_{s,kW}}{\left(0.02\ \frac{kJ\cdot min}{m^3\cdot s\cdot °C}\right)(T_{id,°C} - T_{in,°C})}$$

$$= \frac{60\ kW}{\left(0.02\ \frac{kJ\cdot min}{m^3\cdot s\cdot °C}\right)(23.9°C - 12.8°C)}$$

$$= 270.3\ m^3/min$$

This is a mixing problem.

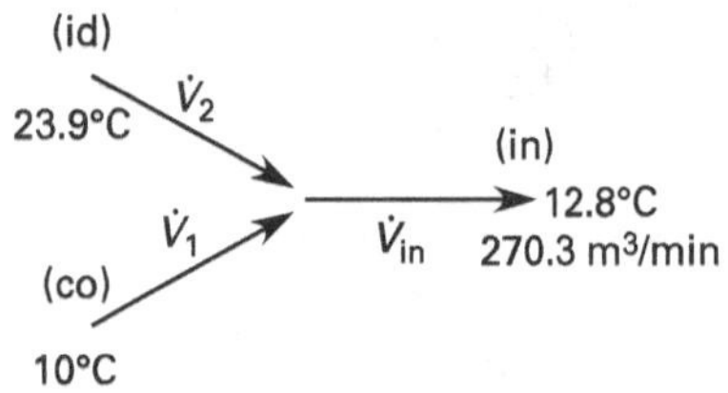

Using the lever rule, and since the temperature scales are all linear, the fraction of air passing through the coil is

$$\frac{T_{id} - T_{in}}{T_{id} - T_{co}} = \frac{23.9°C - 12.8°C}{23.9°C - 10°C} = 0.799$$

$$\dot{V}_1 = (0.799)\left(270.3\ \frac{m^3}{min}\right)$$

$$= 216.0\ m^3/min \quad (220\ m^3/min)$$

The answer is (C).

5. Clean rooms should be maintained at positive pressures with filtered forced air in order to prevent the incursion of airborne particles from the atmosphere. The intake air is filtered and cleaned, and the interior of the clean room is protected. Most laboratories should be maintained at negative pressures in order to keep airborne hazardous materials in the air distribution system. The exhaust is filtered and cleaned, and the environment is protected.

The answer is (C).

31 Engineering Materials

PRACTICE PROBLEMS

1. Approximately how much (in molecules of HCl per gram of PVC) HCl should be used as an initiator in PVC if the efficiency is 20% and an average molecular weight of 7000 g/mol is desired? The final polymer has the structure shown.

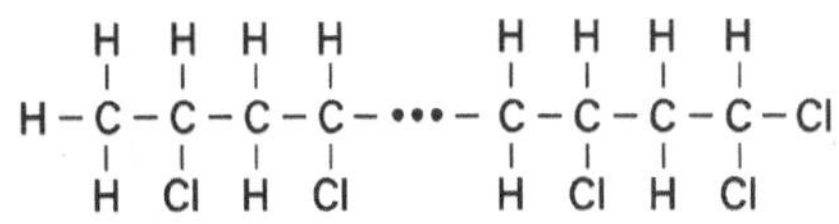

(A) 7.8×10^{18} molecules HCl/gram PVC

(B) 4.3×10^{20} molecules HCl/gram PVC

(C) 9.5×10^{21} molecules HCl/gram PVC

(D) 3.6×10^{23} molecules HCl/gram PVC

2. 10 ml of a 0.2% solution (by weight) of hydrogen peroxide is added to 12 g of ethylene to stabilize the polymer. Hydrogen peroxide breaks down according to

$$H_2O_2 \rightarrow 2(OH^-) + \cdots$$

Assume that the stabilized polymer has the structure shown.

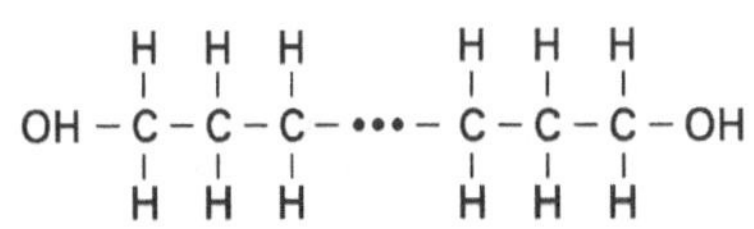

What is most nearly the average degree of polymerization if the hydrogen peroxide is completely utilized?

(A) 180

(B) 730

(C) 910

(D) 1200

SOLUTIONS

1. The vinyl chloride mer is

With 20% efficiency, 5 molecules of HCl per PVC molecule are required to supply each end Cl atom. This is the same as 5 mol HCl per mole PVC. Using Avogadro's number of 6.022×10^{23} molecules per mole from *NCEES Handbook* table "Physical Constants," the number of molecules of HCl per gram of PVC is

$$\frac{\left(5\ \frac{\text{mol HCl}}{\text{mol PVC}}\right)\left(6.022 \times 10^{23}\ \frac{\text{molecules}}{\text{mol}}\right)}{7000\ \frac{\text{g}}{\text{mol}}} = 4.3 \times 10^{20} \text{ molecules HCl/gram PVC}$$

The answer is (B).

2. As in *NCEES Handbook:* Relative Atomic Mass, the molecular weight of hydrogen peroxide, H_2O_2, is

$$(2)\left(1\ \frac{\text{g}}{\text{mol}}\right) + (2)\left(16\ \frac{\text{g}}{\text{mol}}\right) = 34 \text{ g/mol}$$

The weight of H_2O_2 is

$$(10 \text{ mL})\left(1\ \frac{\text{g}}{\text{mL}}\right) = 10 \text{ g}$$

From *NCEES Handbook* table "Physical Constants," the number of H_2O_2 molecules in a 0.2% solution is

$$\frac{(10 \text{ g})\left(\frac{0.2\%}{100\%}\right)\left(6.022 \times 10^{23}\ \frac{\text{molecules}}{\text{mol}}\right)}{34\ \frac{\text{g}}{\text{mol}}} = 3.54 \times 10^{20} \text{ molecules}$$

The molecular weight of ethylene, C_2H_4, is

$$(2)\left(12\ \frac{\text{g}}{\text{mol}}\right) + (4)\left(1\ \frac{\text{g}}{\text{mol}}\right) = 28\ \text{g/mol}$$

The number of ethylene molecules is

$$\frac{(12\ \text{g})\left(6.022\times 10^{23}\ \frac{\text{molecules}}{\text{mol}}\right)}{28\ \frac{\text{g}}{\text{mol}}} = 2.58\times 10^{23}\ \text{molecules}$$

Since it takes one H_2O_2 molecule (i.e., $2OH^-$ radicals) to stabilize a polyethylene molecule, there are 3.54×10^{20} polymers.

The degree of polymerization is

$$\text{DP} = \frac{2.58\times 10^{23}\ C_2H_4\ \text{molecules}}{3.54\times 10^{20}\ \text{polymers}} = 729 \quad (730)$$

The answer is (B).

32 Physical Properties of Construction Materials

PRACTICE PROBLEMS

There are no problems in this book corresponding to Chap. 32 of the *PE Chemical Review*.

33 Analysis of Engineering Systems

PRACTICE PROBLEMS

1. What type of signal is indicated by this line symbol on a piping and instrumentation diagram (P&ID)?

(A) pneumatic signal

(B) hydraulic signal

(C) electromagnetic signal

(D) sonic signal

2. Which of these statements about piping and instrumentation diagram (P&ID) documents is NOT correct?

(A) Issue for construction (IFC) revisions can include sizes of control valves and pressure safety valves as well as HAZOP actions.

(B) Equipment is drawn to scale except for pumps and valves.

(C) An instrumentation line is nearly always represented by a plain, solid line.

(D) An instrument that is visible on a shared video display as part of a distributed control system is represented by a circle inscribed in a square.

3. To assign a robustness indicator figure (RIF) for durability testing, the Arrhenius model can be applied. The Arrhenius model is usually applicable when high temperature tests or lifetime tests with constant temperature are conducted (see *NCEES Handbook:* Temperature Dependence).

A device under test (DUT) is required to work at a maximum temperature of 87°C for 1600 hr. When the DUT is tested at 98°C, the failure at accelerated conditions occurs after 1950 hr. From the Arrhenius model, the acceleration factor is

$$\pi_B = e^{E_A(\text{TF})}$$

The Arrhenius activation energy, E_A, is 0.45 eV. The temperature factor, TF, is defined as

$$\text{TF} = \frac{1}{k}\left(\frac{1}{T_{\text{max}}} - \frac{1}{T_{\text{test}}}\right)$$

$$\text{RIF} = \frac{\tau_2 \pi_B}{\tau_1}$$

The Boltzmann constant, k, is 8.617×10^{-5} eV/K. Most nearly, the RIF is

The robustness can be represented by the RIF, which is calculated from the following equation.

(A) 1.9

(B) 2.2

(C) 2.5

(D) 3.0

SOLUTIONS

1. Pneumatic signals are represented by lines with only TWO axial traversing line segments as represented by the spring-opposed single-acting, double-acting, and I/P actuator symbols in the *NCEES Handbook*. The line symbol shown represents a pneumatic signal. Standard symbols on P&IDs are defined by ANSI/ISA Standard S5.1.

The answer is (A).

2. On process flow drawings (PFDs), which are the simplest P&ID documents, instrumentation lines are represented only by plain, solid lines. On other kinds of P&ID documents, however, instrumentation lines may be dashed, dotted, or marked with other patterns, mainly to indicate the type of signal (pneumatic, sonic, electrical, and so on) that each line transmits.

The answer is (C).

3. The maximum absolute temperature is $87°\text{C} + 273° = 360\text{K}$. The absolute testing temperature is $98°\text{C} + 273° = 371\text{K}$. The temperature factor is

$$\begin{aligned}\text{TF} &= \frac{1}{k}\left(\frac{1}{T_{\text{max}}} - \frac{1}{T_{\text{test}}}\right) \\ &= \left(\frac{1}{8.617\times 10^{-5}\ \dfrac{\text{eV}}{\text{K}}}\right)\left(\frac{1}{360\text{K}} - \frac{1}{371\text{K}}\right) \\ &= 0.96\ 1/\text{eV}\end{aligned}$$

From the Arrhenius model, the acceleration factor is

$$\begin{aligned}\pi_B &= e^{E_A(\text{TF})} \\ &= e^{(0.45\ \text{eV})(0.96\ 1/\text{eV})} \\ &= 1.54\end{aligned}$$

Robustness is defined as the difference between the limits of the design or the product and the specification requirement. The RIF for the device under test is

$$\text{RIF} = \frac{\tau_2\pi_B}{\tau_1} = \frac{(1950\ \text{hr})(1.54)}{1600\ \text{hr}} = 1.88 \quad (1.9)$$

The answer is (A).

34 Illumination and Sound

PRACTICE PROBLEMS

1. A worker is exposed to a sound level of 85 decibels (dB) for the first 3.5 hr of his 8 hr shift and then exposed to a sound level of 89 dB for the remaining 4.5 hr. Most nearly, the worker's 8 hr time-weighted average (TWA) noise level is

(A) 71 dB

(B) 82 dB

(C) 86 dB

(D) 88 dB

SOLUTIONS

1. Calculate the permissible sound level durations from the initial noise level, L_1, and the time of exposure, C_1.

$$\begin{aligned} T_1 &= \frac{8 \text{ hr}}{2^{(L_1 - 90\,\text{dB})/5\,\text{dB}}} \\ &= \frac{8 \text{ hr}}{2^{(85\,\text{dB} - 90\,\text{dB})/5\,\text{dB}}} \\ &= 16 \text{ hr} \end{aligned}$$

Calculate the permissible sound level durations from the second noise level, L_2, and the time of exposure, C_2.

$$\begin{aligned} T_2 &= \frac{8 \text{ hr}}{2^{(L_2 - 90\,\text{dB})/5\,\text{dB}}} \\ &= \frac{8 \text{ hr}}{2^{(89\,\text{dB} - 90\,\text{dB})/5\,\text{dB}}} \\ &= 9.190 \text{ hr} \end{aligned}$$

Calculate the total noise dose over the work day.

$$\begin{aligned} D &= \left(\frac{C_1}{T_1} + \frac{C_2}{T_2}\right) \times 100\% \\ &= \left(\frac{3.5 \text{ hr}}{16 \text{ hr}} + \frac{4.5 \text{ hr}}{9.190 \text{ hr}}\right) \times 100\% \\ &= 70.84\% \end{aligned}$$

Calculate the TWA.

$$\begin{aligned} \text{TWA} &= (16.61 \text{ dB}) \log\left(\frac{D}{100}\right) + 90 \text{ dB} \\ &= (16.61 \text{ dB}) \log \frac{70.84\%}{100\%} + 90 \text{ dB} \\ &= 87.51 \text{ dB} \quad (88 \text{ dB}) \end{aligned}$$

The answer is (D).

35 Construction and Jobsite Safety

PRACTICE PROBLEMS

1. The ACGIH threshold limit value (TLV) by volume for eight hours of carbon monoxide exposure is

(A) 10 ppm
(B) 25 ppm
(C) 35 ppm
(D) 50 ppm

2. Which of the following will NOT form a flammable or explosive mixture with air?

(A) fly ash or portland cement dust
(B) coal dust
(C) whey dust
(D) cellulose

3. Which type of work glove is suitable for working with temperatures up to 500°C and flames?

(A) aluminized gloves
(B) impervious gloves
(C) leather gloves
(D) metal mesh gloves

4. According to the OSHA *Hazard Communication Standard* (OSHA 1910.1200), which of the following data is NOT compulsory on a safety data sheet?

(A) flash point
(B) vapor pressure
(C) enthalpy of sublimation
(D) odor threshold

5. According to OSHA *Hazard Communication Standard* (HCS), which information is NOT required in Section 2 of a safety data sheet?

(A) chemical name
(B) signal word
(C) hazard statement
(D) symbol(s) and precautionary statement(s)

SOLUTIONS

1. The ACGIH TLV values can be found in *NCEES Handbook:* Risk Assessment/Toxicology. For carbon monoxide, the value is 25 ppm.

The answer is (B).

2. Dust explosions require a combustible material. Fly ash and cement dust will not form a flammable or explosive mixture when exposed to air. The explosive thresholds for the other three dusts can be found in *NCEES Handbook:* Combustible Dust.

The answer is (A).

3. Aluminized gloves are resistant to very high temperatures because they reflect most of the radiant heat.

The answer is (A).

4. According to the OSHA *Hazard Communication Standard*, information about the enthalpy of sublimation is not required.

The minimum information given about the physical and chemical properties of a chemical substance consists of

- appearance
- upper/lower flammability or explosive limits
- odor
- vapor pressure
- odor threshold
- vapor density
- pH
- relative density
- melting point/freezing point
- solubility
- initial boiling point and boiling range
- flash point
- evaporation rate
- flammability
- partition coefficient
- auto-ignition temperature
- decomposition temperature
- viscosity

The answer is (C).

5. The chemical name is required in Section 3 of a safety data sheet, not Section 2.

The answer is (A).

36 Management Science

PRACTICE PROBLEMS

1. A chemical plant can process 500 tons/day of raw material. The amount of raw material processed in one day can be modeled as having an exponential distribution with a mean of 500 tons. It is desired that the probability of running out of raw material be 0.05. Most nearly, the amount of raw material that should be stocked each day is

(A) 500 tons

(B) 750 tons

(C) 1000 tons

(D) 1500 tons

2. The physical property of a product within a production line is modeled as time-dependent variable $P(t)$. The time rate of change for the physical property is equal to a constant k multiplied by $P(t)$. The constant k is 0.00001 s^{-1}. The boundary condition is that the physical property initially equals 100 units. Most nearly, what is the physical property's value 24 hours later?

(A) 12

(B) 50

(C) 100

(D) 237

SOLUTIONS

1. Let a be the amount of raw material that should be stocked each day. The probability of running out of raw material in a day is $P(X > a) = 0.05$. By definition, the relationship between the amount of raw material that should be stocked each day and the given probability is

$$\begin{aligned} P(X > a) &= \int_a^\infty f(x)\,dx \\ &= \int_a^\infty \left(\frac{1}{500 \text{ tons}}\right) e^{-x/500 \text{ tons}}\,dx \\ &= -e^{-a/500 \text{ tons}}\Big|_a^\infty \end{aligned}$$

Setting this equal to 0.05 gives

$$\begin{aligned} -e^{-a/500 \text{ ton}} &= 0.05 \\ -\frac{a}{500 \text{ tons}} &= \ln 0.05 \\ a &= -(500 \text{ tons})\ln 0.05 \\ &= 1497 \text{ tons} \quad (1500 \text{ tons}) \end{aligned}$$

The answer is (D).

2. This problem can be modeled as a differential equation. The time rate of change of the physical property is

$$\frac{dP(t)}{dt} = kP(t) = \left(\frac{0.00001}{1 \text{ s}}\right)P(t)$$

$$\frac{dP}{dt} - (0.00001)P = 0$$

The above differential equation's characteristic equation is

$$r - 0.00001 = 0$$

The root, r, is 0.00001. Therefore, the solution to the equation is

$$P(t) = Ce^{rt} = Ce^{0.00001t}$$

To solve for C, apply the boundary condition $P(0) = 100$ units.

$$\begin{aligned} P(0) = 100 &= Ce^{(0.00001)(0)} = C(1) \\ C &= 100 \end{aligned}$$

The physical property, P, 24 hours (or 86,400 seconds) later is

$$\begin{aligned} P(86{,}400) &= 100e^{(0.00001)(86{,}400\,\text{s})} \\ &= (100)(2.373) \\ &= 237.3 \quad (237) \end{aligned}$$

The answer is (D).

37 Engineering Economic Analysis

PRACTICE PROBLEMS

1. At 6% effective annual interest, approximately how much will be accumulated if \$1000 is invested for 10 years?

(A) \$560

(B) \$790

(C) \$1600

(D) \$1800

2. At 6% effective annual interest, the present worth of \$2000 that becomes available in four years is most nearly

(A) \$520

(B) \$580

(C) \$1600

(D) \$2500

3. At 6% effective annual interest, approximately how much should be invested to accumulate \$2000 in 20 years?

(A) \$620

(B) \$1400

(C) \$4400

(D) \$6400

4. At 6% effective annual interest, the year-end annual amount deposited over seven years that is equivalent to \$500 invested now is most nearly

(A) \$90

(B) \$210

(C) \$300

(D) \$710

5. At 6% effective annual interest, the accumulated amount at the end of ten years if \$50 is invested at the end of each year for ten years is most nearly

(A) \$90

(B) \$370

(C) \$660

(D) \$900

6. At 6% effective annual interest, approximately how much should be deposited at the start of each year for 10 years (a total of 10 deposits) in order to empty the fund by drawing out \$200 at the end of each year for 10 years (a total of 10 withdrawals)?

(A) \$190

(B) \$210

(C) \$220

(D) \$250

7. At 6% effective annual interest, approximately how much should be deposited at the start of each year for five years to accumulate \$2000 on the date of the last deposit?

(A) \$350

(B) \$470

(C) \$510

(D) \$680

8. At 6% effective annual interest, approximately how much will be accumulated in 10 years if three payments of \$100 are deposited every other year for four years, with the first payment occurring at $t = 0$?

(A) \$180

(B) \$480

(C) \$510

(D) \$540

9. \$500 is compounded monthly at a 6% nominal annual interest rate. Approximately how much will have accumulated in five years?

(A) \$515

(B) \$530

(C) \$675

(D) \$690

10. The effective annual rate of return on an \$80 investment that pays back \$120 in seven years is most nearly

(A) 4.5%

(B) 5.0%

(C) 5.5%

(D) 6.0%

11. A new machine will cost \$17,000 and will have a resale value of \$14,000 after five years. Special tooling will cost \$5000. The tooling will have a resale value of \$2500 after five years. Maintenance will be \$2000 per year. The effective annual interest rate is 6%. The equivalent uniform annual cost (EUAC) is defined as the sum of the annual cost (A) and the present value (P) minus the resale or salvage value (F). The average annual cost of ownership during the next five years will be most nearly

$$\text{EUAC} = A + P - F$$

(A) \$2000

(B) \$2300

(C) \$4300

(D) \$5500

12. An old covered wooden bridge can be strengthened at a cost of \$9000, or it can be replaced for \$40,000. The present salvage value of the old bridge is \$13,000. It is estimated that the reinforced bridge will last for 20 years, will have an annual cost of \$500, and will have a salvage value of \$10,000 at the end of 20 years. The estimated salvage value of the new bridge after 25 years is \$15,000. Maintenance for the new bridge would cost \$100 annually. The effective annual interest rate is 8%. Which is the best alternative?

(A) Strengthen the old bridge.

(B) Build the new bridge.

(C) Both options are economically identical.

(D) Not enough information is given.

13. A firm expects to receive \$32,000 each year for 15 years from sales of a product. An initial investment of \$150,000 will be required to manufacture the product. Expenses will run \$7530 per year. Salvage value is zero, and straight-line depreciation is used. The income tax rate is 48%. The after-tax rate of return is most nearly

(A) 8.0%

(B) 9.0%

(C) 10%

(D) 11%

14. A public works project has initial costs of \$1,000,000, benefits of \$1,500,000, and disbenefits of \$300,000. The excess of benefits over costs is most nearly

(A) \$200,000

(B) \$500,000

(C) \$700,000

(D) \$800,000

15. A speculator in land pays \$14,000 for property that he expects to hold for 10 years. \$1000 is spent in renovation, and a monthly rent of \$75 is collected from the tenants. (Use the year-end convention.) Taxes are \$150 per year, and maintenance costs are \$250 per year. In 10 years, the sale price needed to realize a 10% rate of return is most nearly

(A) \$26,000

(B) \$31,000

(C) \$34,000

(D) \$36,000

16. The effective annual interest rate for a payment plan of 30 equal payments of \$89.30 per month when a lump sum payment of \$2000 would have been an outright purchase is most nearly

(A) 27%

(B) 35%

(C) 43%

(D) 51%

17. A depreciable item is purchased for \$500,000. The salvage value at the end of 25 years is estimated at \$100,000. The depreciation in each of the first three years using the straight line method is most nearly

(A) \$4000

(B) \$16,000

(C) \$20,000

(D) \$24,000

18. Equipment that is purchased for $12,000 now is expected to be sold after ten years for $2000. The estimated maintenance is $1000 for the first year, but it is expected to increase $200 each year thereafter. The effective annual interest rate is 10%. The present worth (cost) is most nearly

(A) $16,000

(B) $17,000

(C) $21,000

(D) $22,000

19. A new grain combine with a 20-year life can remove seven pounds of rocks from its harvest per hour. Any rocks left in its output hopper will cause $25,000 damage in subsequent processes. Several investments are available to increase the rock-removal capacity, as listed in the table. The effective annual interest rate is 10%. What should be done?

rock removal rate	annual probability of exceeding rock removal rate	required investment to achieve removal rate
7	0.15	0
8	0.10	$15,000
9	0.07	$20,000
10	0.03	$30,000

(A) Do nothing.

(B) Invest $15,000.

(C) Invest $20,000.

(D) Invest $30,000.

20. A mechanism that costs $10,000 has operating costs and salvage values as given. An effective annual interest rate of 20% is to be used.

year	operating cost	salvage value
1	$2000	$8000
2	$3000	$7000
3	$4000	$6000
4	$5000	$5000
5	$6000	$4000

The economic life of the mechanism is most nearly

(A) one year

(B) two years

(C) three years

(D) five years

21. A salesperson intends to purchase a car for $50,000 for personal use, driving 15,000 miles per year. Insurance for personal use costs $2000 per year, and maintenance costs $1500 per year. The car gets 15 miles per gallon, and gasoline costs $1.50 per gallon. The resale value after five years will be $10,000. The salesperson's employer has asked that the car be used for business driving of 50,000 miles per year and has offered a reimbursement of $0.30 per mile. Using the car for business would increase the insurance cost to $3000 per year and maintenance to $2000 per year. The salvage value after five years would be reduced to $5000. If the employer purchased a car for the salesperson to use, the initial cost would be the same, but insurance and maintenance would be $2500 and $2000 respectively. The salesperson's effective annual interest rate is 10%. With a reimbursement of $0.30 per mile, approximately how many miles must the car be driven per year to justify the employer buying the car for the salesperson to use?

(A) 20,000 mi

(B) 55,000 mi

(C) 82,000 mi

(D) 150,000 mi

22. Alternatives A and B are being evaluated. The effective annual interest rate is 10%.

	alternative A	alternative B
first cost	$80,000	$35,000
life	20 years	10 years
salvage value	$7000	0
annual costs		
years 1–5	$1000	$3000
years 6–10	$1500	$4000
years 11–20	$2000	0
additional cost		
year 10	$5000	0

Which alternative is economically superior?

(A) Alternative A is economically superior.

(B) Alternative B is economically superior.

(C) Alternatives A and B are economically equivalent.

(D) Not enough information is provided.

23. A car is needed for three years. Plans A and B for acquiring the car are being evaluated. An effective annual interest rate of 10% is to be used.

Plan A: lease the car for $0.25/mile (all inclusive)

Plan B: purchase the car for $30,000; keep the car for three years; sell the car after three years for $7200; pay $0.14 per mile for oil and gas; pay other costs of $500 per year

Which plan is economically superior?

(A) Plan A is economically superior.

(B) Plan B is economically superior.

(C) Plans A and B are economically equivalent.

(D) Not enough information is provided.

24. Two methods are being considered to meet strict air pollution control requirements over the next ten years. Method A uses equipment with a life of 10 years. Method B uses equipment with a life of five years that will be replaced with new equipment with an additional life of five years. Capacities of the two methods are different, but operating costs do not depend on the throughput. Operation is 24 hours per day, 365 days per year. The effective annual interest rate for this evaluation is 7%.

	method A	method B	
	years 1–10	years 1–5	years 6–10
installation cost	$13,000	$6000	$7000
equipment cost	$10,000	$2000	$2200
operating cost per hour	$10.50	$8.00	$8.00
salvage value	$5000	$2000	$2000
capacity (tons/yr)	50	20	20
life	10 years	5 years	5 years

The uniform annual cost per ton for method B is most nearly

(A) $3500

(B) $3600

(C) $4200

(D) $4300

25. A transit district has asked for assistance in determining the proper fare for its bus system. An effective annual interest rate of 7% is to be used. The following additional information was compiled.

cost per bus	$60,000
bus life	20 years
salvage value	$10,000
miles driven per year	37,440
number of passengers per year	80,000
operating cost	$1.00 per mile in the first year, increasing $0.10 per mile each year thereafter

If the transit district decides to set the per-passenger fare at $0.35 for the first year, approximately how much should the passenger fare go up each year thereafter such that the district can break even in 20 years?

(A) $0.022 increase per year

(B) $0.036 increase per year

(C) $0.067 increase per year

(D) $0.072 increase per year

26. Make a recommendation to your client to accept one of the following alternatives. Use the present worth comparison method. (Initial costs are the same.)

Alternative A: a 25-year annuity paying $4800 at the end of each year, where the interest rate is a nominal 12% per annum

Alternative B: a 25-year annuity paying $1200 every quarter at 12% nominal annual interest

(A) Alternative A is economically superior.

(B) Alternative B is economically superior.

(C) Alternatives A and B are economically equivalent.

(D) Not enough information is provided.

27. A firm has two alternatives for improvement of its existing production line. The data are as follows.

	alternative A	alternative B
initial installment cost	$1500	$2500
annual operating cost	$800	$650
service life	5 years	8 years
salvage value	0	0

Determine the best alternative using an interest rate of 15%.

(A) Alternative A is economically superior.

(B) Alternative B is economically superior.

(C) Alternatives A and B are economically equivalent.

(D) Not enough information is provided.

28. Two mutually exclusive alternatives requiring different investments are being considered. The life of both alternatives is estimated at 20 years with no salvage values. The minimum rate of return that is considered acceptable is 4%. Which alternative is best?

	alternative A	alternative B
investment required	$70,000	$40,000
net income per year	$5620	$4075
rate of return on total investment	5%	8%

(A) Alternative A is economically superior.

(B) Alternative B is economically superior.

(C) Alternatives A and B are economically equivalent.

(D) Not enough information is provided.

29. Compare the costs of two plant renovation schemes, A and B. Assume equal lives of 25 years, no salvage values, and interest at 25%.

	alternative A	alternative B
first cost	$20,000	$25,000
annual expenditure	$3000	$2500

Determine the best alternative using the present worth method.

(A) Alternative A is economically superior.

(B) Alternative B is economically superior.

(C) Alternatives A and B are economically equivalent.

(D) Not enough information is provided.

30. A machine costs $18,000 and has a salvage value of $2000. It has a useful life of eight years. The interest rate is 8%. Using straight line depreciation, its book value at the end of five years is most nearly

(A) $2000

(B) $3000

(C) $6000

(D) $8000

31. A soda ash plant has the water effluent from processing equipment treated in a large settling basin. The settling basin eventually discharges into a river that runs alongside the basin. Recently enacted environmental regulations require all rainfall on the plant to be diverted and treated in the settling basin. A heavy rainfall will cause the entire basin to overflow. An uncontrolled overflow will cause environmental damage and heavy fines. The construction of additional height on the existing basin walls is under consideration.

Data on the costs of construction and expected costs for environmental cleanup and fines are shown. Data on 50 typical winters have been collected. The soda ash plant management considers 12% to be their minimum rate of return, and it is felt that after 15 years the plant will be closed. The company wants to select the alternative that minimizes its total expected costs.

additional basin height (ft)	number of winters with basin overflow	expense for environmental cleanup per year	construction cost
0	24	$550,000	0
5	14	$600,000	$600,000
10	8	$650,000	$710,000
15	3	$700,000	$900,00
20	1	$800,000	$1,000,000
	50		

The additional height the basin should be built to is most nearly

(A) 5.0 ft

(B) 10 ft

(C) 15 ft

(D) 20 ft

32. A wood processing plant installed a waste gas scrubber at a cost of $30,000 to remove pollutants from the exhaust discharged into the atmosphere. The scrubber has no salvage value and will cost $18,700 to operate next year, with operating costs expected to increase at the rate of $1200 per year thereafter. Money can be borrowed at 12%. Approximately when should the company consider replacing the scrubber?

(A) 3 yr

(B) 6 yr

(C) 8 yr

(D) 10 yr

33. Two alternative piping schemes are being considered by a water treatment facility. Head and horsepower are reflected in the hourly cost of operation. On the basis of a 10-year life and an interest rate of 12%, what is most nearly the number of hours of operation for which the two installations will be equivalent?

	alternative A	alternative B
pipe diameter	4 in	6 in
head loss for required flow	48 ft	26 ft
size motor required	20 hp	7 hp
energy cost per hour of operation	$0.30	$0.10
cost of motor installed	$3600	$2800
cost of pipes and fittings	$3050	$5010
salvage value at end of 10 years	$200	$280

(A) 1000 hr

(B) 3000 hr

(C) 5000 hr

(D) 6000 hr

34. A company is considering two alternatives, only one of which can be selected.

alternative	initial investment	salvage value	annual net profit	life
A	$120,000	$15,000	$57,000	5 yr
B	$170,000	$20,000	$67,000	5 yr

The net profit is after operating and maintenance costs, but before taxes. The company pays 45% of its year-end profit as income taxes. Use straight line depreciation. Do not use investment tax credit. Determine whether each alternative has an ROR greater than the MARR.

(A) Alternative A has ROR > MARR.

(B) Alternative B has ROR > MARR.

(C) Both alternatives have ROR > MARR.

(D) Neither alternative has ROR > MARR.

35. A company is considering the purchase of equipment to expand its capacity. The equipment cost is $300,000. The equipment is needed for five years, after which it will be sold for $50,000. The company's before-tax cash flow will be improved $90,000 annually by the purchase of the asset. The corporate tax rate is 48%, and straight line depreciation will be used. The company will take an investment tax credit of 6.67%. What is the after-tax rate of return associated with this equipment purchase?

(A) 10.9%

(B) 11.8%

(C) 12.2%

(D) 13.2%

36. A 120-room hotel is purchased for $2,500,000. A 25-year loan is available for 12%. The year-end convention applies to loan payments. A study was conducted to determine the various occupancy rates.

occupancy	probability
65% full	0.40
70%	0.30
75%	0.20
80%	0.10

The operating costs of the hotel are as follows.

taxes and insurance	$20,000 annually
maintenance	$50,000 annually
operating	$200,000 annually

The life of the hotel is figured to be 25 years when operating 365 days per year. The salvage value after 25 years is $500,000. The new hotel owners want to receive an annual rate of return of 15% on their investment. Neglect tax credit and income taxes. The average rate that should be charged per room per night is most nearly

(A) $27

(B) $29

(C) $30

(D) $31

37. A company is insured for $3,500,000 against fire and the insurance rate is $0.69/$1000. The insurance company will decrease the rate to $0.47/$1000 if fire sprinklers are installed. The initial cost of the sprinklers is $7500. Annual costs are $200; additional taxes are $100 annually. The system life is 25 years. The rate of return is most nearly

(A) 3.8%

(B) 5.0%

(C) 13%

(D) 16%

38. Heat losses through the walls in an existing building cost a company $1,300,000 per year. This amount is considered excessive, and two alternatives are being evaluated. Neither of the alternatives will increase the life of the existing building beyond the current expected life of six years, and neither of the alternatives will produce a salvage value. Improvements can be depreciated.

Alternative A: Do nothing, and continue with current losses.

Alternative B: Spend $2,000,000 immediately to upgrade the building and reduce the loss by 80%. Annual maintenance will cost $150,000.

Alternative C: Spend $1,200,000 immediately. Then, repeat the $1,200,000 expenditure 3 years from now. Heat loss the first year will be reduced 80%. Due to deterioration, the reduction will be 55% and 20% in the second and third years. (The pattern is repeated starting after the second expenditure.) There are no maintenance costs.

All energy and maintenance costs are considered expenses for tax purposes. The company's tax rate is 48%, and straight line depreciation is used. 15% is regarded as the effective annual interest rate. Evaluate each alternative on an after-tax basis. Which alternative should be recommended?

(A) alternative A

(B) alternative B

(C) alternative C

(D) not enough information

39. You have been asked to determine if a seven-year-old machine should be replaced. Base your decision on a before-tax interest rate of 15%.

The existing machine is presumed to have a 10-year life. It has been depreciated on a straight line basis from its original value of $1,250,000 to a current book value of $620,000. Its ultimate salvage value was assumed to be $350,000 for purposes of depreciation. Its present salvage value is estimated at $400,000, and this is not expected to change over the next three years. The current operating costs are not expected to change from $200,000 per year.

A new machine costs $800,000, with operating costs of $40,000 the first year, and increasing by $30,000 each year thereafter. The new machine has an expected life of 10 years. The salvage value depends on the year the new machine is retired.

year retired	salvage
1	$600,000
2	$500,000
3	$450,000
4	$400,000
5	$350,000
6	$300,000
7	$250,000
8	$200,000
9	$150,000
10	$100,000

From an economic perspective, when does it makes sense to buy the new piece of equipment?

(A) during years 1–3

(B) during years 4–6

(C) during years 7–10

(D) never

40. As production facilities move toward just-in-time manufacturing, it is important to minimize

(A) demand rate

(B) production rate

(C) inventory carrying cost

(D) setup cost

SOLUTIONS

1.

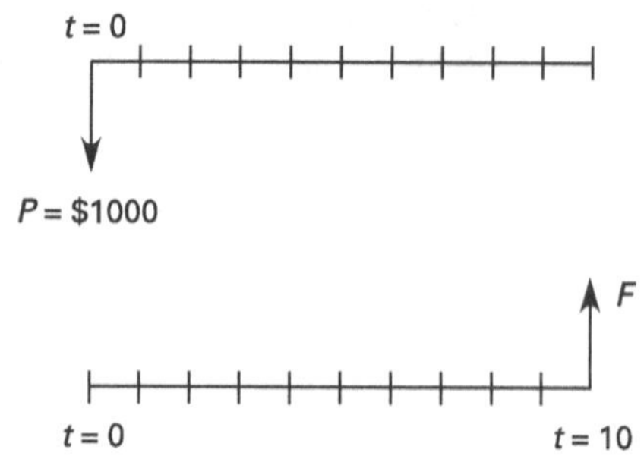

$i = 6\%$ a year

Using the formula from Table 37.1 (also *NCEES Handbook:* Cost Estimation and Project Evaluation),

$$F = P(1+i)^n = (\$1000)(1+0.06)^{10} = \$1790.85$$

Using the factor from App. 37A (also *NCEES Handbook:* Cost Estimation and Project Evaluation), $(F/P, i\%, n) = 1.7908$ for $i = 6\%$ a year and $n = 10$ years.

$$F = P(F/P, 6\%, 10) = (\$1000)(1.7908) = \$1790.80 \quad (\$1800)$$

The answer is (D).

2.

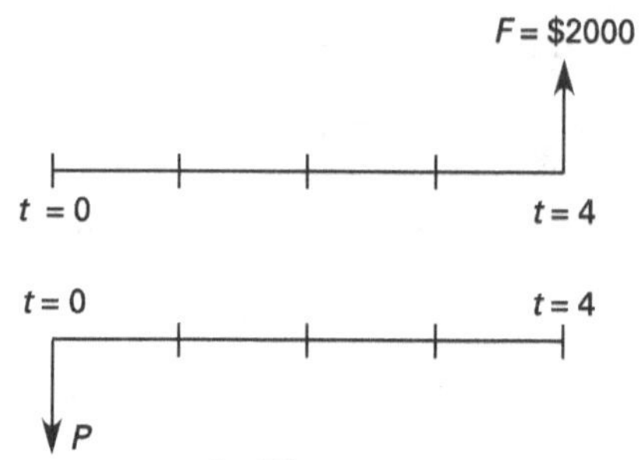

$i = 6\%$ a year

Using the formula from Table 37.1 (also *NCEES Handbook:* Cost Estimation and Project Evaluation),

$$P = \frac{F}{(1+i)^n} = \frac{\$2000}{(1+0.06)^4} = \$1584.19$$

Using the factor from App. 37A (also *NCEES Handbook:* Cost Estimation and Project Evaluation), $(P/F,\ i,\ n) = 0.7921$ for $i = 6\%$ a year and $n = 4$ years.

$$\begin{aligned} P &= F(P/F, 6\%, 4) = (\$2000)(0.7921) \\ &= \$1584.20 \quad (\$1600) \end{aligned}$$

The answer is (C).

3.

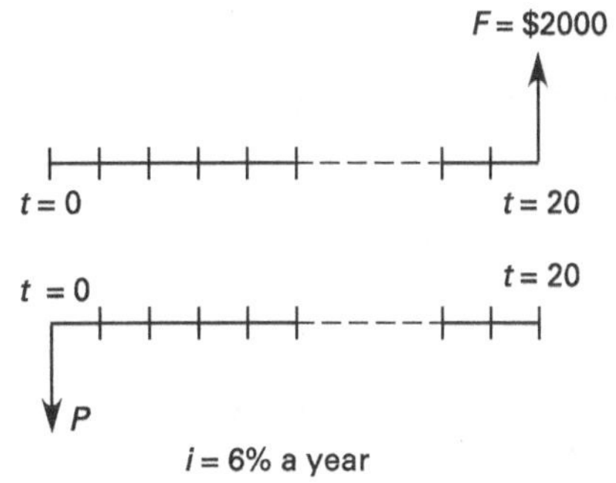

Using the formula from Table 37.1 (also *NCEES Handbook:* Cost Estimation and Project Evaluation),

$$\begin{aligned} P &= \frac{F}{(1+i)^n} = \frac{\$2000}{(1+0.06)^{20}} \\ &= \$623.61 \end{aligned}$$

Using the factor from App. 37A (also *NCEES Handbook:* Cost Estimation and Project Evaluation), $(P/F,\ i,\ n) = 0.3118$ for $i = 6\%$ a year and $n = 20$ years.

$$\begin{aligned} P &= F(P/F, 6\%, 20) = (\$2000)(0.3118) \\ &= \$623.60 \quad (\$620) \end{aligned}$$

The answer is (A).

4.

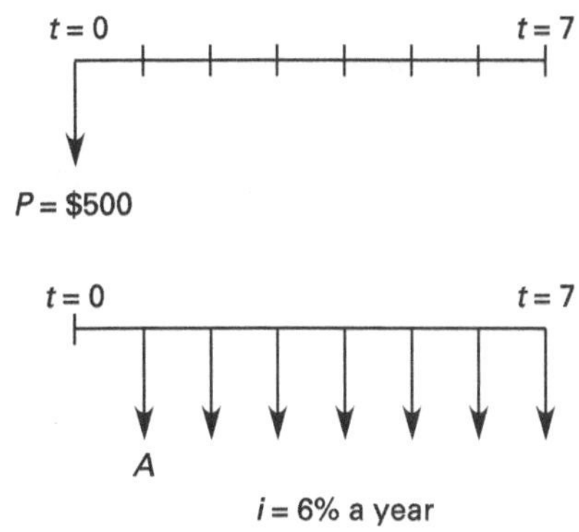

Using the formula from Table 37.1 (also *NCEES Handbook:* Cost Estimation and Project Evaluation),

$$\begin{aligned} A &= P\left(\frac{i(1+i)^n}{(1+i)^n - 1}\right) = (\$500)\left(\frac{(0.06)(1+0.06)^7}{(1+0.06)^7 - 1}\right) \\ &= \$89.57 \end{aligned}$$

Using the factor from App. 37A (also *NCEES Handbook:* Cost Estimation and Project Evaluation), $(A/P,\ i,\ n) = 0.1791$ for $i = 6\%$ a year and $n = 7$ years.

$$\begin{aligned} A &= P(A/P, 6\%, 7) = (\$500)(0.17914) \\ &= \$89.55 \quad (\$90) \end{aligned}$$

The answer is (A).

5.

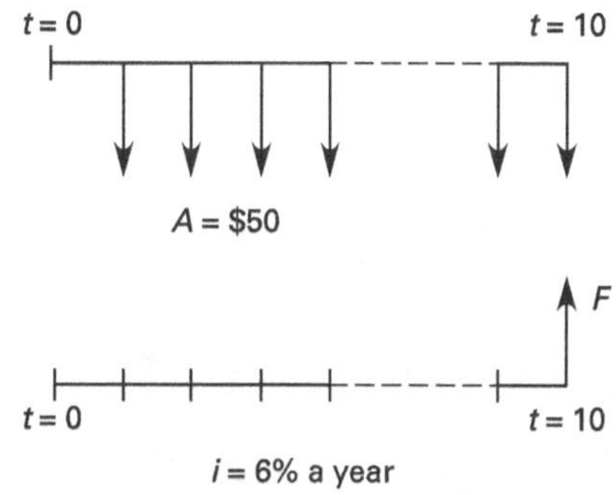

Using the formula from Table 37.1 (also *NCEES Handbook:* Cost Estimation and Project Evaluation),

$$\begin{aligned} F &= A\left(\frac{(1+i)^n - 1}{i}\right) \\ &= (\$50)\left(\frac{(1+0.06)^{10} - 1}{0.06}\right) \\ &= \$659.04 \end{aligned}$$

Using the factor from App. 37A (also *NCEES Handbook:* Cost Estimation and Project Evaluation), $(F/A,\ i,\ n) = 13.1808$ for $i = 6\%$ a year and $n = 10$ years.

$$\begin{aligned} F &= A(F/A, 6\%, 10) \\ &= (\$50)(13.1808) \\ &= \$659.04 \quad (\$660) \end{aligned}$$

The answer is (C).

6.

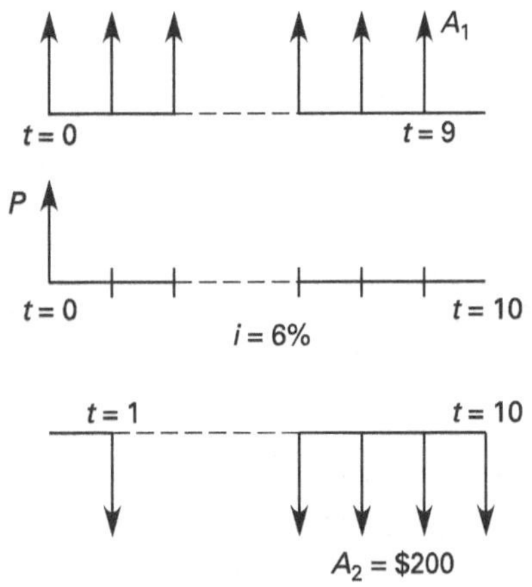

From Table 37.1 (also *NCEES Handbook:* Cost Estimation and Project Evaluation), for each cash flow diagram,

$$\begin{aligned} P &= A_1 + A_1\left(\frac{(1+0.06)^9 - 1}{(0.06)(1+0.06)^9}\right) \\ &= A_2\left(\frac{(1+0.06)^{10} - 1}{(0.06)(1+0.06)^{10}}\right) \end{aligned}$$

Therefore, for $A_2 = \$200$,

$$\begin{aligned} A_1 + A_1\left(\frac{(1+0.06)^9 - 1}{(0.06)(1+0.06)^9}\right) &= (\$200) \\ &\quad \times\left(\frac{(1+0.06)^{10} - 1}{(0.06)(1+0.06)^{10}}\right) \\ 7.80A_1 &= \$1472.02 \\ A_1 &= \$188.72 \end{aligned}$$

Using factors from App. 37A (also *NCEES Handbook:* Cost Estimation and Project Evaluation),

$$\begin{aligned} (P/A, 6\%, 9) &= 6.8017 \\ (P/A, 6\%, 10) &= 7.3601 \\ A_1 + A_1(6.8017) &= (\$200)(7.3601) \\ 7.8017A_1 &= \$1472.02 \\ A_1 &= \frac{\$1472.02}{7.8017} \\ &= \$188.68 \quad (\$190) \end{aligned}$$

The answer is (A).

7.

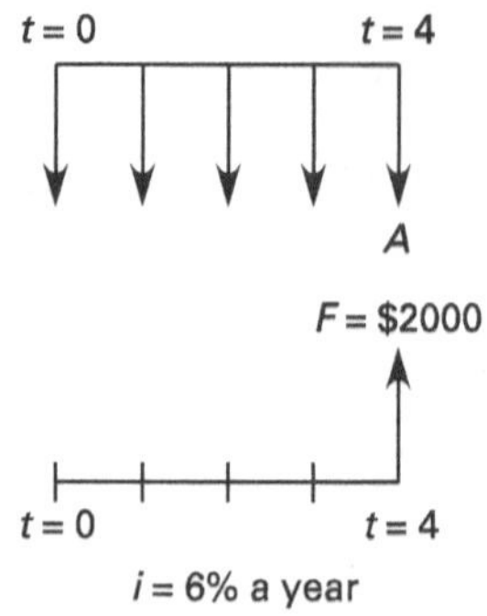

From Table 37.1 (also *NCEES Handbook:* Cost Estimation and Project Evaluation),

$$F = A\left(\frac{(1+i)^n - 1}{i}\right)$$

Since the deposits start at the beginning of each year, five deposits are made that contribute to the final amount. This is equivalent to a cash flow that starts at $t = -1$ without a deposit and has a duration (starting at $t = -1$) of five years.

$$\begin{aligned} F &= A\left(\frac{(1+i)^n - 1}{i}\right) = \$2000 \\ &= A\left(\frac{(1+0.06)^5 - 1}{0.06}\right) \end{aligned}$$

$$\begin{aligned} \$2000 &= 5.6371A \\ A &= \frac{\$2000}{5.6371} \\ &= \$354.79 \end{aligned}$$

Using factors from App. 37A (also *NCEES Handbook:* Cost Estimation and Project Evaluation),

$$\begin{aligned} F &= A\big((F/P, 6\%, 4) + (F/A, 6\%, 4)\big) \\ \$2000 &= A(1.2625 + 4.3746) \\ A &= \$354.79 \quad (\$350) \end{aligned}$$

The answer is (A).

8.

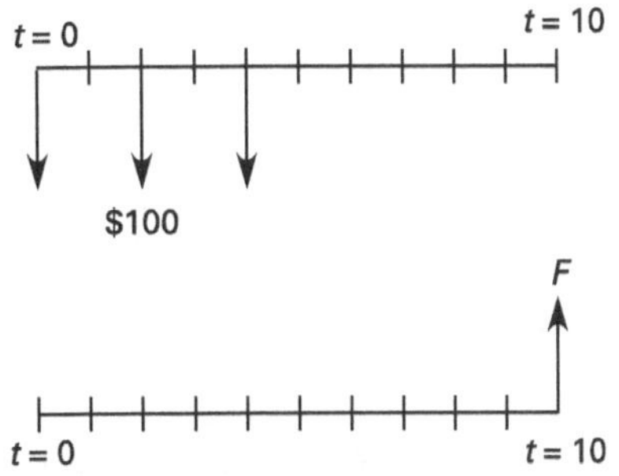

i = 6% a year

From Table 37.1 (also *NCEES Handbook:* Cost Estimation and Project Evaluation), $F = P(1+i)^n$. If each deposit is considered as P, each will accumulate interest for periods of 10, 8, and 6 years. Therefore,

$$\begin{aligned} F &= (\$100)(1+0.06)^{10} + (\$100)(1+0.06)^{8} \\ &\quad + (\$100)(1+0.06)^{6} \\ &= (\$100)(1.7908 + 1.5938 + 1.4185) \\ &= \$480.31 \end{aligned}$$

Using App. 37A (also *NCEES Handbook:* Cost Estimation and Project Evaluation),

$$\begin{aligned} (F/P,i,n) &= 1.7908 \text{ for } i = 6\% \text{ and } n = 10 \\ &= 1.5938 \text{ for } i = 6\% \text{ and } n = 8 \\ &= 1.4185 \text{ for } i = 6\% \text{ and } n = 6 \end{aligned}$$

By summation,

$$\begin{aligned} F &= (\$100)(1.7908 + 1.5938 + 1.4185) \\ &= \$480.31 \quad (\$480) \end{aligned}$$

The answer is (B).

9.

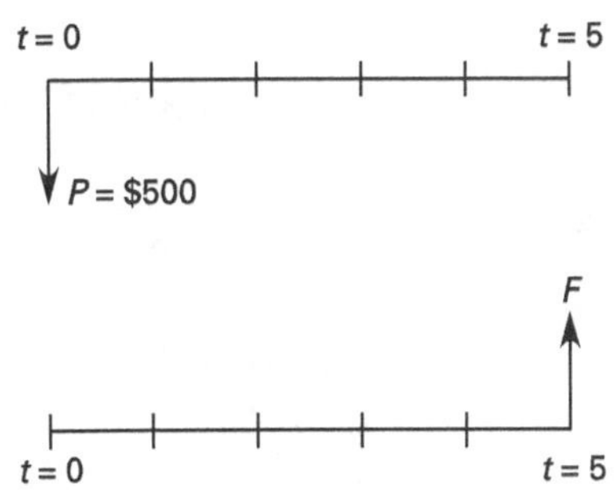

r = 6% a year

Since the deposit is compounded monthly, the effective interest rate should be calculated from Eq. 37.51 (also *NCEES Handbook:* Cost Estimation and Project Evaluation).

$$\begin{aligned} i &= \left(1+\frac{r}{k}\right)^k - 1 = \left(1+\frac{0.06}{12}\right)^{12} - 1 \\ &= 0.061678 \end{aligned}$$

From Table 37.1 (also *NCEES Handbook:* Cost Estimation and Project Evaluation),

$$\begin{aligned} F &= P(1+i)^n = (\$500)(1+0.061678)^5 \\ &= \$674.43 \end{aligned}$$

Use *NCEES Handbook:* Cost Estimation and Project Evaluation.

$$\begin{aligned} F &= P(F/P, 6.1678\%, 5) = (\$500)(1.3490) \\ &= \$674.50 \quad (\$675) \end{aligned}$$

The answer is (C).

10.

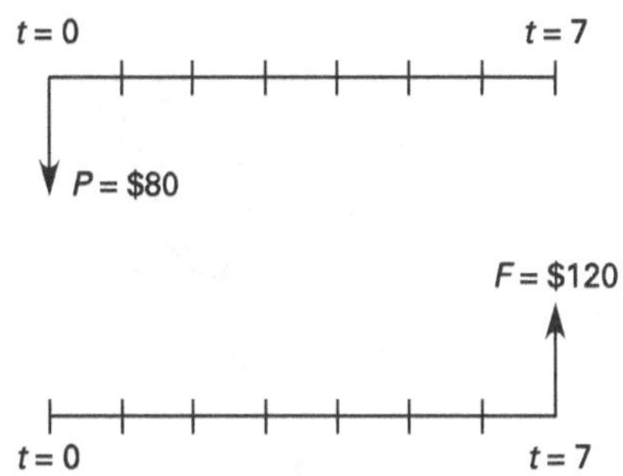

From Table 37.1 (also *NCEES Handbook:* Cost Estimation and Project Evaluation),

$$F = P(1+i)^n$$

Therefore,

$$\begin{aligned} (1+i)^n &= F/P \\ i &= (F/P)^{1/n} - 1 = \left(\frac{\$120}{\$80}\right)^{1/7} - 1 \\ &= 0.0596 \approx 6.0\% \end{aligned}$$

From Table 37.1 (also *NCEES Handbook:* Cost Estimation and Project Evaluation),

$$\begin{aligned} F &= P(F/P, i\%, 7) \\ (F/P, i\%, 7) &= F/P = \frac{\$120}{\$80} = 1.5 \end{aligned}$$

From App. 37A (also *NCEES Handbook:* Cost Estimation and Project Evaluation),

$$
\begin{aligned}
(F/P, i\%, 7) &= 1.4071 \text{ for } i = 5\% \\
&= 1.5036 \text{ for } i = 6\% \\
&= 1.6058 \text{ for } i = 7\%
\end{aligned}
$$

Therefore, $i \approx 6.0\%$.

The answer is (D).

11.

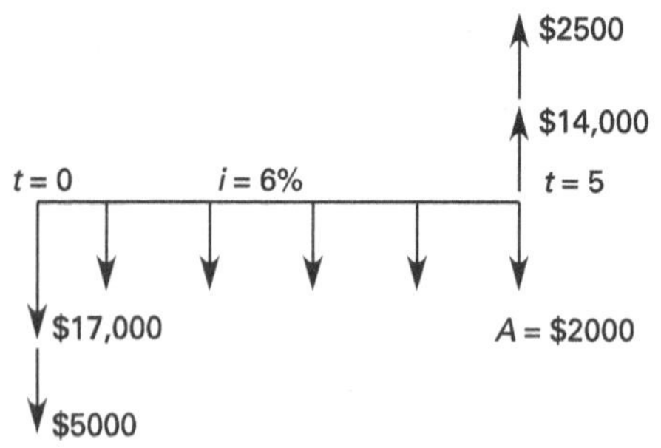

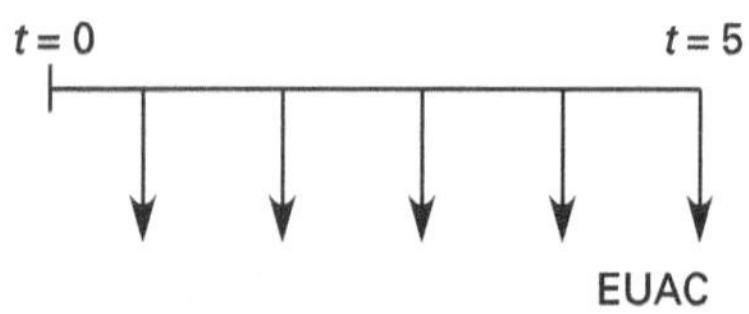

As in *NCEES Handbook:* Cost Estimation and Project Evaluation, annual cost of ownership, EUAC, can be obtained by the factors converting P to A and F to A.

$$
\begin{aligned}
P &= \$17{,}000 + \$5000 \\
&= \$22{,}000 \\
F &= \$14{,}000 + \$2500 \\
&= \$16{,}500 \\
\text{EUAC} &= A + P(A/P, 6\%, 5) - F(A/F, 6\%, 5) \\
(A/P, 6\%, 5) &= 0.2374 \\
(A/F, 6\%, 5) &= 0.1774 \\
\text{EUAC} &= \$2000 + (\$22{,}000)(0.2374) \\
&\quad - (\$16{,}500)(0.1774) \\
&= \$4295.70 \quad (\$4300)
\end{aligned}
$$

The answer is (C).

12.

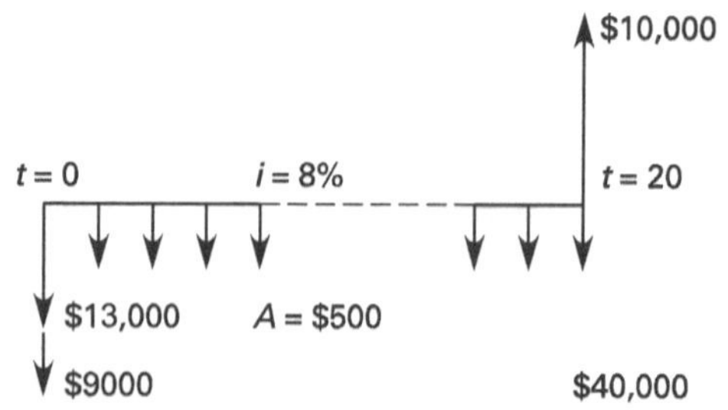

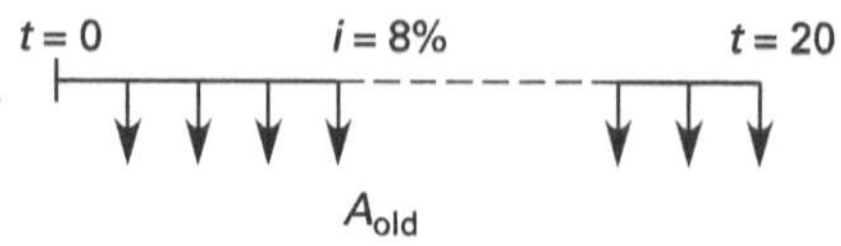

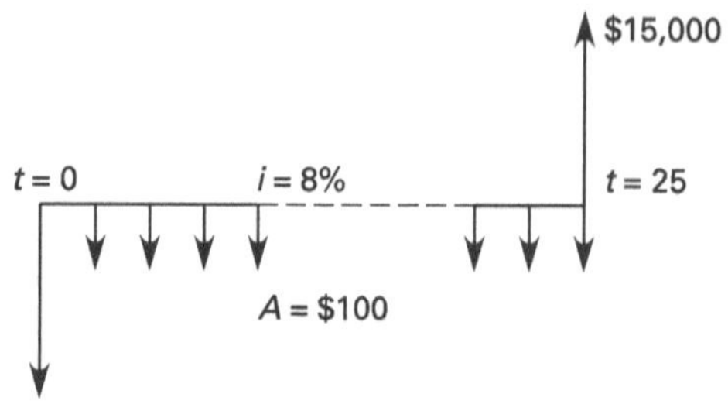

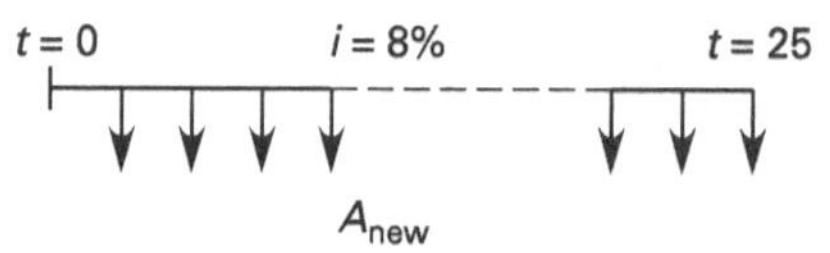

Consider the salvage value as a benefit lost (cost).

$$
\begin{aligned}
\text{EUAC}_{\text{old}} &= \$500 + (\$9000 + \$13{,}000)(A/P, 8\%, 20) \\
&\quad - (\$10{,}000)(A/F, 8\%, 20) \\
(A/P, 8\%, 20) &= 0.1019 \\
(A/F, 8\%, 20) &= 0.0219 \\
\text{EUAC}_{\text{old}} &= \$500 + (\$22{,}000)(0.1019) \\
&\quad - (\$10{,}000)(0.0219) \\
&= \$2522.80
\end{aligned}
$$

Similarly,

$$
\begin{aligned}
\text{EUAC}_{\text{new}} &= \$100 + (\$40{,}000)(A/P, 8\%, 25) \\
&\quad - (\$15{,}000)(A/F, 8\%, 25) \\
(A/P, 8\%, 25) &= 0.0937 \\
(A/F, 8\%, 25) &= 0.0137 \\
\text{EUAC}_{\text{new}} &= \$100 + (\$40{,}000)(0.0937) \\
&\quad - (\$15{,}000)(0.0137) \\
&= \$3642.50
\end{aligned}
$$

Therefore, the new bridge is more costly.

The best alternative is to strengthen the old bridge.

The answer is (A).

13.

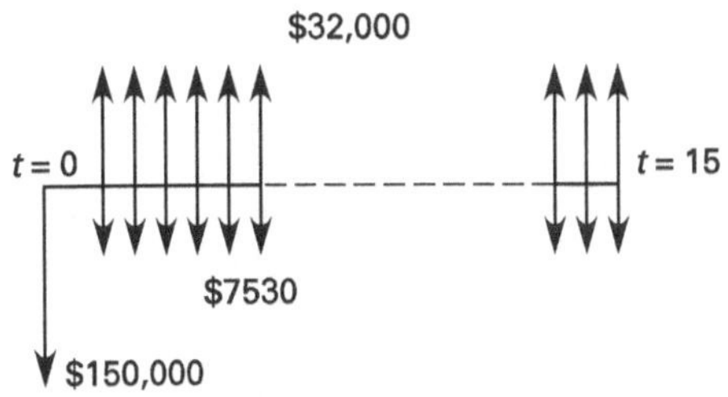

As in *NCEES Handbook* table "Depreciation Methods," the annual depreciation is

$$D = \frac{C - S_n}{n} = \frac{\$150{,}000}{15} = \$10{,}000/\text{year}$$

The taxable income is

$$\$32{,}000 - \$7530 - \$10{,}000 = \$14{,}470/\text{year}$$

Taxes paid are

$$(\$14{,}470)(0.48) = \$6945.60/\text{year}$$

The after-tax cash flow is

$$\$32{,}000 - \$7530 - \$6945.60 = \$17{,}524.40$$

The present worth of the alternate is zero when evaluated at its ROR.

$$0 = -\$150{,}000 + (\$17{,}524.40)(P/A, i\%, 15)$$

Therefore,

$$(P/A, i\%, 15) = \frac{\$150{,}000}{\$17{,}524.40} = 8.55949$$

From App. 37A, this factor matches $i = 8\%$.

$$\text{ROR} = 8.0\%$$

The answer is (A).

14. The conventional benefit/cost ratio is

$$B/C = \frac{B - D}{C}$$

The benefit/cost ratio will be

$$B/C = \frac{\$1{,}500{,}000 - \$300{,}000}{\$1{,}000{,}000} = 1.2$$

The excess of benefits over cost is \$200,000.

The answer is (A).

15. The annual rent is

$$(\$75)\left(12 \ \frac{\text{months}}{\text{year}}\right) = \$900$$

$$P = P_1 + P_2 = \$15{,}000$$
$$A_1 = -\$900$$
$$A_2 = \$250 + \$150 = \$400$$

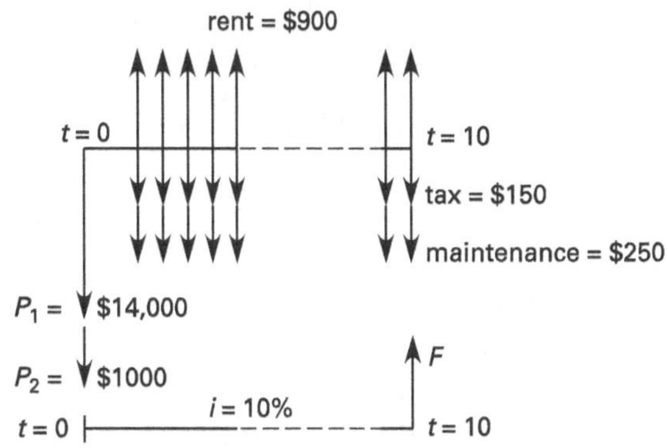

From Table 37.1 (also *NCEES Handbook:* Cost Estimation and Project Evaluation),

$$\begin{aligned} F &= (\$15{,}000)(F/P, 10\%, 10) \\ &\quad + (\$400)(F/A, 10\%, 10) \\ &\quad - (\$900)(F/A, 10\%, 10) \end{aligned}$$

$$(F/P, 10\%, 10) = 2.5937$$
$$(F/A, 10\%, 10) = 15.9374$$

$$\begin{aligned} F &= (\$15{,}000)(2.5937) + (\$400)(15.9374) \\ &\quad - (\$900)(15.9374) \\ &= \$30{,}937 \quad (\$31{,}000) \end{aligned}$$

The answer is (B).

16.

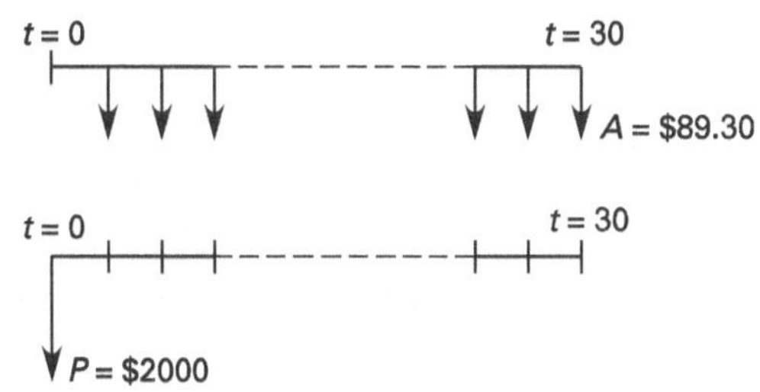

From Table 37.1 (also *NCEES Handbook:* Cost Estimation and Project Evaluation),

$$P = A\left(\frac{(1+i)^n - 1}{i(1+i)^n}\right)$$

$$\frac{(1+i)^{30} - 1}{i(1+i)^{30}} = \frac{\$2000}{\$89.30} = 22.40$$

By trial and error,

$i\%$	$(1+i)^{30}$	$\frac{(1+i)^{30}-1}{i(1+i)^{30}}$
10	17.45	9.43
6	5.74	13.76
4	3.24	17.29
2	1.81	22.40

2% per month is close.

$$\begin{aligned} i &= (1+0.02)^{12} - 1 \\ &= 0.2682 \quad (27\%) \end{aligned}$$

The answer is (A).

17. Use the straight line method, from Eq. 37.25 (also *NCEES Handbook:* Cost Estimation and Project Evaluation).

$$D = \frac{C - S_n}{n}$$

Each year depreciation will be the same.

$$D = \frac{\$500{,}000 - \$100{,}000}{25} = \$16{,}000$$

The answer is (B).

18.

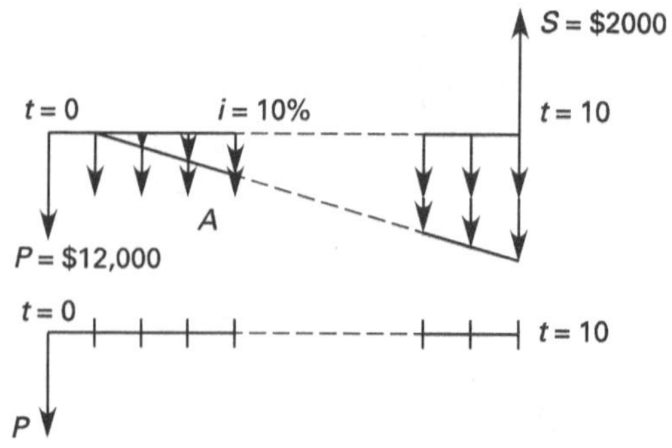

A is \$1000 for t = 10 years, and G is \$200 for $t = n - 1 = 9$ years. With a future value F = the sale value S = \$2,000, the present worth (cost) is

$$\begin{aligned} P &= \$12{,}000 + A(P/A, 10\%, 10) + G(P/G, 10\%, 9) \\ &\quad - F(P/F, 10\%, 10) \\ &= \$12{,}000 + (\$1000)(6.1446) + (\$200)(19.4215) \\ &\quad - (\$2000)(0.3855) \\ &= \$21{,}258 \quad (\$21{,}000) \end{aligned}$$

The answer is (C).

19. An increase in rock removal capacity can be achieved by a 20-year loan (investment). Different cases can be compared by equivalent uniform annual cost (EUAC).

$$\begin{aligned} \text{EUAC} &= \text{annual loan cost} \\ &\quad + \text{expected annual damage} \\ &= \text{cost}\,(A/P, 10\%, 20) \\ &\quad + (\$25{,}000)(\text{probability}) \end{aligned}$$

$$(A/P, 10\%, 20) = 0.1175$$

A table can be prepared for different cases.

rock removal rate	cost ($)	annual loan cost ($)	expected annual damage ($)	EUAC ($)
7	0	0	3750	3750.00
8	15,000	1761.90	2500	4261.90
9	20,000	2349.20	1750	4099.20
10	30,000	3523.80	750	4273.80

It is cheapest to do nothing.

The answer is (A).

20. Calculate the cost of owning and operating for years one and two using *NCEES Handbook:* Cost Estimation and Project Evaluation.

$$\begin{aligned} A_1 &= (\$10{,}000)(A/P,20\%,1) + \$2000 \\ &\quad -(\$8000)(A/F,20\%,1) \end{aligned}$$

$$(A/P,20\%,1) = 1.2$$

$$(A/F,20\%,1) = 1.0$$

$$\begin{aligned} A_1 &= (\$10{,}000)(1.2) + \$2000 - (\$8000)(1.0) \\ &= \$6000 \end{aligned}$$

$$\begin{aligned} A_2 &= (\$10{,}000)(A/P,20\%,2) + \$2000 \\ &\quad +(\$1000)(A/G,20\%,2) \\ &\quad -(\$7000)(A/F,20\%,2) \end{aligned}$$

$$(A/P,20\%,2) = 0.6545$$

$$(A/G,20\%,2) = 0.4545$$

$$(A/F,20\%,2) = 0.4545$$

$$\begin{aligned} A_2 &= (\$10{,}000)(0.6545) + \$2000 \\ &\quad +(\$1000)(0.4545) - (\$7000)(0.4545) \\ &= \$5818 \end{aligned}$$

Calculate the cost of owning and operating for years three through five using *NCEES Handbook:* Cost Estimation and Project Evaluation.

$$\begin{aligned} A_3 &= (\$10{,}000)(A/P,20\%,3) + \$2000 \\ &\quad +(\$1000)(A/G,20\%,3) \\ &\quad -(\$6000)(A/F,20\%,3) \end{aligned}$$

$$(A/P,20\%,3) = 0.4747$$

$$(A/G,20\%,3) = 0.8791$$

$$(A/F,20\%,3) = 0.2747$$

$$\begin{aligned} A_3 &= (\$10{,}000)(0.4747) + \$2000 \\ &\quad +(\$1000)(0.8791) \\ &\quad -(\$6000)(0.2747) \\ &= \$5977.90 \end{aligned}$$

$$\begin{aligned} A_4 &= (\$10{,}000)(A/P,20\%,4) \\ &\quad +\$2000 + (\$1000)(A/G,20\%,4) \\ &\quad -(\$5000)(A/F,20\%,4) \end{aligned}$$

$$(A/F,20\%,4) = 0.3863$$

$$(A/F,20\%,4) = 1.2742$$

$$(A/F,20\%,4) = 0.1863$$

$$\begin{aligned} A_4 &= (\$10{,}000)(0.3863) + \$2000 \\ &\quad +(\$1000)(1.2742) - (\$5000)(0.1863) \\ &= \$6205.70 \end{aligned}$$

$$\begin{aligned} A_5 &= (\$10{,}000)(A/P,20\%,5) + \$2000 \\ &\quad +(\$1000)(A/G,20\%,5) \\ &\quad -(\$4000)(A/F,20\%,5) \end{aligned}$$

$$(A/P,20\%,5) = 0.3344$$

$$(A/G,20\%,5) = 1.6405$$

$$(A/F,20\%,5) = 0.1344$$

$$\begin{aligned} A_5 &= (\$10{,}000)(0.3344) + \$2000 \\ &\quad +(\$1000)(1.6405) \\ &\quad -(\$4000)(0.1344) \\ &= \$6446.90 \end{aligned}$$

Since the annual owning and operating cost is smallest after two years of operation, it is advantageous to sell the mechanism after the second year.

The economic life is two years.

The answer is (B).

21. To find out if the reimbursement is adequate, calculate the business-related expense as in *NCEES Handbook:* Cost Estimation and Project Evaluation.

Charge the company for business travel.

$$\begin{aligned} \text{insurance: } \$3000 - \$2000 &= \$1000 \\ \text{maintenance: } \$2000 - \$1500 &= \$500 \\ \text{drop in salvage value: } \$10{,}000 - \$5000 &= \$5000 \end{aligned}$$

The annual portion of the drop in salvage value is

$$\begin{aligned} A &= (\$5000)(A/F,10\%,5) \\ (A/F,10\%,5) &= 0.1638 \\ A &= (\$5000)(0.1638) \\ &= \$819/\text{yr} \end{aligned}$$

The annual cost of gas is

$$\left(\frac{50{,}000 \text{ mi}}{15 \dfrac{\text{mi}}{\text{gal}}}\right)\left(\frac{\$1.50}{\text{gal}}\right) = \$5000$$

$$\begin{aligned} \text{EUAC per mile} &= \frac{\$1000 + \$500 + \$819 + \$5000}{50{,}000 \text{ mi}} \\ &= \$0.14638/\text{mi} \end{aligned}$$

Since the reimbursement per mile is \$0.30 and since $\$0.30 > \$0.14638$, the reimbursement is adequate.

Determine (with reimbursement) how many miles the car must be driven to break even.

If the car is driven M miles per year,

$$\left(\frac{\$0.30}{1\text{ mi}}\right)M = (\$50{,}000)(A/P,10\%,5) + \$2500 + \$2000 - (\$8000)(A/F,10\%,5) + \left(\frac{M}{15\ \frac{\text{mi}}{\text{gal}}}\right)(\$1.50)$$

$$(A/P,10\%,5) = 0.2638$$
$$(A/F,10\%,5) = 0.1638$$
$$0.3M = (\$50{,}000)(0.2638) + \$2500 + \$2000 - (\$8000)(0.1638) + 0.1M$$
$$0.2M = \$16{,}379.60$$
$$M = \frac{\$16{,}379.60}{\frac{\$0.20}{1\text{ mi}}} = 81{,}898\text{ mi}\quad(82{,}000\text{ mi})$$

The answer is (C).

22. Using *NCEES Handbook:* Cost Estimation and Project Evaluation, the present worth of alternative A is

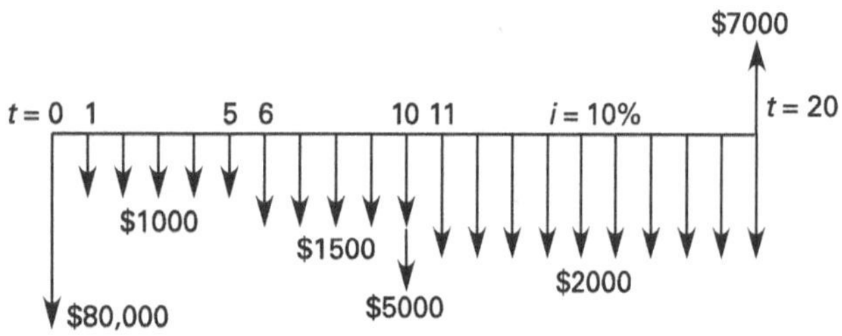

$$P_A = \$80{,}000 + (\$1000)(P/A,10\%,5) + (\$1500)(P/A,10\%,5)(P/F,10\%,5) + (\$2000)(P/A,10\%,10) + (\$5000)(P/F,10\%,10) - (\$7000)(P/F,10\%,20)$$

$$(P/A,10\%,5) = 3.7908$$
$$(P/F,10\%,5) = 0.6209$$
$$(P/A,10\%,10) = 6.1446$$
$$(P/F,10\%,10) = 0.3855$$
$$(P/F,10\%,20) = 0.1486$$

$$P_A = \$80{,}000 + (\$1000)(3.7908) + (\$1500)(3.7908)(0.6209) + (\$2000)(6.1446)(0.3855) + (\$5000)(0.3855) - (\$7000)(0.1486) = \$92{,}946.15$$

Since the lives are different, compare by EUAC.

$$\text{EUAC(A)} = (\$92{,}946.14)(A/P,10\%,20) = (\$92{,}946.14)(0.1175) = \$10{,}921$$

Evaluate alternative B.

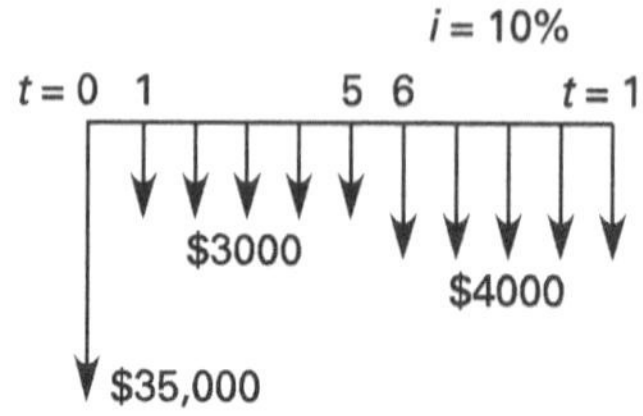

$$P_B = \$35{,}000 + (\$3000)(P/A,10\%,5) + (\$4000)(P/A,10\%,5)(P/F,10\%,5)$$
$$(P/A,10\%,5) = 3.7908$$
$$(P/F,10\%,5) = 0.6209$$
$$P_B = \$35{,}000 + (\$3000)(3.7908) + (\$4000)(3.7908)(0.6209) = \$55{,}787.23$$

Since the lives are different, compare by EUAC.

$$\text{EUAC(B)} = (\$55{,}787.23)(A/P,10\%,10) = (\$55{,}787.23)(0.1627) = \$9077$$

Since EUAC(B) < EUAC(A), Alternative B is economically superior.

The answer is (B).

23. If the annual cost is compared with a total annual mileage of M, for plan A,

$$A_A = \$0.25M$$

Use *NCEES Handbook:* Cost Estimation and Project Evaluation throughout the rest of the problem.

For plan B,

$$A_B = (\$30{,}000)(A/P, 10\%, 3) + \$0.14M + \$500 - (\$7200)(A/F, 10\%, 3)$$

$$(A/P, 10\%, 3) = 0.4021$$

$$(A/F, 10\%, 3) = 0.3021$$

$$A_B = (\$30{,}000)(0.4021) + \$0.14M + \$500 - (\$7200)(0.3021) = \$10{,}387.88 + \$0.14M$$

For an equal annual cost $A_A = A_B$,

$$\$0.25M = \$10{,}387.88 + \$0.14M$$

$$\$0.11M = \$10{,}387.88$$

$$M = 94{,}435 \quad (94{,}000)$$

An annual mileage would be $M = 94{,}000$ mi.

For an annual mileage less than that, $A_A < A_B$.

Plan A is economically superior until 94,000 mi is exceeded.

The answer is (A).

24. Use *NCEES Handbook:* Cost Estimation and Project Evaluation.

Method A:

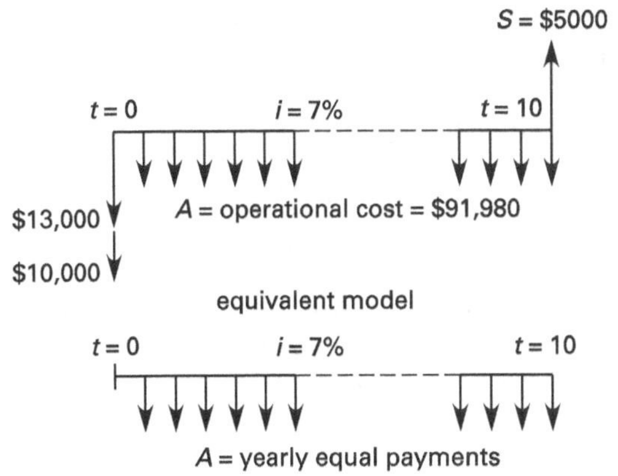

$$24 \text{ hr/day}$$

$$365 \text{ days/yr}$$

$$\text{total of } (24)(365) = 8760 \text{ hr/yr}$$

$$\$10.50 \text{ operational cost/hr}$$

$$\text{total of } (8760)(\$10.50) = \$91{,}980 \text{ operational cost/yr}$$

$$A = \$91{,}980 + (\$23{,}000)(A/P, 7\%, 10) - (\$5000)(A/F, 7\%, 10)$$

$$(A/P, 7\%, 10) = 0.1424$$

$$(A/F, 7\%, 10) = 0.0724$$

$$A = \$91{,}980 + (\$23{,}000)(0.1424) - (\$5000)(0.0724) = \$94{,}893.20/\text{yr}$$

Therefore, the uniform annual cost per ton each year will be

$$\frac{\$94{,}893.20}{50 \text{ ton}} = \$1897.86$$

Method B:

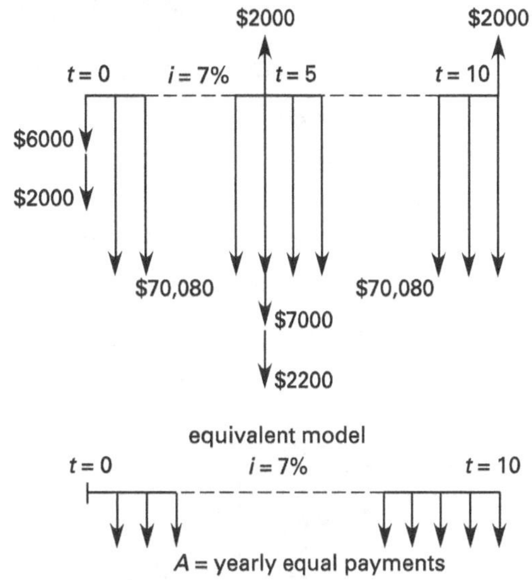

$$8760 \text{ hr/yr}$$

$$\$8 \text{ operational cost/hr}$$

$$\text{total of } \$70{,}080 \text{ operational cost/yr}$$

$$A = \$70{,}080 + (\$6000 + \$2000) \times (A/P, 7\%, 10) + (\$7000 + \$2200 - \$2000) \times (P/F, 7\%, 5)(A/P, 7\%, 10) - (\$2000)(A/F, 7\%, 10)$$

$$(A/P, 7\%, 10) = 0.1424$$

$$(A/F, 7\%, 10) = 0.0724$$

$$(P/F, 7\%, 5) = 0.7130$$

$$A = \$70{,}080 + (\$8000)(0.1424) + (\$7200)(0.7130)(0.1424) - (\$2000)(0.0724) = \$71{,}805.42/\text{yr}$$

Therefore, the uniform annual cost per ton each year will be

$$\frac{\$71{,}805.42}{20\ \text{ton}} = \$3590.27 \quad (\$3600)$$

The answer is (B).

25. As in *NCEES Handbook:* Cost Estimation and Project Evaluation,

$$\begin{aligned} A_e &= (\$60{,}000)(A/P, 7\%, 20) + A \\ &\quad + G(P/G, 7\%, 20)(A/P, 7\%, 20) \\ &\quad - (\$10{,}000)(A/F, 7\%, 20) \\ (A/P, 7\%, 20) &= 0.0944 \\ A &= (37{,}440\ \text{mi})\left(\frac{\$1.0}{1\ \text{mi}}\right) = \$37{,}440 \\ G &= 0.1A = (0.1)(\$37{,}440) = \$3744 \\ (P/G, 7\%, 20) &= 77.5091 \\ (A/F, 7\%, 20) &= 0.0244 \\ A_e &= (\$60{,}000)(0.0944) + \$37{,}440 \\ &\quad + (\$3744)(77.5091)(0.0944) \\ &\quad - (\$10{,}000)(0.0244) \\ &= \$70{,}254.32 \end{aligned}$$

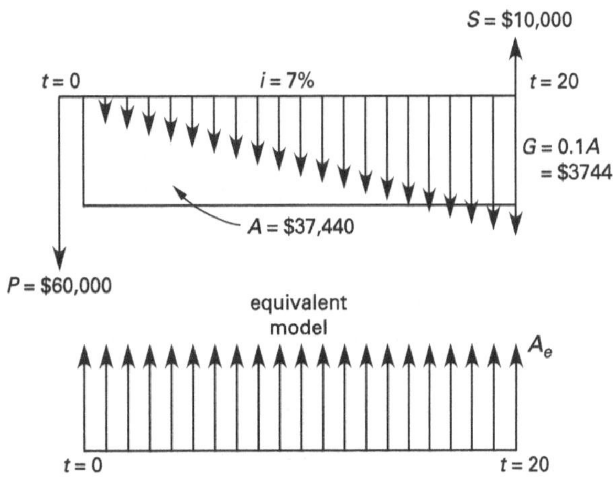

With 80,000 passengers a year, the break-even fare per passenger would be

$$\begin{aligned} \text{fare} &= \frac{A_e}{80{,}000} = \frac{\$70{,}254.32}{80{,}000} \\ &= \$0.878/\text{passenger} \end{aligned}$$

The passenger fare should go up each year by

$$\begin{aligned} \$0.878 &= \$0.35 + G(A/G, 7\%, 20) \\ G &= \frac{\$0.878 - \$0.35}{7.3163} \\ &= \$0.072 \text{ increase per year} \end{aligned}$$

The answer is (D).

26. Use the present worth comparison method (also *NCEES Handbook:* Cost Estimation and Project Evaluation).

$$\begin{aligned} P(\text{A}) &= (\$4800)(P/A, 12\%, 25) \\ &= (\$4800)(7.8431) \\ &= \$37{,}646.88 \end{aligned}$$

(4 quarters)(25 years) = 100 compounding periods

$$\begin{aligned} P(\text{B}) &= (\$1200)(P/A, 3\%, 100) \\ &= (\$1200)(31.5989) \\ &= \$37{,}918.68 \end{aligned}$$

Alternative B is economically superior.

The answer is (B).

27. Use the equivalent uniform annual cost method (also *NCEES Handbook:* Cost Estimation and Project Evaluation).

$$\begin{aligned} \text{EUAC(A)} &= (\$1500)(A/P, 15\%, 5) + \$800 \\ &= (\$1500)(0.2983) + \$800 \\ &= \$1247.45 \end{aligned}$$

$$\begin{aligned} \text{EUAC(B)} &= (\$2500)(A/P, 15\%, 8) + \$650 \\ &= (\$2500)(0.2229) + \$650 \\ &= \$1207.25 \end{aligned}$$

Alternative B is economically superior.

The answer is (B).

28. The data given imply that both investments return 4% or more. However, the increased investment of $30,000 may not be cost effective. Do an incremental analysis (see *NCEES Handbook:* Cost Estimation and Project Evaluation).

$$\text{incremental cost} = \$70{,}000 - \$40{,}000 = \$30{,}000$$

$$\text{incremental income} = \$5620 - \$4075 = \$1545$$

$$\begin{aligned} 0 &= -\$30{,}000 + (\$1545)(P/A, i\%, 20) \\ (P/A, i\%, 20) &= 19.417 \\ i &\approx 0.25\% < 4\% \end{aligned}$$

Alternative B is economically superior.

(The same conclusion could be reached by taking the present worths of both alternatives at 4%.)

The answer is (B).

29. The present worth comparison is

$$\begin{aligned} P(\mathrm{A}) &= (-\$3000)(P/A, 25\%, 25) - \$20{,}000 \\ &= (-\$3000)(3.9849) - \$20{,}000 \\ &= -\$31{,}954.70 \\ P(\mathrm{B}) &= (-\$2500)(3.9849) - \$25{,}000 \\ &= -\$34{,}962.25 \end{aligned}$$

Alternative A is economically superior.

The answer is (A).

30. As in *NCEES Handbook:* Cost Estimation and Project Evaluation, the depreciation in the first five years is

$$\begin{aligned} \mathrm{BV} &= \$18{,}000 - (5)\left(\frac{\$18{,}000 - \$2000}{8}\right) \\ &= \$8000 \end{aligned}$$

The answer is (D).

31. Use the equivalent uniform annual cost method (also *NCEES Handbook:* Cost Estimation and Project Evaluation) to find the best alternative.

$$(A/P, 12\%, 15) = 0.1468$$

$$\begin{aligned} \mathrm{EUAC}_{5\,\mathrm{ft}} &= (\$600{,}000)(0.1468) \\ &\quad + \left(\frac{14}{50}\right)(\$600{,}000) + \left(\frac{8}{50}\right)(\$650{,}000) \\ &\quad + \left(\frac{3}{50}\right)(\$700{,}000) + \left(\frac{1}{50}\right)(\$800{,}000) \\ &= \$418{,}080 \end{aligned}$$

$$\begin{aligned} \mathrm{EUAC}_{10\,\mathrm{ft}} &= (\$710{,}000)(0.1468) \\ &\quad + \left(\frac{8}{50}\right)(\$650{,}000) + \left(\frac{3}{50}\right)(\$700{,}000) \\ &\quad + \left(\frac{1}{50}\right)(\$800{,}000) \\ &= \$266{,}228 \end{aligned}$$

$$\begin{aligned} \mathrm{EUAC}_{15\,\mathrm{ft}} &= (\$900{,}000)(0.1468) \\ &\quad + \left(\frac{3}{50}\right)(\$700{,}000) + \left(\frac{1}{50}\right)(\$800{,}000) \\ &= \$190{,}120 \\ \mathrm{EUAC}_{20\,\mathrm{ft}} &= (\$1{,}000{,}000)(0.1468) \\ &\quad + \left(\frac{1}{50}\right)(\$800{,}000) \\ &= \$162{,}800 \end{aligned}$$

Build to 20 ft.

The answer is (D).

32. Calculate the EUAC for the options given in the multiple choice answers.

Assume replacement after three years.

$$\begin{aligned} \mathrm{EUAC}(3) &= (\$30{,}000)(A/P, 12\%, 3) \\ &\quad + \$18{,}700 + (\$1200)(A/G, 12\%, 3) \\ &= (\$30{,}000)(0.4163) + \$18{,}700 \\ &\quad + (\$1200)(0.9246) \\ &= \$32{,}299 \end{aligned}$$

Similarly, calculate to obtain the numbers in the following table.

years in service	EUAC
3	\$32,299
6	\$28,602
8	\$28,234
10	\$28,312

Replace after eight years.

The answer is (C).

33. Since the head and horsepower data are already reflected in the hourly operating costs, there is no need to work with head and horsepower.

Use N = no. of hours operated each year.

$$\begin{aligned}\text{EUAC(A)} &= (\$3600 + \$3050)(A/P, 12\%, 10)\\ &\quad -(\$200)(A/F, 12\%, 10) + 0.30N\\ &= (\$6650)(0.1770) - (\$200)(0.0570) + 0.30N\\ &= 1165.65 + 0.30N\\ \text{EUAC(B)} &= (\$2800 + \$5010)(A/P, 12\%, 10)\\ &\quad +(\$280)(A/F, 12\%, 10) + 0.10N\\ &= (\$7810)(0.1770) - (\$280)(0.0570) + 0.10N\\ &= 1366.41 + 0.10N\end{aligned}$$

$$\begin{aligned}\text{EUAC(A)} &= \text{EUAC(B)}\\ 1165.65 + 0.30N &= 1366.41 + 0.10N\\ N &= 1003.8 \text{ hr} \quad (1000 \text{ hr})\end{aligned}$$

The answer is (A).

34. First check that both alternatives have an ROR greater than the MARR. Work in thousands of dollars. Evaluate alternative A using *NCEES Handbook:* Cost Estimation and Project Evaluation.

$$\begin{aligned}P(\text{A}) &= -\$120 + (\$15)(P/F, i\%, 5)\\ &\quad +(\$57)(P/A, i\%, 5)(1 - 0.45)\\ &\quad +\left(\frac{\$120 - \$15}{5}\right)(P/A, i\%, 5)(0.45)\\ &= -\$120 + (\$15)(P/F, i\%, 5)\\ &\quad +(\$40.8)(P/A, i\%, 5)\end{aligned}$$

Try 15%.

$$\begin{aligned}P(\text{A}) &= -\$120 + (\$15)(0.4972) + (\$40.8)(3.3522)\\ &= \$24.23\end{aligned}$$

Try 25%.

$$\begin{aligned}P(\text{A}) &= -\$120 + (\$15)(0.3277) + (\$40.8)(2.6893)\\ &= -\$5.36\end{aligned}$$

Since P(A) goes through 0,

$$(\text{ROR})_\text{A} > \text{MARR} = 15\%$$

Next, evaluate alternative B.

$$\begin{aligned}P(\text{B}) &= -\$170 + (\$20)(P/F, i\%, 5)\\ &\quad +(\$67)(P/A, i\%, 5)(1 - 0.45)\\ &\quad +\left(\frac{\$170 - \$20}{5}\right)(P/A, i\%, 5)(0.45)\\ &= -\$170 + (\$20)(P/F, i\%, 5)\\ &\quad +(\$50.35)(P/A, i\%, 5)\end{aligned}$$

Try 15%.

$$\begin{aligned}P(\text{B}) &= -\$170 + (\$20)(0.4972) + (\$50.35)(3.3522)\\ &= \$8.73\end{aligned}$$

Since P(B) > 0 and will decrease as i increases,

$$(\text{ROR})_\text{B} > 15\%$$

ROR > MARR for both alternatives.

The answer is (C).

35. Use the year-end convention with the tax credit. The purchase is made at $t = 0$. However, the credit is received at $t = 1$ and must be multiplied by $(P/F, i\%, 1)$. As in *NCEES Handbook:* Cost Estimation and Project Evaluation,

$$\begin{aligned}P &= -\$300{,}000 + (0.0667)(\$300{,}000)(P/F, i\%, 1)\\ &\quad +(\$90{,}000)(P/A, i\%, 5)(1 - 0.48)\\ &\quad +\left(\frac{\$300{,}000 - \$50{,}000}{5}\right)(P/A, i\%, 5)(0.48)\\ &\quad +(\$50{,}000)(P/F, i\%, 5)\\ &= -\$300{,}000 + (\$20{,}010)(P/F, i\%, 1)\\ &\quad +(\$46{,}800)(P/A, i\%, 5)\\ &\quad +(\$24{,}000)(P/A, i\%, 5)\\ &\quad +(\$50{,}000)(P/F, i\%, 5)\end{aligned}$$

By trial and error,

i	P
10%	\$17,625
15%	−\$20,409
12%	\$1456
13%	−\$6134
12¼%	−\$472

i is between 12% and 12¼%.

The answer is (C).

36. The distributed profit is

$$\begin{aligned}\text{distributed profit} &= (0.15)(\$2{,}500{,}000)\\ &= \$375{,}000\end{aligned}$$

As in *NCEES Handbook:* Cost Estimation and Project Evaluation, find the annual loan payment.

$$\begin{aligned}\text{payment} &= (\$2{,}500{,}000)(A/P, 12\%, 25)\\ &= (\$2{,}500{,}000)(0.1275)\\ &= \$318{,}750\end{aligned}$$

After paying all expenses and distributing the 15% profit, the remainder should be 0.

$$\begin{aligned}0 &= \text{EUAC}\\ &= \$20{,}000 + \$50{,}000 + \$200{,}000\\ &\quad + \$375{,}000 + \$318{,}750 - \text{annual receipts}\\ &\quad - (\$500{,}000)(A/F, 15\%, 25)\\ &= \$963{,}750 - \text{annual receipts}\\ &\quad - (\$500{,}000)(0.0047)\end{aligned}$$

This calculation assumes $i = 15\%$, which equals the desired return. However, this assumption only affects the salvage calculation, and since the number is so small, the analysis is not sensitive to the assumption.

$$\text{annual receipts} = \$961{,}400$$

The average daily receipts are

$$\frac{\$961{,}400}{365} = \$2634$$

Use the expected value approach. The average occupancy is

$$\begin{aligned}&(0.40)(0.65) + (0.30)(0.70) + (0.20)(0.75)\\ &\quad + (0.10)(0.80) = 0.70\end{aligned}$$

The average number of rooms occupied each night is

$$(0.70)(120 \text{ rooms}) = 84 \text{ rooms}$$

The minimum required average daily rate per room is

$$\frac{\$2634}{84} = \$31.36 \quad (\$31)$$

The answer is (D).

37. The annual savings are

$$\begin{array}{c}\text{annual}\\ \text{savings}\end{array} = \left(\frac{0.69 - 0.47}{1000}\right)(\$3{,}500{,}000) = \$770$$

$$\begin{aligned}P &= -\$7500 + (\$770 - \$200 - \$100)\\ &\quad \times (P/A, i\%, 25) = 0\\ (P/A, i\%, 25) &= 15.957\end{aligned}$$

One can calculate P/A using the various i values from the answer selection to find that

$$\begin{aligned}(P/A, 3.8\%, 25) &= 15.957\\ (P/A, 5.0\%, 25) &= 14.094\\ (P/A, 13\%, 25) &= 7.330\\ (P/A, 16\%, 25) &= 6.097\end{aligned}$$

The 3.8% interest rate gives the closest (exact) value to 15.957.

The answer is (A).

38. Evaluate alternative A, working in millions of dollars and using *NCEES Handbook:* Cost Estimation and Project Evaluation.

$$\begin{aligned}P(\text{A}) &= -(\$1.3)(1 - 0.48)(P/A, 15\%, 6)\\ &= -(\$1.3)(0.52)(3.7845)\\ &= -\$2.56 \quad [\text{millions}]\end{aligned}$$

Use straight line depreciation to evaluate alternative B.

$$\begin{aligned}D_j &= \frac{\$2}{6} = \$0.333\\ P(\text{B}) &= -\$2 - (0.20)(\$1.3)(1 - 0.48)(P/A, 15\%, 6)\\ &\quad - (\$0.15)(1 - 0.48)(P/A, 15\%, 6)\\ &\quad + (\$0.333)(0.48)(P/A, 15\%, 6)\\ &= -\$2 - (0.20)(\$1.3)(0.52)(3.7845)\\ &\quad - (\$0.15)(0.52)(3.7845)\\ &\quad + (\$0.333)(0.48)(3.7845)\\ &= -\$2.202 \quad [\text{millions}]\end{aligned}$$

Evaluate alternative C.

$$D_j = \frac{1.2}{3} = 0.4$$

$$\begin{aligned}P(\mathrm{C}) &= -(\$1.2)\big(1+(P/F,15\%,3)\big)\\ &\quad -(\$0.20)(\$1.3)(1-0.48)\\ &\quad \times\big((P/F,15\%,1)+(P/F,15\%,4)\big)\\ &\quad -(\$0.45)(\$1.3)(1-0.48)\\ &\quad \times\big((P/F,15\%,2)+(P/F,15\%,5)\big)\\ &\quad -(\$0.80)(\$1.3)(1-0.48)\\ &\quad \times\big((P/F,15\%,3)+(P/F,15\%,6)\big)\\ &\quad +(\$0.4)(\$0.48)(P/A,15\%,6)\\ &= -(\$1.2)(1.6575)\\ &\quad -(\$0.20)(\$1.3)(0.52)(0.8696+0.5718)\\ &\quad -(\$0.45)(\$1.3)(0.52)(0.7561+0.4972)\\ &\quad -(\$0.80)(\$1.3)(0.52)(0.6575+0.4323)\\ &\quad +(\$0.4)(0.48)(3.7845)\\ &= -\$2.428 \quad \text{[millions]}\end{aligned}$$

Alternative B is superior.

The answer is (B).

39. This is a replacement study. Since production capacity and efficiency are not a problem with the defender, the only question is when to bring in the challenger. Since this is a before-tax problem, depreciation is not a factor, nor is book value. The cost of keeping the defender one more year is

$$\begin{aligned}\text{EUAC(defender)} &= \$200{,}000 + (0.15)(\$400{,}000)\\ &= \$260{,}000\end{aligned}$$

For the challenger, as in *NCEES Handbook:* Cost Estimation and Project Evaluation,

$$\begin{aligned}&\text{EUAC(challenger)}\\ &= (\$800{,}000)(A/P,15\%,10)+\$40{,}000\\ &\quad +(\$30{,}000)(A/G,15\%,10)\\ &\quad -(\$100{,}000)(A/F,15\%,10)\\ &= (\$800{,}000)(0.1993)+\$40{,}000\\ &\quad +(\$30{,}000)(3.3832)\\ &\quad -(\$100{,}000)(0.0493)\\ &= \$296{,}006\end{aligned}$$

Since the defender is cheaper, keep it. The same analysis next year will give identical answers. Therefore, keep the defender for the next three years, at which time the decision to buy the challenger will be automatic.

Having determined that it is less expensive to keep the defender than to maintain the challenger for 10 years, determine whether the challenger is less expensive if retired before 10 years. If retired in nine years,

$$\begin{aligned}\text{EUAC(challenger)} &= (\$800{,}000)(A/P,15\%,9)+\$40{,}000\\ &\quad +(\$30{,}000)(A/G,15\%,9)\\ &\quad -(\$150{,}000)(A/F,15\%,9)\\ &= (\$800{,}000)(0.2096)\\ &\quad +\$40{,}000+(\$30{,}000)(3.0922)\\ &\quad -(\$150{,}000)(0.0596)\\ &= \$291{,}506\end{aligned}$$

Similar calculations yield the following results for all the retirement dates.

n	EUAC
10	\$296,000
9	\$291,506
8	\$287,179
7	\$283,214
6	\$280,016
5	\$278,419
4	\$279,909
3	\$288,013
2	\$313,483
1	\$360,000

Since none of these equivalent uniform annual costs are less than that of the defender, it is not economical to buy and keep the challenger for any length of time.

Keep the defender.

The answer is (D).

40. With just-in-time manufacturing, production is one-at-a-time, according to demand. When one is needed, one is made. In order to make the economic order quantity approach zero, the setup cost must approach zero.

The answer is (D).